T0192212

VOLUME I

Quantum Mechanics

Fundamentals

KURT GOTTFRIED
Cornell University

CRC Press is an imprint of the
Taylor & Francis Group, an **informa** business

Quantum Mechanics Volume I: Fundamentals

Originally published in 1966 by the Benjamin/Cummings Publishing Company, Inc.

Published 1974 by Westview Press

Published 2018 by CRC Press
Taylor & Francis Group
6000 Broken Sound Parkway NW, Suite 300
Boca Raton, FL 33487-2742

CRC Press is an imprint of the Taylor & Francis Group, an informa business

Visit the Taylor & Francis Web site at
http://www.taylorandfrancis.com

and the CRC Press Web site at
http://www.crcpress.com

Library of Congress Cataloging-in-Publication Data

Gottfried, Kurt.
Quantum mechanics / Kurt Gottfried.
p. cm. – (Advanced book classics)
Originally published: Reading, Mass. : W.A. Benjamin, Advanced Book Program, 1966.
Bibliography: p.
Includes index.
1. Quantum theory. I. Title. II. Series.
QC174.12.G68 1989 530.1'2–dc19 89-30696

ISBN 13: 978-0-201-40633-7 (pbk)

Quantum Mechanics

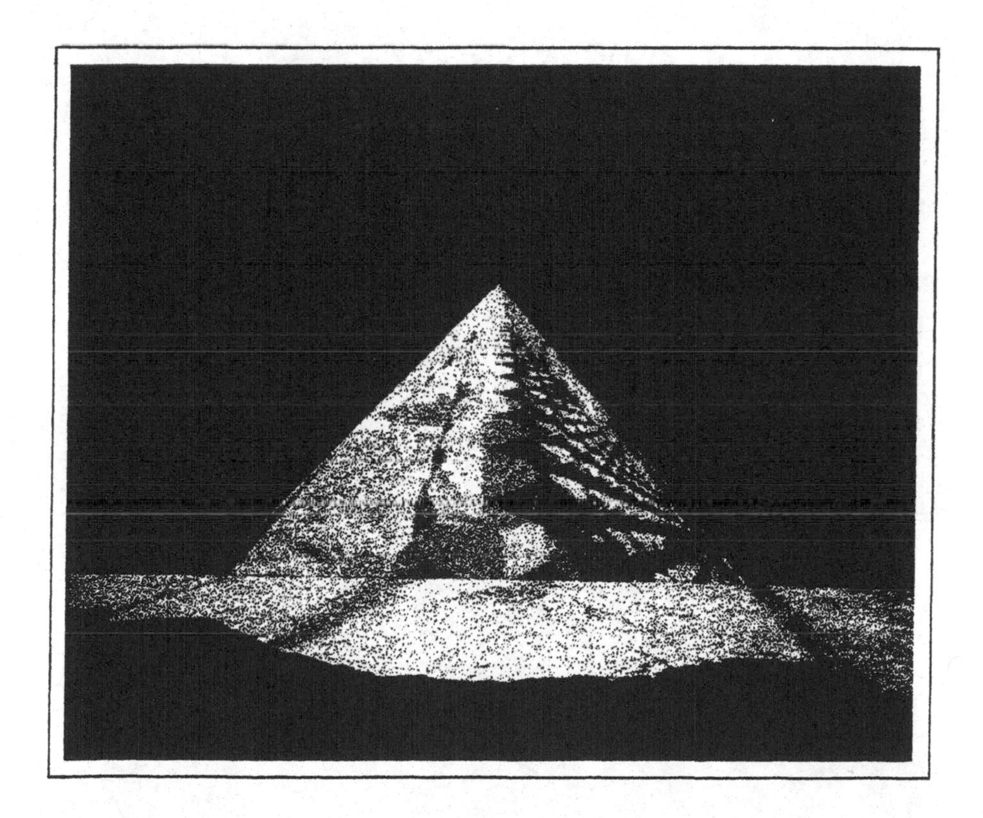

This unique image was created with special-effects photography. Photographs of a broken road, an office building, and a rusted object were superimposed to achieve the effect of a faceted pyramid on a futuristic plain. It originally appeared in a slide show called "Fossils of the Cyborg: From the Ancient to the Future," produced by Synapse Productions, San Francisco. Because this image evokes a fusion of classicism and dynamism, the future and the past, it was chosen as the logo for the Advanced Book Classics series.

Publisher's Foreword

"Advanced Book Classics" is a reprint series which has come into being as a direct result of public demand for the individual volumes in this program. That was our initial criterion for launching the series. Additional criteria for selection of a book's inclusion in the series include:

- Its intrinsic value for the current scholarly buyer. It is not enough for the book to have some historic significance, but rather it must have a timeless quality attached to its content, as well. In a word, "uniqueness."
- The book's global appeal. A survey of our international markets revealed that readers of these volumes comprise a boundaryless, worldwide audience.
- The copyright date and imprint status of the book. Titles in the program are frequently fifteen to twenty years old. Many have gone out of print, some are about to go out of print. Our aim is to sustain the lifespan of these very special volumes.

We have devised an attractive design and trim-size for the "ABC" titles, giving the series a striking appearance, while lending the individual titles unifying identity as part of the "Advanced Book Classics" program. Since "classic" books demand a long-lasting binding, we have made them available in hardcover at an affordable price. We envision them being purchased by individuals for reference and research use, and for personal and public libraries. We also foresee their use as primary and recommended course materials for university level courses in the appropriate subject area.

The "Advanced Book Classics" program is not static. Titles will continue to be added to the series in ensuing years as works meet the criteria for inclusion which we've imposed. As the series grows, we naturally anticipate our book buying audience to grow with it. We welcome your support and your suggestions concerning future volumes in the program and invite you to communicate directly with us.

Advanced Book Classics

1989 Reissues

V.I. Arnold and A. Avez, *Ergodic Problems of Classical Mechanics*

E. Artin and J. Tate, *Class Field Theory*

Michael F. Atiyah, *K-Theory*

David Bohm, *The Special Theory of Relativity*

Ronald C. Davidson, *Theory of Nonneutral Plasmas*

P.G. de Gennes, *Superconductivity of Metals and Alloys*

Bernard d'Espagnat, *Conceptual Foundations of Quantum Mechanics, 2nd Edition*

Richard Feynman, *Photon-Hadron Interactions*

William Fulton, *Algebraic Curves: An Introduction to Algebraic Geometry*

Kurt Gottfried, *Quantum Mechanics*

Leo Kadanoff and Gordon Baym, *Quantum Statistical Mechanics*

I.M. Khalatnikov, *An Introduction to the Theory of Superfluidity*

George W. Mackey, *Unitary Group Representations in Physics, Probability and Number Theory*

A. B. Migdal, *Qualitative Methods in Quantum Theory*

Phillipe Nozières and David Pines, *The Theory of Quantum Liquids, Volume II* - new material, 1989 copyright

David Pines and Phillipe Nozières, *The Theory of Quantum Liquids, Volume I: Normal Fermi Liquids*

David Ruelle, *Statistical Mechanics: Rigorous Results*

Julian Schwinger, *Particles, Source and Fields, Volume I*

Julian Schwinger, *Particles, Sources and Fields, Volume II*

Julian Schwinger, *Particles, Sources and Fields, Volume III* - new material, 1989 copyright

Jean-Pierre Serre, *Abelian ℓ-Adic Representations and Elliptic Curves*

R.F. Streater and A.S. Wightman, *PCT Spin and Statistics and All That*

René Thom, *Structural Stability and Morphogenesis*

Vita

Kurt Gottfried

Born in Vienna, Kurt Gottfried emigrated to Canada in 1939 and received his Ph.D. in theoretical physics from MIT in 1955. He is Professor of Physics at Cornell University, and has been a Junior Fellow and Assistant Professor of Physics at Harvard University, a Visiting Professor at the Massachusetts Institute of Technology, and a Senior Staff Member at CERN in Geneva. A recipient of a John Simon Guggenheim fellowship, Dr. Gottfried is also a member of the American Academy of Arts and Sciences and has served as Chairman, Division of Particles and Fields of the American Physical Society. He is also the co-author, with V.F. Weisskopf, of *Concepts of Particle Physics*. Dr. Gottfried has an active interest in arms control and human rights and has testified before Congress on these issues.

Special Preface

Quantum mechanics was already a venerable subject when this book first appeared in 1966—as solidly grounded in experiment, and as luxuriously developed in its theoretical virtuosity, as Newtonian dynamics 150 years ago. In that sense, therefore, little that would fundamentally change the face of the subject was to be expected. On the whole, that expectation has been fulfilled, and that presumably explains why this book still seems to meet the needs of many serious students of the subject.

This qualification "on the whole" in the preceding sentence is an allusion to the one development of a truly fundamental nature that has occurred. True enough, it did not change the face of the subject, but it did illuminate its soul. I refer, of course, to John Bell's now famous inequality[1], published in 1964, which was conceived (as the first draft of the book was being written) during one of my long stays at John's home institution, CERN.

Bell's accomplishment was remarkable in that it led to laboratory, not just *Gendanken*, experiments. The data has demonstrated[2] that quantum mechanical probabilities *cannot* be explained by assuming the existence of a still undiscovered substructure in nature described by "hidden variables" obeying laws that comply with everyday common sense. In short, whether you, the reader, like it or not, you will have to put up with the exquisite mysteries of quantum mechanics! New developments may, indeed almost certainly will, show that quantum mechanics, like its Newtonian ancestor, is only an approximation of something deeper. That something will, however, be at least as peculiar as quantum mechanics, and in all likelihood, even stranger—a prejudice I held firmly long before Bell's work made it possible to confront such biases with reality.

While I was aware of Bell's work before this book went to press, I had no more than an obtuse inkling of its significance, and only referred to it in an endnote beginning with the second printing of 1969 (see p. 190). Although this book continues to be one of the few that takes a hard look at the theory of measurement, it is seriously deficient because it does not include discussions of the Einstein-Podolsky-Rosen paper, Bell's theorem, and related matters[3].

On the other hand, to my mind, and that of three old friends and colleagues who have also pondered these matters (the late Wendell Furry, David Mermin, and Victor Weisskopf), pp. 185-189 resolves the long-standing "mystery" of the "reduction of the wave packet." This I do by showing that for all observable A that are local on a macroscopic length, the result of any measurement is given by $\mathrm{Tr}\, A\, \hat{\rho}$, where $\hat{\rho}$ is the density matrix obtained from the true and complete density matrix by discarding terms corresponding to interference between macroscopically distinct states of the measurement apparatus. In other words, I divide the set of all Hermitian operators into two categories, those that are macroscopically local and those that are not, and show that the former, which includes ail those conventionally associated with "observables," the reduction postulate is superfluous[4]. Strangely enough, this demonstration that the reduction problem is a red herring has received virtually no notice, good or bad, in the community of scholars that struggles with the foundations[5].

The most important technical gap in the book is Feynman's path integral formulation of quantum mechanics[6], which was already an old topic in 1966. It has become far more important in recent years, thanks in no small way to my colleague Kenneth Wilson.

Over the years hundreds of people (and many a Benjamin and Addison-Wesley editor) have asked about Volume II. The decision to split the manuscript into two volumes was actually made quite late, when it became clear that the book was becoming very large and encompassing topics that might be viewed as too broad a spectrum of difficulty. In retrospect, this was a terrible mistake. Proofreading Volume I was so arduous a task that I could not face the manuscript when it was completed. Then research interests and other matters created an unending sequence of obstacles, and as a result, several hundred pages of manuscript have lain idle for over twenty years, including material that I think is superior to what is in this book, and also more unique. I still intend to return to this task.

Kurt Gottfried
Ithaca, New York
February 1989

REFERENCES

1. John S. Bell, *Physics 1*, 195 (1964). For a collection of papers by Bell on this theorem and other developments concerning the foundations of quantum mechanics, see his *Speakable and unspeakable in quantum mechanics* (Cambridge: Cambridge University Press, 1987). Delightful and instructive discussions of Bell's theorem and related issues can be found in N. David Mermin, *Physics Today 38*, 38 (1985); and *Am. J. Phys. 55*, 585 (1987).
2. A. Aspect, P. Grangier and G. Roger, *Phys. Rev. Let. 47*, 460 (1981); and *49*, 91 (1982); A. Aspect, J. Dalibard and G. Roger, ibid. *49*, 1804 (1982). For earlier experimental work see the review of J.F. Clauser and A. Shimony, *Rep. Prog. Phys. 41*, 1881, (1978).
3. For a text that discusses these topics, see J.J. Sakurai, *Modern Quantum Mechanics* (Menlo Park, CA: Addison-Wesley, 1985); Sec. 3.9.
4. As pointed out on p. 186, the set of macroscopically local observables does not include everything of interest to physics. Superfluids have observables that are macroscopically non-local, and one can devise experiments using superfluids which manifest non-local effects over macroscopic distances. In this circumstance nonsense would result were one to replace the true density matrix by its truncated counterpart; on the other hand, such an apparatus would not perform a "measurement" in the sense that the term is used in measurement theory.
5. I am aware of only one exception, M. Cini, "Quantum Theory of Measurement without Wave Packet Collapse," *Nuovo Cimento 73B*, 27 (1983), who states on p. 29 "that the right answer to the question about the origin of the so-called wave function collapse has been outlined already" in this book, and that this "has remained...unnoticed for more than fifteen years" by the (interminable) journal and monograph literature on the topic. When Cini wrote this, the book was in its sixth printing!
6. Richard P. Feynman and Albert R. Hibbs, *Quantum Mechanics and Path Integrals* (New York: McGraw Hill, 1965). For introductory treatments, see Sakuri, op. cit., Section 2.5; and Sidney Coleman, *Aspects of Symmetry* (Cambridge: Cambridge University Press, 1985), pp. 145-159.

TO SORKY, WITH LOVE

AND GRATITUDE

Preface

In 1961-62 I taught the graduate quantum mechanics course at Harvard, and unwittingly prepared a detailed set of handwritten notes for distribution to the students. Having heard of this, the Pied Piper from Benjamin appeared. Little did he or I suspect that his enticing tune would lead to a two-volume tome.

This volume is intended as a text for a first-year graduate quantum mechanics course. The great majority of students entering graduate school today have had a term or more of introductory quantum theory. The lectures I have given at Cornell and Harvard therefore assumed that the audience had at least a nodding acquaintance with the phenomenological and epistemological background, the Schrödinger equation, and its application to very simple systems. This book is written in the same spirit. It is self-contained, and in principle it can be read by someone who has not heard of Planck's constant. But only a very gifted student could expect to digest it comfortably without any previous exposure to the quantum theory.

As for other prerequisites, I should like to emphasize that I do *not* assume a knowledge of group theory, functional spaces, or classical mechanics beyond Hamilton's equations. Except for Section 3, which can be skipped without destroying the continuity of the argument, only elementary electromagnetic theory is necessary until Chapter VIII. On the other hand, I do assume some knowledge of vector analysis, linear algebra, ordinary differential equations, Fourier analysis, and analytic functions. The general level of mathematical rigor is typical of most of current theoretical physics.

In the courses I have taught, and also in this book, I have tried to meet several, occasionally conflicting, goals. On the one hand, quantum mechanics underlies essentially all current thinking in physics, and the serious student must therefore master the conceptual and mathematical foundations of the subject if he is not to be a mere technician.

On the other hand, quantum mechanics is also an exceedingly complicated and rapidly proliferating technique used by many physicists in their day-to-day work. It is therefore some combination of philosopher and artisan that I have in mind—a "quantum mechanic."

Insofar as "fundamentals" are concerned, I have emphasized the theory of the measurement process, and symmetry principles. The analysis of measurements, and its relationship to the statistical interpretation, are usually overlooked in courses on quantum mechanics. While it is true that measurement theory cannot be appreciated by the novice, it can be understood by hard-thinking students who are able to master the more technical portions of the subject. Here I do not mean to imply that measurement theory is easy. Far from it. As for symmetry, or lack thereof, no apologies are necessary because recent developments have repeatedly shown that this is a subject that one can ill afford to ignore. No matter how much one has studied invariance principles and their consequences, there always appear to be sensational surprises that reveal the superficiality of one's understanding.

In the applications of the general theory I have tried to find examples that really do justice to the technique used. It was also my intention to illustrate every important technique with an application of genuine practical interest. Unfortunately I found it rather inconvenient to meet these objectives completely within the compass of Volume I. For example, the Wigner-Eckart theorem and the angular momentum decomposition of the electromagnetic field are not applied until Volume II (containing Chapters X-XIII). But by and large I believe I have illustrated the theory with realistic applications, or, as in Section 15, with artificial examples of sufficient complexity to reveal the essential ingredients of a practical calculation.

A few remarks concerning the organization of the book are in order here. If logic and efficiency were the only considerations, I would begin straightaway with Dirac's abstract formulation. But most students appear to find this too difficult, and I therefore develop the wave-mechanical formulation first. Once students have had the opportunity to master this approach by applying it to several problems of moderate complexity, they are far more receptive to the abstract approach, and also in a better position to appreciate its power and elegance. In the remainder of the course (i.e., following Chapter IV) I always use Dirac's mode of thought and notation. The applications treated in this volume are almost exclusively concerned with systems having a small number of degrees of freedom. In thinking about such systems one is not forced to make drastic approximations *ab initio*, and much confusion between basic principles and approximation methods is thereby avoided. The theory of systems with many degrees of freedom has, on the whole, been relegated to Volume II. That volume will contain discussions of radiation theory, quantum statistics and the many-body problem, and more advanced topics in collision theory.

Here I should like to interject a few suggestions to those who set out to learn this subject by themselves, without the crutch of a lecture course. There are a number of topics of more than average technical complexity that can be deleted in a first approach to the subject. Some of these have been set in small type and are also marked off by the symbol •. But even larger portions can be saved for later consumption. For example, Section 3 can be read just prior to Chapter VIII. Section 17 on the Coulomb field contains a good deal of rather intricate analysis; readers who do not feel at home off the real axis might do well to substitute a treatment along more conventional lines. The advice in the footnote at the beginning of Section 20 should be heeded by beginners. The bulk of Section 34, and essentially all of Section 49, can safely be skipped. But the problems should not be skipped under any circumstances. It is impossible to know whether one has understood the theory unless one has solved problems. Some of the problems in this book are rather difficult, and one should not be dismayed if several long days are required for their solution.

In citing the literature I have used a set of rather arbitrary rules. The papers from the Heroic Age are not referred to, unless they contain further information that the reader might find useful. On the other hand, more modern articles whose existence may be unknown to the student are frequently cited. On several occasions I have purposely used published material in the problems without stating a reference. The abused authors are R.J. Glauber, J.V. Lepore, J. Schwinger, and Y. Yamaguchi. The wave packet treatment of collisions described in Section 12.2 is due to F.E. Low.

Kurt Gottfried
Ithaca, New York
December 1965

Acknowledgments

By far the most pleasant task in writing a book is the compilation of acknowledgments. Above all I should like to express my debt to the men who have taught me this subject, both by the spoken and written word. As a student I was very fortunate to hear lectures by three outstanding teachers, J.D. Jackson, J. Schwinger, and V.F. Weisskopf. Schwinger's approach made an especially deep impression on me, and I frequently consulted my voluminous notes from his course in preparing my own lectures. A brief glance at this book also reveals that I have been greatly influenced by Dirac's classic monograph, and by Pauli's *Handbuch* article. In truth, portions of Chapters I, II, and IV are merely expanded translations of Pauli's work, "a poor man's Pauli." I have also received advice and criticism from a number of friends, colleagues, and students. On innumerable occasions Jeffrey Goldstone has shared his exceptional understanding of the foundations of this subject with me; these conversations had a marked effect on the final shape of Sections 20 and 27. I have also benefited from many suggestions and comments by David Jackson, Paul Martin, and Charles Schwartz. Wendell Furry kindly read and commented on Chapter IV. Some of the views and ideas expressed in Sections 20.3 and 56.2 grew out of stimulating conversations with Donald Yennie.

I also wish to thank Alan Chodos and Velma Ray for diligent help in preparing the manuscript and correcting proofs, and all the good people at W.A. Benjamin, Inc., for their remarkable patience. A large portion of the manuscript was prepared while I held a John Simon Guggenheim Fellowship at CERN; I should like to thank the Guggenheim Foundation for its generous support, and Leon Van Hove and Victor Weisskopf for the warm hospitality extended to me in Geneva. Finally, I am very indebted to my wife for encouragement and support throughout the endless hours that have gone into the writing of this book.

K.G.

Contents

Quantum Mechanics

I

Uncertainty and Complementarity

During the first decade of this century it became increasingly clear that classical physics could not account for some of the most significant features of the newly discovered atomic phenomena. The inadequacy of classical theory was strikingly emphasized by the partial success of the ideas proposed by Planck, Einstein, and Bohr. This "old quantum theory" was a diabolically clever hodge-podge of classical laws and seemingly unrelated *ad hoc* recipes. The creation of quantum mechanics in the period 1924–1928 restored logical consistency to its rightful place in theoretical physics. Of even greater importance, it provided us with a theory that appears to be in complete accord with our empirical knowledge of all nonrelativistic phenomena. On the other hand, the new theory brought with it a most profound revolution in the concepts, and to some extent even the aims, of physics.

We shall begin our study of quantum mechanics by analyzing some of the microscopic phenomena to which we have alluded. This analysis will force us to the conclusion that some of the most "obvious" and dearly cherished notions abstracted from our vast experience of macroscopic phenomena are simply inapplicable to the microcosm. We shall see that this failure of classical physics is not merely a matter of quantitative disagreements with experiments. The problem is much more fundamental because *classical physics does not even provide an appropriate language for describing certain microscopic phenomena in purely qualitative terms.* Once the extent of this breakdown of classical physics has been carefully delineated, the stage upon which the new theory is to appear will be set. Our train of argument will not follow historical lines; instead it will draw heavily on the wisdom of hindsight. The historical development of the theory is a long and fascinating story that falls outside the scope of this book.*

1. Nonclassical Aspects of the Electromagnetic Field

The study of interaction between light and electrons provided most of the important clues in the development of the quantum theory, and so we shall first address ourselves to this topic. Consider an electromagnetic wave packet of mean wave vector $\mathbf{k}$ having a spatial extension considerably in excess of $1/k$, i.e., a fairly monochromatic packet.† According to classical electrodynamics the energy and momentum carried by this packet depend on the mean intensity of the field. The classical theory asserts that after this packet has interacted with an electron, the field will consist of an outgoing spherical wave, as well as a packet of slightly depleted intensity proceeding in the initial direction $\hat{\mathbf{k}}$. Furthermore, classical theory predicts that if the electron was initially at rest, it will possess a momentum in the direction $\hat{\mathbf{k}}$ after the collision. Let us now compare these predictions with the experimental facts originally obtained by Compton in a study of X-ray scattering. His experiments revealed that (a) electrons frequently acquire a momentum transverse to $\hat{\mathbf{k}}$; (b) no vestige of the spherically scattered wave is observed: on the contrary, the electromagnetic energy and momentum are concentrated in a spatially confined packet after the collision; (c) the propagation of this scattered packet is correlated in direction with the momentum vector of the scattered electron.

We are obviously faced with a number of glaring disagreements with classical electrodynamics. In fact, as Compton himself recognized, one can retrieve some of the experimental results by invoking the photon

* For an account of developments preceding 1926, see Whittaker. Pauli's footnotes provide an outline history for the period 1925–1932. (When only an author's name appears in a citation, see the Bibliography for a detailed reference.)

† If $\hat{\mathbf{k}}$ is a unit vector in the direction of propagation of a plane wave, then $\mathbf{k} = \hat{\mathbf{k}}/\lambda\!\!\!^{-}$, where $2\pi\lambda\!\!\!^{-}$ is the wavelength. We shall always use the notation $\hat{\mathbf{a}}$ for a unit vector in the direction of $\mathbf{a}$.

concept introduced long before by Einstein in his theory of the photo-effect. That is, one treats the incident field as an assembly of particles (photons) that can scatter from the electrons like billiard balls. The energy E and momentum $\mathbf{p}$ which one ascribes to these particles are related to the mean circular frequency ω and wave vector $\mathbf{k}$ of the electromagnetic disturbance by

$$E = \hbar\omega, \qquad \mathbf{p} = \hbar\mathbf{k}, \tag{1}$$

where $\hbar$ ($= 1.054 \times 10^{-27}$ erg sec) is Planck's constant.* The conservation laws of energy and momentum, in conjunction with (1), correctly determine the frequency and propagation vector of the scattered packet in terms of the momentum vector of the scattered electron. We should note here that the electrodynamic dispersion law $\omega = ck$, where c is the velocity of light, implies that $E = pc$. One therefore speaks of the photon as a particle with vanishing rest mass.

Let us examine the novel features of Compton's "explanation" in more detail. In the first instance we note that the relations (1) constitute a most remarkable liaison between two families of concepts that are quite unrelated in classical physics—that of particles and that of waves. Here we may already catch a glimpse of Bohr's *Complementarity Principle*, which, among other things, asserts that in the atomic domain it is not possible to describe phenomena by any single classical concept. Concerning this point, *the wave-particle duality*, we shall have much more to say shortly. Let us first emphasize another striking feature of the Compton experiment. That is, *there is a stochastic aspect to the actually observed course of events*, because there does not appear to be any imaginable way in which the experimenter can control or predict the momentum that an electron will acquire in any single collision. The only reproducible experimental quantity is the probability distribution for the magnitude and direction of the momentum transfer. The naive theory of the Compton effect based on (1) also involves this probabilistic aspect, because it only gives the momentum of the scattered light in terms of the momentum of the scattered electron. In fact, this naive theory is clearly *incomplete*, because it cannot predict the probability distribution referred to above. On the other hand, the classical theory is not only incomplete; it is *wrong*, for it does not contain any stochastic element. Rather, it asserts that the momentum transfer is uniquely determined once the incident light packet is specified.

An immense amount of evidence attesting to the wave nature of light had been collected in the century preceding Compton's work. Can one reconcile the phenomena of interference and diffraction with those of the

* A list of fundamental constants and conversion factors is provided in the appendix.

photo-effect and Compton scattering? In order to answer this question, let us investigate a typical interference experiment. Consider the grating arrangement shown in Fig. 1.1. According to classical theory, the angular separation between maxima is $\Delta\alpha \simeq \lambda/d$, and the angular width of each maximum is $\delta\alpha \simeq (d/D)(\Delta\alpha)$, where λ is the wavelength of the radiation. As a detector we could use a photographic plate, which on detailed examination would reveal a multitude of spots whose density is given by classical wave theory. Each individual spot is actually the

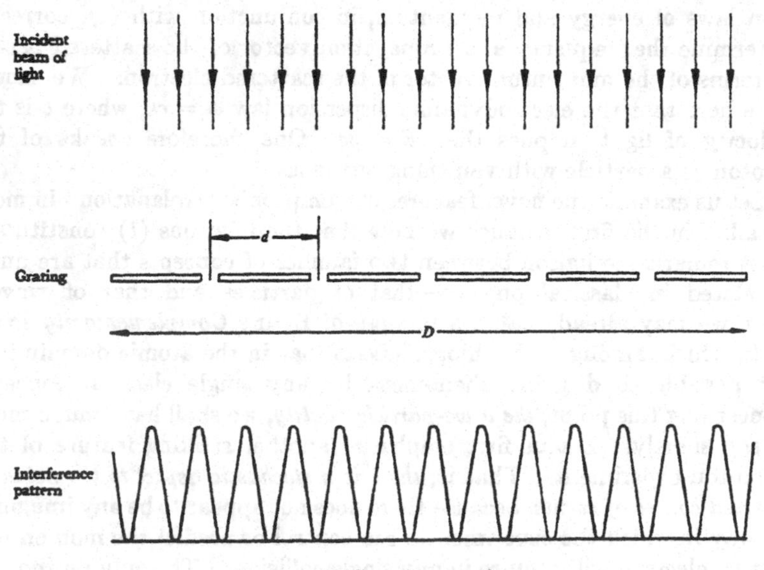

Fig. 1.1. Diffraction grating.

result of a photo-chemical reaction which, as we now know, is triggered by a *single* quantum. This can also be shown by reducing the beam intensity to a point where, on the average, only one quantum is passing through the apparatus at a time. Thus only one chemical reaction in the detector would be triggered at a time.* This is in complete disagreement with the classical theory which predicts that the interference pattern remains unaltered, and that the total intensity is reduced. In the one-photon-at-a-time experiment, we only see one spot at a time. If we make a long exposure so that many photons pass through the apparatus, and fail to resolve the different spots on the photographic plate, we

* Or we could put a cloud chamber behind the screen and observe Compton scattering of the diffracted light.

regain the classical wave pattern. In other words, *the wave theory only predicts the probability of where a light quantum will hit the detector.*

Now it is natural to ask how seriously the particle picture of light is to be taken. We could, for example, ask whether a trajectory through the apparatus can be ascribed to the photons. That is, can the photon be localized not only in the act of absorption by the detector, but throughout its course in the apparatus? Since the illumination of the grating is uniform, we might expect that if we allowed only one section of length $D' \ll D$ to be exposed at a time, and moved the exposed part about so that all parts received equal exposure, then the interference pattern would be unchanged, except possibly for a reduction in intensity. This is not so, however, since the width of each maximum increases by D/D'. In fact, if we cover all the rulings but one, on the assumption that the photon has atomic dimensions and only goes through one ruling, we completely lose the interference pattern. Thus we may conclude quite generally that in any arrangement that causes light to traverse different paths followed by recombination, either we may observe an interference pattern, and remain ignorant as to which of the paths the quanta have traversed, or we may experimentally determine which path the quanta followed, but thereby destroy the interference pattern. This reveals another aspect of the *Complementarity Principle: an experimental arrangement designed to manifest one of the classical attributes (e.g., wave- or particle-like aspects) precludes the possibility of observing at least some of the other classical attributes.*

We have just seen that in the interference experiments the statistical distribution of quanta is correctly given by the intensity of the wave field as computed from classical electromagnetic theory. We should point out that under special circumstances the classical theory also makes contact with some of the empirical facts in Compton's experiment. In particular, one finds that in the long wavelength limit

$$\lambdabar \gg \frac{\hbar}{mc}, \tag{2}$$

where m is the mass of the electron and $\lambdabar = \lambda/2\pi$, the angular distribution of the scattered photons agrees with the angular distribution of intensity of the scattered wave as computed from the classical theory by J. J. Thomson. Consider now an electron immersed in an intense beam of such soft radiation. In each Compton event it will acquire some longitudinal and some transverse momentum, $\mathbf{p}_\parallel$ and $\mathbf{p}_\perp$. The orientations of $\mathbf{p}_\perp$ are random, and will therefore tend to cancel, but $\mathbf{p}_\parallel$ is always in the direction of the incident wave. After many collisions the electron's momentum vector will therefore be in the incident direction to a high degree of accuracy (which improves as λ increases). Hence in the long wavelength limit an experiment using detectors that cannot distinguish between the arrivals of individual photons, and only measure time-averaged values of field strengths and electron momentum, would yield results in complete accordance with classical electrodynamics. This argument is an example of another fundamental insight which we also owe to Bohr, known as the *Correspondence Principle.* This doctrine states that from a satisfactory quantum theory one must be able to deduce classical mechanics and electrodynamics by taking an appropriate limit. The existence of such a limit is clearly necessary, because we know that classical mechanics and electrodynamics describe a vast body of data with great accuracy. As we shall see, the Correspondence Principle plays a very important role in the mathematical formulation of quantum mechanics.

It appears from (2), and our discussion of the grating experiment, that classical electrodynamics gives the correct statistical distribution of quanta (intensity of wave equals probability of absorbing photons) whenever the material bodies with which the light interacts are sufficiently massive. Thus whenever we use macroscopic equipment to localize or confine electromagnetic disturbances, we may predict the probability for detecting quanta from the wave equation

$$\nabla^2\phi = \frac{1}{c^2}\frac{\partial^2\phi}{\partial t^2}, \tag{3}$$

together with the boundary conditions appropriate to the apparatus in question. (Here we ignore polarization, and therefore use a scalar electromagnetic field.)

The most general solution of (3) is

$$\phi(\mathbf{x}, t) = \left(\frac{1}{2\pi}\right)^{\frac{3}{2}} \int d^3k\, e^{i(\mathbf{k}\cdot\mathbf{x} - \omega t)} A(\mathbf{k}), \tag{4}$$

where $\omega = c|\mathbf{k}|$. If we are to find the light quantum in some region of space, $|\phi|^2$ must be nonzero in this region, and must nearly vanish elsewhere. Let us say that the region has the dimensions Δx_i $(i = 1, 2, 3)$. It then follows from the theory of Fourier integrals that $A(\mathbf{k})$ is nonzero in a finite domain of $\mathbf{k}$-space whose size is related to Δx_i by *

$$\Delta x_i\, \Delta k_j \gtrsim \delta_{ij}. \tag{5}$$

Similarly the time Δt taken by the packet to pass any point is related to

* A derivation of (5) will be found in Sec. 4.3.

the width in frequency by

$$\Delta\omega\,\Delta t \gtrsim 1. \tag{5'}$$

If we now use the relations (1) we have

$$\Delta x_i\,\Delta p_j \gtrsim \delta_{ij}\,\hbar, \qquad \Delta E\,\Delta t \gtrsim \hbar. \tag{6}$$

In other words, it is not possible to simultaneously ascribe a position *and* momentum to a photon with arbitrary accuracy, nor is it possible to simultaneously give the energy *and* time of passage of a photon past a fixed point. These are the ***Heisenberg uncertainty relations;*** they were first discovered by him in 1927 for particle mechanics, not for light.

Strangely enough, the uncertainty relations (6) achieve a certain reconciliation of the particle and wave picture, in that they say that the quantum can display only one of the attributes of a particle at a time ($\mathbf{p}$ or $\mathbf{x}$), but not both, i.e., it is not really a particle in the classical sense. On the other hand it is not a wave either, because of the original Einstein relations (1).

2. Wave-Particle Duality for Systems with Finite Rest Mass

The discussion of the Compton effect in the foregoing section leaned heavily on the conservation of energy and momentum, and we shall therefore assume that these quantities are rigorously conserved even in atomic physics. Consider now the consequences of assuming that massive systems (e.g., the electron in the Compton effect), in contrast to the electromagnetic field, were accurately described by classical mechanics and therefore violated the Heisenberg relations * (1.6). In the Compton experiment we could then place the electron at rest at a known point and use the knowledge of its final position and momentum to determine the final position and momentum of the photon to arbitrary precision, in violation of (1.6). We may therefore conclude that ***consistency requires the uncertainty relations (1.6) to hold equally well for systems having a finite rest mass.*** In fact, by similar considerations, we can conclude that $\hbar$ must be a universal constant which does not depend on the

* The system of equation numbering is as follows:

1. Within each section equations are numbered starting from one and are referred to in the text as (1), (2), etc.

2. Equations belonging to another section will be cited by section number followed by equation number. Thus (14.3) indicates the third equation of Sec. 14. To facilitate the use of this system, section numbers appear at the top of the page.

particular system in question. If $\hbar$ did not have this character we could always use the system with the smallest Planck constant to violate the uncertainty relations for all other systems. Hence $\hbar$ plays a role similar to that of the limiting velocity c in relativity theory: it gives a measure of the utmost accuracy to which two properties of a system can be determined simultaneously.

The most direct way of assuring a Heisenberg type of uncertainty for particles is to attribute wave-like properties to them as well. This was first done by de Broglie in 1924. He conjectured that a wavelength λ (now called the de Broglie wavelength) should be attributed to particles in a fashion quite analogous to (1.1), i.e.,

$$\lambda = \frac{\hbar}{|\mathbf{p}|}, \tag{1}$$

where $\mathbf{p}$ is the momentum of the particle. We now know from electron and neutron diffraction that (1) is a correct and universal association.

There are a number of *Gedanken* experiments whose analysis confirms the universal nature of the uncertainty relations (1.6). These are not proofs, because they all depend on the use of the de Broglie relation (1) which was set up to insure the existence of the uncertainty relations. These experiments merely serve to elucidate the content of the uncertainty principle.

We begin with a very unsophisticated position measurement (Fig. 2.1) that determines the location of an electron in the x-direction to within an uncertainty of $\Delta x = d$. Because of the wave properties, the beam is diffracted, with

$$\sin\theta \simeq \frac{\lambda}{d} \simeq \frac{\hbar}{p_z d}.$$

Therefore p_x is not certain after the measurement. To be precise, $\Delta p_x \simeq p_z \sin\theta$, and therefore $d \cdot \Delta p_x \sim \hbar$, which agrees with (1.6).

Here we have ignored the question of any momentum transferred from the screen to the electron. It is clear that if the screen's momentum were known more precisely than p_x, we could violate the uncertainty principle by using the conservation laws. However, the position of the screen must be known to an accuracy $\delta x \ll d$, i.e., it has an uncertainty $\delta p_x \gtrsim \hbar/\delta x \gg \Delta p_x$. (We will, of course, make the screen very heavy so that it will remain essentially stationary and not affect the position measurement.) Thus we have the important conclusion that the apparatus itself must obey (1.6), or we would easily deduce violations of the uncertainty principle. This is really a repetition of the argument which forced

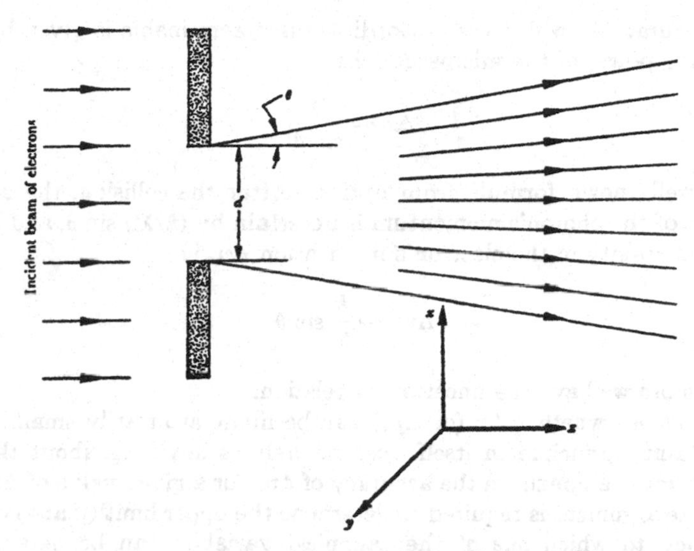

Fig. 2.1. An experiment for measuring the x-coordinate.

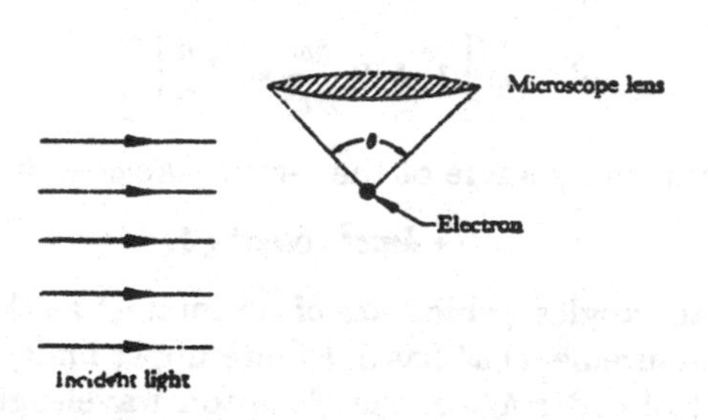

Fig. 2.2. The Heisenberg microscope.

us to extend the uncertainty principle from photons to microscopic particles having a rest mass.

Next we discuss a famous example due to Heisenberg, in which Compton scattered γ-rays are brought to focus by a microscope (Fig. 2.2). Let (λ, ω) and (λ', ω') characterize the incident and scattered radiation.

The accuracy to which the x-coordinate is determinable is given by the resolving power of the microscope, i.e.,

$$\Delta x \sim \frac{\lambda'}{\sin\theta}, \tag{2}$$

by a well-known formula from optics. After the collision, the x-component of the photon's momentum is uncertain by $(\hbar/\lambda\!\!\!^{-}{}')\sin\theta$, and hence the uncertainty in the electron's momentum Δp_x is

$$\Delta p_x \sim \frac{\hbar}{\lambda\!\!\!^{-}{}'}\sin\theta. \tag{3}$$

Once more we have the uncertainty relation.

Let us see whether Δx (or Δp_x) can be made arbitrarily small. The uncertainty principle in itself does not tell us anything about this; it merely gives a bound on the accuracy of Δp_x for a given value of Δx. A separate argument is required to determine the upper limit (if any) on the accuracy to which *one* of the canonical variables can be determined when the uncertainty in the conjugate variable is allowed to be arbitrarily large.

To study the limits on Δx, we return to Heisenberg's microscope. According to (2) we could reduce Δx to zero if λ' could be made to vanish. One might hope to achieve this by letting the incident frequency go to infinity, but the situation is not quite so simple. To see this, consider the Compton effect in more detail. The frequency shift is given by

$$\omega' = \omega\left[1 + 2\frac{\hbar\omega}{mc^2}\sin^2\frac{\vartheta}{2}\right]^{-1},$$

where ϑ is the scattering angle of the γ-ray. As $\omega \to \infty$ we have

$$\hbar\omega' \to \tfrac{1}{2}mc^2 \operatorname{cosec}^2 \tfrac{1}{2}\vartheta.$$

Except at forward angles (which are of no interest as they would interfere with the measurement) ω' has a definite upper limit, and so (2) gives Δx a lower limit of order $\hbar/mc$, the Compton wavelength. For a nonrelativistic particle, this is far smaller than its de Broglie wavelength (1), because

$$\frac{\lambda\!\!\!^{-}_{\text{Compton}}}{\lambda\!\!\!^{-}_{\text{de Broglie}}} = \frac{p}{mc} \simeq \frac{v}{c}, \tag{4}$$

which we can consider as zero in nonrelativistic quantum mechanics ($c \to \infty$). Completely new phenomena (such as pair creation) occur when one attempts to reduce Δx below $\hbar/mc$.

We should also like to know to what extent one can assign a time to this precise position measurement. The uncertainty in this time is given by the time the scattered packet takes to pass any given point. This is given by

$$\Delta t \gtrsim \frac{1}{\omega'} \geqslant \hbar/mc^2,$$

i.e., the time it takes light to traverse the distance $\hbar/mc$. This time is short compared to any time of significance in nonrelativistic physics, and we may therefore consider it to be negligible.

Consider next the possibility of repeating a coordinate measurement. After a time δt the struck particle will have moved from the first observation point to $\delta t(\mathbf{p} + \Delta\mathbf{p})/m$. Hence we cannot tell precisely where the particle will be found in the future, and we must be satisfied with statistical predictions. Nevertheless, by decreasing δt sufficiently, we can reduce this stochastic drift of the particle as much as we please. In principle, therefore, the position measurement is reproducible, and the notion that the particle is at a definite point in space at a given instant seems to be defined within our framework of ideas.

We are now in a position to make the fundamental assumption: *To every state of a nonrelativistic particle there corresponds at every time t a probability* $W(\mathbf{x}, t)\, d^3x$ *of finding the particle in the interval* $(\mathbf{x}, \mathbf{x} + d\mathbf{x})$. $W(\mathbf{x}, t)$ is called the coordinate probability density. This assumption is also reasonable for a particle in a potential, provided the potential is small compared to mc^2 or, better put, provided the external forces cannot supply an appreciable momentum transfer to the particle during the time $\hbar/mc^2$ required to measure the position.

The existence of a probability distribution $W(\mathbf{x}, t)$ does not follow from $\Delta x\, \Delta p \gtrsim \hbar$. Thus a photon cannot be localized to a distance smaller than its wavelength. It is therefore not surprising that in quantum electrodynamics a configuration-space probability cannot be defined for light quanta.

Since the uncertainty principle implies some sort of symmetry between $\mathbf{p}$ and $\mathbf{x}$ we should also look into the question of the existence * of a probability distribution for the momentum, $W(\mathbf{p}, t)$. Once more we investigate whether $\mathbf{p}$, in the nonrelativistic limit, can be measured to

* In doing so we do not mean to imply that $W(\mathbf{p}, t)$ can be computed from $W(\mathbf{x}, t)$, or vice versa; we shall soon see that no unique connection between these functions exists (see (4.21)). The most striking illustration of the subtlety of the situation is again supplied by the electromagnetic field: whereas a physically acceptable (i.e., positive definite) $W(\mathbf{x}, t)$ cannot be defined for photons, a p-space probability can. The best known example of the latter is the Planck distribution law for photons in thermal equilibrium with their source.

arbitrary precision at an instant of time. To measure $\mathbf{p}$ we would roughly localize a particle at a point $\mathbf{x}_1$ and again at $\mathbf{x}_2$ (with large errors $\Delta\mathbf{x}_1$ and $\Delta\mathbf{x}_2$ so as not to affect the momentum too much), and measure the time taken for the traversal of the length $|\mathbf{x}_1 - \mathbf{x}_2| = L$. In order to minimize the importance of $\Delta\mathbf{x}_1$ and $\Delta\mathbf{x}_2$ in the determination of L, the latter should be large, and hence the time taken for the measurement will be long. Note that after such a measurement, $\mathbf{p}$ is still known with considerable accuracy, which means that, given enough time, one can measure the momentum with great precision and cause only a "small change of state." This is simply a consequence of the fact that for an almost free particle, $\mathbf{p}$ is almost a constant of the motion. On the other hand, the coordinate is not at all conserved, and hence a state in which the position is well defined must be very nonstationary.

The method just outlined does not provide us with a particle having a certain value of $\mathbf{p}$ at a definite time t. To prepare particles in a state of fairly definite $\mathbf{p}$ at a well-defined time, we can use the Compton effect once more. We first use the "slow" momentum measurement of the preceding paragraph to prepare a target of very low momentum $p_0 \ll \hbar\omega/c$, where ω is the frequency of the γ-rays used in the subsequent Compton experiment. The momentum of the Compton-scattered electrons is then determined to within an error of order p_0 in a time of order $1/\omega'$, where ω' is again the frequency of the scattered light. Thus we conclude that the momentum can be assigned at a time which is definite to within $\Delta t \gtrsim \hbar/mc^2$. The identical lower limit occurred in the position measurement.

The situation is thus the same as for the x-measurement. It is therefore natural to conjecture that *for every state of a nonrelativistic particle there exists at every time t a probability $W(\mathbf{p}, t)\, d^3p$ of finding the particle's momentum in the interval* $(\mathbf{p}, \mathbf{p} + d\mathbf{p})$.

3. *Uncertainty Relations for the Electromagnetic Field Strengths*

We have deduced the uncertainty relations for the dynamical variables of a particle from the conservation laws and the properties of photons. Because the motion of charged particles can be deduced from their attendant fields, it is clear that fields cannot be measured to arbitrary precision. Furthermore, the fact that the electromagnetic field has certain corpuscular aspects would lead one to expect some essential modification of the very notion of a field. Now we close the circle by using our new knowledge concerning particles to derive uncertainty relations for the electromagnetic field strengths. Our discussion is a highly simplified version of the original argument due to Bohr and Rosenfeld.*

To measure a field strength we require a test charge. Let this be a body of mass m and charge Q, which extends over the volume Ω_1. By measuring the momentum of the test body at times t_0 and t_1 we can determine the value of the electric field $\mathbf{\mathcal{E}}$ averaged over Ω_1 and $T_1 = t_1 - t_0$:

$$\langle \mathbf{\mathcal{E}} \rangle_{V_1} = \frac{\mathbf{p}_1 - \mathbf{p}_0}{QT_1}, \tag{1}$$

where $V_1 = \Omega_1 T_1$ is a volume in space-time. If the momentum of the test body is uncertain by $\Delta\mathbf{p}$, then the average field strength is uncertain by the amount

$$\Delta\mathcal{E}(V_1) \gtrsim \frac{\Delta\mathbf{p}}{QT_1} \gtrsim \frac{\hbar}{QT_1\,\Delta\mathbf{x}}, \tag{2}$$

where $\Delta\mathbf{x}$ is the uncertainty in the test body's position. We note that (2) can be made as small as we please by increasing Q.

Consider now the problem of measuring both $\mathcal{E}_x$ in the space-time region V_1, and some other field-component in the region $V_2 = \Omega_2 T_2$. We shall assume that V_2 lies entirely in the future relative to V_1. Hence the measurement of $\mathcal{E}_x$ in V_1 can disturb the measurement in V_2, but not vice versa. Thus an uncertainty in the motion of the first test body will lead to unknown fields which propagate into V_2 and confuse observations there. In particular, the uncertain position Δx means that we have an uncertain dipole-moment density $Q\,\Delta x/\Omega_1$ in the x-direction. At distances large compared to Δx the leading contribution to $\Delta\mathcal{E}$ arises from the dipole field. When $\Delta p \ll mc$ the uncertainty in the retarded dipole potential is

$$\Delta\phi(\mathbf{r}_2, t_2) \simeq \frac{Q\,\Delta x}{\Omega_1} \int_{V_1} dV_1 \frac{\partial}{\partial x_1} \frac{\delta(t_2 - t_1 - |\mathbf{r}_2 - \mathbf{r}_1|c^{-1})}{|\mathbf{r}_2 - \mathbf{r}_1|}, \tag{3}$$

where we have introduced Dirac's δ-function. There is also a current in the x-direction due to the undetermined motion of the test charge, which gives rise to an uncertainty in the x-component of the vector potential,† $\Delta A_x(\mathbf{r}_2, t_2)$. Hence a measurement in V_1 of $\mathcal{E}_x$ leads to un-

* N. Bohr and L. Rosenfeld, *Det. Kgl. dansk. Vid. Selskab.* **12,** No. 8 (1933); L. Rosenfeld, in *Niels Bohr and the Development of Physics,* edited by W. Pauli. McGraw-Hill (New York, 1955). See also Heitler, pp. 81–86.

† This is evaluated in Heitler, p. 85.

certainties in $\mathcal{E}$, $\mathcal{H}_y$, and $\mathcal{H}_z$ in V_2, because

$$\Delta\mathcal{E} = -\operatorname{grad}(\Delta\phi) - \frac{1}{c}\frac{\partial}{\partial t}\Delta\mathbf{A}, \qquad \Delta\mathcal{H} = \operatorname{curl}(\Delta\mathbf{A});$$

where only ΔA_y and ΔA_z may vanish. Therefore we have

$$\Delta\mathcal{E}_x(V_1)\,\Delta\mathcal{H}_x(V_2) \gtrsim 0, \tag{4}$$

but all other uncertainty products have a nonzero lower bound. If we assume that $\Delta A_y = 0$, we find that $\Delta\mathcal{E}_y$ averaged over V_2 is

$$\begin{aligned}\Delta\mathcal{E}_y(V_2) &\simeq -\frac{1}{V_2}\int_{V_2} dV_2 \frac{\partial}{\partial y_2}\Delta\phi(\mathbf{r}_2, t_2)\\ &= -\frac{Q\,\Delta x}{\Omega_1 V_2}\int_{V_1} dV_1 \int_{V_2} dV_2 \frac{\partial^2}{\partial x_1\,\partial y_2}\cdot\frac{\delta(t - r/c)}{r},\end{aligned} \tag{5}$$

with $r = |\mathbf{r}_1 - \mathbf{r}_2|$, $t = t_2 - t_1$. According to (2), therefore,

$$\Delta\mathcal{E}_x(V_1)\,\Delta\mathcal{E}_y(V_2) \gtrsim -\frac{\hbar}{V_1 V_2}\int_{V_1} dV_1 \int_{V_2} dV_2 \frac{\partial^2}{\partial x_1\,\partial y_2}\frac{\delta(t - r/c)}{r}. \tag{6}$$

Eventually (see Sec. 53) we shall also derive this uncertainty relation from the quantum theory of the electromagnetic field.

We observe that in this result the properties of the test body (Q, m) and the latter's uncertainty in position Δx have all disappeared; only $\hbar$, c and geometrical factors remain. *This shows that the uncertainty relationship is to be viewed as an intrinsic property of the field which does not depend on the actual measurement technique used* (and of course vanishes when $\hbar \to 0$). We may also note that the δ-function in (6) merely insures that the measurement of $\mathcal{E}_y(V_2)$ will be uninfluenced by that of $\mathcal{E}_x(V_1)$ if V_2 lies outside the union of all the future light cones originating from V_1.

It is now clear that in a complete quantum theory not only the particles but also the fields must be "quantized." If we were to treat part of the universe with classical physics, we could, in principle, arrive at inconsistencies. In any particular practical calculation, it may not be necessary to treat part of the system (e.g., an electromagnetic field) with quantum mechanics, but in the last analysis, all components of a system will have certain quantal aspects. We shall begin our construction of the formal theory by treating the electromagnetic field classically. For many purposes this is, in fact, permissible. Eventually, however, we shall also incorporate the field into the theory. We shall then find that in addition to the uncertainty relations (6), it is also impossible to determine the field strengths and the number of photons simultaneously. This shows the mutual exclusiveness of the photon and field-strength descriptions of electromagnetic phenomena and is a very illuminating example of Bohr's principle of complementarity.

REFERENCES

Bohm, Chapts. 5 & 6.
Bohr.
Heisenberg.
Pauli, Secs. 1, 2.

II

Wave Mechanics

Our next task is to set up the equations that will enable us to predict the (statistical) results of measurements on a system subsequent to its preparation in some particular state. We shall restrict ourselves to nonrelativistic systems. Needless to say, the axioms that form the basis of the theory cannot be inferred in a strictly logical fashion from the experimental data. At this juncture the creators of the theory had to resort to intuition and Occam's razor. We shall confine ourselves here to a brief discussion of the most essential points, and not even attempt an enumeration of the axioms.*

4. The Free Particle

4.1 THE SCHRÖDINGER EQUATION

We begin with the simplest case, that of a particle of mass m which moves in a force-free region between measurement acts. As we already know,

* The form of the Schrödinger equation for a system of free particles can be inferred from rather weak assumptions by invariance arguments. Even in the case of interactions the structure of the equation is rather limited by symmetry requirements. These considerations will be presented in detail in Chapt. VI. Here we shall proceed in a more pedestrian manner.

wave-like properties must be attributed to this system, the wave vector $\mathbf{k}$ and frequency ω being related to the momentum and energy by

$$\mathbf{p} = \hbar\mathbf{k}, \qquad E = \hbar\omega. \tag{1}$$

Experiments on electron diffraction show that the interference phenomena associated with these "matter waves" obey the familiar laws of physical optics. We recall that the laws of conventional optics are a consequence of the linearity of Maxwell's equations, which permit the superposition of different waves (the Huygens construction). We shall take this to mean that the interference laws for de Broglie waves also arise from a fundamental equation that is *linear* in a wave function $\psi(\mathbf{x}, t)$. This basic assumption is known as the *Superposition Principle.*

In Sec. 2 we inferred the existence of probability distributions for the coordinate and momentum, $W(\mathbf{x}, t)$ and $W(\mathbf{p}, t)$. In virtue of their physical significance, these functions must have the following properties:

(a) positive definiteness,

$$W(\mathbf{x}, t) \geq 0, \qquad W(\mathbf{p}, t) \geq 0; \tag{2}$$

(b) conservation of probability,

$$\frac{d}{dt}\int d^3x\, W(\mathbf{x}, t) = \frac{d}{dt}\int d^3p\, W(\mathbf{p}, t) = 0; \tag{3}$$

(c) normalizability, which requires the existence of the integrals appearing in (b). We may then arrange the definitions of the W's so as to have

$$\int d^3x\, W(\mathbf{x}, t) = 1, \qquad \int d^3p\, W(\mathbf{p}, t) = 1. \tag{4}$$

Because we have already agreed to attempt a formulation in terms of a linear wave equation, we cannot identify the wave function $\psi(\mathbf{x}, t)$ with the probability $W(\mathbf{x}, t)$; a linear equation will not, in general, guarantee the positive-definite character of ψ. We must, therefore, suppose that W is some nonlinear functional of ψ, and we shall furthermore guess that this relationship is local in space and time, i.e., $W(\mathbf{x}, t)$ can be computed if ψ is given at the point $\mathbf{x}$. These assumptions are reminiscent of classical electrodynamics, where the fields $\mathcal{E}$ and $\mathcal{H}$ satisfy linear equations, whereas the energy density is quadratic in the field strengths and, therefore, positive definite. At this point in the argument the connection between $\psi(\mathbf{x}, t)$ and the $\mathbf{p}$-space probability is still obscure, however. Nevertheless, *we shall suppose that a wave function* $\psi(\mathbf{x}, t)$ *and an attendant linear equation exist, and that this* ψ *is the funda-*

mental mathematical object in the theory from which all observable quantities may be computed.

The next assumption we make is that the state of the system at any instant of time suffices to completely determine its evolution until the next measurement act. As we have already assumed that ψ uniquely determines the state of the system (in the sense that it allows the computation of all observable quantities), we are therefore asserting that the fundamental equation that determines ψ contains, at most, first-order derivatives with respect to the time. Finally, we shall assume a homogeneous equation, because we would expect inhomogeneous terms to destroy the conservation of probability, much as the inhomogeneous source terms in Maxwell's equations lead to changes in the total energy of the field.

All these assumptions having been made, we are now ready to set down the fundamental equations that govern the motion of free particles. In what follows, $\mathbf{x}$ is the position of a free "elementary" particle (i.e., one we assume to have no internal structure), or the center-of-mass coordinate of a complex system not subject to external forces. In this case, the de Broglie hypothesis associates a wave $e^{i(\mathbf{k}\cdot\mathbf{x}-\omega t)}$ with the system, the dispersion law being given by (1) and the relation between E and p, which remains to be specified. The assumed linearity of the fundamental equation therefore means that the most general ψ has the form

$$\psi(\mathbf{x}, t) = \frac{1}{(2\pi)^{\frac{3}{2}}} \int A(\mathbf{k}) e^{i(\mathbf{k}\cdot\mathbf{x}-\omega t)}\, d^3k \tag{5}$$

$$= \frac{1}{(2\pi)^{\frac{3}{2}}} \int A(\mathbf{k}) e^{i(\mathbf{p}\cdot\mathbf{x}-Et)/\hbar}\, d^3k. \tag{6}$$

Before proceeding any further, we had better make sure that the recipe (6) does not conflict with classical mechanics in the limit where the de Broglie wavelength is small. If this correspondence with classical mechanics were not contained in (6), we would know that we are on the wrong track. In order to investigate this point, we assume that ψ in (6) is confined to a spatial region V_x (a so-called wave packet). Our hypothesis concerning the relationship between the wave function and the probability distribution implies that $W(\mathbf{x}, t)$ vanishes outside V_x in this case. We may construct such a packet with a function $A(\mathbf{k})$ which is concentrated in a volume V_k of k-space whose dimensions are inversely proportional to V_x. Furthermore, we are free to center V_k at some mean propagation vector $\mathbf{k}_0$. Let us now ask how this packet moves in time. Although we shall answer this question in considerable detail in the latter parts of this section, we shall, for the moment, again take recourse to optics and recall that such a packet will move with the group velocity. As usual, the group velocity is evaluated by finding the point of stationary phase of the integrand in (6). This phase φ is $\mathbf{k}\cdot\mathbf{x} - \omega t + \alpha(\mathbf{k})$, where $\alpha(\mathbf{k})$ is the phase of the complex function $A(\mathbf{k})$. Thus the stationary phase point is

$$\left.\frac{\partial\varphi}{\partial k_i}\right|_{\mathbf{k}_0} = x_i - t\left.\frac{\partial\omega}{\partial k_i}\right|_{\mathbf{k}_0} + \left.\frac{\partial\alpha}{\partial k_i}\right|_{\mathbf{k}_0} = 0, \tag{7}$$

where i labels the three Cartesian coordinates. The only important contributions to the integral in (5) [or equivalently, in (6)] arise when (7) is satisfied. When (7) is violated the integrand oscillates violently and gives a small net result. This means that the values of $\mathbf{x}$ for which (5) is appreciable change with time, because $\partial\omega/\partial\mathbf{k}$ and $\partial\alpha/\partial\mathbf{k}$ in (7) are independent of t. The region V_x in which $W(\mathbf{x}, t)$ is significantly different from zero therefore moves about in space. The velocity of this motion can be read off from (7), viz.

$$\mathbf{v} = \frac{\partial\omega}{\partial\mathbf{k}} = \frac{\partial E}{\partial\mathbf{p}}, \tag{8}$$

the last equality being a consequence of (1). In the classical limit we require $\mathbf{v} = \mathbf{p}/m$; the dependence of the energy on the momentum must therefore be given by the familiar classical relationship

$$E = \frac{p^2}{2m}. \tag{9}$$

The dispersion law for de Broglie waves in the force-free case is therefore

$$\omega = \frac{\hbar|\mathbf{k}|^2}{2m}. \tag{9'}$$

By construction, the location of our packet as a function of time will be given by Newton's laws. On the other hand, at the moment we do not yet know whether the spatial extent of the packet will spread in time and thereby preclude a sensible classical limit. Nevertheless, we can feel confident that we have not committed any gross error in setting down (6) and (9). We shall therefore proceed to the construction of the basic wave equation from (6). Shortly thereafter we shall return to a careful examination of the motion of wave packets.

The fundamental wave equation of the theory is now a direct consequence of (6) and (9), and of the hypothesis that it be of first order in

$\partial/\partial t$. Namely (6), with E as given in (9), is the general solution of

$$i\hbar \frac{\partial}{\partial t} \psi(\mathbf{x}, t) = -\frac{\hbar^2}{2m} \nabla^2 \psi(\mathbf{x}, t). \tag{10}$$

This is *the Schrödinger wave equation for a free particle.* The explicit appearance of i is of very great significance, as it implies that ψ is not real in general; this is due to the fact that ψ^* satisfies a *different* equation, viz.,

$$i\hbar \frac{\partial}{\partial t} \psi^*(\mathbf{x}, t) = \frac{\hbar^2}{2m} \nabla^2 \psi^*(\mathbf{x}, t). \tag{10'}$$

4.2 PROBABILITY DISTRIBUTIONS

We now turn to the construction of the function $W(\mathbf{x}, t)$. Aside from the requirements of (2)–(4), we must again be guided by simplicity and experiment. Because of (2), Re ψ^2 and Im ψ^2 will not do. Neither will Re ψ nor Im ψ. The simplest expression that has a fighting chance is $\psi^*\psi$. Now we must see if (3) is satisfied by this expression. Using (10) and (10′) we have

$$i\hbar \frac{\partial}{\partial t} \psi^*\psi = \frac{\hbar^2}{2m} [(\nabla^2\psi^*)\psi - \psi^*\nabla^2\psi].$$

This can be cast into a continuity equation,

$$\frac{\partial}{\partial t} (\psi^*\psi) + \nabla \cdot \mathbf{j} = 0, \tag{11}$$

with the current being given by

$$\mathbf{j}(\mathbf{x}, t) = \frac{\hbar}{2mi} \{\psi^*\nabla\psi - \psi\nabla\psi^*\}. \tag{12}$$

Applying Gauss' theorem to (11), we obtain

$$\frac{\partial}{\partial t} \int d^3x \, \psi^*\psi = 0, \tag{13}$$

provided $\mathbf{j}$ tends to zero faster than $1/r^2$ at infinity. Such a fall-off is also necessary if the normalization integral (4) is to exist. We are therefore justified in calling *

$$W(\mathbf{x}, t) = |\psi(\mathbf{x}, t)|^2 \tag{14}$$

* There may be more complicated functionals W of ψ which satisfy (2)–(4). We shall not discuss other possibilities, however, because (14) agrees completely with experiment.

the coordinate-space probability distribution. As ψ satisfies a linear equation, we may always choose the arbitrary multiplicative constant in the solution of (9) in such a way as to satisfy the normalization condition

$$\int \psi^* \psi \, d^3x = 1. \tag{15}$$

Hence ψ has the dimension $(\text{length})^{-\frac{3}{2}}$. The vector field $\mathbf{j}(\mathbf{x}, t)$ is the probability current or flux; for example, j_z gives the probability per unit time that the particle has passed through the element of area $dx\,dy$.

Our next task is to deduce an expression for the momentum-space probability distribution, $W(\mathbf{p}, t)$. To this end we note that Parseval's relation tells us that

$$\int |\psi|^2 \, d^3x = \int d^3k |A(\mathbf{k})|^2 \equiv \int \frac{d^3p}{\hbar^3} \left| A\left(\frac{\mathbf{p}}{\hbar}\right)\right|^2. \tag{16}$$

The quantity $|A(\mathbf{p}/\hbar)|^2/\hbar^3$ is therefore a likely candidate. In virtue of (15), it also satisfies the normalization condition. It is independent of time and therefore obeys (3) trivially, and it is certainly positive definite. Let us therefore take $W(\mathbf{p}, t)$ to be *

$$W(\mathbf{p}, t) = |\phi(\mathbf{p}, t)|^2, \tag{17}$$

where †

$$\phi(\mathbf{p}, t) = \frac{1}{\hbar^{\frac{3}{2}}} A\left(\frac{\mathbf{p}}{\hbar}\right) e^{-iE_pt/\hbar}; \tag{18}$$

this quantity will henceforth be referred to as the momentum-space wave function. In virtue of (6) and the fundamental theorem of Fourier transforms, ϕ is related to ψ by

$$\phi(\mathbf{p}, t) = (2\pi\hbar)^{-\frac{3}{2}} \int d^3x \, \psi(\mathbf{x}, t) e^{-i\mathbf{p}\cdot\mathbf{x}/\hbar}. \tag{19}$$

For future reference we also rewrite (6) in terms of ϕ:

$$\psi(\mathbf{x}, t) = (2\pi\hbar)^{-\frac{3}{2}} \int d^3p \, \phi(\mathbf{p}, t) e^{i\mathbf{p}\cdot\mathbf{x}/\hbar}. \tag{20}$$

We may pause here to note that ψ does indeed determine all the observable quantities discussed so far. Thus $W(\mathbf{x}, t)$ is given by the simple

* It will be noticed that $W(\mathbf{p}, t)$ actually does not depend on time in this case. This is hardly surprising because there are no forces acting.

† Here E_p is given by (9); we will occasionally use this notation instead of E to emphasize that the energy depends on the momentum.

relation (14), while $W(\mathbf{p}, t)$ is given by the more complicated nonlocal relation

$$W(\mathbf{p}, t) = (2\pi\hbar)^{-3} \int d^3x\, d^3x'\, \psi(\mathbf{x}, t)\psi^*(\mathbf{x}', t)e^{-i\mathbf{p}\cdot(\mathbf{x}-\mathbf{x}')/\hbar}. \tag{21}$$

The more complicated observable quantities which we shall discuss momentarily will also be completely determined by ψ.

It will be recalled that one of the basic assumptions that entered into the theory was that the wave function at any instant would determine the future state of the system. To show that this is actually so we shall now demonstrate explicitly that given a wave function at some instant t_0 throughout all space, we can compute ψ at any other time. This follows directly from (20) and (18), according to which

$$\begin{aligned}\psi(\mathbf{x}, t) &= (2\pi\hbar)^{-\frac{3}{2}} \int d^3p\, e^{i[\mathbf{p}\cdot\mathbf{x} - E(t-t_0)]/\hbar}\phi(\mathbf{p}, t_0)\\ &= (2\pi\hbar)^{-3} \int d^3p\, d^3x_0\, e^{i[\mathbf{p}\cdot(\mathbf{x}-\mathbf{x}_0) - E(t-t_0)]/\hbar}\,\psi(\mathbf{x}_0, t_0).\end{aligned}$$

Therefore

$$\psi(\mathbf{x}, t) = \int d^3x_0\, G(\mathbf{x} - \mathbf{x}_0, t - t_0)\psi(\mathbf{x}_0, t_0), \tag{22}$$

where

$$\begin{aligned}G(\mathbf{x}, t) &= (2\pi\hbar)^{-3} \int d^3p\, e^{i(\mathbf{p}\cdot\mathbf{x} - E_p t)/\hbar}\\ &= \left(\frac{m}{2\pi i\hbar t}\right)^{\frac{3}{2}} \exp\left(\frac{im|\mathbf{x}|^2}{2\hbar t}\right).\end{aligned} \tag{23}$$

These equations provide us with the Huygens construction for the Schrödinger equation. $G(\mathbf{x}, t)$ is seen to be the solution which evolves out of $\delta(\mathbf{x})$ at $t = 0$. It is interesting to note that no relationship like (22) exists for the probabilities. In order to compute $W(\mathbf{x}, t)$ in terms of information at some other instant one needs the wave function at that time, because

$$\begin{aligned}W(\mathbf{x}, t) = \int d^3x_0\, d^3x_0'\, G^*(\mathbf{x} - \mathbf{x}_0, t - t_0)&\\ \times G(\mathbf{x} - \mathbf{x}_0', t - t_0)\psi^*(\mathbf{x}_0, t_0)\psi(\mathbf{x}_0', t_0)&.\end{aligned}$$

Hence the probability distribution at time t_0 does not determine W at some other time, and a theory that works purely in terms of probabilities appears to be impossible. This is equivalent to the statement that not only the magnitude, but also the phase, of the wave function contains information of physical significance. Only an $\mathbf{x}$-dependent phase matters in this connection, of course; multiplication of ψ (or ϕ) by an arbitrary complex constant of modulus one gives a wave function that describes the same physical state.

4.3 EXPECTATION VALUES, OPERATORS

We can now use our probability distributions to calculate the average values of various quantities in a state of the system represented by ψ (or ϕ). Thus the mean or *expectation value* of the position is*

$$\langle \mathbf{x} \rangle \equiv \int \mathbf{x} W(\mathbf{x}, t)\, d^3x$$
$$= \int \psi^*(\mathbf{x}, t)\mathbf{x}\psi(\mathbf{x}, t)\, d^3x. \tag{24}$$

By using (20) and partial integration we can write

$$\langle \mathbf{x} \rangle = (2\pi\hbar)^{-3} \int d^3p\, d^3p'\, d^3x\, \phi^*(\mathbf{p}, t) e^{-i\mathbf{p}\cdot\mathbf{x}/\hbar}\, \mathbf{x}\, e^{i\mathbf{p}'\cdot\mathbf{x}/\hbar}\, \phi(\mathbf{p}', t)$$
$$= (2\pi\hbar)^{-3} \int d^3p\, d^3p'\, d^3x\, \phi^*(\mathbf{p}, t) e^{i(\mathbf{p}'-\mathbf{p})\cdot\mathbf{x}/\hbar} i\hbar \frac{\partial}{\partial \mathbf{p}'}\, \phi(\mathbf{p}', t).$$

Here we have discarded a surface term in p-space under the assumption that ϕ vanishes sufficiently rapidly as $|\mathbf{p}| \to \infty$. We shall, for now, take this property of ϕ for granted, as well as the analogous property for ψ as $|\mathbf{x}| \to \infty$. The Dirac delta function $\delta(\mathbf{p})$ is given by †

$$(2\pi\hbar)^{-3} \int e^{i\mathbf{p}\cdot\mathbf{x}/\hbar}\, d^3x = \delta(\mathbf{p});$$

we therefore obtain

$$\langle \mathbf{x} \rangle = \int d^3p\, \phi^*(\mathbf{p}, t) i\hbar \frac{\partial}{\partial \mathbf{p}} \phi(\mathbf{p}, t). \tag{25}$$

Similarly the mean value of $\mathbf{p}$ is

$$\langle \mathbf{p} \rangle = \int d^3p\, \phi^*(\mathbf{p}, t)\, \mathbf{p}\, \phi(\mathbf{p}, t)$$
$$= \int \psi^*(\mathbf{x}, t) \frac{\hbar}{i} \frac{\partial}{\partial \mathbf{x}} \psi(\mathbf{x}, t)\, d^3x. \tag{26}$$

* We use the peculiar notation $\psi^*\mathbf{x}\,\psi$ instead of $|\psi|^2\mathbf{x}$ because the former is closely related to developments in the formalism yet to come.

† Readers not familiar with delta functions should consult Sec. 7.3 and the references quoted there.

In looking back over the preceding pages, especially Eqs. (19), (20), (24), (25), and (26), we note the great symmetry that exists between the p- and x-descriptions. In fact it is clear that we can give a theoretical description in either the coordinate or momentum "representation"; both are complete. To take advantage of this symmetry between the x- and p-representations we introduce the notion of an *operator*. Although this is merely a convenience at this point, the operator concept will become increasingly important as the theory is developed, and eventually it will be far more than a bit of technical shorthand. The first operator that we introduce is that associated with the momentum. In the x-representation it is the differential operator *

$$\tilde{\mathbf{p}} = \frac{\hbar}{i} \frac{\partial}{\partial \mathbf{x}}, \tag{27}$$

but in the p-representation it is merely multiplication by $\mathbf{p}$. In terms of $\tilde{\mathbf{p}}$ the Schrödinger equation reads

$$i\hbar \frac{\partial \psi}{\partial t} = \frac{\tilde{p}^2}{2m} \psi. \tag{28}$$

Note that the differential operator on the right-hand side of (28) has the same formal structure as the Hamiltonian of a free particle in classical mechanics. This is, of course, a consequence of our use of $E = p^2/2m$ in the de Broglie relations.

With the help of $\tilde{\mathbf{p}}$ we can write the expectation value of the momentum as

$$\langle \tilde{\mathbf{p}} \rangle = \int \psi^*(\mathbf{x}, t)\, \tilde{\mathbf{p}}\, \psi(\mathbf{x}, t)\, d^3x \tag{29}$$

in the x-representation. Since $\tilde{\mathbf{p}}$ is merely multiplication by $\mathbf{p}$ in the p-representation, we also have

$$\langle \tilde{\mathbf{p}} \rangle = \int \phi^*(\mathbf{p}, t)\, \tilde{\mathbf{p}}\, \phi(\mathbf{p}, t)\, d^3p.$$

In a similar fashion we also introduce the position operator $\tilde{\mathbf{x}}$, which is

$$\tilde{\mathbf{x}} = i\hbar \frac{\partial}{\partial \mathbf{p}} \tag{30}$$

* We shall designate operators by a tilde (~) for the time being. Soon experience will teach us to recognize whether a quantity is an operator or not without a distinguishing notation, and at that point we shall drop the tilde.

in the p-representation, and multiplication by $\mathbf{x}$ in the x-representation. The expectation value of $\bar{\mathbf{x}}$ can then be written in either of the following forms:

$$\langle \bar{\mathbf{x}} \rangle = \int \phi^*(\mathbf{p}, t)\, \bar{\mathbf{x}}\, \phi(\mathbf{p}, t)\, d^3p, \tag{31}$$

$$\langle \bar{\mathbf{x}} \rangle = \int \psi^*(\mathbf{x}, t)\, \bar{\mathbf{x}}\, \psi(\mathbf{x}, t)\, d^3x. \tag{32}$$

Let $F(\bar{\mathbf{x}})$ be any function of $\bar{\mathbf{x}}$ that can be expanded in a power series. Then

$$\begin{aligned}\langle F(\bar{\mathbf{x}}) \rangle &= \int \psi^*(\mathbf{x}, t) F(\mathbf{x}) \psi(\mathbf{x}, t)\, d^3x \\ &= \int \phi^*(\mathbf{p}, t) F\left(i\hbar \frac{\partial}{\partial \mathbf{p}}\right) \phi(\mathbf{p}, t)\, d^3p\end{aligned} \tag{33}$$

For a similarly defined function of $\mathbf{p}$, $G(\mathbf{p})$, we have

$$\begin{aligned}\langle G(\bar{\mathbf{p}}) \rangle &= \int \phi^*(\mathbf{p}, t) G(\mathbf{p}) \phi(\mathbf{p}, t)\, d^3p \\ &= \int \psi^*(\mathbf{x}, t) G\left(\frac{\hbar}{i} \frac{\partial}{\partial \mathbf{x}}\right) \psi(\mathbf{x}, t)\, d^3x.\end{aligned} \tag{34}$$

When G is not a polynomial, (34) is rather useless. It is then more appropriate to introduce the Fourier transform of G,

$$\mathcal{G}(\mathbf{x}) \equiv (2\pi\hbar)^{-3} \int d^3p\, e^{i\mathbf{p}\cdot\mathbf{x}/\hbar} G(\mathbf{p}),$$

in terms of which (34) can be written as

$$\langle G(\bar{\mathbf{p}}) \rangle = \int d^3x\, d^3x'\, \psi^*(\mathbf{x}, t) \mathcal{G}(\mathbf{x} - \mathbf{x}') \psi(\mathbf{x}', t). \tag{34'}$$

A similar relationship can easily be obtained for $F(\bar{\mathbf{x}})$:

$$\langle F(\bar{\mathbf{x}}) \rangle = \int d^3p\, d^3p'\, \phi^*(\mathbf{p}, t) \mathcal{F}(\mathbf{p} - \mathbf{p}') \phi(\mathbf{p}', t), \tag{33'}$$

with

$$\mathcal{F}(\mathbf{p}) \equiv (2\pi\hbar)^{-3} \int d^3x\, e^{-i\mathbf{p}\cdot\mathbf{x}/\hbar} F(\mathbf{x}).$$

When F is a polynomial, (33′) becomes (33), because $\mathcal{F}$ then involves derivatives of the Dirac δ-function. The same remark applies to (34′) when G is a polynomial.

We may now use these techniques to make a precise statement of the *uncertainty principle*. Naturally, in order to do so we must have a precise definition of the uncertainties $\Delta\mathbf{x}$ and $\Delta\mathbf{p}$. We shall adopt the root-mean-square definition of these quantities, i.e.,

$$(\Delta x_i)^2 = \langle(\bar{x}_i - \langle\bar{x}_i\rangle)^2\rangle = \langle\bar{x}_i^2\rangle - \langle\bar{x}_i\rangle^2, \tag{35}$$

and

$$(\Delta p_i)^2 = \langle(\bar{p}_i - \langle\bar{p}_i\rangle)^2\rangle = \langle\bar{p}_i^2\rangle - \langle\bar{p}_i\rangle^2. \tag{36}$$

In Sec. 24 we shall give a general way of deriving uncertainty relationships, but at the moment we shall use a special device. Define

$$D(\alpha) = \int d^3x |(\bar{x}_i - \langle\bar{x}_i\rangle)\psi + i\alpha(\bar{p}_i - \langle\bar{p}_i\rangle)\psi|^2 \geqslant 0; \tag{37}$$

a straightforward calculation then gives

$$\begin{aligned} D(\alpha) = (\Delta x_i)^2 &+ \alpha^2 \int d^3x [(\bar{p}_i - \langle\bar{p}_i\rangle)\psi]^*(\bar{p}_i - \langle\bar{p}_i\rangle)\psi \\ &+ i\alpha \int d^3x \{\psi^*(\bar{x}_i - \langle\bar{x}_i\rangle)(\bar{p}_i - \langle\bar{p}_i\rangle)\psi \\ &\qquad - [(\bar{p}_i - \langle\bar{p}_i\rangle)\psi]^*(\bar{x}_i - \langle\bar{x}_i\rangle)\psi\}. \end{aligned}$$

If $|\mathbf{x}|^2\psi \to 0$ as $|\mathbf{x}| \to \infty$, integrating by parts yields

$$D(\alpha) = (\Delta x_i)^2 + \alpha^2(\Delta p_i)^2 - \alpha\hbar.$$

Because this quadratic form in α is positive semi-definite, it can only have complex roots. Its discriminant is therefore negative, $\hbar^2 - 4(\Delta x_i)^2(\Delta p_i)^2 \leqslant 0$, which gives us the desired precise form of the uncertainty relation:

$$\Delta x_i \, \Delta p_i \geqslant \frac{\hbar}{2}. \tag{38}$$

4.4 THE MOTION OF FREE WAVE PACKETS

Let us now examine the time development of free packets more closely. From (18) and (25) we have

$$\begin{aligned} \langle\bar{x}_i\rangle &= \int d^3p \, e^{iE_pt/\hbar}\phi^*(\mathbf{p}, 0) i\hbar \frac{\partial}{\partial p_i} \phi(\mathbf{p}, 0) e^{-iE_pt/\hbar} \\ &= \int d^3p \, \phi^*(\mathbf{p}, 0) i\hbar \frac{\partial}{\partial p_i} \phi \, (\mathbf{p}, 0) + t \int d^3p \, \phi^*(\mathbf{p}, 0) \frac{\partial E}{\partial p_i} \phi(\mathbf{p}, 0), \end{aligned}$$

or

$$\frac{d}{dt}\langle\bar{x}_i\rangle = \left\langle\frac{\partial E}{\partial p_i}\right\rangle = \frac{1}{m}\langle\bar{p}_i\rangle, \tag{39}$$

and

$$\frac{d^2}{dt^2}\langle\bar{x}_i\rangle = 0. \tag{40}$$

The first equation just restates the fact that the group velocity is the "classical" velocity. Together (39) and (40) show that the mean value of $\bar{x}_i$ moves in accordance with the laws of classical mechanics.

In order to understand the limitations on the validity of classical mechanics, one must also study the time dependence of Δx_i (Δp_i is t-independent, of course). We have already seen that the mean position $\langle \bar{x}_i \rangle$ moves with constant velocity. Hence by going to a uniformly moving coordinate system we can always make $\langle \bar{x}_i \rangle = \langle \bar{p}_i \rangle = 0$. Let us therefore see what happens to $\Delta x_i(t)$ in this system. Because $(\Delta x_i)^2 = \langle \bar{x}_i^2 \rangle$ in this frame, we have

$$\langle \bar{x}_i^2 \rangle_t = \int \phi^*(\mathbf{p}, t) \left(i\hbar \frac{\partial}{\partial p_i} \right)^2 \phi(\mathbf{p}, t)\, d^3p$$
$$= \int \left| i\hbar \frac{\partial}{\partial p_i} \phi(\mathbf{p}, t) \right|^2 d^3p.$$

But

$$i\hbar \frac{\partial}{\partial p_i} \phi(\mathbf{p}, t) = i\hbar e^{-iE_p t/\hbar} \frac{\partial}{\partial p_i} \phi(\mathbf{p}, 0) + t \frac{\partial E}{\partial p_i} \phi(\mathbf{p}, t),$$

or

$$\langle \bar{x}_i^2 \rangle_t = \hbar^2 \int \left| \frac{\partial \phi(\mathbf{p}, 0)}{\partial p_i} \right|^2 d^3p + \frac{t^2}{m^2} \langle p_i^2 \rangle$$
$$+ \frac{i\hbar t}{m} \int p_i \left\{ \phi^*(\mathbf{p}, 0) \frac{\partial}{\partial p_i} \phi(\mathbf{p}, 0) - \phi(\mathbf{p}, 0) \frac{\partial}{\partial p_i} \phi^*(\mathbf{p}, 0) \right\} d^3p.$$

The first term is just $\langle \bar{x}_i^2 \rangle_0$. The last term can easily be cast into the form

$$\frac{\hbar t}{im} \int x_i \left\{ \psi^*(\mathbf{x}, 0) \frac{\partial}{\partial x_i} \psi(\mathbf{x}, 0) - \frac{\partial \psi^*(\mathbf{x}, 0)}{\partial x_i} \psi(\mathbf{x}, 0) \right\} d^3x$$
$$= 2t \int x_i j_i(\mathbf{x}, 0)\, d^3x$$

by use of (12) and (20), and therefore

$$\langle \bar{x}_i^2 \rangle_t = \langle \bar{x}_i^2 \rangle_0 + 2t \int x_i j_i(\mathbf{x}, 0)\, d^3x + \frac{t^2}{m^2} \langle \bar{p}_i^2 \rangle. \tag{41}$$

This shows that $\langle \bar{x}_i^2 \rangle$ is a quadratic form in t when there are no forces acting.

It must be remembered that (41) describes the spreading of the packet in the frame where $\Delta x_i = \sqrt{\langle \bar{x}_i^2 \rangle}$. For large values of t, $(\Delta x_i)^2$ is seen

to be a quadratically growing function of time. We may estimate the importance of the quadratic term in (41) by considering the ratio

$$\frac{t\,\Delta p}{m\,\Delta x_{t=0}} \sim \frac{\hbar t}{m(\Delta x_{t=0})^2},$$

where $\Delta x_{t=0}$ is the initial uncertainty in position. If we put $m \equiv N m_P$, where m_P is the mass of the proton, we obtain

$$\frac{t\,\Delta p}{m\,\Delta x_{t=0}} \sim 0.6 \times 10^{-4}\,\frac{t}{N(\Delta x_{t=0})^2}.$$

Here t is in seconds and $\Delta x_{t=0}$ in centimeters. For a macroscopic body $N \sim 10^{23}$, and therefore the spreading of the packet is totally insignificant on an astronomic time scale even for the smallest values of $\Delta x_{t=0}$ of interest. In the case of an alpha particle initially localized within a distance of order 10^{-11} cm, the ratio attains unity in 10^{-17} sec. Nevertheless, under the appropriate conditions, one can still correctly predict the motion of the α-particle with classical mechanics. To make this plausible, consider the scattering by a nucleus of such a localized α-particle moving with velocity $c/30$. This particle would traverse $\sim 10^{-8}$ cm before the spreading is significant. But this distance is large compared to nuclear dimensions (*circa* 10^{-12} cm), and it may then be possible to ascribe a classical trajectory to the α-particle during the collision. The precise criteria that must be met if such a classical description is to be accurate will be derived in Sec. 8.

Although the spreading of the wave packet may, under certain circumstances, be exceedingly small, it can never be completely eliminated. The Schrödinger equation is not Lorentz invariant, and contains no maximum velocity of propagation. It is therefore to be expected that a wave function $\psi(\mathbf{x}, 0)$ which vanishes identically outside some region will be nonzero everywhere for all $t > 0$. An example which illustrates this fact is the subject of Prob. 2. That there is an instantaneous spreading can also be seen from the fundamental solution $G(\mathbf{x}, t)$ of Eq. 23. In relativistic field theories such fundamental solutions always possess sharply defined wave fronts outside of which the wave function vanishes.

In spite of these remarks, it is possible to form packets with wave fronts which remain very sharp over long time intervals. For such packets the instantaneous spreading, although present, is negligible in the sense that the probability of finding the particle before the arrival of the wave front is exceedingly small. For our purposes the phrase "exceedingly small" means "smaller than relativistic corrections." To the extent that the Schrödinger equation is a nonrelativistic theory in any case, such errors need not concern us.

•For the sake of simplicity we shall construct a packet with a well-defined wave front only in one space dimension. At $t = 0$ we write the wave function as

$$\psi(x, 0) = \frac{1}{\sqrt{2\pi}} \int_{-\infty}^{\infty} dk\, e^{ikx} A(k). \tag{42}$$

Assume that $A(k)$ is analytic in the upper half of the complex k-plane, and grows less rapidly than an exponential as k tends to infinity in the upper half plane. When $x > 0$, we can add an infinite semicircle in the upper half plane to the integration path, and thereby show that $\psi(x, 0) = 0$ if $x > 0$. An example of such a function is

$$A(k) = A\,\frac{1}{(k - k_0 + i\kappa)^n}. \tag{43}$$

If $\kappa \ll k_0$, and $n \gg 1$, this packet is sharply peaked in momentum space about the mean value $\hbar k_0$, with a momentum uncertainty of order $\hbar\kappa$. By construction, (43) will insure $\psi(x, 0) = 0$ when $x > 0$. For $t > 0$, (42) becomes

$$\psi(x, t) = \frac{1}{\sqrt{2\pi}} \int_{-\infty}^{\infty} e^{i(kx - \hbar k^2 t/2m)} A(k)\, dk. \tag{44}$$

The presence of k^2 in the exponent prevents us from closing the contour* and using Cauchy's theorem when $t > 0$; strictly speaking, $\psi(x, t)$ is nonzero almost everywhere unless $t = 0$. Nevertheless, if $A(k)$ is sharply peaked near some value k_0 (for example, as in (43)), we may approximate k^2 by $k_0^2 + 2k_0(k - k_0)$ in the exponent of (44), and thereby obtain a sharp front:

$$\begin{aligned} \psi(x, t) &= \frac{1}{\sqrt{2\pi}} e^{i\hbar k_0^2 t/2m} \int_{-\infty}^{\infty} e^{ik(x - v_0 t)} A(k)\, dk \\ &= e^{i\hbar k_0^2 t/2m} \psi(x - v_0 t, 0), \end{aligned} \tag{45}$$

where $v_0 = \hbar k_0/m$ is the mean velocity of the packet. Clearly (45) vanishes if $x > v_0 t$. The discarded terms in the exponent describe the spreading of the packet. Their influence on the time development of ψ can be evaluated by expanding the $(\Delta k)^2$-term in the exponent as a power series:

$$\begin{aligned} \psi(x, t) &= e^{i\hbar k_0^2 t/2m} \psi(x - v_0 t, 0) \\ &\quad + \frac{1}{\sqrt{2\pi}} e^{i\hbar k_0^2 t/2m} \int_{-\infty}^{\infty} dk\, e^{i(kx - v_0 t)} A(k) \sum_{n=1}^{\infty} \frac{(k - k_0)^{2n}}{n!} \left(-\frac{i\hbar t}{2m}\right)^n \\ &= e^{i\hbar k_0^2 t/2m} \Bigg\{ \psi(x - v_0 t, 0) \\ &\quad + \sum_{n=1}^{\infty} \frac{1}{n!} \left(-\frac{it}{2m\hbar}\right)^n \left(\frac{\hbar}{i}\frac{\partial}{\partial x} - \hbar k_0\right)^{2n} \psi(x - v_0 t, 0) \Bigg\}. \end{aligned}$$

* When the energy is proportional to the momentum, the situation is completely different. Thus for light the exponent in (44) is $ik(x - ct)$, and Cauchy's theorem then shows that $\psi(x, t) = 0$ unless $x < ct$.

The correction terms are small if $(\Delta p)^2 t/m\hbar \ll 1$. In view of $\Delta p \sim \hbar/\Delta x_{t=0}$, the last inequality can also be written as $t\,\Delta p/m \ll \Delta x_{t=0}$. This is the condition required by (41) if the spreading of the packet is to be small after the time t has elapsed.•

5. Schrödinger's Equation for an Interacting System of Particles

We now turn to the problem of extending the theory to a system of particles which interact with each other and with any external fields that may be present. Two basic questions arise immediately:

(a) How are we to generalize the wave function, i.e., how much information do we expect it to contain?

(b) What is to replace the wave equation

$$i\hbar \frac{\partial}{\partial t} \psi(\mathbf{x}, t) = -\frac{\hbar^2}{2m} \nabla^2 \psi(\mathbf{x}, t)? \tag{1}$$

5.1 ONE PARTICLE IN AN EXTERNAL FIELD

Before tackling the more difficult problem of an interacting many-particle system, we consider a single particle in an external field. In this case we need not generalize ψ, because we already know that $W(\mathbf{x}, t)$ and $W(\mathbf{p}, t)$ are still well defined notions (see Sec. 2). A knowledge of ψ at one time previously determined ψ for all times, and it would seem that this should not depend on the presence of forces. It is therefore natural to conjecture that the basic equation is again of first order in $\partial/\partial t$. As electric and magnetic fields only affect the details of electron diffraction patterns, but do not cause interference phenomena to disappear, we assume that the wave equation remains linear under the present circumstance. The assumed form of the Schrödinger equation is thus

$$i\hbar \frac{\partial}{\partial t} \psi(\mathbf{x}, t) = \mathfrak{H}\psi(\mathbf{x}, t), \tag{2}$$

where, by hypothesis, $\mathfrak{H}$ is a linear operator. Assuming that $W(\mathbf{x}, t)$ is still $|\psi|^2$, we compute

$$i\hbar \frac{\partial}{\partial t} \int \psi^*\psi \, d^3x = -\int [(\mathfrak{H}\psi)^*\psi - \psi^*\mathfrak{H}\psi] \, d^3x, \tag{3}$$

which must be zero if this is a sensible choice of W; we therefore require

$$\int [(\mathfrak{H}\psi)^*\psi - \psi^*\mathfrak{H}\psi] \, d^3x = 0. \tag{4}$$

$\tilde{O}$ is linear, i.e.,

$$\tilde{O}(c_1\psi_1 + c_2\psi_2) = c_1\tilde{O}\psi_1 + c_2\tilde{O}\psi_2, \tag{5}$$

and therefore (4) implies that

$$0 = \sum_{i=1,2} c_i^* c_j \int [(\tilde{O}\psi_i)^*\psi_j - \psi_i^*\tilde{O}\psi_j]\, d^3x$$

for arbitrary complex constants c_1 and c_2. Thus for any two solutions ψ_1 and ψ_2, $\tilde{O}$ must satisfy

$$\int (\tilde{O}\psi_1)^*\psi_2\, d^3x = \int \psi_1^*\tilde{O}\psi_2\, d^3x; \tag{6}$$

such an operator is said to be *Hermitian*. (Note that $(\hbar/i)\partial/\partial\mathbf{x}$ is Hermitian in the space of normalizable functions, i.e., those that can be integrated by parts without surface terms.) To go the rest of the way in determining $\tilde{O}$, we must do a little more guessing. We note that in (1) $\tilde{O}$ is obtained from the classical Hamiltonian by the replacement

$$p_i \rightarrow \tilde{p}_i = \frac{\hbar}{i}\frac{\partial}{\partial x_i}. \tag{7}$$

In classical transformation theory, the Hamiltonian H is the generator of displacements in time. In the case of a free particle, the Schrödinger equation (1) tells us that the Hamiltonian operator $\tilde{H} = \tilde{p}^2/2m$ plays an analogous role, because the time evolution is completely determined by $\tilde{H}$. In fact, a formal solution of (1) can be written as*

$$\psi(\mathbf{x}, t) = e^{-i\tilde{H}t/\hbar}\psi(\mathbf{x}, 0). \tag{8}$$

Once one has recognized this relationship between the differential operator $-\hbar^2\nabla^2/2m$ and the classical kinetic energy, it is only natural to assume that the operator in the case where the particle is influenced by forces is also related to the classical Hamiltonian. We therefore write (2) as

$$i\hbar\frac{\partial}{\partial t}\psi(\mathbf{x}, t) = \tilde{H}\psi(\mathbf{x}, t), \tag{9}$$

where $\tilde{H}$ is obtained from the classical Hamiltonian by the substitution (7). For example, a particle in an external electromagnetic field, as described by the potentials $(\mathbf{A}, V)$, has the classical Hamiltonian

$$H = \frac{1}{2m}\left(\mathbf{p} - \frac{e}{c}\mathbf{A}\right)^2 + eV, \tag{10}$$

* That (8) is a solution of (1) can be verified by expanding the exponential in a power series. A more elegant derivation of (8) can be found in Sec. 28.

where **A** and V are given functions of **x** and t, and **p** is the canonical momentum. In this case the *Ansatz* reads *

$$i\hbar \frac{\partial}{\partial t}\psi(\mathbf{x}, t) = \left\{\frac{1}{2m}\left(\frac{\hbar}{i}\frac{\partial}{\partial \mathbf{x}} - \frac{e}{c}\mathbf{A}\right)^2 + eV\right\}\psi. \tag{11}$$

We note that the operator $(\hbar/i)(\partial/\partial\mathbf{x}) - (e/c)\mathbf{A}$ is Hermitian; this is so because $(\hbar/i)(\partial/\partial\mathbf{x})$ is Hermitian, and **A** is real. We again have a continuity equation:

$$\frac{\partial}{\partial t} W(\mathbf{x}, t) + \nabla \cdot \mathbf{j}(\mathbf{x}, t) = 0, \tag{12}$$

where the probability current is

$$\begin{aligned}\mathbf{j}(\mathbf{x}, t) &= \frac{1}{2m}\left\{\psi^*\left(\mathbf{\bar{p}} - \frac{e}{c}\mathbf{A}\right)\psi + \left[\left(\mathbf{\bar{p}} - \frac{e}{c}\mathbf{A}\right)\psi\right]^*\psi\right\} \\ &= \operatorname{Re}\left\{\psi^*\frac{1}{m}\left(\mathbf{\bar{p}} - \frac{e}{c}\mathbf{A}\right)\psi\right\}\end{aligned} \tag{13}$$

instead of $\operatorname{Re}\psi^*(\mathbf{\bar{p}}/m)\psi$. The current must always have the form (probability density) $\times$ (velocity operator). In the presence of an electromagnetic field, the classical velocity is $(1/m)(\mathbf{p} - (e/c)\mathbf{A})$, if **p** is the momentum canonically conjugate to **x**, and therefore (13) has a very plausible structure.

We may just point out that the existence of a continuity equation, i.e., a *local* conservation law, depends on the local nature of the Hamiltonian. The Hermiticity and linearity of the Hamiltonian guarantee that $\int\psi^*\psi\, d^3x$ is conserved, but they do *not* assure the existence of a probability current. (An illustration of this fact will be found in Prob. 4.)

5.2 THE MANY-PARTICLE WAVE EQUATION

Now we come to the question of how a many-particle system is to be described. Clearly the probability distribution will have to depend on the coordinates of all the N particles. The only question is whether we should make this distribution a function of N times, i.e., $W(\mathbf{x}_1, t_1, \ldots, x_N t_N) \cdot d^3x_1 \cdots d^3x_N$. The existence of such a W presupposes the possibility of measuring simultaneously and without mutual disturbance the position $\mathbf{x}_1$ of particle 1 at time t_1, that of particle 2 at time t_2, and so forth. On the other hand, when there are interactions, a coordinate measurement of particle 1 at time t_1 will affect particle 2 at space-time

* The form of this Schrödinger equation is intimately connected with the requirement of *gauge invariance*. This question is taken up in Prob. 7. The significance of gauge invariance in quantum mechanics is discussed in an exceptionally interesting paper by Y. Aharanov and D. Bohm, *Phys. Rev.* **115**, 485 (1959).

points $(\mathbf{x}_2, t_2)$ that satisfy $|t_2 - t_1| \geqslant |\mathbf{x}_1 - \mathbf{x}_2|/c$. In a nonrelativistic theory c must be treated as infinite, and particle 2 will therefore be influenced instantly when a coordinate measurement is performed on particle 1. Hence we can only measure all the coordinates at a given instant of time, and we therefore postulate the existence of a probability distribution with only a single time argument, $W(\mathbf{x}_1, \ldots, \mathbf{x}_N, t)$. This function must possess the usual reality and positive definiteness properties required of all probability distributions.

Some guidance in the construction of the theory can easily be obtained by considering a system of N noninteracting particles. In this case we would expect an uncorrelated probability distribution for the coordinates of the various particles, i.e., a distribution having the form *

$$W(\mathbf{x}_1, \ldots, \mathbf{x}_N, t) = \prod_{\alpha=1}^{N} W_\alpha(\mathbf{x}_\alpha, t). \tag{14}$$

This uncorrelated form will emerge if we continue to identify the probability distribution as $|\psi|^2$, and use the wave equation

$$\left(i\hbar \frac{\partial}{\partial t} - \sum_\alpha \tilde{H}_\alpha\right)\psi(\mathbf{x}_1, \ldots, \mathbf{x}_N, t) = 0, \tag{15}$$

where the Hamiltonian operator $\tilde{H}_\alpha$ only contains the coordinates of particle α. The kinetic energy operator of a system of free particles has this form, viz.,

$$\tilde{H} = \sum_{\alpha=1}^{N} \frac{\tilde{\mathbf{p}}_\alpha^2}{2m}. \tag{16}$$

That there are solutions of (15) which lead to (14) then follows from the fact that the product wave function

$$\psi(\mathbf{x}_1, \ldots, \mathbf{x}_N, t) = \prod_{\alpha=1}^{N} \psi_\alpha(\mathbf{x}_\alpha, t) \tag{17}$$

is a solution of (15) provided ψ_α satisfies the one-particle Schrödinger equation

$$\left(i\hbar \frac{\partial}{\partial t} - \tilde{H}_\alpha\right)\psi_\alpha(\mathbf{x}_\alpha, t) = 0. \tag{18}$$

* When the particles are indistinguishable, this intuitive argument fails, and there are correlations even when there are no interactions. These correlations arise from the fact that the correct wave function for a system of indistinguishable particles is a linear combination of solutions of the type (17). For such combinations the probability distribution does not have the factorized form (14). These questions are discussed in detail in Sec. 41.

Encouraged by this (slight) success, we postulate the following:

(a) The N-body system is described by a wave function $\psi(\mathbf{x}_1, \ldots, \mathbf{x}_N, t)$ which is the solution of a linear equation of first order in $\partial/\partial t$:

$$i\hbar \frac{\partial}{\partial t} \psi(\mathbf{x}_1, \ldots, \mathbf{x}_N, t) = \bar{H}\psi(\mathbf{x}_1, \ldots, \mathbf{x}_N, t);$$

(b) the operator $\bar{H}$ in this equation is constructed from the classical Hamiltonian by the substitution $\mathbf{p}_\alpha \to \bar{\mathbf{p}}_\alpha = (\hbar/i)\, \partial/\partial \mathbf{x}_\alpha$, where $\mathbf{p}_\alpha$ is the canonical momentum; and

(c) the joint probability distribution $W(\mathbf{x}_1, \ldots, \mathbf{x}_N, t)$ is still $|\psi|^2$.

As we shall see in Sec. 8, postulate (b) guarantees the correct classical limit. On the other hand, it leaves us free to add terms that have no classical counterpart, i.e., terms that vanish explicitly when $\hbar \to 0$. The presence of such terms in $\bar{H}$ can only be inferred from the experimental data—or from logical necessity. An illustration of the latter is provided by the requirement that $\bar{H}$ be Hermitian, because

$$i\hbar \frac{\partial}{\partial t} \int |\psi|^2 (dx) = \int (dx)[\psi^* \bar{H}\psi - (\bar{H}\psi)^*\psi],$$

must vanish. Here we use the abbreviation

$$(dx) \equiv \prod_{\alpha=1}^{N} d^3x.$$

The substitution involving $\mathbf{p}_\alpha$ in postulate (b) may lead to a non-Hermitian $\bar{H}$, and then terms (that vanish as $\hbar \to 0$) must be added so as to render it Hermitian. Examples of this procedure occur in Sec. 52.

It frequently happens that microscopic systems possess degrees of freedom that do not have any classical counterpart. The most familiar example of this situation is provided by spin. When this occurs, our recipes (a) to (c) fail. Guidance in the construction of the theory must then come from the experimental data, and more powerful theoretical concepts than those which we have already introduced. In this and the subsequent two chapters, we shall confine our attention to systems which have only the familiar classical degrees of freedom.

On the basis of assumptions (a)–(c) we may now write the Hamiltonian operator for a system of charged particles (having charges e_α and masses m_α) which interact through their Coulomb fields and are subjected to an applied electromagnetic field described by the potentials $A(\mathbf{x}, t)$ and

$V(\mathbf{x}, t)$:

$$\bar{H} = \sum_{\alpha=1}^{N} \left\{ \frac{1}{2m_\alpha} \left(\frac{\hbar}{i} \frac{\partial}{\partial \mathbf{x}_\alpha} - \frac{e_\alpha}{c} \mathbf{A}(\mathbf{x}_\alpha, t) \right)^2 + e_\alpha V(\mathbf{x}_\alpha, t) \right\} + \frac{1}{2} \sum_{\alpha \neq \beta} \frac{e_\alpha e_\beta}{|\mathbf{x}_\alpha - \mathbf{x}_\beta|}. \quad (19)$$

Let us set up a continuity equation for the system described by this Hamiltonian. As usual, we compute

$$\begin{aligned} i\hbar \frac{\partial}{\partial t} |\psi|^2 &= \psi^* \bar{H} \psi - (\bar{H}\psi)^* \psi \\ &= -\sum_\alpha \frac{\partial}{\partial \mathbf{x}_\alpha} \cdot \left[\frac{\hbar^2}{2m_\alpha} \left(\psi^* \frac{\partial}{\partial \mathbf{x}_\alpha} \psi - \psi \frac{\partial}{\partial \mathbf{x}_\alpha} \psi^* \right) \right] \\ &\quad - \sum_\alpha \frac{e_\alpha}{m_\alpha c} \frac{\hbar}{i} \left\{ \mathbf{A}_\alpha \cdot \left(\psi^* \frac{\partial}{\partial \mathbf{x}_\alpha} \psi + \psi \frac{\partial}{\partial \mathbf{x}_\alpha} \psi^* \right) + \psi^* \psi \nabla \cdot \mathbf{A}_\alpha \right\}, \end{aligned}$$

where $\mathbf{A}_\alpha \equiv \mathbf{A}(\mathbf{x}_\alpha, t)$. This equation can be cast into the form

$$\frac{\partial}{\partial t} W(\mathbf{x}_1, \mathbf{x}_2, \ldots, \mathbf{x}_N, t) + \sum_{\alpha=1}^{N} \frac{\partial}{\partial \mathbf{x}_\alpha} \cdot \mathbf{j}_\alpha(\mathbf{x}_1, \ldots, \mathbf{x}_N, t) = 0, \quad (20)$$

where

$$\mathbf{j}_\alpha(\mathbf{x}_1, \ldots, \mathbf{x}_N, t) = \mathrm{Re}\, \psi^* \tilde{\mathbf{v}}_\alpha \psi. \quad (21)$$

Here

$$\tilde{\mathbf{v}}_\alpha = \frac{1}{m_\alpha} \left(\tilde{\mathbf{p}}_\alpha - \frac{e_\alpha}{c} \mathbf{A}_\alpha \right)$$

is the velocity operator for partile α.

Note that (20) is a continuity equation in a $3N$-dimensional space. Although the current $\mathbf{j}_\alpha$ transforms like a 3-vector under a spatial rotation, the continuity equation states that the rate of change of W in a $3N$-volume is given by the flow through a $(3N - 1)$-dimensional hypersurface. One should never forget that the Schrödinger equation describes waves in a $3N$-dimensional configuration space, *not* in the three-dimensional physical space. In spite of this, it is possible to define electric charge and current densities of the conventional type. To this end, we ask for the probability of finding particle α at $\mathbf{r}$; this is clearly

$$\int (dx)\, \delta(\mathbf{r} - \mathbf{x}_\alpha) W(\mathbf{x}_1, \ldots, \mathbf{x}_N, t).$$

We therefore define the charge density as

$$\rho(\mathbf{r}, t) = \sum_{\alpha=1}^{N} e_\alpha \int (dx)\, \delta(\mathbf{r} - \mathbf{x}_\alpha) W(\mathbf{x}_1, \ldots, \mathbf{x}_N, t); \tag{22}$$

note that the total charge is correctly given by

$$\int \rho(\mathbf{r}, t)\, d^3r = \sum_\alpha e_\alpha.$$

In a similar fashion we define the electric current as

$$\mathbf{i}(\mathbf{r}, t) = \sum_{\alpha=1}^{N} e_\alpha \int (dx)\, \delta(\mathbf{r} - \mathbf{x}_\alpha) \mathbf{j}_\alpha(\mathbf{x}_1, \ldots, \mathbf{x}_N, t). \tag{23}$$

Because

$$\frac{\partial}{\partial \mathbf{r}} \delta(\mathbf{r} - \mathbf{x}_\alpha) = - \frac{\partial}{\partial \mathbf{x}_\alpha} \delta(\mathbf{r} - \mathbf{x}_\alpha),$$

we easily deduce the more conventional continuity equation

$$\frac{\partial}{\partial t} \rho(\mathbf{r}, t) + \nabla \cdot \mathbf{i}(\mathbf{r}, t) = 0 \tag{20'}$$

from (20).

The transcription of the theory into the momentum representation is done by means of a straightforward generalization of the formulas of Sec. 4. The p-space wave function is defined as

$$\phi(\mathbf{p}_1, \ldots, \mathbf{p}_N, t) = \left(\frac{1}{2\pi\hbar}\right)^{\frac{3}{2}N} \int (dx) \exp\left\{-\frac{i}{\hbar} \sum_\alpha \mathbf{p}_\alpha \cdot \mathbf{x}_\alpha\right\} \psi(\mathbf{x}_1, \ldots, \mathbf{x}_N, t), \tag{24}$$

and the inverse of this relation is

$$\psi(\mathbf{x}_1, \ldots, \mathbf{x}_N, t) = \left(\frac{1}{2\pi\hbar}\right)^{\frac{3}{2}N} \int (dp) \exp\left\{\frac{i}{\hbar} \sum_\alpha \mathbf{p}_\alpha \cdot \mathbf{x}_\alpha\right\} \phi(\mathbf{p}_1, \ldots, \mathbf{p}_N, t). \tag{25}$$

Again

$$\int |\psi|^2 (dx) = \int |\phi|^2 (dp), \tag{26}$$

where $(dp) = \Pi_\alpha\, d^3p_\alpha$, whence

$$\frac{\partial}{\partial t} \int (dp) |\phi(\mathbf{p}_1, \ldots, \mathbf{p}_N, t)|^2 = 0.$$

We are therefore justified in calling

$$W(\mathbf{p}_1, \ldots, \mathbf{p}_N, t) = |\phi(\mathbf{p}_1, \ldots, \mathbf{p}_N, t)|^2 \tag{27}$$

the p-space probability density.

•A warning is in order here. The transcription $\mathbf{p} \to (\hbar/i)\, \partial/\partial\mathbf{x}$ seems to work in general only when Cartesian coordinates are used. A counter example will show this. The 2-dimensional Laplacian in Cartesian coordinates is

$$\nabla^2 = \frac{\partial^2}{\partial x^2} + \frac{\partial^2}{\partial y^2},$$

whereas in cylindrical coordinates it becomes

$$\nabla^2 = \frac{\partial^2}{\partial r^2} + \frac{1}{r}\frac{\partial}{\partial r} + \frac{1}{r^2}\frac{\partial^2}{\partial \varphi^2}. \tag{28}$$

The classical Hamiltonian in cylindrical coordinates is

$$\frac{1}{2m}\left(p_r^2 + \frac{1}{r^2}\, p_\varphi^2\right), \tag{29}$$

and the obvious substitution would therefore lead to

$$-\frac{\hbar^2}{2m}\left[\frac{\partial^2}{\partial r^2} + \frac{1}{r^2}\frac{\partial^2}{\partial \varphi^2}\right] \tag{30}$$

for the Hamiltonian operator, which differs markedly from (28). Equation (30) is incorrect because it does not have all the invariances of (28); in particular, it is not invariant under space translations. In general, one should use Cartesian coordinates to make the transcription $\mathbf{p} \to (\hbar/i)\, \partial/\partial\mathbf{x}$, and make coordinate transformation afterwards. A general discussion of this question using the theory of curvilinear coordinates is given by Pauli, pp. 39–40.•

6. *Constants of the Motion and Symmetries of the Hamiltonian*

6.1 EXPECTATION VALUES OF OBSERVABLES AND THEIR TIME DEPENDENCE

In Sec. 4.3 we determined the time development of the expectation value of the position operator for a free particle. It is our purpose to extend this discussion to operators pertaining to interacting many-particle systems. Following Dirac, we shall call an operator which is associated with a physical attribute of a system an *observable*. The position, momentum, and Hamiltonian operators are all examples of observables.

By definition, the expectation value of an observable F which depends only on the coordinate operators * $\mathbf{x}_\alpha$ ($\alpha = 1, 2, \ldots, N$) can be evaluated with the help of the x-space wave function as

$$\langle F(\mathbf{x}_1, \ldots, \mathbf{x}_N)\rangle = \int \psi^*(\mathbf{x}_1, \ldots, \mathbf{x}_N, t) F \psi(\mathbf{x}_1, \ldots, \mathbf{x}_N, t)(dx). \quad (1)$$

This quantity must be real for any conceivable solution of the Schrödinger equation. F must, therefore, be a real function of the coordinates.

The expectation value of an observable $G(\mathbf{p}_1, \ldots, \mathbf{p}_N)$ is easiest to evaluate in terms of the p-space wave function (5.24):

$$\langle G(\mathbf{p}_1, \ldots, \mathbf{p}_N)\rangle = \int \phi^*(\mathbf{p}_1, \ldots, \mathbf{p}_N, t) G \phi(\mathbf{p}_1, \ldots, \mathbf{p}_N, t)(dp). \quad (2)$$

G must be a real function of the momenta. Just as for the free particle (see Eq. 4.34), we can, if we wish, compute $\langle G\rangle$ in the coordinate representation:

$$\langle G\rangle = \int \psi^* G\left(\frac{\hbar}{i}\frac{\partial}{\partial \mathbf{x}_1}, \ldots, \frac{\hbar}{i}\frac{\partial}{\partial \mathbf{x}_N}\right)\psi(dx). \quad (2')$$

But $\langle G\rangle$ is real for an arbitrary linear superposition $c_1\psi_1 + c_2\psi_2$ of linearly independent solutions of the Schrödinger equation. Thus we require

$$c_1^* c_2 \int \psi_1^* G \psi_2(dx) + c_2^* c_1 \int \psi_2^* G \psi_1(dx)$$

to be real, or

$$\int (\psi_1^* G \psi_2)^*(dx) = \int (G\psi_2)^* \psi_1(dx) = \int \psi_2^* G \psi_1(dx).$$

This shows that G, considered as a differential operator, is Hermitian (recall the definition (5.6) of Hermiticity).

Frequently we shall require the expectation values of observables that are functions of the coordinates and the momenta. The most important example of such an observable is the angular momentum. Such observables cannot be reduced to numerical functions by working in either the x- or the p-representation. Nevertheless, we shall postulate that for an arbitrary observable Θ which is a function both of the canonical coordinates and the momenta, the expectation value is given by the simplest generalization of (1) and (2′), i.e.,

$$\langle \Theta(\mathbf{x}_1, \ldots, \mathbf{x}_N; \mathbf{p}_1, \ldots, \mathbf{p}_N; t)\rangle = \int \psi^* \Theta\left(\mathbf{x}_1, \ldots, \mathbf{x}_N; \frac{\hbar}{i}\frac{\partial}{\partial \mathbf{x}_1}, \ldots, \frac{\hbar}{i}\frac{\partial}{\partial \mathbf{x}_N}; t\right)\psi(dx). \quad (3)$$

* Henceforth we shall only designate operators by a tilde if confusion might otherwise result.

The requirement that $\langle \mathcal{O} \rangle$ be real again implies that $\mathcal{O}$ must be Hermitian. We therefore conclude that *all observables are represented by Hermitian operators* in the mathematical formalism.

A reminder as to the empirical significance of these expectation values may be in order here. The assertion of the theory is: *If we take a number n of identically prepared systems in the state represented by the wave function* $\psi(\mathbf{x}_1, \ldots, \mathbf{x}_N, t)$, *and in each, at time t, measure the value of the physical quantity represented by the operator* $\mathcal{O}$, *then* $\int \psi^*(\mathbf{x}_1, \ldots, \mathbf{x}_N, t)\mathcal{O}\psi(\mathbf{x}_1, \ldots, \mathbf{x}_N, t)(dx)$ *equals the mean value of the results of these n measurements in the limit* $n \to \infty$. A little reflection quickly reveals that this is actually a very obscure statement. Except for the brief and rather superficial discussion of Chap. I, we have not explained how we are to prepare systems so that we can be sure that they are in the state ψ, nor have we shown how a measurement will give us the values with which we are to compute $\langle \mathcal{O} \rangle$. Unfortunately, it is virtually impossible to treat these questions in a cogent and undeceptive fashion unless one uses the mathematical formalism of quantum mechanics. The reader will therefore have to be patient and set aside some of the misgivings to which the rather formalistic treatment presented here will undoubtedly give rise. Once we have established more of the techniques of the theory, we shall return, in Chap. IV, to the question of how one prepares states and carries out measurements upon them.

The time dependence of the expectation value of an observable $\mathcal{O}$ cannot be evaluated explicitly unless the operator has special properties, or the Hamiltonian appearing in the wave equation is especially simple. We can, however, obtain useful expressions for the time derivatives of the expectation value. Assuming that ψ in (3) satisfies a Schrödinger equation with Hamiltonian H, we easily compute

$$i\hbar \frac{d}{dt} \langle \mathcal{O} \rangle = i\hbar \int \psi^* \frac{\partial \mathcal{O}}{\partial t} \psi(dx) + \int [\psi^* \mathcal{O} H \psi - (H\psi)^* \mathcal{O}\psi](dx).$$

But H is Hermitian,* and therefore

$$i\hbar \frac{d}{dt} \langle \mathcal{O} \rangle = i\hbar \left\langle \frac{\partial \mathcal{O}}{\partial t} \right\rangle + \langle \mathcal{O}H - H\mathcal{O} \rangle. \tag{4}$$

* Here we have claimed that $\int (H\psi)^* \mathcal{O}\psi(dx) = \int \psi^*(H\mathcal{O}\psi)(dx)$, but previously we had only required Hermiticity for linearly independent pairs of solutions of the Schrödinger equation. In general $\mathcal{O}\psi$ will not be a solution of the Schrödinger equation, however. Hence there is a gap in the argument leading to (4). We shall see in Sec. 7 that the linearly independent solutions of the Schrödinger equation form a complete set of functions. We may therefore write $\mathcal{O}\psi$ as a linear superposition of solutions, and thereby show that the step leading to (4) is legitimate.

This very important identity is sometimes referred to as Heisenberg's equation of motion. We shall actually reserve this name for a related, but somewhat more general, equation of motion (see Sec. 28). The combination of operators $AB-BA$ appears very frequently in quantum mechanics, and we therefore give it the special name of *commutator*, and designate it by

$$[A, B] \equiv AB - BA. \tag{5}$$

With this notation (4) becomes

$$i\hbar \frac{d}{dt}\langle \mathcal{O} \rangle = i\hbar \left\langle \frac{\partial \mathcal{O}}{\partial t} \right\rangle + \langle [\mathcal{O}, H] \rangle. \tag{6}$$

Most of the observables we shall deal with are not explicitly functions of the time; when this is the case (6) simplifies to

$$i\hbar \frac{d}{dt}\langle \mathcal{O} \rangle = \langle [\mathcal{O}, H] \rangle. \tag{7}$$

Higher derivatives can be computed by a repetition of the calculation; thus

$$\left(i\hbar \frac{d}{dt} \right)^2 \langle \mathcal{O} \rangle = \langle [[\mathcal{O}, H], H] \rangle, \tag{8}$$

and so forth. These equations give us the very important theorem that *the expectation value of an observable which commutes with the Hamiltonian is a constant of the motion.* This is the crudest type of quantum mechanical conservation law.* We observe that when H is not an explicit function of the time, H is a constant of the motion: *the expectation value of the energy is conserved.*

6.2 COMMUTATION RULES

It is now abundantly clear that we shall have to develop techniques for the evaluation of commutators. We begin with the canonical coordinates, which we shall designate by x_α^i, where the subscript refers to the particle and the superscript ($i = 1, 2, 3$) to the Cartesian components. In the x-representation these operators are simply numbers, and they therefore commute in this representation. In the p-representation

$$[x_\alpha^i, x_\beta^j]\phi(\mathbf{p}_1, \ldots, \mathbf{p}_N, t) = (i\hbar)^2 \left[\frac{\partial}{\partial p_\alpha^i}, \frac{\partial}{\partial p_\beta^j} \right] \phi(\mathbf{p}_1, \ldots, \mathbf{p}_N, t),$$

* The statement is crude because it only refers to expectation values. It will be generalized in Sec. 28.

which also vanishes. Thus we have

$$[x_\alpha^i, x_\beta^j] = 0. \tag{9}$$

By essentially the same argument we obtain

$$[p_\alpha^i, p_\beta^j] = 0. \tag{10}$$

The commutator $[x_\alpha^i, p_\beta^j]$ does not vanish, on the other hand. In the x-representation

$$[\bar{x}_\alpha^i, \bar{p}_\beta^j]\psi = \frac{\hbar}{i}\left(x_\alpha^i \frac{\partial \psi}{\partial x_\beta^j} - \frac{\partial}{\partial x_\beta^j} x_\alpha^i \psi\right) = i\hbar\, \delta_{ij}\, \delta_{\alpha\beta}\, \psi,$$

and in the p-representation

$$[\bar{x}_\alpha^i, \bar{p}_\beta^j]\phi = i\hbar\left(\frac{\partial}{\partial p_\alpha^i} p_\beta^j \phi - p_\beta^j \frac{\partial \phi}{\partial p_\alpha^i}\right) = i\hbar\, \delta_{ij}\, \delta_{\alpha\beta}\, \phi.$$

Thus we have

$$[x_\alpha^i, p_\beta^j] = i\hbar\, \delta_{ij}\, \delta_{\alpha\beta}. \tag{11}$$

Equations (9)–(11) are known as *the canonical commutation rules.* As we see, they take the same form in both of the representations already familiar to us. They are, in fact, not only invariant under transformations from the x- to the p-representation, and vice versa, but also under a far wider class of transformations. This fact will be demonstrated in Sec. 23.3. In the sequel we shall not require this theorem, but we shall only compute commutators in either the x- or p-representation, and leave it to the reader to convince himself that the results hold in both of these representations.

Let F again be any observable which is a function of the coordinates only. Clearly $[\mathbf{x}_\alpha, F] = 0$. On the other hand $[p_\alpha^i, F]\psi = -i\hbar(\partial F/\partial x_\alpha^i)\psi$, or

$$[p_\alpha^i, F(\mathbf{x}_1, \ldots, \mathbf{x}_N)] = \frac{\hbar}{i}\frac{\partial F}{\partial x_\alpha^i}. \tag{12}$$

Similarly

$$[x_\alpha^i, G(\mathbf{p}_1, \ldots, \mathbf{p}_N)] = i\hbar \frac{\partial G}{\partial p_\alpha^i}. \tag{13}$$

The angular momentum operator for particle α is defined in the classical manner as

$$\hbar \mathbf{L}_\alpha = \mathbf{x}_\alpha \times \mathbf{p}_\alpha. \tag{14}$$

We observe that parallel components of the vectors $\mathbf{x}_\alpha$ and $\mathbf{p}_\alpha$ are never multiplied together in $\mathbf{L}_\alpha$, and we therefore do not have to specify the order of the operator factors in (14). The explicit factor of $\hbar$ that appears in (14) has been inserted so as to make $\mathbf{L}_\alpha$ dimensionless. In order to simplify the formulas we shall *always* use dimensionless angular momenta in this book.

The evaluation of more complicated commutators is greatly facilitated by the two identities

$$\begin{aligned}[A, BC] &= [A, B]C + B[A, C],\\ [AB, C] &= A[B, C] + [A, C]B.\end{aligned} \tag{15}$$

With their help we can reduce the angular momentum commutator $[L^1, L^2]$ to $x^2[p^3, x^3]p^1 + x^1[x^3, p^3]p^2$. Using the canonical commutation rules we therefore have $[L^1, L^2] = iL^3$. This result is easily generalized to the very important *angular momentum commutation rule*

$$[L^i, L^j] = iL^k, \tag{16}$$

where (i, j, k) is a cyclic permutation of (1, 2, 3). In writing (16) we have deleted the particle index α, since it is perfectly obvious that $\mathbf{L}_\alpha$ and $\mathbf{L}_\beta$ commute unless $\alpha = \beta$.

The commutator of an angular with a linear momentum can be evaluated with equal ease. As x^i does not appear in L^i, it is obvious that

$$[L^i, p^i] = 0. \tag{17}$$

On the other hand, $\hbar[L^1, p^2] = [x^2, p^2]p^3$, and therefore

$$[L^i, p^j] = ip^k \qquad (i, j, k, \text{cyclic}). \tag{18}$$

We shall also require the commutator of $\mathbf{L}_\alpha$ with a function F of the $\mathbf{x}_\alpha$ and G of the $\mathbf{p}_\alpha$. These follow immediately from (12), (13) and (15):

$$i[\mathbf{L}_\alpha, F] = \mathbf{x}_\alpha \times \frac{\partial F}{\partial \mathbf{x}_\alpha}, \tag{19}$$

$$i[\mathbf{L}_\alpha, G] = \mathbf{p}_\alpha \times \frac{\partial G}{\partial \mathbf{p}_\alpha}. \tag{20}$$

Let us finally define the total linear and angular momentum operators by

$$\mathbf{P} = \sum_\alpha \mathbf{p}_\alpha, \tag{21}$$

and

$$\mathbf{J} = \sum_\alpha \mathbf{L}_\alpha. \tag{22}$$

As these are linear functions of the one-particle operators, it follows immediately that they satisfy the same commutation rules. In detail

$$[J^i, J^j] = iJ^k \qquad (i, j, k, \text{ cyclic}), \tag{23}$$
$$[J^i, P^j] = iP^k \qquad (i, j, k, \text{ cyclic}), \tag{24}$$
$$[J^i, P^i] = 0. \tag{25}$$

6.3 INFINITESIMAL TRANSLATIONS AND ROTATIONS. CONSERVATION OF LINEAR AND ANGULAR MOMENTA

In classical mechanics the total linear and angular momenta are constants of the motion if the Hamiltonian is invariant under translations and rotations, respectively. We shall now show that this statement continues to hold in quantum mechanics.

Although our argument could be generalized, we shall, for the sake of simplicity, assume that H has the form $T(\mathbf{p}_1, \ldots, \mathbf{p}_N) + V(\mathbf{x}_1, \ldots, \mathbf{x}_N)$. This is the case for a system with conservative interactions; T and V are then the kinetic and potential energies, respectively.

The time rate of change of the expectation value of the total linear momentum is given by

$$i\hbar \frac{d}{dt}\langle \mathbf{P} \rangle = \langle [\mathbf{P}, H] \rangle. \tag{26}$$

According to (12) and (21)

$$[\mathbf{P}, H] = \frac{\hbar}{i} \sum_\alpha \frac{\partial H}{\partial \mathbf{x}_\alpha}.$$

On the other hand, Taylor's theorem tells us that an infinitesimal translation $\delta\mathbf{a}$ of the coordinates can be expressed as

$$H(\mathbf{x}_1 + \delta\mathbf{a}, \ldots; \mathbf{p}_1, \ldots) - H(\mathbf{x}_1, \ldots; \mathbf{p}_1 \ldots) = \delta\mathbf{a} \cdot \sum_\alpha \frac{\partial H}{\partial \mathbf{x}_\alpha} = \frac{i}{\hbar}\, \delta\mathbf{a} \cdot [\mathbf{P}, H]. \tag{27}$$

If H is translationally invariant, the left-hand side of (27) vanishes. This proves that *the total linear momentum commutes with translation invariant observables.* One consequence of this fact is that *the expectation value of the total linear momentum is conserved if the Hamiltonian is translation invariant.*

Under an infinitesimal rotation $\mathbf{x}_\alpha \to \mathbf{x}_\alpha + \delta\boldsymbol{\omega} \times \mathbf{x}_\alpha$, $\mathbf{p}_\alpha \to \mathbf{p}_\alpha + \delta\boldsymbol{\omega} \times \mathbf{p}_\alpha$, the Hamiltonian changes by

$$\delta H = \sum_\alpha \left(\delta\boldsymbol{\omega} \times \mathbf{x}_\alpha \cdot \frac{\partial}{\partial \mathbf{x}_\alpha} + \delta\boldsymbol{\omega} \times \mathbf{p}_\alpha \cdot \frac{\partial}{\partial \mathbf{p}_\alpha} \right) H. \tag{28}$$

Equations (19) and (20) now apply directly, and allow us to rewrite the right-hand side of (28) as

$$\delta\omega \cdot \sum_{\alpha} \left(\mathbf{x}_\alpha \times \frac{\partial V}{\partial \mathbf{x}_\alpha} + \mathbf{p}_\alpha \times \frac{\partial T}{\partial \mathbf{p}_\alpha} \right) = i\, \delta\omega \cdot [\mathbf{J}, H]. \tag{29}$$

Therefore the total angular momentum commutes with rotation invariant observables, and the expectation value of the total angular momentum is a constant of the motion if H is invariant under rotation.

6.4 THE TWO-BODY PROBLEM

As an important illustration of the results just derived, we consider the isolated two-body problem. The Hamiltonian is

$$H = \frac{p_1^2}{2m_1} + \frac{p_2^2}{2m_2} + V(|\mathbf{x}_1 - \mathbf{x}_2|). \tag{30}$$

The interaction potential V is seen to be invariant under translations and rotations. We introduce center-of-mass and relative coordinates by

$$\begin{aligned} \mathbf{R} &= \frac{1}{M}(m_1\mathbf{x}_1 + m_2\mathbf{x}_2), \\ \mathbf{r} &= \mathbf{x}_1 - \mathbf{x}_2, \end{aligned} \tag{31}$$

where $M = m_1 + m_2$. From the rules of differential calculus one readily shows that the total momentum operator $\mathbf{P} = \mathbf{p}_1 + \mathbf{p}_2$ has the coordinate representation

$$\mathbf{P} = \frac{\hbar}{i}\left(\frac{\partial}{\partial \mathbf{x}_1} + \frac{\partial}{\partial \mathbf{x}_2}\right) = \frac{\hbar}{i}\frac{\partial}{\partial \mathbf{R}}. \tag{32}$$

Furthermore, the relative momentum $\mathbf{p} = (m_2\mathbf{p}_1 - m_1\mathbf{p}_2)/M$ is represented by

$$\mathbf{p} = \frac{\hbar}{iM}\left(m_2\frac{\partial}{\partial \mathbf{x}_1} - m_1\frac{\partial}{\partial \mathbf{x}_2}\right) = \frac{\hbar}{i}\frac{\partial}{\partial \mathbf{r}}. \tag{33}$$

The nonzero commutators involving the new variables are obviously

$$\begin{aligned} [R^i, P^j] &= i\hbar\, \delta_{ij}, \\ [r^i, p^j] &= i\hbar\, \delta_{ij}. \end{aligned} \tag{34}$$

The commutators still have the canonical structure.

The utility of the variables $(\mathbf{R}, \mathbf{P}, \mathbf{r}, \mathbf{p})$ is that in terms of them the Hamiltonian can be written as

$$H = \frac{1}{2M} P^2 + \frac{1}{2\mu} p^2 + V(r), \tag{35}$$

where $\mu = m_1 m_2/M$ is the reduced mass. The Hamiltonian therefore breaks up into two commuting pieces. The first, $P^2/2M$, is just the kinetic energy for the center-of-mass motion. The second portion is the Hamiltonian of a single particle of mass μ moving in the static potential $V(r)$.

The total angular momentum $\hbar\mathbf{J} = \mathbf{x}_1 \times \mathbf{p}_1 + \mathbf{x}_2 \times \mathbf{p}_2$ can also be split into operators relating separately to the center-of-mass and relative motion:

$$\mathbf{J} = \mathbf{L}_R + \mathbf{L}_r, \tag{36}$$

where

$$\hbar\mathbf{L}_R = \mathbf{R} \times \mathbf{P}, \tag{37}$$

$$\hbar\mathbf{L}_r = \mathbf{r} \times \mathbf{p}. \tag{38}$$

The observables which possess time-independent expectation values are those which commute with (30):

$$H, \quad \frac{P^2}{2M}, \quad \frac{p^2}{2\mu} + V, \quad \mathbf{P}, \quad \mathbf{L}_R, \quad \mathbf{L}_r. \tag{39}$$

It should be noted that in classical mechanics all of these quantities are also constants of the motion.*

7. *Eigenfunctions*

7.1 STATIONARY STATES

When the Hamiltonian is not a function of time, the Schrödinger equation

$$\left(i\hbar \frac{\partial}{\partial t} - H\right) \Psi(\mathbf{x}_1, \ldots, \mathbf{x}_N, t) = 0 \tag{1}$$

can be reduced to a time-independent equation. The latter is obtained by substituting

$$\Psi(\mathbf{x}_1, \ldots, \mathbf{x}_N, t) = e^{-iEt/\hbar} \psi_E(\mathbf{x}_1, \ldots, \mathbf{x}_N) \tag{2}$$

* The analogy with classical mechanics is actually very close, at least from a formal point of view. This can be seen by replacing the expression $[A, B]/i\hbar$ by the Poisson bracket of A and B in the equations of this section.

into (1):

$$(H - E)\psi_E = 0. \tag{3}$$

This is known as *the time-independent Schrödinger equation.* An equation of type (3) is called an *eigenvalue problem*, the number E being the *eigenvalue* and ψ_E the *eigenfunction* of the operator H. A number of very important properties of these eigenfunctions and eigenvalues can easily be derived.

(a) *The eigenvalues E are real.* To prove this we multiply (3) on the left by ψ_E^* and integrate over the whole of configuration space. We then obtain

$$E = \|\psi_E\|^{-1} \int \psi_E^* H \psi_E (dx), \tag{4}$$

where

$$\|\psi_E\| \equiv \int |\psi_E|^2 (dx) \tag{5}$$

is called the norm of ψ_E. The complex conjugate of (4) is

$$E^* = \|\psi_E\|^{-1} \int (H\psi_E)^* \psi_E (dx).$$

But H is Hermitian, and therefore $E = E^*$. *QED.*

(b) *In a state described by the wave function* $e^{-iEt/\hbar}\psi_E(\mathbf{x}_1, \ldots, \mathbf{x}_N)$, *the expectation value of any time-independent observables is stationary.* This is an immediate consequence of the reality of the eigenvalue E. By the same token, the x-space and p-space probability distributions, and the currents associated with them, are time independent. In view of these facts, one calls a state described by a wave function of type (2) *a stationary state.*

(c) *In a stationary state there is no dispersion in energy.* The dispersion in energy is defined in the usual way as

$$\sqrt{\langle H^2 \rangle - \langle H \rangle^2} = \Delta E. \tag{6}$$

But according to (3)

$$\langle H^n \rangle = \|\psi_E\|^{-1} \int \psi_E^* H^n \psi_E (dx) = E^n,$$

and therefore $\Delta E = 0$. *QED.* Furthermore, all the cumulants $\langle (H - \langle H \rangle)^n \rangle$ vanish. We therefore conclude that in a stationary state there is no uncertainty whatsoever in the energy, and the precise or sharp value of the energy is equal to the relevant eigenvalue of the Hamiltonian.

We shall only consider solutions of (3) that are physically sensible. By physically sensible we mean that $|\psi_E(\mathbf{x}_1, \ldots, \mathbf{x}_N)|^2$ must have the properties of a probability distribution. There are two aspects to this requirement: (1) The current (5.21) must be free of sources or sinks of probability, which is always satisfied by Hermitian Hamiltonians which do not possess pathological potentials.* (2) As long as $\|\psi_E\| < \infty$, we can always take advantage of the linearity of the Schrödinger equation to redefine ψ_E so that the probability distribution $|\psi_E|^2$ is normalized to unity. At first sight one might therefore demand $\|\psi_E\| < \infty$.

This last requirement is actually too stringent and eliminates solutions of the Schrödinger equation which we must retain. That this is so can be seen from the familiar example of a free particle. Here the eigenfunctions are $\psi_E(\mathbf{x}) = e^{i\mathbf{k}\cdot\mathbf{x}}$, and the corresponding eigenvalues E are $\hbar^2k^2/2m$. Let us call

$$\|\psi_E\|_\Omega \equiv \int_\Omega d^3x|\psi_E|^2,$$

where Ω is a finite integration volume. Clearly $\|\psi_E\|_\Omega \propto \Omega$, and therefore $\|\psi_E\| = \lim_{\Omega\to\infty} \|\psi_E\|_\Omega$ does not exist. The trouble is that the state described by the wave function $e^{i\mathbf{k}\cdot\mathbf{x}}$ is not localized; the probability of finding the particle in the volume element d^3x is independent of $\mathbf{x}$, and ψ_E must therefore vanish if we insist on normalizing $|\psi_E|^2$ to unity. In any realistic situation, e.g., an accelerator, the particles are always localized to some extent. As we know, such localized states are represented by wave packets which are linear combinations of stationary states belonging to various energy eigenvalues, and therefore nonstationary. Were we to banish solutions like $e^{i\mathbf{k}\cdot\mathbf{x}}$ because they are not normalizable, we would lose the possibility of building wave packets. Clearly we cannot afford to do this. We shall therefore admit solutions of infinite norm and consider them as the mathematical idealizations of wave packets. Whenever we wish to be realistic, we shall have to build nonstationary packets with these eigenfunctions. In admitting stationary wave functions of infinite norm we cannot be completely indiscriminate, however. In the free particle case, for example, we do not admit solutions of the type $e^{\boldsymbol{\gamma}\cdot\mathbf{x}}$, where γ is real. This function places the particle at infinity with overwhelming probability, and must be eliminated as nonphysical.† To be

* An example of a pathological potential would be one containing the derivative of a δ-function.

† It must be emphasized that there is nothing novel in our discarding certain solutions of infinite norm, while retaining others. Precisely the same thing must be done in the continuum theories of classical physics. In electromagnetic theory we frequently use plane waves such as $\mathcal{E} = \mathcal{E}_0 \cos(\mathbf{k}\cdot\mathbf{x} - ckt)$, even though the total energy and momentum of such disturbances are infinite. In any realistic situation $\int(\mathcal{E}^2 + \mathcal{H}^2)\,d^3x$ is finite; the plane waves are used as mathematical aids in the construction of such "normalizable" solutions. On the other hand, even though $\mathcal{E} = \mathcal{E}_0 \exp(\mathbf{k}\cdot\mathbf{x} - ckt)$, together with the associated magnetic field, constitutes a solution of Maxwell's equations, we must discard it, since it leads to an energy density which becomes infinite as $|\mathbf{x}| \to \infty$. This is to be contrasted with the plane waves, which have a finite energy density everywhere.

precise, *we shall demand that solutions of the time-independent N-particle Schrödinger equation satisfy*

$$\|\psi_E(\mathbf{x}_1, \ldots, \mathbf{x}_N)\|_\Omega \propto \Omega^n, \tag{7}$$

with $n \leqslant N$. The equality sign holds for the case of N free particles. At the other extreme, if $n = 0$, $\|\psi_E\|$ is finite, in which case all the particles must be localized. This applies to the ground state of an N-electron atom. We call such a localized state a *bound state*. When $0 < n < N$, the probability distribution for certain particles will tend to a constant as their coordinate tends to infinity, whereas others will be localized. For example, in the scattering of an electron by a helium atom, $N = 3$ but $n = 1$, because two of the electrons are bound to the nucleus. This last example brings out the general point that whenever $n \neq 0$ in (7), the energy eigenfunction describes a collision phenomenon *; we therefore call such states *scattering states*.

7.2 THE SPECTRUM OF THE HAMILTONIAN

The conditions on the norm which we have just spelled out can never be met for all values of E. In fact, we must demand that the energy eigenvalues have a lower bound. This is a stability requirement which the Hamiltonian must surely satisfy. Assuming that H is such that its eigenvalues have a lower bound E_0, the next question is whether all $E > E_0$ are eigenvalues. This question cannot be answered unless H is at least partially specified. Sometimes the requirement (7) on the norm will eliminate all solutions of (3) except for those belonging to a discrete set of values. The most common situation is that up to some value $E_c > E_0$ only a discrete set of eigenvalues correspond to sensible solutions, but for $E > E_c$ all values of E are admissible. In quantum mechanics it is this property of eigenvalue problems which leads to quantization.†

The set of all eigenvalues of the Hamiltonian is called *the energy spectrum*. Where this set is made up of isolated points, the spectrum is said to be discrete; otherwise it is called continuous.

Under almost all circumstances one can tell what type of spectrum a Hamiltonian possesses by considering the identical physical system from the classical point of view. There are a number of different cases which we take up separately.

* This statement does not apply to large systems such as one treats in statistical mechanics or solid state physics. It does hold as long as N is finite, however.

† The title of Schrödinger's first paper is "Quantization as an Eigenvalue Problem."

(a) If all the classical orbits are confined to a finite region of configuration space, the spectrum is discrete. The only realistic problem of this type occurs in statistical physics where we confine a system with an astronomic number of degrees of freedom in an enclosure. We leave such systems, and also the periodic potentials of solid state physics, out of our discussion at this time. Once this is done, the only systems where all orbits are confined are artificial constructs like the perfectly harmonic oscillator; the analogous quantum mechanical oscillator then has a purely discrete energy spectrum.

(b) If the classical system has only confined orbits for energy $E_m < E < E_c$, and only orbits of infinite extent for $E > E_c$, then the spectrum of H will be continuous for $E > E_c$, and discrete for $E_m < E < E_c$. In classical mechanics an attractive potential can bind states no matter how weak it is, because there are classical orbits in which the particles have no kinetic energy and take full advantage of the attraction. The situation is entirely different in quantum mechanics because the uncertainty principle guarantees that a great deal of localization leads to a correspondingly large dispersion in momentum, i.e., a large kinetic energy.* Consequently it usually takes a finite attraction to bind a state in quantum mechanics.† The upshot of all this is that the discrete spectrum below E_c that we would expect on the basis of the first sentence of this paragraph may be entirely absent.

(c) If the classical system only possesses unconfined ("hyperbolic") orbits, the spectrum is continuous.

(d) If there are classical orbits with finite turning points that have the same energy as unconfined orbits, these confined orbits will not appear as bound state solutions of the Schrödinger equation. This is one facet of the phenomenon of barrier penetration.

A set of four one-dimensional potentials which illustrate the possibilities (a)–(d) are shown in Fig. 7.1. For one particle in one dimension, and also in three dimensions when the potential is spherically symmetric, one can prove these statements concerning the spectrum which we have just made directly from the differential equation. Instead of discussing general proofs we shall, throughout this book, consider various specific problems which illustrate the general validity of these statements.

* It is easy to see that by the same argument the lowest energy eigenvalue E_0 satisfies $E_0 > E_m$, where E_m is the lowest possible energy of the classical problem. This classical minimum energy may, of course, be negatively infinite, whereas E_0 is finite. This is the case for the Coulomb field, for example.

† Exceptions to this statement are provided by one-dimensional problems, and by the Coulomb field which, owing to its infinite range, leads to binding for all values of the electric charge.

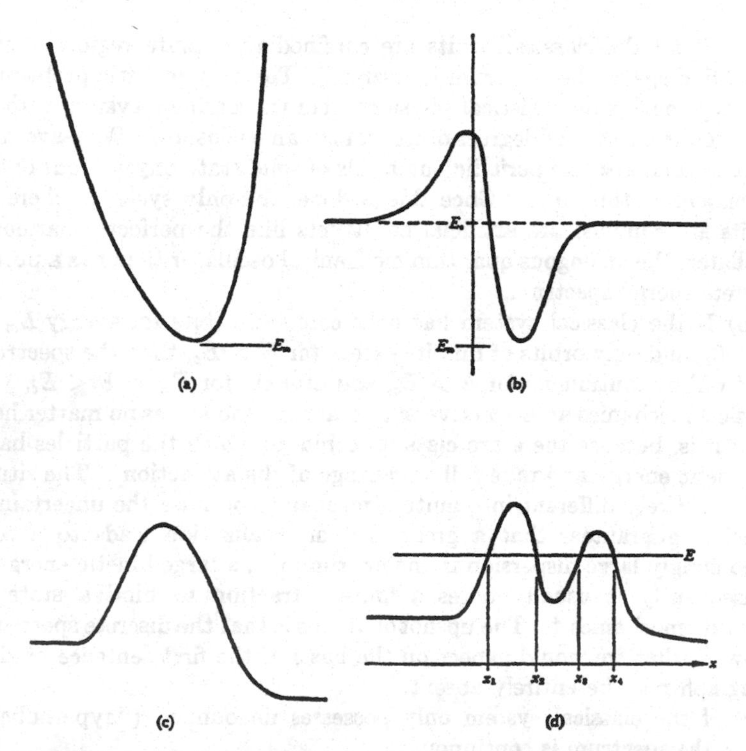

Fig. 7.1. Various one-dimensional potentials $V(x)$ illustrating statements (a)–(d) in the text. In case (a) the classical orbits are all confined, and the energy spectrum is discrete. In case (b) there may be bound states having energy eigenvalues between E_m and E_c. In case (c) there are no bound states, nor are there in (d). In classical mechanics the potential (d) would have bound states of energy E confined in the interval $x_2 < x < x_3$, and unbound states of energy E for $x < x_1$ and $x > x_4$. The wave function does not vanish in the classically forbidden regions $x_1 < x < x_2$ and $x_3 < x < x_4$, however. As a consequence, the energy spectrum associated with case (d) is purely continuous. For the same reason a particle incident from the right on the potential (b), and having an energy $E > E_c$ but smaller than the top of the barrier, may eventually be found on the left of the barrier.

A simple example will serve to show that the bound states, as defined above, belong to the discrete part of the spectrum, whereas the scattering states are in the continuum. Consider the one-dimensional problem of a particle of mass m interacting with the potential * $-\gamma\hbar^2\,\delta(x)/2m$. The

* This δ-function potential may be obtained from the potential $V(x) = -\hbar^2\gamma/4am$ for $|x| < a$, $V(x) = 0$ for $|x| > a$, by taking the limit $a \to 0$.

Schrödinger equation for this problem reads

$$\left(\frac{d^2}{dx^2} + \gamma\,\delta(x) + k^2\right)\psi_k(x) = 0, \tag{8}$$

where we have replaced the energy eigenvalue E by $\hbar^2k^2/2m$. When $x \neq 0$, this equation reduces to $\psi_k'' + k^2\psi_k = 0$, and therefore the solution must be a linear combination of $e^{\pm ikx}$. Furthermore, by integrating (8) from $x = -\epsilon$ to $x = +\epsilon$, we obtain

$$\psi_k'(\epsilon) - \psi_k'(-\epsilon) + \gamma\psi_k(0) + k^2\int_{-\epsilon}^{\epsilon}\psi_k(x)\,dx = 0.$$

If we now let $\epsilon \to 0$, we obtain

$$\lim_{\epsilon\to 0}\,[\psi_k'(\epsilon) - \psi_k'(-\epsilon)] = -\gamma\psi_k(0), \tag{9}$$

provided $\psi_k(x)$ is continuous at $x = 0$. But if $|\psi_k(x)|^2$ were discontinuous, there would be sources and sinks of probability, and therefore the continuity assumption implicit in (9) is unavoidable.*

When the energy E is negative, we replace k by $i\alpha$, where $\alpha > 0$ by definition. The form of $\psi_k(x)$ must then be $e^{-\alpha|x|}$ if it is to be normalizable. Substitution into (9) then yields $2\alpha = \gamma$. Thus for $\gamma < 0$ there are no solutions with negative energy. This was to be expected because $\gamma < 0$ gives a purely repulsive potential. On the other hand, when $\gamma > 0$ the potential is attractive and there is exactly one solution with $\alpha = \frac{1}{2}\gamma$, having energy $E_b = -\hbar^2\gamma^2/8m$:

$$\psi_b(x) = \sqrt{\tfrac{1}{2}\gamma}\; e^{-\frac{1}{2}\gamma|x|}. \tag{10}$$

This function has been arranged so that $\|\psi_b\| = 1$; it is obviously a localized, bound state. In this example E_b is the lower bound to the spectrum referred to above. Hence the spectrum below $E = 0$ is discrete.

For positive E there is no quantization, as one readily verifies by use of (9): all values of $E > 0$ lead to solutions. As we shall have further occasion to refer to these solutions, we shall write them out here.† For

* Furthermore, one easily shows that (9) guarantees that the current $(\hbar/2im)(\psi^*\psi' - \psi\psi'^*)$ is continuous at $x = 0$.

† The reader can easily verify that (9), together with the boundary conditions and the fact that $\psi_k(x)$ must be a linear combination of e^{ikx} and e^{-ikx}, suffices to determine (11) to within a factor. Alternatively, one can substitute (11) and (14) into (8), and show by direct calculation that they are solutions. In doing so one should bear in mind that $d\theta(x)/dx = \delta(x)$. A more elegant method of solving this problem is taken up in Prob. 6 of Chapt. III.

$k > 0$ we impose the boundary condition that we are to have a plane wave incident from the left (i.e., $x < 0$), and only a transmitted wave propagating to the right when $x > 0$. (These statements concerning the direction of propagation depend on our convention that the time dependence of a wave function is always $e^{-iEt/\hbar}$, and not $e^{iEt/\hbar}$.) The solution which satisfies these boundary conditions for $k > 0$ is

$$\psi_{k\rightarrow}(x) = \frac{1}{\sqrt{2\pi}}\{e^{ikx} + R(k)e^{-ikx}\theta(-x) + [T(k) - 1]e^{ikx}\theta(x)\}, \quad (11)$$

where $\theta(x)$ is the step function,

$$\begin{aligned} \theta(x) &= 1 \qquad (x > 0) \\ &= 0 \qquad (x < 0) \end{aligned}. \quad (12)$$

$|R|^2$ and $|T|^2$ are the reflection and transmission coefficients; they are given by

$$R(k) = \frac{i\gamma}{2|k|D(k)}, \qquad T(k) = 1 + \frac{i\gamma}{2|k|D(k)},$$
$$D(k) = 1 - \frac{i\gamma}{2|k|}. \quad (13)$$

If we replace k by $-k$ in (11), we obtain a solution which has a wave proceeding to the right on the left of the potential, even though the incident wave is coming in from the right. Hence (11), with $k < 0$, does not satisfy the physically sensible boundary condition. The solution which does have the correct behavior for negative values of k is

$$\psi_{k\leftarrow}(x) = \frac{1}{\sqrt{2\pi}}\{e^{ikx} + R(k)e^{-ikx}\theta(x) + [T(k) - 1]e^{ikx}\theta(-x)\}. \quad (14)$$

We see immediately that these scattering solutions, which belong to the continuous spectrum, are not normalizable, i.e.,

$$\int_{-L/2}^{L/2} |\psi_{k\rightleftarrows}|^2 \, dx$$

is of order L. They therefore satisfy the one-dimensional counterpart of (7).

7.3 ORTHOGONALITY AND COMPLETENESS

There are two further basic properties of the energy eigenfunctions which we must now discuss. We took advantage of these properties in our discussion of the free particle in Sec. 4, but did not state them explicitly there.

•A brief pseudo-mathematical digression will be necessary here. In order to simplify the notation we begin with the one-dimensional case; the generalization to three dimensions and N particles will be straightforward. We recall that Fourier's theorem states that a square integrable function $f(x)$ can be represented by the integral

$$f(x) = \frac{1}{\sqrt{2\pi}} \int_{-\infty}^{\infty} e^{ikx} A(k)\, dk, \tag{15}$$

where

$$A(k) = \frac{1}{\sqrt{2\pi}} \int_{-\infty}^{\infty} e^{-ikx} f(x)\, dx. \tag{16}$$

The equality sign here means that

$$\left\| f(x) - \frac{1}{\sqrt{2\pi}} \int_{-\infty}^{\infty} e^{ikx} A(k)\, dk \right\| = 0. \tag{17}$$

If we substitute (16) into (15) we obtain

$$f(x) = \int_{-\infty}^{\infty} \frac{dk}{\sqrt{2\pi}} e^{ikx} \int_{-\infty}^{\infty} \frac{dx'}{\sqrt{2\pi}} e^{-ikx'} f(x'). \tag{18}$$

We now introduce the symbolic or generalized function $\delta(x - x')$, which we shall always refer to as Dirac's δ-function, by

$$f(x) = \int \delta(x - x') f(x')\, dx'. \tag{19}$$

From a rigorous point of view, the δ-function is not a function, and (19) is nonsensical. Mathematicians * prefer to introduce $\delta(x - x')$ as the linear functional that assigns the number $f(x)$ to any sufficiently smooth function $f(x')$, and they therefore write (19) as $\delta_x[f(x')] = f(x)$. We shall not dwell on these refinements here. Suffice it to say that the δ-function never enters into any final result which we actually compare with experimental data. In the last analysis, we shall always integrate the δ-function multiplied by a smooth function (called a testing function by the mathematicians) over some finite interval, and thereby in fact use the δ-function as the linear functional described above.

We shall also have occasion to use derivatives of the δ-function. The nth derivative will be written as $\delta^{(n)}(x - x')$, and it is also a linear functional which assigns the number $d^n f(x)/dx^n$ to the n-fold differentiable function $f(x')$. In our more cryptic notation, we shall write this as

$$\frac{d^n f(x)}{dx^n} = \int_{-\infty}^{\infty} \delta^{(n)}(x - x') f(x')\, dx'. \tag{20}$$

* See Lighthill; Schwartz; Gel'fand and Shilov; Courant and Hilbert, Vol. II, Appendix.

In the remainder of this book we shall encounter many different representations of the δ-function and its derivatives. These are all equivalent in the sense that when they are multiplied by a sufficiently smooth function and integrated, they produce the function (or one of its derivatives) at some specified point. The most common such representation can be obtained by comparing (18) and (19):

$$\delta(x - x') = \int_{-\infty}^{\infty} \frac{dk}{2\pi} e^{ik(x-x')}. \tag{21}$$

It should be realized that this is nothing but a shorthand for Fourier's theorem.* In the same vein the representation

$$\delta^{(n)}(x - x') = \int_{-\infty}^{\infty} \frac{dk}{2\pi} (ik)^n e^{ik(x-x')} \tag{22}$$

is another way of stating Fourier's theorem and the fact that the Fourier transform of $d^n f/dx^n$ is $(ik)^n A(k)$.

The multi-dimensional counterparts of these representations are simply trivial generalizations of (21) *et seq.* Thus the three-dimensional δ-function is

$$\delta(\mathbf{x} - \mathbf{x}') \equiv \delta(x - x')\,\delta(y - y')\,\delta(z - z') = \int \frac{d^3k}{(2\pi)^3} e^{i\mathbf{k}\cdot(\mathbf{x}-\mathbf{x}')}.$$

The gradient of the δ-function, which we designate by $\boldsymbol{\delta}^{(1)}(\mathbf{x} - \mathbf{x}')$, is

$$\boldsymbol{\delta}^{(1)}(\mathbf{x} - \mathbf{x}') = i \int \frac{d^3k}{(2\pi)^3} \mathbf{k}\, e^{i\mathbf{k}\cdot(\mathbf{x}-\mathbf{x}')}; \tag{23}$$

clearly

$$\nabla f(\mathbf{x}) = \int d^3x'\, \boldsymbol{\delta}^{(1)}(\mathbf{x} - \mathbf{x}')\, f(\mathbf{x}').^{*} \tag{24}$$

We may now return to quantum mechanics. The energy eigenfunctions of a free particle in one dimension may be chosen as

$$\phi_k(x) = \frac{1}{\sqrt{2\pi}} e^{ikx}. \tag{25}$$

The energy eigenvalue associated with (25) is $\hbar^2 k^2/2m$. Comparing with (21) we observe that

$$\int \phi_k(x)\phi_k^*(x')\, dk = \delta(x - x'), \tag{26}$$

* In the orthodox theory of Fourier transforms, the functions $f(x)$ and $A(k)$ which appear in the basic relation (15) must be square integrable. This restriction can be removed if we allow $A(k)$, for example, to be a generalized function. The most trivial example is obtained by setting $A(k) = \sqrt{2\pi}\,\delta(k - k_0)$, in which case (15) produces e^{ik_0x}, which is not square integrable.

and also

$$\int \phi_k^*(x)\phi_{k'}(x)\, dx = \delta(k - k'). \tag{27}$$

By definition, one calls two functions $f(x)$ and $g(x)$ *orthogonal* * if $\int_{-\infty}^{\infty} f(x)^* g(x)\, dx = 0$. Equation (27) therefore tells us that two solutions of the free particle Schrödinger equation belonging to different energy eigenvalues are orthogonal. When the eigenvalues are equal (i.e., $k = k'$), (27) gives us $\|\phi_k(x)\|$. Our earlier discussion already had led us to the conclusion that $\|\phi_k(x)\|$ is infinite when the eigenfunction lies in the continuous spectrum, and (27) merely tells "how infinite it is." This last statement is not completely inane, for as we now know, there are a great number of these generalized functions (cf. (22)), and (27) tells precisely which one enters into the orthogonality relation. The constant which multiplies the δ-function in (27) is, however, at our disposal; we chose it to be unity by setting the multiplicative constant in (25) equal to $(2\pi)^{-\frac{1}{2}}$.

We shall now show that (27) generalizes to the following theorem: *Eigenfunctions of the Hamiltonian belonging to different eigenvalues are orthogonal.* To prove this we consider two eigenfunctions ψ_E and $\psi_{E'}$ of H, and note that

$$\psi_{E'}^*(H - E)\psi_E = 0,$$
$$\psi_E[(H - E')\psi_{E'}]^* = 0.$$

If we integrate these expressions over the N-particle configuration space, and then subtract, we obtain

$$(E - E') \int \psi_{E'}^* \psi_E (dx) = \int [\psi_{E'}^* H \psi_E - (H\psi_{E'})^* \psi_E](dx).$$

Because of the Hermitian character of H, the right side of this expression vanishes. *QED.*

The norm of ψ_E will be either finite or infinite depending on whether ψ_E is in the discrete or continuous spectrum. In the latter case δ-functions always appear in the orthonormality integral. The precise way in which these δ-functions enter will be deduced when we formulate scattering theory. We shall then show (see (12.30)) that if the incident particles in a collision state are represented by the plane waves $\phi_{\mathbf{k}}(\mathbf{x}) = (2\pi)^{-\frac{3}{2}} e^{i\mathbf{k}\cdot\mathbf{x}}$, then the continuum eigenfunctions of the Hamiltonian (including interactions) possess the orthonormality integral $\delta(\mathbf{k} - \mathbf{k}')$. The one-dimensional counterpart of this theorem, in the special case of scattering by the δ-potential, can be demonstrated by direct integration of

* The motivation behind this terminology will be given in Sec. 7.5.

the continuum eigenfunctions (11) and (14) *:

$$\int dx\, \psi_k^*(x)\psi_{k'}(x) = \delta(k - k'), \tag{28}$$

where one is to use (11) when $k > 0$, (14) when $k < 0$, and similarly for k'.

Aside from the orthonormality relation (27), the one-dimensional eigenfunctions $\phi_k(x)$ satisfy the further identity (26). We recall that (26) is merely a restatement of Fourier's theorem, for it tells us that an arbitrary function $\chi(x)$ can be expressed as the linear superposition of the functions $\phi_k(x)$:

$$\chi(x) = \int_{-\infty}^{\infty} dk\, \phi_k(x) \int_{-\infty}^{\infty} dx'\, \phi_k^*(x')\chi(x'). \tag{29}$$

We already took advantage of this relation when we discussed free particle wave packets in Sec. 4. When an arbitrary † function $\chi(x)$ in some domain $\mathfrak{D}$ of the variable x can be expressed as a linear superposition of a set of orthonormal functions, that set is said to be *complete* in $\mathfrak{D}$. Hence the set of one-dimensional plane waves $\{\phi_k(x)\}$, with $-\infty < k < \infty$, is complete on the real line. An arbitrary orthonormal set of functions may be denumerable, nondenumerable, or both. We shall only consider the latter case for the moment, as it includes the former two as special cases. Let us then designate a one-dimensional orthonormal set by $\{u_n(x), u_\lambda(x)\}$, where $n = 1, 2, \ldots$ is a discrete index, and λ is continuous. These functions, by definition, satisfy the orthonormality relations

$$\begin{aligned} \int u_n^*(x)u_{n'}(x)\, dx &= \delta_{nn'}, \\ \int u_\lambda^*(x)u_{\lambda'}(x)\, dx &= \delta(\lambda - \lambda'), \\ \int u_n^*(x)u_\lambda(x)\, dx &= 0. \end{aligned} \tag{30}$$

By hypothesis, the arbitrary function $\chi(x)$ can be expressed as

$$\chi(x) = \sum_n c_n u_n(x) + \int d\lambda\, c(\lambda)u_\lambda(x). \tag{31}$$

The expansion coefficients can be obtained by multiplying (31) by u^*, integrating, and using (30):

$$\begin{aligned} c_n &= \int dx\, u_n^*(x)\chi(x), \\ c(\lambda) &= \int dx\, u_\lambda^*(x)\chi(x). \end{aligned} \tag{32}$$

* A more elegant derivation is given in Prob. 6 of Chapt. III.

† By "arbitrary" we shall always mean that χ has a norm that satisfies (7).

A much more compact way of writing (30) and (31) is the so-called *completeness relationship:*

$$\delta(x - x') = \sum_n u_n(x)u_n^*(x') + \int d\lambda\, u_\lambda(x)u_\lambda^*(x'). \tag{33}$$

Just as (26) implies (29), so does multiplication of (33) by $\chi(x')$ and integration over x' reproduce (31) and (32). Or put another way, the completeness relation (33) is the generalization of Fourier's theorem from the special set $\{\phi_k\}$ to the set of functions $\{u_n, u_\lambda\}$.

The plane wave set $\{\phi_k(x)\}$ is, at the moment, the only one that we know to be complete. By explicit calculation (see Prob. 3.6, p. 163) one can also show that the solutions of the δ-potential form a complete (and orthonormal) set of functions, viz.,

$$\delta(x - x') = \psi_b(x)\psi_b(x') + \int_{-\infty}^{\infty} dk\{\theta(k)\psi_{k\rightarrow}(x)\psi_{k\rightarrow}^*(x') + \theta(-k)\psi_{k\leftarrow}(x)\psi_{k\leftarrow}^*(x')\}, \tag{34}$$

in the case where there is a bound state ψ_b (i.e., $\gamma > 0$). In this example there is just one function of the type u_n. When the potential is repulsive ($\gamma < 0$), there is no bound state, and the term $\psi_b(x)\psi_b(x')$ is missing from (34). Thus for all values of γ the set of all the linearly independent solutions of the time-independent Schrödinger equation (8) forms a complete set of functions. In this one-dimensional example there are two independent solutions, $\psi_{k\rightarrow}$ and $\psi_{k\leftarrow}$, for each energy eigenvalue $\hbar^2k^2/2m$. Both of these must be included in the completeness relationship, otherwise the set is not complete. This is hardly surprising; when $\gamma = 0$, (34) reduces to Fourier's theorem, which certainly requires that plane waves propagating in both directions be included in the set. If one wishes to single out the energy eigenvalue in the completeness relationship, one can always rewrite (34) as

$$\delta(x - x') = \psi_b(x)\psi_b(x') + \sum_{\nu=\pm1} \int_0^\infty dE\, \psi_{E\nu}(x)\psi_{E\nu}^*(x'), \tag{35}$$

where

$$\psi_{E,\pm1}(x) = \sqrt{dk/dE}\, \psi_{k\rightleftarrows}(x). \tag{36}$$

These new functions possess the orthonormality integral

$$\int \psi_{E\nu}^*(x)\psi_{E'\nu'}(x)\, dx = \delta_{\nu\nu'}\, \delta(E - E'), \tag{37}$$

which is to be compared with the completely equivalent relation (28). Nevertheless, (35) has the advantage of emphasizing more explicitly the fact that a complete set usually contains more than one linearly independent wave function for each energy eigenvalue. In fact, under many circumstances there are an infinity of eigenfunctions belonging to one energy eigenvalue. The most familiar illustration of this fact is provided by the three-dimensional free-particle wave functions $\phi_{\mathbf{k}}(\mathbf{x}) = (2\pi)^{-\frac{3}{2}}e^{i\mathbf{k}\cdot\mathbf{x}}$.

We have now quoted several examples where the totality of the linearly independent eigenfunctions of a Hamiltonian constitutes a complete orthonormal set. It is natural to ask whether this is true in general. In the case of a single particle in three dimensions the affirmative answer is one of the theorems of the Sturm-Liouville theory of differential equations. In the general case of N particles proofs of the analogous theorem require much more powerful methods (functional analysis *). *We shall always suppose that the totality of all the linearly independent solutions of the time-independent Schrödinger equation forms a complete set of functions.*

•Occasionally it is convenient to avoid the continuous spectrum. This can be achieved by a variety of devices, some clumsier than others. We illustrate the most important of these with the example of a free particle in three dimensions. Let us imagine that the whole of 3-space is divided into cubes of volume Ω, where $\Omega^{\frac{1}{3}}$ is large compared to all distances of interest to us. We then impose the requirement that all wave functions $\psi(x, y, z)$ satisfy the periodicity condition

$$\psi(x, y, z) = \psi(x + n_x\Omega^{\frac{1}{3}}, y + n_y\Omega^{\frac{1}{3}}, z + n_z\Omega^{\frac{1}{3}}), \tag{38}$$

where n_x, etc. are integers. We shall refer to (38) as *periodic boundary conditions*. Once the wave functions satisfy (38), we can confine our attention to one cube. The configuration space is now finite and eigenfunctions of infinite norms cannot arise. We may then use these functions of finite norm to carry out all computations, and let $\Omega \to \infty$ at the end of the calculation. Clearly all results of physical interest must be independent of Ω in this limit if this procedure is sensible. We shall use this device when we discuss the electromagnetic field (Chapt. VIII) and the many-body problem (Volume II). A somewhat similar artifice is to enclose the system in a large rigid enclosure of volume Ω. This requires the wave functions to vanish on the surface of the enclosure, and renders the spectrum discrete. We shall use this procedure in a treatment of scattering in Sec. 49.

A complete set of orthonormal wave functions which satisfy (38) are the periodic running waves

$$u_{\mathbf{k}}(\mathbf{x}) = \Omega^{-\frac{1}{2}}e^{i\mathbf{k}\cdot\mathbf{x}}, \tag{39}$$

with

$$\mathbf{k} = (2n_x\pi\Omega^{-\frac{1}{3}}, 2n_y\pi\Omega^{-\frac{1}{3}}, 2n_z\pi\Omega^{-\frac{1}{3}}), \tag{40}$$

where each n may take on any of the values $0, \pm 1, \pm 2, \ldots.$ The orthogo-

* See Riesz and Nagy, Sec. 33; Vulikh, Sec. 6.6.

nality and completeness relations are

$$\int_\Omega u_{\mathbf{k}}^*(\mathbf{x})u_{\mathbf{k}'}(\mathbf{x})\,d^3x = \delta_{\mathbf{k},\mathbf{k}'}, \tag{41}$$

$$\sum_{\mathbf{k}} u_{\mathbf{k}}(\mathbf{x})u_{\mathbf{k}}^*(\mathbf{x}') = \delta(\mathbf{x} - \mathbf{x}'). \tag{42}$$

It should of course be realized that only functions that satisfy the periodicity condition (38) can be expanded in terms of the set (39). The transformation to the nondenumerable basis $\phi_{\mathbf{k}}(\mathbf{x})$ can be made by means of the rather cavalier relationship $\lim_{\Omega\to\infty} \Omega^{\frac{1}{2}}(2\pi)^{-\frac{3}{2}}u_{\mathbf{k}}(\mathbf{x}) \to \phi_{\mathbf{k}}(\mathbf{x})$. When this is substituted into (41) and (42) we obtain the recipes

$$\lim_{\Omega\to\infty} \frac{\Omega}{(2\pi)^3}\delta_{\mathbf{k},\mathbf{k}'} \to \delta(\mathbf{k} - \mathbf{k}'), \tag{43}$$

$$\lim_{\Omega\to\infty} \frac{1}{\Omega}\sum_{\mathbf{k}} \to \int \frac{d^3k}{(2\pi)^3}. \tag{44}$$

7.4 LINEAR VECTOR SPACES

The various properties of the energy eigenfunctions which we have discovered in the preceding portions of this section can be formulated in a more compact and visualizable manner. Consider an arbitrary function $\chi(x)$. Unless otherwise stated, x will stand for all the variables that are required, and $\int dx$ will indicate integration over all of these variables. In fact, it does not affect the general results which follow if we consider χ to be a function of the momenta $\mathbf{p}_1, \ldots$. For definiteness we shall continue to use the x-representation, however. We may think of χ as a vector and the totality of its values $\chi(x)$ as its components labeled by the continuous index x, just as a three-vector $\mathbf{V}$ will have components V_i labeled by the index $i = 1, 2, 3$. If $\varphi(x)$ is another function, we define the scalar product (φ, χ) between the vectors φ and χ by

$$(\varphi, \chi) \equiv \int \varphi^*(x)\chi(x)\,dx = (\chi, \varphi)^*. \tag{45}$$

In analogy with finite-dimensional vectors, two vectors (functions) are orthogonal to each other if their scalar product vanishes. The norm of a vector $\|\chi\| = (\chi, \chi)$ is seen to be the square of its "length." The vector space is said to be linear in the sense that $(\varphi, c_1\chi_1 + c_2\chi_2) = c_1(\varphi, \chi_1) + c_2(\varphi, \chi_2)$ for any two functions χ_1 and χ_2 and any two complex numbers c_1 and c_2.

A complete orthonormal set of functions will be said to form an orthonormal *basis* in the vector space. The existence of various complete orthonormal sets of functions, which we found explicitly in the one-dimensional case, is now seen to be quite natural. Surely we must expect a vast manifold of different orthonormal basis systems in an infinite-dimensional space. We shall occasionally refer to such a vector space as a *Hilbert space.*

Let us work once more with the basis $\{u_n, u_\lambda\}$ of the preceding subsection. The expansion (31) of the arbitrary function χ in terms of this complete set is seen to be the statement that we can write a vector in terms of its components. The components are "directed" along the basis vectors u_n and u_λ, and the "lengths" of these components are given by the scalar products between the arbitrary vector χ and the various members of the basis. Thus in our new notation (31) reads

$$\chi = \sum_n u_n(u_n, \chi) + \int d\lambda\, u_\lambda(u_\lambda, \chi). \tag{46}$$

An operator is an object which maps a vector in the space into another vector. Heretofore we have mainly dealt with multiplicative and differential linear operators. As we saw in (4.34′), however, linear integral operators may also be of interest to us. These are defined by

$$A[\chi(x)] = \int A(x, x')\chi(x')\, dx', \tag{47}$$

where $A(x, x')$ is said to be the kernel of the operator A. The most important properties of the operators with which we shall now be concerned are that they are linear and Hermitian. These properties were already defined in Sec. 5. As in the case of the Hamiltonian, a function φ_a is an eigenfunction of the operator A if it is left in the "direction" φ_a by the operator A, i.e.,

$$A\varphi_a = a\varphi_a,$$

where a is a number called the eigenvalue. The argument previously applied to the Hamiltonian and its eigenfunctions leads directly to the following theorem: *If A is an Hermitian operator, its eigenvalues a are real, and eigenfunctions belonging to different eigenvalues are orthogonal.*

It can be shown that *for any Hermitian operator A there exists at least one basis in the Hilbert space such that each function in the basis is an eigenfunction of A.* Although we shall frequently make use of this theorem, we shall not prove it here.*

* See Riesz and Nagy, Chapt. 6. There must obviously be restrictions on the operator. For example, if A is an integral operator, the kernel $A(x, x')$ must have certain continuity and boundedness properties. These questions are also discussed in Riesz and Nagy.

7.5 SIMULTANEOUS EIGENFUNCTIONS AND COMPATIBLE OBSERVABLES

We have already seen several illustrations of the fact that there frequently are more than one linearly independent energy eigenfunction belonging to an energy eigenvalue. An analogous situation in classical mechanics is that the orientation of the orbit in a central field does not affect the energy. In order to specify the orbit completely, we must state more than the energy. Usually it is most convenient to specify the values of the constants of the motion, for example, the angular momentum vector. It is natural to ask whether one can also distinguish the linearly independent eigenfunctions belonging to the same energy by their properties with respect to other observables. (Linearly independent functions with the same energy eigenvalue are said to be *degenerate*.)

To be sure, the theorem at the end of Sec. 7.4 assures us that there exist eigenfunctions φ_a belonging to any observable A. Furthermore, in a state described by φ_a, the observable A has no dispersion whatsoever. (The proof of this statement is identical to the one given for ψ_E in Sec. 7.1.) The eigenvalue a is therefore the value of the observable A in the state described by φ_a. The question, however, is whether it is possible to have states where A and the energy both have precise values. This obviously requires the existence of a function ψ_{Ea} which is a simultaneous eigenfunction of H and A:

$$(A - a)\psi_{Ea} = 0, \tag{48}$$

$$(H - E)\psi_{Ea} = 0. \tag{49}$$

Assuming this to be so, we multiply (48) by H and (49) by A, and obtain

$$[H, A]\psi_{Ea} = 0. \tag{50}$$

As the energy eigenfunctions form a complete set, (50) can only be satisfied if $[H, A] = 0$. The identical argument can of course be applied to any pair of observables. We therefore have the following theorems:

(I) *The necessary and sufficient condition for the existence of a simultaneous system of eigenfunctions of H and A is that A be a constant of the motion.**

(II) *Only commuting observables can have a system of simultaneous eigenfunctions.**

We illustrate these theorems by some simple examples. In the case of a free particle in three dimensions, the plane waves are simultaneous

* We have inserted the word "system" on purpose here. When ψ_0 is an eigenfunction of A with eigenvalue zero, it may happen that it is also an eigenfunction of B even though $[A, B] \neq 0$. Thus a spherically symmetric wave function is a simultaneous eigenfunction (with eigenvalues zero) of all components of the angular momentum, even though $[L^i, L^j] = iL^k$.

eigenfunctions of all three components of the momentum, and *ipso facto*, of the Hamiltonian:

$$\left(\frac{\hbar}{i}\frac{\partial}{\partial \mathbf{x}} - \mathbf{p}\right) e^{i\mathbf{p}\cdot\mathbf{x}/\hbar} = 0,$$

$$\left(\frac{\hbar^2}{2m}\nabla^2 + \frac{p^2}{2m}\right) e^{i\mathbf{p}\cdot\mathbf{x}/\hbar} = 0.$$

These results are possible because the different components of the momentum operator commute with each other. A somewhat more interesting example is the one-particle wave function $\chi = e^{ipz/\hbar}e^{i\mu\varphi}$, where p and μ are numbers, z is the z-coordinate, and φ the azimuthal angle in a system of polar coordinates with z as the polar axis. χ is seen to be an eigenfunction of p_z:

$$\left(\frac{\hbar}{i}\frac{\partial}{\partial z} - p\right)\chi = 0.$$

It is also an eigenfunction of the z-component of angular momentum, $\hbar L^z = xp_y - yp_x$. In polar coordinates $L^z = -i\,\partial/\partial\varphi$. But

$$\left(\frac{1}{i}\frac{\partial}{\partial\varphi} - \mu\right)\chi = 0;$$

therefore μ is the eigenvalue of L^z. That functions like χ should exist is hardly surprising, because we had already shown that parallel components of $\mathbf{p}$ and $\mathbf{L}$ commute (see (6.17)). On the other hand, χ is not an eigenfunction of any component of $\mathbf{p}$ except p_z, as one immediately realizes by writing χ in terms of Cartesian coordinates, viz., $\chi = e^{ipz/\hbar}\exp[i\mu \arctan(y/x)]$. This was also to be expected, because perpendicular components of $\mathbf{L}$ and $\mathbf{p}$ fail to commute (see (6.18)).

Let us now see what further consequences we can draw from Theorems I and II above. Our first observation concerns the uncertainty principle. According to II, if $[A, B] = 0$, one can construct states in which the observables A and B both have no dispersion whatsoever. We therefore have the following generalization of the position-momentum uncertainty principle: *if two observables commute, it is possible to simultaneously measure these observables to arbitrary accuracy.* For this reason observables that commute with each other are said to be *compatible.* When A and B are incompatible, we shall show in Sec. 24 that the lower bound on $\Delta A\,\Delta B$ (where $\Delta A = \sqrt{\langle (A - \langle A\rangle)^2\rangle}$, etc.) is of order $|\langle [A, B]\rangle|$.

Finally, consider the situation where we have a number of constants of the motion, $A_1, A_2, \ldots, B_1, B_2, \ldots$, where $[A_i, A_j] = [B_i, B_j] = 0$, but $[A_i, B_j] \neq 0$. According to I we can construct simultaneous eigenfunctions of H and all the A_i, or alternatively of H and all the B_i. On the other hand, II informs us that we cannot construct simultaneous eigenfunctions of (H, A_1, B_2), for example. In other words, we can only construct simultaneous eigenfunctions of a set of mutually compatible observables. As an illustration of these statements, consider the angular momentum operator $\mathbf{L}$. As the different components of $\mathbf{L}$ do not commute with each other, only a single component (L^z, say) of any angular momentum vector can appear in a compatible set of observables. On the other hand, in Sec. 6.3 we proved that all rotationally invariant quantities commute with $\mathbf{L}$. Hence $[\mathbf{L}^2, \mathbf{L}] = 0$, and we may therefore incorporate $\mathbf{L}^2$ and L^z into a compatible set.

Let us now return to the two-body problem of Sec. 6.4. We may select from (6.39) different sets of compatible observables:

$$\begin{aligned}
&\text{(a)} \qquad \mathbf{P},\ \ \frac{p^2}{2\mu} + V,\ \ (\mathbf{L}_r)^2,\ \ L_r^z; \\
&\text{(b)} \qquad P^2/2M,\ \ (\mathbf{L}_R)^2,\ \ L_R^z,\ \ \frac{p^2}{2\mu} + V,\ \ (\mathbf{L}_r)^2,\ \ L_r^z; \qquad (51) \\
&\text{(c)} \qquad \mathbf{x}_1,\ \ \mathbf{x}_2; \\
&\text{(d)} \qquad \mathbf{p}_1,\ \ \mathbf{p}_2;
\end{aligned}$$

and many others. The different sets do not commute with each other. Only (a) and (b) include the total Hamiltonian and have stationary states as their eigenfunctions. Nevertheless, (c) and (d) are a perfectly fine compatible set of observables. Inspection of (a)–(d) reveals that in no instance can we add further compatible observables to these sets, unless they are merely functions of the observables listed. (Thus we can always add $P^2/2M$ to (a), or V to (c).) We shall refer to such an unaugmentable set as *a complete set of compatible observables* (or commuting operators).

All of the sets (a)–(d) listed here have six independent compatible observables, a number that equals the number of classical degrees of freedom. This result holds quite generally for systems that may be completely described in terms of the canonical observables $\mathbf{x}_\alpha$ and $\mathbf{p}_\alpha$. When such a description turns out to be incomplete, we must introduce further observables that commute with the canonical observables. Such observables do not have any counterpart in classical physics, and they do not possess a classical limit in the sense of the correspondence principle. As far as we are concerned here, these nonclassical observables can only be discovered experimentally. The most familiar example of such an observable is the intrinsic spin of, say, the electron. Its existence was first inferred from the fact that certain atoms had *more* degenerate eigenstates than one could distinguish by the eigenvalues of the conventional observables. In view of these remarks we shall always demand that *the eigenvalues of a complete set of compatible observables serve to uniquely specify each and every linearly independent solution of the time-independent Schrödinger equation.*

7.6 PROBABILITY AMPLITUDES

In the foregoing discussion we have placed great emphasis on eigenfunctions, and in particular, on stationary states. We must not lose sight of the fact that a linear superposition of such wave function also describes a state of the system. In fact, only under the most unusual circumstances can we hope to find a system in a state described by a single eigenfunction. Should we be concerned with observables with a discrete spectrum, it may be relatively easy to prepare states in which only one eigenfunction of such an observable is present. As soon as we have continuous spectra, however, we must in the last analysis always deal with superpositions of eigenfunctions.*

Let $\{\psi_{E[a]}\}$ be a complete set of eigenfunctions belonging to the complete compatible set $(H, A_1, A_2, \ldots)$, where E is the eigenvalue of H, and $[a] \equiv a_1, a_2, \ldots$, stands for the eigenvalues of the observables $A_1, A_2, \ldots$. The most general solution χ of the Schrödinger equation can then be written as

$$\chi = \sum_{[a]} \int dE \, c_{[a]}(E) \psi_{E[a]} e^{-iEt/\hbar}. \tag{52}$$

Here we have assumed that the energy spectrum is purely continuous, whereas that of A_i is purely discrete. If this is not the case, there must also be a sum over the discrete energy eigenvalues, and an integral over $[a]$. We shall refrain from spelling out all these trivial complications; the assumed spectrum will illustrate the essential points. The functions $\psi_{E[a]}$ are adjusted so as to satisfy the orthonormality condition

$$(\psi_{E[a]}, \psi_{E'[a']}) = \delta_{[a],[a']} \, \delta(E - E'), \tag{53}$$

which is the generalization of (37) to the present case. The (time-independent) expansion coefficients in (52) are therefore

$$c_{[a]}(E) = (\psi_{E[a]}, \chi) e^{iEt/\hbar} \tag{54}$$

* Even this statement is too restrictive. The most general state of a quantum mechanical system cannot be described by a single linear superposition of wave functions. A detailed discussion of this point will be given in Secs. 20 and 26.

We shall also demand that χ is a normalized packet, i.e., $\|\chi\| = 1$; hence

$$\sum_{[a]} \int dE |c_{[a]}(E)|^2 = 1. \tag{55}$$

The observables H and A_i do not have sharply defined values in the state χ. Their expectation values can be evaluated with the help of (52)–(55):

$$(\chi, H\chi) = \int dE \cdot E \sum_{[a]} |c_{[a]}(E)|^2, \tag{56}$$

$$(\chi, A_i\chi) = \sum_{a_i} a_i \sum_{[a_j](j \neq i)} \int dE\, |c_{[a]}(E)|^2. \tag{57}$$

By hypothesis, $(\chi, H\chi)$ is the mean value of the energy that emerges from repeated energy measurements on an ensemble of systems prepared to be in the state described by the wave function χ. We observe that (56) can be written as $\int E W_\chi(E)\, dE$, where $\int W_\chi(E)\, dE = 1$. It is therefore natural to interpret

$$W_\chi(E) = \sum_{[a]} |c_{[a]}(E)|^2$$

as the probability that the energy lies between E and $E + dE$ when the system is in the state described by χ. Precisely the same reasoning can be applied to (57). Thus (57) can be written as $\Sigma a_i W_\chi(a_i)$, where

$$W_\chi(a_i) = \sum_{[a_j](j \neq i)} \int dE\, |c_{[a]}(E)|^2,$$

and $\Sigma W_\chi(a_i) = 1$. We may therefore interpret $W_\chi(a_i)$ as the probability that a measurement of the observable A will yield the value a_i.

Equations (55)–(57) therefore lead us to the following interpretation of the expansion coefficients in the linear superposition (52): $|(\chi, \psi_{E[a]})|^2$ *is the probability density that a simultaneous measurement of all the observables* $H, A_1, A_2, \ldots$ *in the state* χ *will yield the result* E, $[a]$. Needless to say, if $\varphi_{[b]}$ is a simultaneous eigenfunction of any complete set of observables $B_1, B_2, \ldots$ (which need not * contain H), then by the identical argument $|(\chi, \varphi_{[b]})|^2$ is the probability that the observables $B_1, B_2, \ldots,$ have the values $[b]$ in the state χ. In view of this interpretation, one frequently refers to the scalar product as a *probability amplitude*. The interpretation of the scalar product as a probability amplitude is perhaps the most basic feature of quantum mechanics. As the argument which led us to this concept is really very formal, we shall, in Chapt. IV, return

* If the set of observables $\{B_i\}$ does not contain H, $|(\chi, \varphi_{[b]})|^2$ will be time dependent.

to the problem of justifying this interpretation from a more physical point of view.

Let us just comment on an obvious, but nonetheless remarkable, feature of (56) and (57): only the eigenvalues of H and A_i appear in the summand. Hence *any and all measurements of an observable A can only yield a result belonging to the spectrum of A.* When the spectrum of the observable is discrete, this is a very startling statement from the classical viewpoint.

Special examples of probability amplitudes are already known to us in the guise of x- and p-space wave functions. How do these familiar amplitudes fit into the general framework? According to this framework, the p-space probability density associated with the state χ should be $|(\phi_{\mathbf{p}}, \chi)|^2$, where $\phi_{\mathbf{p}}$ is an eigenfunction of the momentum. But

$$(\phi_{\mathbf{p}}, \chi) \equiv \int d^3x \, \phi_{\mathbf{p}}^*(\mathbf{x})\chi(\mathbf{x}) = (2\pi\hbar)^{-\frac{3}{2}} \int d^3x \, e^{-i\mathbf{p}\cdot\mathbf{x}/\hbar}\chi(\mathbf{x}),$$

which agrees with the formulas of Sec. 4. (We have restricted ourselves to a single particle here; as usual the generalization is trivial.) The x-space wave function should also fit into this framework. The x-space eigenfunction of the observable $\mathbf{\hat{x}}$ with eigenvalue $\mathbf{x}_0$ is $\delta(\mathbf{x} - \mathbf{x}_0) \equiv u_{\mathbf{x}_0}(\mathbf{x})$. Thus $(u_{\mathbf{x}_0}, \chi)$ should be the x-space wave function; it is, because of the trivial identity

$$(u_{\mathbf{x}_0}, \chi) = \int u_{\mathbf{x}_0}^*(\mathbf{x})\chi(\mathbf{x}) \, d^3x = \chi(\mathbf{x}_0).$$

8. The Classical Limit

We must still verify that the theory we have constructed actually reduces to classical mechanics under appropriate circumstances. Although we have used the correspondence principle in a formal way up to now, the manner in which the classical limit is to be taken and its domain of validity remain unclear.

We already have a quite clear picture of the circumstances under which classical mechanics is applicable: We must be able to construct wave packets whose spatial extent is small compared to the distances in which the forces change appreciably, and which retain this localization for a macroscopic time. To be more precise, in the case of a bound system we demand that this time be much longer than the classical period of the motion, while in a scattering problem this time would have to be long compared to the classical collision time. Note that the spatial localization does not have to be small compared to the size of the particle itself;

a 100 kev α-particle scattering from Pb can be treated by classical mechanics even though the de Broglie wavelength ($\sim 2 \times 10^{-13}$ cm) is of the order of the radius of the α-particle ($\sim 10^{-13}$ cm). In this example the classical distance of closest approach (3×10^{-10} cm) is much larger than the de Broglie wavelength, and so the projectile can be represented by a well-defined wave packet moving along a classical trajectory.

8.1 EHRENFEST'S THEOREM

We return to the study of wave packets begun in Sec. 4, and seek equations that are the natural extension of (4.39) and (4.40) to the case of an interacting system. We shall assume that the Hamiltonian has the form

$$H = T(\mathbf{p}_1, \ldots, \mathbf{p}_N) + V(\mathbf{x}_1, \ldots, \mathbf{x}_N) + \sum_{\alpha=1}^{N} \gamma_\alpha [\mathbf{p}_\alpha \cdot \mathbf{A}(\mathbf{x}_\alpha, t) + \mathbf{A}(\mathbf{x}_\alpha, t) \cdot \mathbf{p}_\alpha], \quad (1)$$

where the γ_α are constants. This form includes, as the special case $\gamma_\alpha = 0$, all types of velocity-independent interactions between the particles and externally applied fields. The last term, with $\gamma_\alpha = -e_\alpha/2m_\alpha c$, allows us to treat the motion of a system of charged particles in an electromagnetic field described by the vector potential $\mathbf{A}(\mathbf{r}, t)$.

Our first objective is to obtain equations for the expectation values of the dynamical variables. These are given by (6.7):

$$i\hbar \frac{d}{dt} \langle p_\alpha^i \rangle = \langle [p_\alpha^i, H] \rangle,$$

$$i\hbar \frac{d}{dt} \langle x_\alpha^i \rangle = \langle [x_\alpha^i, H] \rangle.$$

Recalling the commutation rules (6.12) and (6.13), and using the form of H as given in (1), we obtain

$$\frac{i}{\hbar} [p_\alpha^i, H] = \frac{\partial V}{\partial x_\alpha^i} + \sum_j \gamma_\alpha \left(p_\alpha^j \frac{\partial A_\alpha^j}{\partial x_\alpha^i} + \frac{\partial A_\alpha^j}{\partial x_\alpha^i} p_\alpha^j \right),$$

$$\frac{1}{i\hbar} [x_\alpha^i, H] = \frac{\partial T}{\partial p_\alpha^i} + 2\gamma_\alpha A_\alpha^i.$$

Therefore we find

$$\frac{d}{dt} \langle p_\alpha^i \rangle = -\left\langle \frac{\partial H}{\partial x_\alpha^i} \right\rangle,$$

$$\frac{d}{dt} \langle x_\alpha^i \rangle = \left\langle \frac{\partial H}{\partial p_\alpha^i} \right\rangle = \frac{1}{m_\alpha} \left\langle p_\alpha^i - \frac{e_\alpha}{c} A_\alpha^i \right\rangle. \quad (2)$$

This is *Ehrenfest's theorem:* The laws of classical mechanics (i.e., Hamilton's equations) hold for the expectation values. This theorem is actually a special case of Dirac's earliest discovery: The operator equations of quantum mechanics become the laws of classical mechanics if, for any pair of observables A and B, one formally replaces $[A, B]/i\hbar$ by the Poisson bracket of A and B. Readers familiar with Poisson brackets will agree to the truth of this statement by glancing at the formulas of Sec. 6.

We should note that Eqs. (2) do *not* constitute the classical limit, because they hold for *any* ψ. Classical mechanics is only accurate if $\Delta\mathbf{x}_\alpha$ and $\Delta\mathbf{p}_\alpha$ are negligible, and this requires the validity of certain approximations that we shall spell out below. The physical nature of these approximations was already discussed at the beginning of this section. At the formal level we note that if we could replace $H(x, p, t)$ by $H(\langle x\rangle, \langle p\rangle, t)$ in (2), we would obtain the classical equations

$$\begin{aligned}\frac{d}{dt}\langle p_\alpha^i\rangle &= -\frac{\partial H(\langle x\rangle, \langle p\rangle, t)}{\partial\langle x_\alpha^i\rangle},\\ \frac{d}{dt}\langle x_\alpha^i\rangle &= \frac{\partial H(\langle x\rangle, \langle p\rangle, t)}{\partial\langle p_\alpha^i\rangle}.\end{aligned}\tag{3}$$

When these equations hold, $\langle p_\alpha^i\rangle$ and $\langle x_\alpha^i\rangle$ are functions of time that evolve according to the classical laws.

To make the step from (2) to (3) clearer, consider the case of a single particle in the absence of a magnetic field. Then

$$-\frac{\partial H}{\partial\mathbf{x}} = \mathbf{F}(\mathbf{x})\tag{4}$$

is the operator representing the force exerted on the particle. In the coordinate representation, $\mathbf{F}$ is just a number and we can expand it about the (time-dependent) expectation value $\langle\mathbf{x}\rangle$:

$$\begin{aligned}\mathbf{F}(\mathbf{x}) = \mathbf{F}(\langle\mathbf{x}\rangle) &+ \sum_{i=1}^{3}(x^i - \langle x^i\rangle)\left.\frac{\partial\mathbf{F}}{\partial x^i}\right|_{\mathbf{x}=\langle\mathbf{x}\rangle}\\ &+ \frac{1}{2}\sum_{i,j=1}^{3}(x^i - \langle x^i\rangle)(x^j - \langle x^j\rangle)\left.\frac{\partial^2\mathbf{F}}{\partial x^i\partial x^j}\right|_{\mathbf{x}=\langle\mathbf{x}\rangle} + \cdots.\end{aligned}\tag{5}$$

Substituting into (2) we obtain

$$\frac{d}{dt}\langle\mathbf{p}\rangle \simeq \mathbf{F}(\langle\mathbf{x}\rangle) + \frac{1}{2}\sum_{ij}\Delta_{ij}\left.\frac{\partial^2\mathbf{F}}{\partial x^i\,\partial x^j}\right|_{\mathbf{x}=\langle\mathbf{x}\rangle}\tag{6}$$

The quantity Δ_{ij}, defined by

$$\Delta_{ij} = \langle x^i x^j \rangle - \langle x^i \rangle \langle x^j \rangle,$$

is of order d^2, where d is the diameter of the packet. The nonclassical term in (6) will therefore be negligible whenever*

$$\left| \frac{d^2\, \partial^3 V / \partial x^i\, \partial x^j\, \partial x^k}{\partial V / \partial x^l} \right| \ll 1. \tag{7}$$

It is obvious that this inequality requires V to be a slowly varying function of $\mathbf{x}$.

In order to gain some appreciation for the content of (7), let us consider a one-dimensional example for the moment, in which case (7) reduces to $d^2 |\partial^3 V/\partial x^3| \ll |\partial V/\partial x|$. When $V(x)$ is a slowly varying function, it is useful to introduce the notion of a local de Broglie wavelength $\lambda\!\!\!^{-}(x)$. This quantity may be defined in a natural manner by treating $V(x)$ as if it were a constant in the Schrödinger equation $-\hbar^2\psi''/2m = (E - V)\psi$. Once this is done, ψ has the form

$$\psi(x) = N e^{\pm ix/\lambda\!\!\!^{-}(x)}, \tag{8}$$

with

$$\lambda\!\!\!^{-}(x) = \frac{\hbar}{\sqrt{2m(E - V(x))}}. \tag{9}$$

The form (8) is actually rather crude and does not provide us with a really consistent approximation. In Sec. 8.3 we shall show that the correct stationary state wave function under quasi-classical circumstances is

$$\psi(x) = N' \exp\left\{ \pm i \int^x \frac{dx'}{\lambda\!\!\!^{-}(x')} \right\}$$

instead of (8). Hence (8) is only valid for a wave packet localized in a region in which the variation of $\lambda\!\!\!^{-}$ with x may be ignored. By superposing wave functions of type (8), we can build a packet localized about the point x. If the mean energy of this packet is to be E, the minimum value of d is of order $\lambda\!\!\!^{-}(x)$. Hence (7) becomes

$$\left| \frac{\hbar^2}{2m(E - V(x))} \frac{\partial^3 V}{\partial x^3} \right| \ll \left| \frac{\partial V}{\partial x} \right| \tag{10}$$

* Note that if the potential is a second-degree polynomial in $\mathbf{x}$ (harmonic oscillator), the expansion (5) of $\mathbf{F}$ does not lead to any correction terms, i.e., $\langle \mathbf{x} \rangle$ and $\langle \mathbf{p} \rangle$ follow the classical path for all t. This does not mean that the quantum oscillator can be described completely by classical mechanics: the energy spectrum of the oscillator Hamiltonian is discrete (see Sec. 31).

in the one dimensional case. (Obviously one can derive a similar expression in three dimensions.) According to (10):

(a) the classical approximation always breaks down at a classical turning point (i.e., where $E = V(x)$);

(b) for a given average energy and shape of potential, the quantum mechanical corrections to the classical trajectory are inversely proportional to the mass of the particle.

Aside from these qualitative remarks, (10) also provides us with a quantitative estimate of the error made when one uses the classical equations of motion.

As it stands, our argument still has one important deficiency: it only tells us under what circumstances we can drop the quantum mechanical corrections to $d\langle \mathbf{p}\rangle/dt = \mathbf{F}(\langle \mathbf{x}\rangle)$ at *one* instant. If the classical approximation is to give accurate results, Δ_{ij} must grow sufficiently slowly in time so as to permit us to ignore the nonclassical terms in (6) over a time span of order the classical period. One can estimate the spreading of the packet by developing approximate equations for $\Delta_{ij}(t)$, equations that are the generalizations of (4.41) to the case where forces are present. Instead of pursuing this route, we shall develop a more elegant technique for taking the classical limit.

8.2 THE RELATIONSHIP BETWEEN THE SCHRÖDINGER AND HAMILTON-JACOBI EQUATIONS

On the basis of a crude but essentially correct argument, we have already concluded that in a situation where classical mechanics is expected to be accurate, the phase of the wave function is, to a first approximation, a linear function of position (see Eq. (8) and the remarks immediately following it). This is reminiscent of the propagation of light through a medium of slowly varying index of refraction. It is therefore natural to expect a strong analogy between the reductions (quantum mechanics) → (classical mechanics) and (wave optics) → (geometrical optics). Inspired by these thoughts, Brillouin and Wentzel made the same *Ansatz* for the Schrödinger wave function as one makes for the electromagnetic quantities when deriving geometrical optics,* i.e.,

$$\Psi(\mathbf{x}, t) = \exp\left[\frac{i}{\hbar} S(\mathbf{x}, t)\right], \tag{11}$$

where S is supposed to be a slowly varying function of $\mathbf{x}$. Upon substituting (11) into the Schrödinger equation, we find

$$-\frac{\partial S}{\partial t} = \frac{1}{2m}\left\{(\nabla S)^2 + \frac{\hbar}{i}\nabla^2 S\right\} + V. \tag{12}$$

* Cf., e.g., Born and Wolf, Chapt. 3; Landau and Lifshitz, *CTF*, Chapt. 7.

Assuming that the second derivative of S is small compared to the first, we obtain an approximate S, which we call S_0, and which satisfies

$$\frac{\partial S_0}{\partial t} + \frac{1}{2m}(\nabla S_0)^2 + V = 0; \tag{13}$$

this is the Hamilton-Jacobi equation of classical mechanics.

Let us briefly review the content of (13) in its classical context. $S_0(\mathbf{x}, t)$ is called Hamilton's principal function. It is parameterized by six constants of integration in our one-body example; these parameters define the initial values of the canonical coordinates and the constants of the motion when V is time independent. When V is a function of time, three of the parameters define the canonical momenta at a given instant. A particular solution S_0 of (13) provides a complete description of a single orbit. The momentum of the particle when it is at $\mathbf{x}(t)$ is given by

$$\mathbf{p}(t) = \nabla S_0(\mathbf{x}, t), \tag{14}$$

and $\mathbf{x}(t)$ is determined by a somewhat complicated procedure which we shall not enter into here.*

In 1834 Hamilton discovered that in the case of a time-independent V, Eq. (13) allows one to think of classical mechanics as the geometric-optics limit of a wave motion in configuration space, the trajectories being the normals to the surfaces of constant S_0. This venerable observation was the starting point in Schrödinger's formulation of quantum mechanics.

In order to systematize the step from (12) to (13), we note that it is obtained by formally letting $\hbar \to 0$. We therefore expand S as a power series in $\hbar$,

$$S = S_0 + \frac{\hbar}{i} S_1 + \left(\frac{\hbar}{i}\right)^2 S_2 + \cdots; \tag{15}$$

after substituting into (12) we equate the coefficients of $\hbar^n$. The $\hbar$-independent terms give (13), and the coefficient of the term linear in $\hbar$ imposes the condition

$$-\frac{\partial S_1}{\partial t} = \frac{1}{m}(\nabla S_0 \cdot \nabla S_1 + \tfrac{1}{2}\nabla^2 S_0). \tag{16}$$

Like all the equations of classical mechanics, (13) does not contain any complex numbers. Hamilton's principal function S_0 is therefore a real function in those regions of configuration space accessible to the manifold of classical orbits. Given a real S_0, (16) then manufactures a real function S_1. In the classically accessible portions of configuration space, and to leading order in $\hbar$, the phase and modulus of Ψ are therefore given

* See Goldstein, Chapt. 9.

by $iS_0/\hbar$ and e^{S_1}, respectively: $\Psi \simeq \exp[S_1 + iS_0/\hbar]$. To the same accuracy the probability density is

$$W(\mathbf{x}, t) \cong e^{2S_1(\mathbf{x},t)}, \tag{17}$$

while the associated probability current is

$$\mathbf{j}(\mathbf{x}, t) = (\hbar/m) \operatorname{Im} \Psi^* \nabla \Psi \cong \frac{1}{m} e^{2S_1} \nabla S_0. \tag{18}$$

One easily verifies that W and $\mathbf{j}$ satisfy the continuity equation in virtue of (16).

Observe that W and $\mathbf{j}$ are both independent of $\hbar$. The classical character of $\mathbf{j}$ becomes more apparent if we take advantage of (14) and (17) to write

$$\mathbf{j}(\mathbf{x}, t) = W(\mathbf{x}, t)\mathbf{v}(\mathbf{x}, t). \tag{18'}$$

Here $\mathbf{v}(\mathbf{x}, t)$ is the velocity at the point $\mathbf{x}$ on the classical trajectory described by Hamilton's principal function S_0. This S_0 is related to the wave function in question by

$$S_0(\mathbf{x}, t) = \lim_{\hbar \to 0} \left[\left(\frac{\hbar}{i} \right) \ln \Psi(\mathbf{x}, t) \right]. \tag{19}$$

As we see, the classical limit may be visualized as the motion of a continuous mass distribution of total mass m. At time t_0 the mass density $\rho(\mathbf{x}, t_0)$ is

$$\rho(\mathbf{x}, t_0) = m \lim_{\hbar \to 0} |\Psi(\mathbf{x}, t)|^2 = m e^{2S_1(\mathbf{x},t_0)},$$

and the local velocity is given by $\mathbf{v}(\mathbf{x}, t_0)$. For $t > t_0$ this mass distribution moves according to the classical equations of motion appropriate to the Hamiltonian in question.

The nonclassical aspects of the motion (which include effects associated with the spreading of the wave packet) are described by the term S_n in the expansion (15) having $n \geqslant 2$. These higher-order terms will be small provided the nonclassical term in (12) is small, i.e.,

$$\hbar |\nabla^2 S_0| \ll (\nabla S_0)^2. \tag{20}$$

8.3 THE SEMICLASSICAL APPROXIMATION FOR STATIONARY STATES

The classical limit for stationary states is of considerable interest because it is easiest to formulate approximation methods for such states. By superposing stationary wave functions we may always return to the general situation studied in Sec. 8.2.

When the Hamiltonian is not a function of time, one may introduce a time independent function $\hat{S}(\mathbf{x})$ by

$$S(\mathbf{x}, t) = \hat{S}(\mathbf{x}) - Et. \tag{21}$$

Instead of (12), one then obtains

$$\frac{1}{2m}\{(\nabla\hat{S})^2 + \frac{\hbar}{i}\nabla^2\hat{S}\} = E - V. \tag{22}$$

The classical limit is again reached by formally setting $\hbar = 0$:

$$\frac{1}{2m}(\nabla\hat{S}_0)^2 = E - V. \tag{23}$$

In classical mechanics $\hat{S}_0$ is called Hamilton's characteristic function. The step from (22) to (23) is only legitimate if

$$\hbar|\nabla^2\hat{S}_0| \ll (\nabla\hat{S}_0)^2, \tag{24}$$

which is the same condition as (20).

It is easiest to appreciate what has been done here by again considering the one-dimensional example. In that case (23) can be integrated:

$$\hat{S}_0(x) = \pm \int^x dx' \sqrt{2m[E - V(x')]}. \tag{25}$$

The constant of integration is then determined by appropriately normalizing $|\Psi|^2$. We observe that the local de Broglie wavelength (9) is related to $\hat{S}_0$ by

$$\frac{d\hat{S}_0}{dx} = \pm \frac{\hbar}{\lambda\!\!\!^{-}(x)}. \tag{26}$$

The approximate wave function resulting from the substitution of (25) into (11),

$$\Psi(x, t) = \exp\left(\pm i \int^x \frac{dx'}{\lambda\!\!\!^{-}(x')}\right) e^{-iEt/\hbar}, \tag{27}$$

is usually referred to as the semiclassical wave function.

In the one-dimensional case, the inequality (24) can be written in terms of the local de Broglie wavelength:

$$\hbar^2|d\lambda\!\!\!^{-}/dx\,(\lambda\!\!\!^{-}(x))^{-2}| \ll \hbar^2|\lambda\!\!\!^{-}(x)|^{-2}.$$

Hence the semiclassical wavefunction (27) provides an accurate approximation whenever

$$\lambda\!\!\!^{-}(x) \neq \infty, \qquad |d\lambda\!\!\!^{-}/dx| \ll 1. \tag{28}$$

The condition $\lambda(x) \neq \infty$ merely states that x cannot be too close to a classical turning point. Both of the requirements stated in (28) were already surmised in Sec. 8.1.

If the energy E and potential V are such as to satisfy (28) everywhere, the simple form (27) will serve as a satisfactory approximation to the true stationary wave function, and a superposition of such functions can then be used to build a wave packet that behaves like a continuous, classical, mass distribution. This shows that when (28) is satisfied the classical trajectory is actually followed. We have therefore completed the argument begun in Sec. 8.1; there we only determined the circumstances under which the classical equations are valid instantaneously.

At low energies classical turning points usually occur. These points, designated by $x(E)$, are determined by the implicit equation $E = V(x(E))$. There are then regions of space, defined by $E < V(x)$, that the classical trajectories can never reach. The wave function does not vanish in such regions, however. As we know, phenomena such as alpha-decay and electron tunneling occur because the quantum mechanical probability distribution does not vanish in the classically forbidden portions of configuration space.

Provided $|d\lambda/dx| \ll 1$ in the classically forbidden region, (27) is also a good approximation to the true Ψ there. The phase will be imaginary in such regions, and will therefore decrease or increase exponentially for such values of x. As (27) always breaks down at the turning points $x(E)$, one must use a special technique due to Kramers to interpolate between the semiclassical wave functions in the classically forbidden and allowed regions.* This interpolation is based on approximating $V(x)$ by $E + [x - x(E)]V'$ near $x(E)$. The Schrödinger equation for a potential linear in x can be solved analytically, and these solutions can then be joined smoothly to the semiclassical forms on either side of $x(E)$.

PROBLEMS

1. The uncertainty principle was derived in Eq. (4.37) *et seq.* From this proof, deduce the wave function for which the equal sign obtains simultaneously for all three values of the index i. Does this wave packet preserve the minimum uncertainty product in the course of time?

2. Consider a free particle of mass m which, at $t = 0$, is prepared in the state

$$\begin{aligned} \psi(\mathbf{x}, 0) &= V^{-\frac{1}{2}} \qquad (x \leqslant R), \\ &= 0 \qquad (x > R), \end{aligned}$$

* See Kemble, p. 572; W. H. Furry, *Phys. Rev.* **71**, 360 (1947); Merzbacher, Chapt. 7; Landau and Lifshitz, *QM*, Chapt. 7.

where $V = 4\pi R^3/3$. Show that for all x, but large t,

$$\psi(\mathbf{x}, t) = \left(\frac{2}{\pi V}\right)^{\frac{1}{2}} \left(\frac{aR}{ix^2}\right)^{\frac{3}{2}} \left[\sin\frac{x}{a} - \frac{x}{a}\cos\frac{x}{a}\right] e^{ix^2m/2\hbar t},$$

where $a = \hbar t/mR$. What is the significance of the length a? Determine the behavior of the probability distribution in the limits $x \gg a$ and $x \ll a$.

3. If ψ_1 and ψ_2 are two solutions of the Schrödinger equation, show that

$$\int \psi_1^*(\mathbf{x}_1, \ldots, \mathbf{x}_N, t)\psi_2(\mathbf{x}_1, \ldots, \mathbf{x}_N, t)(dx)$$

does not depend on t.

4. Consider the one-particle Schrödinger equation

$$i\hbar\frac{\partial}{\partial t}\psi(\mathbf{x}, t) = -\frac{\hbar^2}{2m}\nabla^2\psi(\mathbf{x}, t) + \int V(\mathbf{x}, \mathbf{x}')\psi(\mathbf{x}', t)\, d^3x'.$$

Derive the condition on the nonlocal potential $V(\mathbf{x}, \mathbf{x}')$ that insures the conservation of probability. Does a current that depends only on ψ at a point exist in this example?

5. Consider the one-particle Schrödinger equation

$$i\hbar\frac{\partial}{\partial t}\psi(\mathbf{x}, t) = -\frac{\hbar^2}{2m}\nabla^2\psi(\mathbf{x}, t) + [V_1(\mathbf{x}) + iV_2(\mathbf{x})]\psi(\mathbf{x}, t),$$

where V_1 and V_2 are real functions. Show that probability is not conserved, and give an expression for the rate at which probability is "lost" or "gained" in a spatial volume Ω.

6. In the coordinate representation, the Hamiltonian for an electron moving in the field of a nucleus of charge Ze fixed at the origin is

$$H = -\frac{\hbar^2}{2m}\nabla^2 - \frac{Ze^2}{|\mathbf{x}|}.$$

(a) Derive the Schrödinger equation for the momentum-space wave function $\phi(\mathbf{p}, t)$. [Hint: use (4.33').]

(b) By working in the p-representation, show that all the components of the angular momentum (referred to the origin) are constants of the motion.

7. In classical electromagnetic theory, the field strengths are unchanged when the gauge transformation

$$\begin{aligned} \mathbf{A} &\to \mathbf{A} + \nabla\Lambda, \\ \phi &\to \phi - \frac{1}{c}\frac{\partial\Lambda}{\partial t} \end{aligned} \tag{1}$$

is made, where Λ is a scalar solution of the wave equation. If, in the quantum theory, we desire the physical results to be invariant under the gauge transformation, show that (I) must be augmented by the transformation

$$\psi(\mathbf{x}_1, \ldots, \mathbf{x}_N, t) \rightarrow \psi(\mathbf{x}_1, \ldots, \mathbf{x}_N, t) \exp\left\{\frac{ie}{\hbar c}\sum_{\alpha=1}^{N} \Lambda(\mathbf{x}_\alpha, t)\right\}$$

on the wave function.

8. Consider a particle in a magnetic field $\mathcal{H}$, and let $\mathbf{v}$ be its velocity operator. Show that

$$[v_x, v_y] = \frac{ie\hbar}{m^2c}\mathcal{H}_z.$$

III

Illustrative Solutions of Schrödinger's Equation

This chapter is devoted to a detailed examination of a number of relatively simple problems pertaining to the scattering and bound states of two structureless particles interacting through a central field. Some of the results that we shall obtain have a much broader significance than first meets the eye. In Sec. 10 we shall determine the spectrum of the angular momentum operators by a method that does not refer at all to the structure of the system. Consequently the eigenvalues that we shall find there pertain to any angular momentum: they apply equally well to a complex molecule or the quantized electromagnetic field. Many of the techniques that we shall develop here can be elaborated in a fairly straightforward fashion and applied to problems of great complexity. For example, the formulation of scattering theory in terms of an integral equation in Sec. 12 also serves as an introduction to the theory of collisions between complex systems (Chapt. IX). The phase shift analysis described in Sec. 14 holds not only in potential scattering: With only minor modifications it applies to any collision process where two particles collide and two others emerge, e.g., to $k^- + p \to \pi^- + \Sigma^+$.

9. *Separation of Variables in the Two-Body Problem*

The observables pertaining to the two-body problem were already introduced and discussed in Sec. 6.4. We shall now investigate the eigenvalues and eigenfunctions of these observables.

With a slight change of notation we write the Hamiltonian (6.35) as

$$\mathcal{H} = \frac{P^2}{2M} + H, \tag{1}$$

where H is the Hamiltonian in the center-of-mass system:

$$H = \frac{p^2}{2\mu} + V(\mathbf{r}). \tag{2}$$

9.1 SEPARATION OF VARIABLES

The center-of-mass and relative coordinates can be separated by introducing a wave function of the form

$$\Psi(\mathbf{r}, \mathbf{R}, t) = e^{i\mathbf{P}\cdot\mathbf{R}/\hbar}\psi_E(\mathbf{r}) \exp\left[-i\left(\frac{P^2}{2M} + E\right)t/\hbar\right]\cdot \tag{3}$$

This will be a stationary state solution of $i\hbar\dot{\Psi} = \mathcal{H}\Psi$ provided

$$(H - E)\psi_E(\mathbf{r}) = 0. \tag{4}$$

Clearly Ψ is an eigenfunction of the total momentum operator with eigenvalue $\mathbf{P}$, and also an eigenfunction of $\mathcal{H}$ with eigenvalue $E + P^2/2M$. When ψ_E describes a bound state one can construct localized two-body states (or wave packets) by superposing solutions of type (3) with different values of $\mathbf{P}$.

We confine ourselves to the important case of a central field $V(r)$. The relative angular momentum is then conserved:

$$[\mathbf{L}_r, H] = 0. \tag{5}$$

(We shall delete the subscript r from $\mathbf{L}_r$ henceforth, because we need only concern ourselves with the motion in the center-of-mass system.) As we already demonstrated in (7.51), the only complete set of compatible observables pertaining to the center-of-mass motion that contains H consists of H, L^2 and L_z. Hence we may choose ψ_E to be a simultaneous eigenfunction of L^2 and L_z, as well as of H:

$$[L^2 - l(l + 1)]\psi_{Elm}(\mathbf{r}) = 0, \tag{6}$$
$$(L_z - m)\psi_{Elm}(\mathbf{r}) = 0. \tag{7}$$

The eigenvalue of L^2 has been written in this form for our later convenience because in the following section we shall discover that l is a positive integer or zero. At the moment we shall not make use of this fact.

For our purpose we require the angular momentum operators in spherical coordinates (r, θ, ϕ), with the z-direction as the polar axis. By means of elementary calculus one finds that $L_z = -i(x\,\partial/\partial y - y\,\partial/\partial x)$ is simply

$$L_z = \frac{1}{i}\frac{\partial}{\partial\phi}. \tag{8}$$

A lengthier calculation results in

$$L_\pm = \pm e^{\pm i\phi}\left(\frac{\partial}{\partial\theta} \pm i\cot\theta\frac{\partial}{\partial\phi}\right), \tag{9}$$

where L_+ and L_- are the non-Hermitian operators

$$L_\pm = L_x \pm iL_y. \tag{10}$$

As we shall see, these non-Hermitian operators have somewhat simpler properties than L_x and L_y. In terms of L_z, L_+ and L_-, the operator L^2 is

$$L^2 = L_z^2 + \tfrac{1}{2}(L_+L_- + L_-L_+). \tag{11}$$

Direct substitution then reveals that

$$L^2 = -\left(\frac{1}{\sin\theta}\frac{\partial}{\partial\theta}\sin\theta\frac{\partial}{\partial\theta} + \frac{1}{\sin^2\theta}\frac{\partial^2}{\partial\phi^2}\right).$$

In spherical coordinates the Laplacian is

$$\nabla^2 = \frac{1}{r^2}\frac{\partial}{\partial r}r^2\frac{\partial}{\partial r} - \frac{L^2}{r^2},$$

and the kinetic energy is therefore

$$\frac{p^2}{2\mu} = \frac{1}{2\mu}\left(p_r^2 + \frac{\hbar^2L^2}{r^2}\right), \tag{12}$$

where

$$p_r = \frac{\hbar}{i}\left(\frac{1}{r} + \frac{\partial}{\partial r}\right). \tag{13}$$

Equation (12) has the same form as a familiar expression for the kinetic energy in classical mechanics. The differential operator p_r has been arranged to be Hermitian in the interval $0 \leq r < \infty$, i.e., it satisfies

$$\int_0^\infty r^2\,dr\,f^*(r)p_r g(r) = \int_0^\infty r^2\,dr[p_r f(r)]^* g(r).$$

Because of (12) and (6), the Schrödinger equation (4) becomes

$$\left[\frac{1}{2\mu}\left(p_r^2 + \frac{\hbar^2 l(l+1)}{r^2}\right) + V(r)\right]\psi_{Elm} = E\psi_{Elm}. \tag{14}$$

This differential operator depends only on the single variable r, while L^2 and L_z depend only on θ and ϕ. We can therefore separate variables. In other words, a solution of the system of differential equations can be written as a product of two functions,

$$\psi_{Elm}(\mathbf{r}) = R_{El}(r)Y_{lm}(\theta\phi), \tag{15}$$

provided $R_{El}(r)$ satisfies the ordinary differential equation

$$\left[\frac{\hbar^2}{2\mu}\left(-\frac{1}{r^2}\frac{d}{dr}r^2\frac{d}{dr} + \frac{l(l+1)}{r^2}\right) + V(r) - E\right]R_{El}(r) = 0, \tag{16}$$

and $Y_{lm}(\theta\phi)$ satisfies the differential equations

$$[L^2 - l(l+1)]Y_{lm} = 0, \tag{17}$$
$$(L_z - m)Y_{lm} = 0. \tag{18}$$

9.2 IMPLICATIONS OF ROTATIONAL INVARIANCE

There are several features of these equations that deserve special emphasis. Note that the energy eigenvalue E, the potential V, and the mass μ only appear in the radial equation (16). This means that the angular momentum eigenfunctions $Y_{lm}(\theta\phi)$, and the associated eigenvalues $l(l+1)$ and m, do not depend on the particular dynamical system that we are considering. Equations (17) and (18) are completely free of any physical parameters or constants of nature, and the spectrum and eigenfunctions of the angular momentum observables therefore involve only pure numbers.* As we shall see in Chapt. VI, this is one facet of the fact that all information concerning the angular momentum observables can be deduced from symmetry considerations. The radial equation also demonstrates that the energy eigenvalue E can only depend on l, because the eigenvalue m does not appear in (16). This result is a consequence of the rotational invariance of H. Under the same circumstances, the orientation of a classical orbit—as given by the direction of its angular momentum vector—does not affect its energy, whereas the magnitude of the angular momentum does. Here the number m specifies the projec-

* Provided all angular momenta are measured in units of $\hbar$, of course.

tion of the angular momentum along the z-axis, but as there is no preferred direction in space, the energy cannot depend on m.

The rotational invariance of the Hamiltonian has several other consequences of great importance. These can be deduced by applying the infinitesimal rotation operators of Sec. 6.3 to equations (4), (6) and (7). A function $f(\mathbf{r})$ can be rotated through the infinitesimal angle $\delta\omega$ about the axis $\hat{\mathbf{n}}$ as follows:

$$(1 - i\,\delta\boldsymbol{\omega}\cdot\mathbf{L})f(\mathbf{r}) = f(\mathbf{r} - \delta\boldsymbol{\omega}\times\mathbf{r}),$$

where $\delta\boldsymbol{\omega} = \delta\omega\,\hat{\mathbf{n}}$. Taking advantage of $[\mathbf{L}, H] = [\mathbf{L}, L^2] = 0$, we see that

$$(H - E)(1 - i\,\delta\boldsymbol{\omega}\cdot\mathbf{L})\psi_{Elm}(\mathbf{r}) = 0, \tag{19}$$

$$[L^2 - l(l+1)](1 - i\,\delta\boldsymbol{\omega}\cdot\mathbf{L})\psi_{Elm}(\mathbf{r}) = 0. \tag{20}$$

Hence the rotated function $(1 - i\,\delta\boldsymbol{\omega}\cdot\mathbf{L})\psi_{Elm}(\mathbf{r})$ is also a simultaneous eigenfunction of the energy and L^2 with eigenvalues E and $l(l+1)$, respectively. On the other hand, if the rotation axis $\hat{\mathbf{n}}$ does not coincide with the z-direction, L_z will fail to commute with $(1 - i\,\delta\boldsymbol{\omega}\cdot\mathbf{L})$, and $(1 - i\,\delta\boldsymbol{\omega}\cdot\mathbf{L})\psi_{Elm}$ will not, in general, be an eigenfunction of L_z with eigenvalue m. When $\psi_{Elm}(\mathbf{r})$ is not a spherically symmetric function (i.e., not a function of $|\mathbf{r}|$ alone), $(1 - i\,\delta\boldsymbol{\omega}\cdot\mathbf{L})\psi_{Elm}$ and ψ_{Elm} are obviously different functions, and therefore linearly independent. But they are both energy eigenfunctions with the *same* eigenvalue E, and the energy level is therefore degenerate. Although we do not yet know how many linearly independent stationary states belong to the energy level E, we can make the following statement:

*If a stationary state of a spherically symmetric Hamiltonian is not itself spherically symmetric, the energy level in question is necessarily degenerate.**

By precisely the same argument we also conclude that for every nontrivial solution Y_{lm} of (17) there exists at least one other eigenfunction with the same eigenvalue $l(l+1)$ of L^2, and a different eigenvalue m of L_z. The qualification "nontrivial" is meant to exclude the exceptional—though important—case where Y_{lm} is independent of θ and ϕ.

Since the degeneracies of the energy spectrum are a consequence of the symmetries of the Hamiltonian, they can be determined experimentally by destroying this symmetry. In the laboratory this can often be

* One may well ask whether the converse is true: Does a degenerate energy spectrum necessarily imply a symmetry of the Hamiltonian? It would take us too far afield to analyze this question here, but the answer appears to be "yes." In atomic and molecular physics, these symmetries are associated with ordinary geometric concepts. The degeneracies in the mass spectrum of the "elementary" particles (e.g., the equality $m_{\pi^+} = m_{\pi^-}$, and $(m_p - m_n)/(m_p + m_n) \ll 1$) seem to be manifestations of symmetries that have no macroscopic counterparts. (For details, consult Källén and Sakurai.)

achieved by allowing the system under study to interact with forces arranged so as to introduce a preferred spatial direction. For example, the radiation emitted by an atom in a weak externally applied magnetic field $\mathcal{H}$ reveals groups of energy levels (multiplets) that are nearly degenerate. This is known as the Zeeman effect. As $\mathcal{H} \to 0$, the symmetry of the problem is restored and the energy differences between the members of a Zeeman multiplet tend to zero.

The Hilbert space $\mathfrak{H}$ spanned by the totality of eigenfunctions $\{\psi_{Elm}\}$ can be decomposed into orthogonal subspaces that are invariant under rotations.* Let $\mathfrak{H}_{El}$ be a subspace of $\mathfrak{H}$ spanned by all the eigenfunctions belonging to the indicated eigenvalues of H and L^2. The dimensionality of $\mathfrak{H}_{El}$ is given by the number of orthogonal eigenfunctions ψ_{Elm} of L_z that belong to the *same* eigenvalue $l(l + 1)$ of L^2. Let $\chi(\mathbf{r})$ be an arbitrary function belonging to $\mathfrak{H}_{El}$:

$$\chi(\mathbf{r}) = \sum_m c_m \psi_{Elm}(\mathbf{r}).$$

Under an infinitesimal rotation χ transforms as follows:

$$\chi(\mathbf{r}) \to \chi'(\mathbf{r}) = \sum_m c_m(1 - i\,\delta\omega \cdot \mathbf{L})\psi_{Elm}(\mathbf{r}).$$

Every term on the right-hand side is an eigenfunction of H and L^2 with eigenvalue E and $l(l + 1)$, respectively, and hence $\chi(\mathbf{r})$ still belongs to $\mathfrak{H}_{El}$. In particular, if χ is the function $Y_{lm}(\theta\phi)$ itself, we must have

$$(1 - i\,\delta\omega\,\hat{\mathbf{n}} \cdot \mathbf{L})Y_{lm}(\theta\phi) = Y_{lm}(\theta\phi) + \delta\omega \sum_{m'} c_{mm'}^{(l)} Y_{lm'}(\theta\phi), \qquad (21)$$

where the coefficients $c_{mm'}^{(l)}$ depend only on the orientation of $\hat{\mathbf{n}}$. Although our proofs have been confined to infinitesimal rotations, they can be generalized to finite rotations by repeated applications of (21). A more elegant way of doing this will be discussed in Sec. 32, but from the foregoing discussion it is already clear that *the result of an arbitrary rotation applied to the function Y_{lm} can be expressed as a linear superposition of eigenfunctions belonging to the same eigenvalue of L^2.*

* This terminology has the following meaning. Let $(u_1, \ldots, u_N)$ and $(v_1, \ldots, v_M)$ be two sets of linearly independent functions. The set of all functions of the type $\Sigma_n c_n u_n$, where the c_n are arbitrary constants, constitute the space $\mathfrak{H}_u$. In a similar fashion the set of all functions $\Sigma_m c'_m v_m$ constitute the space $\mathfrak{H}_v$, and the set of all functions $\Sigma_n c_n u_n + \Sigma_m c'_m v_m$ constitute the space $\mathfrak{H}$. Then $\mathfrak{H}_u$ and $\mathfrak{H}_v$ are orthogonal subspaces of $\mathfrak{H}$ if $(u_n, v_m) = 0$ for all n and m. The subspace $\mathfrak{H}_u$ is invariant under the operation A if Au_n is a vector in $\mathfrak{H}_u$ for all u_n.

10. Eigenvalues and Eigenfunctions of the Angular Momentum Operators

In this section we shall determine the spectra of the operators L_z and L^2, and evaluate their eigenfunctions Y_{lm} (the so-called spherical harmonics). Instead of approaching this problem by techniques drawn from the standard theory of differential equations, we shall use operator methods.

10.1 THE EIGENVALUES

We normalize the eigenfunction according to

$$(Y_{lm}, Y_{l'm'}) \equiv \int_0^\pi \sin\theta\, d\theta \int_{-\pi}^{\pi} d\phi\, Y^*_{lm}(\theta\phi) Y_{l'm'}(\theta\phi) = \delta_{ll'}\,\delta_{mm'}. \quad (1)$$

The basic commutation rules are given by (6.16):

$$[L_x, L_y] = iL_z, \quad (2)$$

and cyclic permutations. The commutators involving L_z and $L_\pm$ are evaluated with the help of (2). Thus $[L_x \pm iL_y, L_z] = -iL_y \pm i(iL_x)$, or

$$[L_\pm, L_z] = \mp L_\pm, \quad (3)$$

and $[L_+, L_-] = i[L_y, L_x] - i[L_x, L_y]$, or

$$[L_+, L_-] = 2L_z; \quad (4)$$

naturally

$$[L^2, L_\pm] = [L^2, L_z] = 0. \quad (5)$$

Using (4) in (9.11), we obtain

$$\begin{aligned} L_+L_- &= L^2 - L_z(L_z - 1), \\ L_-L_+ &= L^2 - L_z(L_z + 1). \end{aligned} \quad (6)$$

Next we compute

$$\begin{aligned} \|L_xY_{lm}\| + \|L_yY_{lm}\| &= (Y_{lm}, (L^2 - L_z^2)Y_{lm}) \\ &= l(l+1) - m^2. \end{aligned}$$

This must be positive, and therefore

$$m^2 \leqslant l(l+1), \quad (7)$$

i.e., m is bounded both above and below. (Note that $l(l+1) \geqslant 0$ by the same argument.) Let $m_>$ and $m_<$ be these least upper and greatest lower bounds.

Consider now the functions $L_\pm Y_{lm}$. Because of (5)

$$L^2 L_\pm Y_{lm} = L_\pm L^2 Y_{lm} = l(l+1) L_\pm Y_{lm}, \tag{8}$$

while (3) leads to

$$L_z L_\pm Y_{lm} = (L_\pm L_z \pm L_\pm) Y_{lm} = L_\pm (m \pm 1) Y_{lm}; \tag{9}$$

thus $L_\pm Y_{lm}$ is an eigenfunction of L^2 and L_z with eigenvalues $l(l+1)$ and $(m \pm 1)$, respectively. However, as a consequence of (7), $L_\pm Y_{lm_\gtrless}$ cannot be eigenfunctions! This can only occur if these are null vectors. But the norms $\|L_\pm Y_{lm}\|$ are

$$\begin{aligned} \|L_\pm Y_{lm}\| &= (Y_{lm}, L_\mp L_\pm Y_{lm}) \\ &= (Y_{lm}, [L^2 - L_z(L_z \pm 1)] Y_{lm}) \\ &= l(l+1) - m(m \pm 1), \end{aligned} \tag{10}$$

and so we require

$$l(l+1) = m_>(m_> + 1) = m_<(m_< - 1), \tag{11}$$

or

$$(m_> + m_<)(m_> - m_< + 1) = 0.$$

Hence $m_< = -m_>$, because $(m_> - m_< + 1) > 0$.

Assume that m_0 is one of the eigenvalues of L_z. By repeatedly applying L_+ to Y_{lm_0} we can generate the sequence of eigenfunctions Y_{lm_0+1}, Y_{lm_0+2}, etc. This sequence must terminate with $Y_{lm_>}$, because $L_+ Y_{lm}$ only vanishes when $m = m_>$. That is to say, if the sequence of functions $(L_+)^k Y_{lm_0}$, where $k = 0, 1, 2, \ldots$, did not lead to $Y_{lm_>}$, we could increase the eigenvalue of L_z interminably and thereby violate (7). Having established that $Y_{lm_>}$ is one of the eigenfunctions, we can then apply successively higher powers of L_- to this function, and descend the chain of eigenfunctions. But $L_- Y_{lm}$ only vanishes if $m = m_<$, and (7) therefore implies that $Y_{lm_<}$ must be the eigenfunction with the smallest eigenvalue of L_z. Because each step in the chain of L_z-eigenvalues is of unit length, $m_> - m_<$ must either be zero or a positive integer. But $m_> - m_< = 2m_>$, and $m_> = l$. Hence $2l$ is either zero or a positive integer.

To summarize:

(a) *the eigenvalues of L^2 have the form $l(l+1)$, where $l = 0, \frac{1}{2}, 1, \frac{3}{2}, \ldots$;*
(b) *the eigenvalues of L_z are $-l, -l+1, \ldots, l-1, l$.*

The arguments that led us to these conclusions did not make use of the differential operators of Sec. 9. Only the commutation rules (2), and the requirement that the norm of any wave function must be positive (or zero), entered into the derivation. Consequently any angular momentum—such as the total angular momentum of an N-particle system—will possess the spectrum given by (a) and (b) above.

10.2 THE EIGENFUNCTIONS

The eigenfunctions $Y_{lm}(\theta\phi)$ are constructed by using the differential form of the angular momentum operators. Because of this the results that we shall obtain in this subsection are not as general as those concerning the eigenvalues of L^2 and L_z. The spherical harmonics are the eigenfunctions appropriate to a single particle that can be described by the three classical degrees of freedom, or for a pair of such particles in their center-of-mass frame. They are not the angular momentum eigenfunctions of a many-particle system, nor do they describe a particle with intrinsic spin.

The ϕ-dependence of $Y_{lm}(\theta\phi)$ follows immediately from (9.8):

$$\left(\frac{1}{i}\frac{\partial}{\partial\phi} - m\right) Y_{lm}(\theta\phi) = 0,$$

or

$$Y_{lm}(\theta\phi) = y_{lm}(\theta)e^{im\phi}. \tag{12}$$

The functions at the top of and bottom of the chain, $Y_{l,\pm l}(\theta\phi)$, are readily determined from $L_\pm Y_{l,\pm l} = 0$. With the help of (9.9) we can translate this into the differential equations

$$\left(\frac{d}{d\theta} - l\cot\theta\right) y_{l,\pm l}(\theta) = 0.$$

Hence

$$y_{l,\pm l}(\theta) = c_l^\pm \sin^l\theta; \tag{13}$$

the constants $c_l^\pm$ will be determined by the normalization condition (1) and a phase convention.

At this point we can eliminate the half-integral eigenvalues $l = \frac{1}{2}, \frac{3}{2}$, etc. It is frequently stated that these eigenvalues do not occur because $Y_{lm}(\theta\phi)$ must equal $Y_{lm}(\theta, \phi + 2\pi)$. This is a somewhat misleading argument because we need only require the probability distribution to be single valued. Thus

$$|aY_{lm} + bY_{l'm'}|^2 = |a|^2|y_{lm}(\theta)|^2 + |b|^2|y_{l'm'}(\theta)|^2 + 2\,\mathrm{Re}\,[ab^* y_{lm}(\theta)y^*_{l'm'}(\theta)e^{i(m-m')\phi}]$$

must be single valued, which requires $m - m'$ to be an integer. Hence a wave function that is a linear combination of eigenfunctions belonging to *either* integral values of l *or* half-integral values of l will do. The spherical harmonics of half-integral order can be eliminated on other grounds, however. Consider the example of $l = \frac{1}{2}$, for which the eigenfunctions are $Y_{\frac{1}{2},\pm\frac{1}{2}} = c^{\pm}\sqrt{\sin\theta}\, e^{\pm i\phi/2}$. If we perform an arbitrary rotation on either of these functions, we should be able to express the result as a linear combination of $Y_{\frac{1}{2},\pm\frac{1}{2}}$, as discussed in (9.21) *et seq.* It is a simple matter to show that these functions do not satisfy this general requirement. One can do this by carrying out a particular rotation (see Prob. 1(iii)), or by noting that

$$L_{-}Y_{\frac{1}{2}\frac{1}{2}} = -c^{+}e^{-i\phi/2}\frac{\cos\theta}{\sqrt{\sin\theta}}$$

is not proportional to $Y_{\frac{1}{2},-\frac{1}{2}}$, as it should be. These difficulties are common to all the spherical harmonics of half-integer l. On the other hand, for integer values of l, the spherical harmonics transform linearly among themselves under a rotation, and satisfy all the conditions set out in Sec. 10.1. We therefore conclude that the angular momentum of a system that can be described by the three classical degrees of freedom necessarily has integer eigenvalues. The half-integer eigenvalues ($l = \frac{1}{2}, \frac{3}{2}$, etc.) also occur in nature, but they can only be incorporated into the theory by introducing a further, nonclassical, degree of freedom—the spin (see Sec. 33).

We proceed to the construction of the functions $Y_{lm}(\theta, \phi)$ when l (and hence m) are integers. First of all, (10) and $L_{\pm}Y_{lm} = c_{\pm}Y_{l,m\pm1}$ imply that we can choose the arbitrary phases so that*

$$L_{\pm}Y_{lm} = \sqrt{l(l+1) - m(m\pm1)}\; Y_{l,m\pm1}. \tag{14}$$

By repeated use of this identity we find

$$Y_{lm} = \sqrt{\frac{(l+m)!}{(2l)!(l-m)!}}\, L_{-}^{l-m}Y_{ll}. \tag{15}$$

Now recall (9.9), according to which

$$\begin{aligned} L_{-}e^{im\phi}f(\theta) &= -e^{i(m-1)\phi}\left(\frac{\partial}{\partial\theta} + m\cot\theta\right)f(\theta) \\ &= e^{i(m-1)\phi}\sin^{1-m}\theta\,\frac{d}{d\cos\theta}\left[\sin^{m}\theta\cdot f(\theta)\right]. \end{aligned}$$

* Our phase convention is the one employed by Condon and Shortley.

Note that this is again a function of the form $e^{i\lambda\phi}f(\theta)$, and therefore we immediately have

$$L_-^2\, e^{im\phi}f(\theta) = e^{i(m-2)\phi}\sin^{2-m}\theta\,\frac{d}{d\cos\theta}\left\{\sin^{m-1}\theta \times \sin^{1-m}\theta\,\frac{d}{d\cos\theta}\sin^m\theta f(\theta)\right\},$$

or

$$L_-^k\, e^{im\phi}f(\theta) = e^{i(m-k)\phi}\sin^{k-m}\theta\left(\frac{d}{d\cos\theta}\right)^k\sin^m\theta f(\theta). \tag{16}$$

We now apply this identity to $Y_{ll} = c_l e^{il\phi}\sin^l\theta$, where the constant c_l follows from normalization:

$$(Y_{ll}, Y_{ll}) = |c_l|^2 4\pi\,\frac{(2^l l!)^2}{(2l+1)!}.$$

By convention the phase of c_l is chosen so that

$$c_l = (-1)^l\,\frac{1}{2^l l!}\sqrt{\frac{(2l+1)!}{4\pi}}.$$

Combining this with (15), and putting $k = l - m$, $m = l$ in (16), we finally obtain

$$Y_{lm}(\theta\phi) = \frac{(-1)^l}{2^l l!}\sqrt{\frac{2l+1}{4\pi}\frac{(l+m)!}{(l-m)!}}\,e^{im\phi}\sin^{-m}\theta\left(\frac{d}{d\cos\theta}\right)^{l-m}\sin^{2l}\theta. \tag{17}$$

When $m = 0$ this becomes

$$Y_{l0}(\theta\phi) = \sqrt{\frac{2l+1}{4\pi}}\,P_l(\cos\theta), \tag{18}$$

where P_l is the Legendre polynomial.

It is frequently convenient to relate (17) to the associated Legendre function

$$P_l^{|m|}(\theta) = \frac{(-1)^{m+l}}{2^l l!}\frac{(l+|m|)!}{(l-|m|)!}\sin^{-|m|}\theta\left(\frac{d}{d\cos\theta}\right)^{l-|m|}\sin^{2l}\theta.$$

For $m \geqslant 0$ the desired relation is seen to be

$$Y_{lm}(\theta\phi) = (-1)^m\sqrt{\frac{2l+1}{4\pi}\frac{(l-m)!}{(l+m)!}}\,e^{im\phi}P_l^m(\theta); \tag{19}$$

for negative m one uses

$$Y_{l,-m}(\theta\phi) = (-1)^m Y_{lm}^*(\theta\phi). \tag{19}$$

Occasionally we shall require the explicit form of the spherical harmonics. In tabulating the formulas it is convenient to write $Y_{lm}(\theta\phi) = \Theta_{lm}(\theta)e^{im\phi}/\sqrt{2\pi}$. Working from (17) one then finds

$$\Theta_{11} = -\sqrt{\tfrac{3}{4}}\sin\theta, \qquad \Theta_{10} = \sqrt{\tfrac{3}{2}}\cos\theta;$$

$$\Theta_{22} = \sqrt{\tfrac{15}{16}}\sin^2\theta, \qquad \Theta_{21} = -\sqrt{\tfrac{15}{4}}\cos\theta\sin\theta,$$
$$\Theta_{20} = \sqrt{\tfrac{5}{8}}\,(2\cos^2\theta - \sin^2\theta). \tag{20}$$

Finally we quote two very important relations involving the spherical harmonics. The first of these states that $\{Y_{lm}(\theta\phi)\}$ is a complete set of single-valued functions on the unit sphere:

$$\sum_{l=0}^{\infty}\sum_{m=-l}^{l} Y_{lm}(\theta\phi)Y_{lm}^*(\theta'\phi') = \frac{\delta(\theta-\theta')\,\delta(\phi-\phi')}{\sin\theta}. \tag{21}$$

The proof of this formula is the subject of Prob. 2. The other relationship is the *addition theorem:* Let $\hat{\mathbf{n}}_1$ and $\hat{\mathbf{n}}_2$ have the orientations specified by $(\theta_1\phi_1)$ and $(\theta_2\phi_2)$, respectively; then

$$P_l(\hat{\mathbf{n}}_1\cdot\hat{\mathbf{n}}_2) = \frac{4\pi}{2l+1}\sum_{m=-l} Y_{lm}(\theta_1\phi_1)Y_{lm}^*(\theta_2\phi_2). \tag{22}$$

This theorem should be familiar from electrostatics (see Jackson, p. 67), and therefore we shall not prove it now. A derivation of (22) that involves the theory of the rotation group will be given in Sec. 35.1.

11. Free-Particle Wave Functions

In the absence of forces the radial Schrödinger equation (9.16) reduces to

$$\left(\frac{p_r^2}{2\mu} + \frac{\hbar^2 l(l+1)}{2\mu r^2} - E\right)R_{El}(r) = 0. \tag{1}$$

This equation can be cast into dimensionless form by setting $E = \hbar^2k^2/2\mu$ and $\rho = kr$:

$$\left(\frac{1}{\rho^2}\frac{d}{d\rho}\rho^2\frac{d}{d\rho} - \frac{l(l+1)}{\rho^2} + 1\right)R_l(\rho) = 0. \tag{2}$$

The energy eigenvalue E now makes its appearance only through the dependence of ρ on k. Equation (2) is one of Bessel's equations, and the general solution of (2) is related to the cylinder functions $Z_\nu(\rho)$ by

$$R_l(\rho) = \text{const.}\ \frac{Z_{l+\frac{1}{2}}(\rho)}{\sqrt{\rho}}. \tag{3}$$

Functions of this kind are called spherical cylinder functions. As we know, any second-order equation has two linearly independent solutions. For any equation having a singular point at $r = 0$, the solutions fall into two classes: those that are regular (analytic) at $r = 0$, and the others that are not. In the case of (2), the regular solution is the spherical Bessel function:

$$j_l(\rho) = \sqrt{\frac{\pi}{2\rho}}\, J_{l+\frac{1}{2}}(\rho). \tag{4}$$

It has the asymptotic behavior

$$j_l(\rho) \xrightarrow[\rho\to\infty]{} \frac{1}{\rho} \sin\left(\rho - \frac{\pi l}{2}\right) + O\left(\frac{1}{\rho^2}\right), \tag{5}$$

and at the origin it tends to *

$$j_l(\rho) \xrightarrow[\rho\to 0]{} \frac{\rho^l}{(2l+1)!!}. \tag{6}$$

A (linearly independent) irregular solution is the spherical Neumann function

$$n_l(\rho) = \sqrt{\frac{\pi}{2\rho}}\, N_{l+\frac{1}{2}}(\rho). \tag{7}$$

Its properties are

$$n_l(\rho) \xrightarrow[\rho\to\infty]{} -\frac{1}{\rho} \cos\left(\rho - \frac{l\pi}{2}\right), \tag{8}$$

$$n_l(\rho) \xrightarrow[\rho\to 0]{} -(2l-1)!!\left(\frac{1}{\rho}\right)^{l+1} \tag{9}$$

From the spherical Bessel and Neumann functions we can form two other irregular solutions of great utility. These are the spherical Hankel function †

$$h_l(\rho) = j_l(\rho) + i n_l(\rho), \tag{10}$$

* $(2l+1)!! = (2l+1)(2l-1)\cdots 3\cdot 1$.
† Many authors use the notation $h_l^{(1)}$ for what we call h_l, and $h_l^{(2)}$ for h_l^*.

and its complex conjugate $h_l^*(\rho)$. Their asymptotic form is

$$h_l(\rho) \xrightarrow[\rho\to\infty]{} \frac{1}{i^{l+1}\rho} e^{i\rho}. \tag{11}$$

For small angular momenta the various radial functions are

$$\begin{aligned}
j_0(\rho) &= \frac{1}{\rho}\sin\rho, \qquad j_1(\rho) = \frac{\sin\rho}{\rho^2} - \frac{\cos\rho}{\rho},\\
j_2(\rho) &= \left(\frac{3}{\rho^3} - \frac{1}{\rho}\right)\sin\rho - \frac{3\cos\rho}{\rho^2};\\
n_0(\rho) &= -\frac{1}{\rho}\cos\rho, \qquad n_1(\rho) = -\frac{\cos\rho}{\rho^2} - \frac{\sin\rho}{\rho},\\
n_2(\rho) &= -\left(\frac{3}{\rho^3} - \frac{1}{\rho}\right)\cos\rho - \frac{3\sin\rho}{\rho^2};
\end{aligned} \tag{12}$$

$$h_0(\rho) = \frac{1}{i\rho} e^{i\rho}, \qquad h_1(\rho) = -\left(1 + \frac{i}{\rho}\right)\frac{e^{i\rho}}{\rho}, \qquad h_2(\rho) = \left(1 + \frac{3i}{\rho} - \frac{3}{\rho^2}\right)\frac{ie^{i\rho}}{\rho}.$$

The function $n_0(\rho)$ cannot be the wave function of a free particle because the operator p_r^2 is not Hermitian with respect to it: differentiation of $1/\rho$ leads to unacceptable singularities. The functions n_l, with $l > 0$, cannot be normalized. Therefore the $j_l(\rho)$ are the only acceptable functions.

For the sake of consistency it is important to show that $\{j_l(\rho)\}$ forms a complete set. We make use of the indefinite integral

$$\int x^2\,dx\, z_l(\alpha x)\bar{z}_l(\beta x) = \frac{x^2}{\alpha^2 - \beta^2}\{\beta z_l(\alpha x)\bar{z}_{l-1}(\beta x) - \alpha z_{l-1}(\alpha x)\bar{z}_l(\beta x)\},$$

which holds for any two solutions z_l and $\bar{z}_l$ of (2). The asymptotic expansion (5) then implies that

$$\int_0^K j_l(kr)j_l(kr')k^2\,dk = \frac{1}{2rr'}\left\{\frac{(-1)^l}{r+r'}\sin K(r+r') + \frac{1}{r-r'}\sin K(r-r')\right\}$$

as $K \to \infty$. In this limit the first term always oscillates because r and r' are positive; it is therefore negligible. The contribution of the second term can be evaluated with the help of

$$\frac{1}{\pi}\lim_{K\to\infty}\frac{\sin Kx}{x} = \delta(x), \tag{13}$$

and therefore

$$\int_0^\infty j_l(kr)j_l(kr')k^2\,dk = \frac{\pi}{2r^2}\,\delta(r-r'). \tag{14}$$

Obviously,

$$\int_0^\infty j_l(kr)j_l(k'r)r^2\,dr = \frac{\pi}{2k^2}\,\delta(k-k'). \tag{14'}$$

The normalization of the simultaneous eigenfunctions ψ^0_{Elm} of H, L^2, and L_z shall be chosen so that the orthogonality and completeness relations conform with the format originally set out in Sec. 7.6. Bearing (14) and (10.21) in mind, we set

$$\psi^0_{Elm}(\mathbf{r}) = \left(\frac{2}{\pi}\frac{k^2\,dk}{dE}\right)^{\frac{1}{2}} j_l(kr)Y_{lm}(\theta\phi). \tag{15}$$

The completeness and orthogonality relations then have the desired form:

$$\int_0^\infty dE \sum_{l=0}^{\infty}\sum_{m=-l}^{l} \psi^0_{Elm}(\mathbf{r})\psi^0_{Elm}(\mathbf{r}')^* = \delta(\mathbf{r}-\mathbf{r}'), \tag{16}$$

$$\int d^3r\,\psi^0_{Elm}(\mathbf{r})^*\psi^0_{E'l'm'}(\mathbf{r}) = \delta_{ll'}\,\delta_{mm'}\,\delta(E-E'). \tag{17}$$

In deriving (16) we have made use of the formula

$$\delta(\mathbf{r}-\mathbf{r}') = \frac{\delta(r-r')\,\delta(\theta-\theta')\,\delta(\phi-\phi')}{r^2\sin\theta}. \tag{18}$$

There is another set of free particle wave functions with which we are already very familiar, the plane waves

$$\phi_{\mathbf{k}}(\mathbf{r}) = (2\pi)^{-\frac{3}{2}}e^{i\mathbf{k}\cdot\mathbf{r}}. \tag{19}$$

These are eigenfunctions of the momentum with eigenvalue $\hbar\mathbf{k}$, and consequently they possess the energy eigenvalue $E = \hbar^2k^2/2\mu$. They are not eigenfunctions of L^2 and L_z, however. The set $\{\phi_{\mathbf{k}}\}$ is both complete and orthonormal:

$$\int d^3\mathbf{k}\;\phi_{\mathbf{k}}(\mathbf{r})\phi^*_{\mathbf{k}}(\mathbf{r}') = \delta(\mathbf{r}-\mathbf{r}'), \tag{20}$$

$$\int d^3\mathbf{r}\;\phi^*_{\mathbf{k}}(\mathbf{r})\phi_{\mathbf{k}'}(\mathbf{r}) = \delta(\mathbf{k}-\mathbf{k}'). \tag{21}$$

Because the sets $\{\phi_{\mathbf{k}}\}$ and $\{\psi^0_{Elm}\}$ are both complete, they may be expanded in terms of each other. The functions $\phi_{\mathbf{k}}$ and ψ^0_{Elm} have the same energy eigenvalue $\hbar^2k^2/2\mu$, and therefore the expansion of the former in terms of the latter only involves a sum over the angular momentum quantum numbers l and m. Furthermore, we note that if the z-axis is chosen along $\hat{\mathbf{k}}$,

$$L_z\phi_{\mathbf{k}} = \frac{1}{i(2\pi)^{\frac{3}{2}}}\frac{\partial}{\partial\phi}e^{ikr\cos\theta},$$

and therefore $\phi_{\mathbf{k}}(\mathbf{r})$ is already an eigenfunction of L_z with eigenvalue zero. (This illustrates the theorem that parallel components of the linear and angular momenta can be measured simultaneously.) In view of these observations the relationship between the two sets must have the form

$$\phi_{\mathbf{k}}(\mathbf{r}) = \sum_l c_l\psi^0_{El0}(\mathbf{r})$$

provided the z-axis is along $\mathbf{k}$. The constant c_l is determined with the help of (17) and (15):

$$\begin{aligned} c_l\,\delta(E-E') &= (\psi^0_{E'l0},\phi_{\mathbf{k}}) \\ &= (2\pi)^{-\frac{3}{2}}\left(\frac{2}{\pi}\frac{k'^2\,dk'}{dE'}\right)^{\frac{1}{2}}\int_0^\infty r^2\,dr\,j_l(k'r)\int d\Omega\, e^{ikr\cos\theta}Y_{l0}(\theta) \\ &= \left(\frac{2l+1}{4\pi}\frac{k'^2\,dk'}{dE'}\right)^{\frac{1}{2}}\frac{1}{\pi}\int_0^\infty r^2\,dr\,j_l(k'r)\int_{-1}^{1}dx\,e^{ikrx}P_l(x). \end{aligned}$$

By brute-force expansion of the exponential, or by consulting Morse and Feshbach (p. 622), one finds

$$j_l(\rho) = \frac{1}{2i^l}\int_{-1}^{1}e^{i\rho x}P_l(x)\,dx,$$

and therefore

$$c_l\,\delta(E-E') = i^l\left(\frac{2l+1}{4\pi}\frac{k^2\,dk}{dE}\right)^{\frac{1}{2}}\frac{\delta(k-k')}{k^2}.$$

The expansion of a momentum eigenfunction in terms of angular momentum eigenfunctions is therefore

$$\phi_{\mathbf{k}}(\mathbf{r}) = \sum_{l=0}^{\infty}\left(\frac{2l+1}{4\pi}\frac{dE}{k^2\,dk}\right)^{\frac{1}{2}} i^l\psi^0_{El0}(\mathbf{r}). \tag{22}$$

This relation is frequently written in the completely equivalent forms

$$e^{ikr\cos\theta} = \sum_{l=0}^{\infty}(2l+1)i^l j_l(kr)P_l(\cos\theta) \tag{23}$$

$$= \sum_{l=0}^{\infty}\sqrt{4\pi(2l+1)}\;i^l j_l(kr)Y_{l0}(\theta). \tag{24}$$

In (22)–(24) the angle θ is measured with respect to $\hat{\mathbf{k}}$. We can use the addition theorem (10.22) to free ourselves of this restriction:

$$e^{i\mathbf{k}\cdot\mathbf{r}} = 4\pi \sum_{lm} i^l j_l(kr) Y_{lm}^*(\hat{\mathbf{k}}) Y_{lm}(\hat{\mathbf{r}}), \tag{25}$$

or equivalently,

$$\phi_{\mathbf{k}}(\mathbf{r}) = \sum_{lm} i^l \sqrt{dE/k^2\,dk}\; Y_{lm}^*(\hat{\mathbf{k}}) \psi_{Elm}^0(\mathbf{r}). \tag{26}$$

To repeat, the axis along which the projection m of $\mathbf{L}$ is measured is arbitrary in (25) and (26). The inverse relation to (26) follows immediately from the orthogonality of the spherical harmonics:

$$\psi_{Elm}^0(\mathbf{r}) = i^{-l} \sqrt{k^2\,dk/dE} \int d\hat{\mathbf{k}}\; Y_{lm}(\hat{\mathbf{k}}) \phi_{\mathbf{k}}(\mathbf{r}). \tag{27}$$

Here $\int d\hat{\mathbf{k}}$ is an integration over the orientation of $\hat{\mathbf{k}}$, i.e., $d^3k = k^2\,dk\,d\hat{\mathbf{k}}$. The scalar product between the linear and angular momentum eigenfunctions is therefore

$$(\phi_{\mathbf{k}'}, \psi_{Elm}^0) = i^{-l} \sqrt{dk/k^2\,dE}\; \delta(k - k') Y_{lm}(\hat{\mathbf{k}}'), \tag{28}$$

where we have used (18), i.e., $\delta(\mathbf{k} - \mathbf{k}) = \delta(\hat{\mathbf{k}} - \hat{\mathbf{k}}')\delta(k - k')/k^2$.

From a mathematical point of view, this section has been rather superfluous. The expansion of a plane wave in terms of spherical waves was well known to the nineteenth century physicist. But quantum mechanics sheds quite a new light on such relationships. For example, (28) tells us that the probability of finding the linear momentum along $\hat{\mathbf{k}}$ in a state having the angular momentum quantum numbers l and m is proportional to $|Y_{lm}(\hat{\mathbf{k}})|^2$. This can be made more precise in terms of the normalized wave packet

$$\chi_{lm} = \int c(E) \psi_{Elm}^0\, dE.$$

The probability of finding the momentum $\hbar\mathbf{k}$ in a state described by χ_{lm} is therefore

$$|(\phi_{\mathbf{k}}, \chi_{lm})|^2 \hbar^{-3} = \frac{1}{\mu\hbar k} |c(E)|^2 |Y_{lm}(\hat{\mathbf{k}})|^2. \tag{29}$$

This formula has some immediate applications. Consider a system initially at rest at the origin in a state having the angular momentum quantum numbers l and m. If this system can decay into two spinless particles, we might be interested in the angular distribution of the decay products. (This type of problem arises frequently in nuclear physics. For example, the $l = 1$ ϕ-meson decays into $K + \bar{K}$.) Assuming that the interaction is invariant under rotations, the final state of the system has the same angular momentum quantum numbers. Momentum conservation also requires that the decay products have equal but opposite momenta. The system is therefore in a state of the type χ_{lm} after the decay products have left the interaction region. The probability that the relative momentum of the decay products has the orientation $\hat{\mathbf{k}}$ is therefore given by (29).

12. Scattering Theory

Before entering into a detailed discussion of the radial Schrödinger equation we shall formulate the three-dimensional problem in a way that will incorporate the boundary condition. This is especially convenient for collision problems.

In all our work we shall assume that the force vanishes as $r \to \infty$. We are thereby excluding the very important class of problems connected with electrons in the periodic potential field present in solids. By choosing the origin of the energy scale appropriately, we can always arrange

$$\lim_{r\to\infty} V(r) = 0. \tag{1}$$

The spectrum now breaks up into two distinct parts:

(a) The continuum: $E \geqslant 0$. These solutions describe scattering; the wave functions are not localized about $r = 0$. Such motions correspond to classical open (or nonperiodic) orbits.

(b) There may also be solutions for $E < 0$. As we shall see, condition (1) implies that there is only a discrete set of such solutions, in general, and they correspond to bound states (the closed or periodic orbits of classical mechanics), because the wave functions are localized about $r = 0$.

Let us first write the wave equation as

$$(\nabla^2 + k^2)\psi(\mathbf{r}) = U(r)\psi(\mathbf{r}), \tag{2}$$

where

$$E = \frac{\hbar^2 k^2}{2\mu}, \tag{3}$$

$$U(r) = \frac{2\mu}{\hbar^2} V(r). \tag{4}$$

If $E < 0$, (3) implies that k is pure imaginary.

The situations (a) and (b) envisaged above correspond to the following boundary conditions, which must be stipulated in addition to (2):

b.c.(a): A plane wave $\phi_{\mathbf{k}}$ incident on the scatterer, and an outgoing scattered wave e^{ikr}/r at infinite separation. These solutions are normalized to δ-functions, as, for example, in (11.21) and (7.28);

b.c.(b): Solutions that fall off sufficiently as $r \to \infty$ so as to yield normalizable wave functions.

12.1 INTEGRAL FORM OF SCHRÖDINGER'S EQUATION

The standard way of incorporating boundary conditions into a differential equation is to convert the equation into an integral equation. With this end in mind, consider the Green's function for the Helmholtz equation,

$$(\nabla^2 + k^2)G_k(\mathbf{r}, \mathbf{r}') = \delta(\mathbf{r} - \mathbf{r}'). \tag{5}$$

By construction

$$\psi(\mathbf{r}) = \int G_k(\mathbf{r}, \mathbf{r}')U(r')\psi(\mathbf{r}')\, d^3r' \tag{6}$$

is a "solution" of (2). However, if we wish we may add to (6) any solution ϕ of the "homogeneous" equation

$$(\nabla^2 + k^2)\phi = 0. \tag{7}$$

The general "solution" of (2) is therefore the inhomogeneous integral equation

$$\psi(\mathbf{r}) = \phi(\mathbf{r}) + \int G_k(\mathbf{r}, \mathbf{r}')U(r')\psi(\mathbf{r}')\, d^3r'. \tag{8}$$

Note that (7) only has acceptable* solutions for real k, which means that the integral equation is necessarily homogeneous if $E < 0$, whereas for $E > 0$ we are, in general, faced with an inhomogeneous equation.

In order to incorporate the boundary conditions, we must choose the appropriate ϕ and G. Let us construct the latter. Clearly it is a function of $|\mathbf{r} - \mathbf{r}'|$ only, and so we put

$$G_k(\mathbf{r}, \mathbf{r}') = \int e^{i\mathbf{q}\cdot(\mathbf{r}-\mathbf{r}')}\, g_k(\mathbf{q})\, \frac{d^3q}{(2\pi)^3}. \tag{9}$$

Upon substituting into (5) we have

$$(k^2 - q^2)g_k(\mathbf{q}) = 1. \tag{10}$$

Because we shall have to integrate over all positive q^2 in (9), including the singular point $k^2 = q^2$, we must learn how to invert (10). To do this we shall consider the generalization of (5) to complex k. We shall see that the solutions as $\text{Im}\, k \to 0^{\pm}$ are not equal, and that the two different limits

* By "acceptable" we here mean solutions that satisfy the condition (7.7).

just correspond to different spatial boundary conditions. Once Im $k \neq 0$, (10) can be inverted; we obtain

$$G_k(r) = \frac{1}{8\pi^2 ri}\int_{-\infty}^{+\infty} q\,dq\,\frac{e^{iqr} - e^{-iqr}}{(k-q)(k+q)} \qquad (r = |\mathbf{r} - \mathbf{r}'|)$$

after integrating over the orientation of $\mathbf{q}$. Since $r > 0$, the contour can be closed in the upper half-plane in the term with e^{iqr}, and vice versa in the e^{-iqr} term. The residue theorem then yields

$$G_k(r) = -\frac{1}{4\pi}\frac{e^{ikr}}{r} \qquad (\text{Im } k > 0), \tag{11}$$

$$G_k(r) = -\frac{1}{4\pi}\frac{e^{-ikr}}{r} \qquad (\text{Im } k < 0). \tag{12}$$

As a function of complex k, G_k is seen to be multivalued, with a branch cut along the real axis. As we shall see, the boundary value of this function above the cut is the physically interesting solution, because it gives *outgoing* scattered waves when inserted into (8). The other Green's function, e^{-ikr}/r, describes incoming waves, while the linear combination

$$\tfrac{1}{2}\lim_{\epsilon\to 0}\,[G_{k+i\epsilon} + G_{k-i\epsilon}] = -\frac{1}{4\pi}\frac{\cos kr}{r} \tag{13}$$

corresponds to standing waves. As we have said, the Green's function which will prove to be the physically useful one corresponds to Im $k = 0^+$, and it can therefore be written as

$$G_k(r) \equiv -\frac{1}{4\pi}\frac{e^{ik|\mathbf{r}-\mathbf{r}'|}}{|\mathbf{r} - \mathbf{r}'|} = \int d^3q\,\frac{\phi_{\mathbf{q}}(\mathbf{r})\phi_{\mathbf{q}}^*(\mathbf{r}')}{k^2 - q^2 + i\epsilon}, \tag{14}$$

where it is understood that $\epsilon \to 0^+$ *after* the evaluation of the integral.

Because we wish to describe the situation where, in addition to scattered waves, we also have an incident beam, we shall choose for ϕ in (8) the momentum eigenfunction $\phi_{\mathbf{k}}(\mathbf{r})$. This gives us the basic integral equation of scattering theory,

$$\psi_{\mathbf{k}}(\mathbf{r}) = \phi_{\mathbf{k}}(\mathbf{r}) - \frac{1}{4\pi}\int \frac{e^{ik|\mathbf{r}-\mathbf{r}'|}}{|\mathbf{r} - \mathbf{r}'|}\,U(r')\psi_{\mathbf{k}}(\mathbf{r}')\,d^3r'. \tag{15}$$

We shall now verify that (15) has the correct asymptotic form. Let $r \to \infty$, and assume that U decreases rapidly so that $r \gg r'$ everywhere. Then

$$k|\mathbf{r} - \mathbf{r}'| = kr\sqrt{1 + \left(\frac{r'}{r}\right)^2 - \frac{2\mathbf{r}\cdot\mathbf{r}'}{r^2}} \approx kr - \mathbf{k}'\cdot\mathbf{r}',$$

where $\mathbf{k}' \equiv k\hat{\mathbf{r}}$. Note that

$$|\mathbf{k}| = |\mathbf{k}'|. \tag{16}$$

Thus the wave function has the desired asymptotic form,

$$\psi_{\mathbf{k}}(\mathbf{r}) \sim \phi_{\mathbf{k}}(\mathbf{r}) - \frac{1}{4\pi}\frac{e^{ikr}}{r}\int e^{-i\mathbf{k}'\cdot\mathbf{r}'}U(r')\psi_{\mathbf{k}}(\mathbf{r}')\,d^3r', \tag{17}$$

i.e., the sum of a plane wave and an *outgoing* spherical wave. If we had used (12) instead, the scattering term would have had the wrong* spatial dependence, i.e., e^{-ikr}/r. For reasons that will soon be clear, one frequently writes (17) as

$$\psi_{\mathbf{k}}(\mathbf{r}) \sim \left(\frac{1}{2\pi}\right)^{\frac{3}{2}}\left\{e^{i\mathbf{k}\cdot\mathbf{r}} + \frac{e^{ikr}}{r}f(\mathbf{k}',\mathbf{k})\right\}, \tag{18}$$

where

$$f(\mathbf{k}',\mathbf{k}) = -2\pi^2(\phi_{\mathbf{k}'}, U\psi_{\mathbf{k}}) = -\frac{4\pi^2\mu}{\hbar^2}(\phi_{\mathbf{k}'}, V\psi_{\mathbf{k}}). \tag{19}$$

The quantity f is called the *scattering amplitude*. It plays a central role in scattering theory, because as we shall see, $|f|^2$ is the differential scattering cross section. We note that $\hbar\mathbf{k}'$ is a vector pointing from the scattering center to the observation point $\mathbf{r}$. The observer looking towards the scatterer at great distances sees particles of momentum $\hbar\mathbf{k}'$ coming towards him, i.e., $\hbar\mathbf{k}'$ is the momentum of particles scattered into the direction $\hat{\mathbf{k}}'$ (see Fig. 12.1). The equality (16) just expresses the fact that

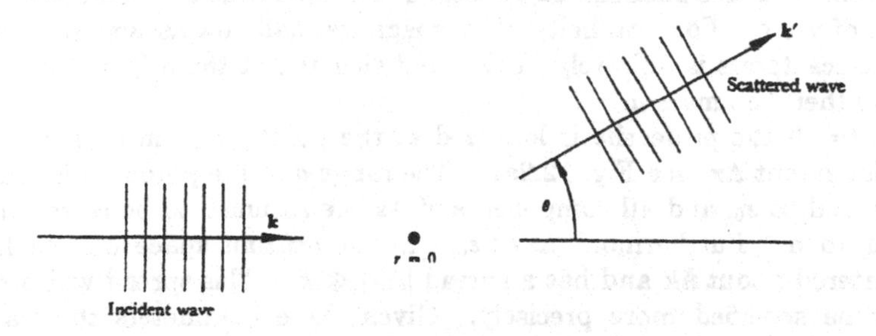

Fig. 12.1. Incident and scattered waves.

we only have elastic scattering since our particles are structureless. We may also note that the incident beam's cylindrical symmetry cannot be destroyed by scattering off a spherically symmetric target. Consequently f can only depend on k and $\hat{\mathbf{k}}\cdot\hat{\mathbf{k}}'$, or equivalently, on k and θ.

* In saying that e^{ikr}/r is an outgoing wave, one must bear in mind that the time dependence is given by $\exp(-iEt/\hbar)$.

A brief remark about bound states is now in order. As already mentioned, for these states k is pure imaginary, say, $i\alpha$. The Green's function (11) with $k = i\alpha$ can be used in the homogeneous integral equation provided $\alpha > 0$, because the solutions must decrease as $r \to \infty$. The bound-state equation is therefore

$$\psi_E(\mathbf{r}) = -\frac{1}{4\pi}\int \frac{e^{-\alpha|\mathbf{r}-\mathbf{r}'|}}{|\mathbf{r}-\mathbf{r}'|} U(\mathbf{r}')\psi_E(\mathbf{r}')\, d^3r', \tag{20}$$

where $E = -\hbar^2\alpha^2/2\mu$. Note that

$$\psi_E \xrightarrow[r\to\infty]{} Ce^{-\alpha r}/r. \tag{20'}$$

A crude understanding of the difference between the $E > 0$ and $E < 0$ spectra can be gained by thinking of the integral equations (15) and (20) as large (though finite) systems of linear equations. A homogeneous system of equations only possesses nontrivial solutions if the determinant of the coefficients vanishes. This leads to special conditions on the parameter E, the eigenvalue condition. On the other hand, an inhomogeneous system always has solutions, and for this reason there is no quantization of the energy for those values of E for which the free particle equation $(\nabla^2 + k^2)\phi = 0$ has solutions.

12.2 TIME-DEPENDENT DESCRIPTION OF COLLISION PHENOMENA

We now turn to a detailed examination of the physical significance of the continuum solutions, and their precise connection with collision phenomena.* These phenomena are best understood from a time-dependent point of view. For simplicity of language we shall now assume that one of the scatterers is infinitely heavy, and situated at the origin; the projectile then has mass μ.

At $t = 0$ the projectile is localized at the point $-z_0\hat{\mathbf{k}}$ in a packet of spatial extent $\Delta\mathbf{x}$ (see Fig. 12.2a). The range a of the potential is small compared to z_0, and all components of $\Delta\mathbf{x}$ are assumed to be large compared to a. Furthermore $\Delta z \lll z_0$. In momentum space our packet is centered about $\hbar\mathbf{k}$ and has a spread $|\Delta\mathbf{k}| \ll k$. This spread will presently be specified more precisely. Given these parameters the wave packet will actually hit the target at time $t \simeq z_0\mu/\hbar k \equiv t_0$.

At $t = 0$ the wave packet just described may be written as

$$\Phi(\mathbf{r}, 0) = \int d^3q\, \chi(\mathbf{q})\phi_{\mathbf{q}}(\mathbf{r}), \tag{21}$$

* The central purpose of this subsection is the derivation of Eq. (38). If one is willing to accept the validity of this formula as intuitively obvious, one can proceed directly to Sec. 12.3. Familiarity with Sec. 12.2 will be assumed in Sec. 16, however.

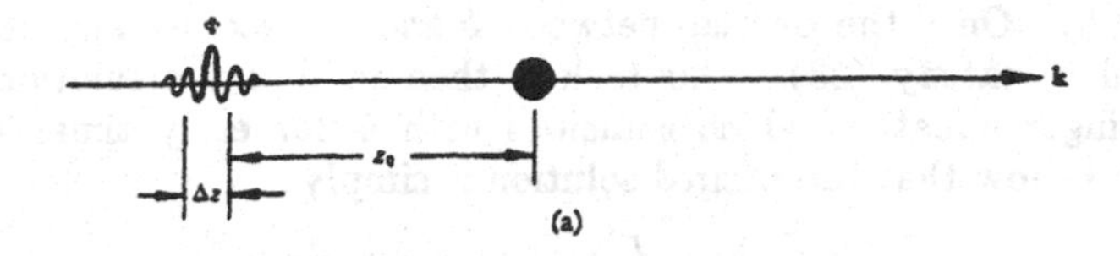

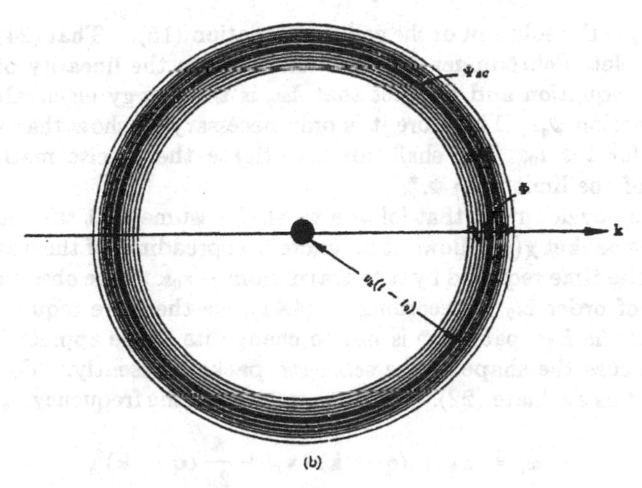

Fig. 12.2. (a) The wave packet before it arrives at the scattering center. (b) The wave packet following the collision.

where the function $\chi(\mathbf{q})$ is chosen so as to make Φ have the properties stipulated above. The time-dependent wave function

$$\Phi(\mathbf{r}, t) = \int d^3q\, \chi(\mathbf{q})\phi_{\mathbf{q}}(\mathbf{r})e^{-i\omega_q t} \tag{22}$$

(with $\omega_q = \hbar q^2/2\mu$) which evolves out of (21) is a packet that moves to the right along the direction $\hat{\mathbf{k}}$. For times small compared to t_0, Φ will have virtually no overlap * with the potential centered at the origin, and (22) therefore satisfies

$$i\hbar\frac{\partial\Phi}{\partial t} = \left(\frac{p^2}{2\mu} + V\right)\Phi \tag{23}$$

* Throughout this discussion our arguments will not be made with complete precision. To do so is really quite straightforward, but rather tedious. Thus instead of the loose statement "virtually no overlap" we should say that given a value of $\epsilon(t_1)$ no matter how small, there exists a length z_0 such that $\int d^3r|\Phi(\mathbf{r}, t)|^2 V(r) < \epsilon(t_1)$ for all $t < t_1$, provided $t_1 \ll t_0$. Using the assumption $rV(r) \to 0$ as $r \to \infty$, one can show that for realistic values of ϵ, z_0 is a small length on a macroscopic scale. The interested reader should consult W. Tobocman and L. L. Foldy, *Am. J. Phys.* **27**, 483 (1959), and Goldberger and Watson, Chapts. 3 and 4.

for $t \ll t_0$. Once the overlap between Φ and V becomes appreciable, Φ will fail to satisfy (23). Our task is then to find the solution of the Schrödinger equation which coincides with Φ for early times ($t \ll t_0$). We shall show that the desired solution is simply

$$\Psi(\mathbf{r}, t) = \int d^3q\, \chi(\mathbf{q}) \psi_{\mathbf{q}}(\mathbf{r}) e^{-i\omega_q t}, \tag{24}$$

where $\psi_{\mathbf{q}}$ is the solution of the collision equation (15). That (24) satisfies the complete Schrödinger equation follows from the linearity of the differential equation and the fact that $\hbar\omega_q$ is the energy eigenvalue of the eigenfunction $\psi_{\mathbf{q}}$. Therefore it is only necessary to show that $\Psi(\mathbf{r}, t) \to \Phi(\mathbf{r}, t)$ for $t \ll t_0$. We shall not investigate the precise mathematical nature of the limit $\Psi \to \Phi$.*

In the development that follows we shall assume that the small width Δk of the packet $\chi(\mathbf{q})$ allows us to ignore the spreadings of the wave packet during the time required by it to travel from $-z_0\hat{\mathbf{k}}$ to the observer. This time is of order $2t_0$. According to (4.41), we therefore require $(\Delta k)^2 \ll \mu/2\hbar t_0$ if the free packet Φ is not to change its shape appreciably. We shall discuss the shape of the scattered packet presently. For the moment, let us evaluate (22). To do so we expand the frequency ω_q about $\mathbf{k}$:

$$\omega_q = \omega_k + (\mathbf{q} - \mathbf{k}) \cdot \mathbf{v}_k + \frac{\hbar}{2\mu} (\mathbf{q} - \mathbf{k})^2,$$

where $\mathbf{v}_k = \hbar\mathbf{k}/\mu$ is the group velocity of the packet. The last term in ω_q is of order $\Delta k/k$ compared to the second, and we may therefore write

$$\Phi(\mathbf{r}, t) \simeq e^{i\omega_k t} \int \frac{d^3q}{(2\pi)^{\frac{3}{2}}} e^{i\mathbf{q}\cdot(\mathbf{r}-\mathbf{v}_k t)} \chi(\mathbf{q}) \left[1 - \frac{i\hbar t}{2\mu} (\mathbf{q} - \mathbf{k})^2\right].$$

If we recall (21), this becomes

$$\Phi(\mathbf{r}, t) \simeq e^{i\omega_k t} \left[1 - \frac{i\hbar t}{2\mu} (i\nabla + \mathbf{k})^2\right] \Phi(\mathbf{r} - \mathbf{v}_k t, 0).$$

The term involving the gradient describes the spreading of the packet in the lowest approximation (recall (4.45) *et seq.*). For the time interval with which we are concerned, it may, as we have said, be dropped, and we then obtain

$$\Phi(\mathbf{r}, t) \simeq e^{i\omega_k t} \Phi(\mathbf{r} - \mathbf{v}_k t, 0). \tag{25}$$

This demonstrates that the packet does indeed move along the classical trajectory without distortion provided $\hbar t(\Delta k)^2/2\mu \ll 1$.

*For a discussion of this question, see W. Brenig and R. Haag, *Fortschritte d. Physik* 7, 183 (1959).

We now turn to the wave packet (24) composed of scattering solutions. When we substitute the integral equation (15) into (24), we immediately evaluate the contribution of the plane wave term as (25). Hence we need only concern ourselves with the term Ψ_{sc} involving the potential:

$$\Psi_{sc}(\mathbf{r}, t) = -\frac{1}{4\pi}\int d^3q\, \chi(\mathbf{q})e^{-i\omega_q t}\int d^3r'\, \frac{e^{iq|\mathbf{r}-\mathbf{r}'|}}{|\mathbf{r}-\mathbf{r}'|}\, U(r')\psi_{\mathbf{q}}(\mathbf{r}'). \tag{26}$$

We shall refer to Ψ_{sc} as the scattered packet. The q-integration in (26) will give a negligible contribution unless we are at the point of stationary phase, which is determined by

$$\frac{\partial}{\partial \mathbf{q}}\left[u(\mathbf{q}) - \omega_q t + q|\mathbf{r}-\mathbf{r}'| + \lambda(\mathbf{q}, \mathbf{r}')\right]\Big|_{\mathbf{q}=\mathbf{k}} = 0, \tag{27}$$

where u is the phase of $\chi(\mathbf{q})$ and $\lambda(\mathbf{q}, \mathbf{r}')$ is the phase of $\psi_{\mathbf{q}}(\mathbf{r}')$. Since $\Phi(\mathbf{r}, 0)$ supposedly has its maximum at $-z_0\hat{\mathbf{k}}$ at $t = 0$, we require the stationary phase point of the integrand of (21) to be

$$\frac{\partial u}{\partial \mathbf{q}}\Big|_{\mathbf{q}=\mathbf{k}} = z_0\hat{\mathbf{k}}. \tag{28}$$

The stationary phase point of Ψ_{sc} is therefore determined from

$$(z_0 - v_k t + |\mathbf{r}-\mathbf{r}'|)\hat{\mathbf{k}} + \frac{\partial\lambda(\mathbf{q}, \mathbf{r}')}{\partial \mathbf{q}}\Big|_{\mathbf{q}=\mathbf{k}} = 0. \tag{29}$$

As we shall now show, this equation succinctly characterizes the region of space-time in which $\Psi_{sc}(\mathbf{r}, t)$ departs significantly from zero. In drawing conclusions from (29), we must remember that a further integration over $\mathbf{r}'$ is necessary in order to evaluate Ψ_{sc} (see (26)). Because of the presence of $U(r')$ in the integrand, the variable $\mathbf{r}'$ is confined to the region specified by $|\mathbf{r}'| \lesssim a$, where a is the range of the potential. In analyzing (29), we must therefore bear in mind that $|\mathbf{r}'| < a$. But if $\mathbf{r}$ in the defining equation for $\psi_{\mathbf{q}}(\mathbf{r})$ is microscopic, there are simply no macroscopic lengths in (15), and hence *all* lengths characterizing $\psi_{\mathbf{q}}(\mathbf{r})$ for such values of $\mathbf{r}$ must themselves be microscopic. In particular, therefore, the length $|\partial\lambda(\mathbf{q}, \mathbf{r}')/\partial\mathbf{q}|$ is microscopic. In most instances this length will be of order a, but under special circumstances it may exceed a by as much as several orders of magnitude. As we shall see in Sec. 16, the latter situation occurs when there is a very narrow resonance in the scattering cross section for incident energy $\hbar\omega_k$.* Nevertheless, even in this case, $|\partial\lambda/\partial\mathbf{q}|$ is a microscopic length, albeit surprisingly large.

We are now ready to draw conclusions from (29). For $t \ll t_0$, $z_0 - v_k t + |\mathbf{r} - \mathbf{r}'|$ is a macroscopic length for all values of $\mathbf{r}$. Since $\partial\lambda/\partial\mathbf{q}$ is microscopic, (29) cannot be satisfied for such values of t throughout all space. Hence $\Psi_{sc}(\mathbf{r}, t)$ vanishes † for $t \ll t_0$, and $\Psi(\mathbf{r}, t) \to \Phi(\mathbf{r}, t)$ for such early times. We have therefore shown that $\Psi(\mathbf{r}, t)$ does satisfy the required initial conditions. When t reaches the value t_0, $z_0 - v_k t \approx 0$, and then (29) *can* be satisfied for microscopic values of $\mathbf{r}$. This simply says that when $\Phi(\mathbf{r}, t)$ actually strikes the target, the scattered wave begins to form in the region of the potential. By the same token, when $t \gg t_0$, $z_0 - v_k t$ is a macroscopic negative length, and $|\mathbf{r} - \mathbf{r}'|$ (with $r' \lesssim a$) must be correspondingly large if (29) is to be satisfied. Thus for $t \gg t_0$, Ψ_{sc} is nonvanishing in a spherical shell of approximate radius $v_k t - z_0 = v_k(t - t_0)$. The evolution of Ψ_{sc} is therefore precisely what we would have expected on intuitive grounds (see Fig. 12.2b). Had we used the incoming wave Green's function (12) in the integral equation that defines $\psi_{\mathbf{q}}$, we would have found that $\Psi_{sc} \neq 0$ *before* the incident packet arrives at the potential.‡

•The orthogonality relation satisfied by the continuum solutions $\psi_{\mathbf{q}}(\mathbf{r})$ can be obtained as a side-product of our proof that $\Psi \to \Phi$ as $t \to -\infty$. Let Ψ_1 and Ψ_2 be two different wave packets of the type just discussed. If these are specified by the Fourier amplitudes χ_1 and χ_2, respectively,

$$(\Psi_1(t), \Psi_2(t)) = \int d^3q\, d^3q'\, \chi_1^*(\mathbf{q})\chi_2(\mathbf{q}')e^{-i(\omega_{q'}-\omega_q)t}(\psi_{\mathbf{q}}, \psi_{\mathbf{q}'}).$$

In Prob. 3 of Chapt. II it was shown that such a scalar product is independent of time. We may, equally well, evaluate it as $t \to -\infty$, in which case it can be written as $(\Phi_1(t), \Phi_2(t))$. With the help of (11.21), we therefore conclude that

$$(\Psi_1(t), \Psi_2(t)) = \int d^3q\, \chi_1^*(\mathbf{q})\chi_2(\mathbf{q}).$$

We therefore infer

$$(\psi_{\mathbf{q}}, \psi_{\mathbf{q}'}) = \delta(\mathbf{q} - \mathbf{q}'). \bullet \tag{30}$$

* One may justify this statement by the following crude plausibility argument. The distance $|\partial\lambda/\partial\mathbf{q}|$ is of order $v_k t_c$, where t_c is the collision time. For nonresonant scattering $t_c \sim a/v_k$. For a resonance whose width (in energy) is Γ, $t_c \sim \hbar/\Gamma$, and $|\partial\lambda/\partial\mathbf{q}| \sim \hbar v_k/\Gamma$.

† This is too strong a statement since we have only shown that the integrand in (26) oscillates exceedingly rapidly when $t \ll t_0$. A precise argument would show that $\lim_{t\to-\infty} ||\Psi - \Phi|| = 0$. See Brenig and Haag, *loc. cit.*

‡ Strangely enough, we shall find that the stationary state solutions of the integral equation wherein the Green's function is given by (12) also play an important role in collision theory. See Sec. 58.2.

In order to determine the cross section, we must evaluate Ψ_{sc} when $r \gg a$. Employing (18), we have

$$\Psi_{sc}(\mathbf{r}, t) \underset{r\to\infty}{\sim} \frac{1}{r} \int \frac{d^3q}{(2\pi)^{3/2}} f(q\hat{\mathbf{r}}, \mathbf{q})\chi(\mathbf{q})e^{i(qr-\omega_q t)}. \tag{31}$$

We shall now make the additional assumption that $f(q\hat{\mathbf{r}}, \mathbf{q})$ varies sufficiently slowly in the interval $\Delta\mathbf{k}$ to permit its replacement by $f(k\hat{\mathbf{r}}, \mathbf{k})$. When this is not the case, the argument that follows breaks down because the shape of the scattered packet is then altered significantly by the collision process.* In principle, one can always reduce Δk until the stated requirement is met, but in the case of a sharp resonance, f may vary too rapidly for this to be practical. When Δk is large compared to the interval in which f changes appreciably, the shape of Ψ_{sc} may differ drastically from the form which we shall now derive. Since this situation is of very great importance, we shall devote a separate section (Sec. 16) to it.

Taking advantage of the peak in $\chi(\mathbf{q})$, we then remove f from the integrand, replace qr by $\mathbf{q}\cdot\hat{\mathbf{k}}r$, and approximate ω_q in the now familiar manner. Then (31) becomes

$$\begin{aligned}\Psi_{sc}(\mathbf{r}, t) &\sim \frac{e^{i\omega_k t}}{r} f(k\hat{\mathbf{r}}, \mathbf{k}) \int \frac{d^3q}{(2\pi)^{3/2}} \chi(\mathbf{q})e^{i\mathbf{q}\cdot(\hat{\mathbf{k}}r - \mathbf{v}_k t)} \\ &\sim e^{i\omega_k t}\frac{f(k\hat{\mathbf{r}}, \mathbf{k})}{r}\Phi(\hat{\mathbf{k}}r - \mathbf{v}_k t, 0).\end{aligned} \tag{32}$$

Note carefully that the orientation of $\mathbf{r}$ only appears in the scattering amplitude; the remainder of Ψ_{sc} is spherically symmetric. Provided $r \approx v_k(t - t_0)$, (32) shows that $|\Psi_{sc}|^2$ falls off as r^{-2}.

The differential cross section is defined as

$$\frac{d\sigma}{d\Omega} = \frac{dN_{sc}/d\Omega}{dN_{inc}/dA}. \tag{33}$$

Here dN_{inc} is the number of particles which traverse the element of area dA normal to $\mathbf{k}$ in the incident packet, and dN_{sc} is the number of particles scattered into the cone subtended by the solid angle $d\Omega \equiv \sin\theta\, d\theta\, d\phi$. In order to calculate the quotient (33), we must find the currents in the incident and scattered wave packets. The incident current is simply

$$\mathbf{j}_{inc}(\mathbf{r}, t) = \mathbf{v}_k|\Phi(\mathbf{r} - \mathbf{v}_k t, 0)|^2$$

* Note that the condition $2(\Delta k)^2\hbar t_0/\mu \ll 1$, which permitted the derivation of (25), does *not* guarantee that Ψ_{sc} will attain the shape given by (32). The aforementioned condition is, in general, only applicable to free-particle wave packets.

to within terms of relative order $\Delta k/k$. Hence

$$dN_{\text{inc}}/dA = \int_{-\infty}^{\infty} dt\, \hat{\mathbf{k}} \cdot \mathbf{j}_{\text{inc}} = v_k \int_{-\infty}^{\infty} dt |\Phi(x, y, z - v_k t, 0)|^2. \quad (34)$$

The wave packet subsequent to the collision was just shown to be

$$\Psi(\mathbf{r}, t) \simeq e^{i\omega_k t}\Phi(\mathbf{r} - \mathbf{v}_k t, 0) + \frac{f(k\hat{\mathbf{r}}, \mathbf{k})}{r} e^{i\omega_k t}\Phi(\hat{\mathbf{k}}r - \mathbf{v}_k t, 0). \quad (35)$$

There are, therefore, three distinct terms in the current following the collision ($t \gg t_0$). The term independent of f, and the interference term linear in f, are both only present when the observation point $\mathbf{r}$ is near the forward direction $\hat{\mathbf{k}}$, i.e., near zero scattering angle (see Fig. 12.2b). The angular interval in which these two terms contribute shrinks as $t \rightarrow \infty$ (or equivalently, $r \rightarrow \infty$), and except in the immediate vicinity of the forward direction we may therefore confine ourselves to the contribution to the current which is quadratic in f. The situation at $\theta = 0$ is more complicated, and will be analyzed below in Sec. 12.3. Bearing the above proviso in mind, we compute the scattered current $\mathbf{j}_{sc}$ from Ψ_{sc}. When we apply the velocity operator $(\hbar/i\mu)\nabla$ to (32), we find that the angular derivatives, and the term arising from ∇r^{-1}, are both of order r^{-2}. As $r \rightarrow \infty$, we therefore have

$$\begin{aligned}
\frac{\hbar}{i\mu} \nabla \frac{f}{r} \Phi(\hat{\mathbf{k}}r - \mathbf{v}_k t, 0) &= \frac{f}{r}\frac{\hbar}{i\mu} \nabla\Phi(\hat{\mathbf{k}}r - \mathbf{v}_k t, 0) + O(r^{-2}) \\
&= \frac{f}{r}\hat{\mathbf{r}}\frac{\hbar}{i\mu}\frac{\partial}{\partial z}\Phi(0, 0, z, 0,)\Big|_{z=r-v_k t} \\
&\simeq \hat{\mathbf{r}} v_k \frac{f}{r}\Phi(\hat{\mathbf{k}}r - \mathbf{v}_k t, 0)
\end{aligned}$$

to within terms of order $\Delta k/k$. Hence

$$\mathbf{j}_{sc}(\mathbf{r}, t) = \hat{\mathbf{r}} v_k \frac{|f(k\hat{\mathbf{r}}, \mathbf{k})|^2}{r^2} |\Phi(\hat{\mathbf{k}}r - \mathbf{v}_k t, 0)|^2 + O(r^{-3}).$$

This shows that the scattered wave is progressing outward radially, and that an observer at the point $\mathbf{r}$ sees particles with velocity $\hat{\mathbf{r}}v_k$ coming towards him. The interpretation of the vector $\hbar\mathbf{k}'$ given after (19) is therefore justified. The number of particles scattered into the cone subtended by $d\Omega$ per unit time is $r^2\, d\Omega\, j_{sc}$, or

$$dN_{sc}/d\Omega = v_k |f(k, \theta)|^2 \int_{-\infty}^{\infty} dt |\Phi(0, 0, r - v_k t, 0)|^2. \quad (36)$$

Therefore

$$\frac{dN_{sc}/d\Omega}{dN_{\text{inc}}/dA} = |f(k, \theta)|^2 \frac{\int_{-\infty}^{\infty} |\Phi(0, 0, z, 0)|^2\, dz.}{\int_{-\infty}^{\infty} |\Phi(x, y, z, 0|^2\, dz}. \quad (37)$$

The number of incident particles dN_{inc} is a function of where the element of area dA is located, needless to say. Hence the appearance of the coordinates x and y in (34). When the experimenter calculates dN_{inc}/dA, he divides the total number of incident particles by the area of the beam. In the plane transverse to **k**, the probability distribution $\int dz|\Phi(x, y, z, 0)|^2$ has the form shown in Fig. 12.3. Under these circumstances, the laboratory definition of dN_{inc}/dA coincides with (34) for any $|x| \ll \Delta x$, $|y| \ll \Delta y$,

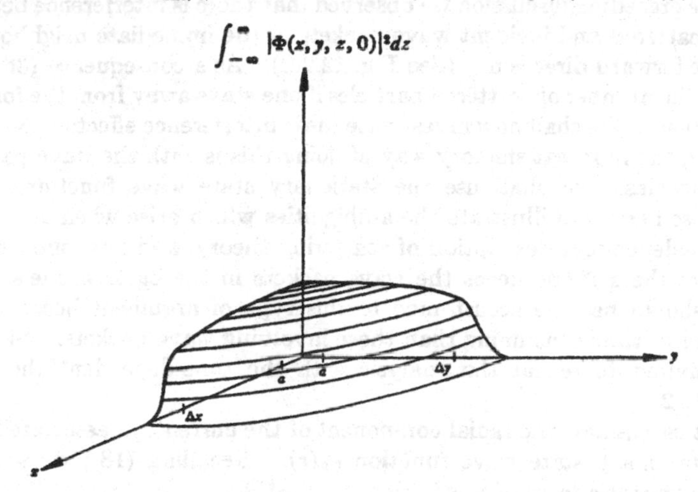

Fig. 12.3. The probability distribution in the incident beam.

and in particular for $x = y = 0$. The quotient in (37) is therefore one. According to the definition (33) *the differential cross section is therefore* *†

$$\frac{d\sigma}{d\Omega} = |f(k, \theta)|^2. \tag{38}$$

* If one is willing to swallow a number of superficially obvious but really somewhat mysterious steps, one can infer (38) very quickly from (18). The reader should examine these well-known arguments critically, and then determine how the treatment given here overcomes the difficulties alluded to.

† If one of the masses is not infinite, then (38) gives the scattering cross section in the center-of-mass system. The transformation to the laboratory system (i.e., where the target is at rest) is the same as in classical mechanics (see Goldstein, p. 85). If Θ is the laboratory scattering angle, and m_1 the incident mass, we have

$$\tan\Theta = \frac{\sin\theta}{\cos\theta + m_1/m_2}$$

and

$$\sigma_{\text{lab}}(\Theta) = \sigma_{\text{c.m.}}(\theta)\,\frac{d(\cos\theta)}{d(\cos\Theta)}.$$

In particular, if $m_1 = m_2$, $\Theta = \frac{1}{2}\theta$, and hence there is no scattering beyond 90° in the laboratory; furthermore $\sigma_{\text{lab}}(\Theta) = 4\cos\Theta\,\sigma_{\text{c.m.}}(2\Theta)$.

We shall frequently write $\sigma(k, \theta)$ instead of $d\sigma/d\Omega$. The total cross section is, by definition,

$$\sigma(k) = \int_{-1}^{1} d(\cos\theta) \int_0^{2\pi} d\phi \, \frac{d\sigma}{d\Omega}. \tag{39}$$

12.3 THE OPTICAL THEOREM

In the preceding discussion we observed that there is interference between the scattered and incident wave packets in the immediate neighborhood of the forward direction. (See Fig. 12.2b.) As a consequence (36) only gives the number of scattered particles if one stays away from the forward direction. We shall now investigate these interference effects. Needless to say, the most satisfactory way of doing this is with the wave packets. Nevertheless, we shall use the stationary state wave functions. Our purpose here is to illustrate the ambiguities which arise when one uses a time-independent description of scattering theory, and how one can dispose of them if one keeps the wave packets in the back of one's mind. One should become accustomed to this type of argument because it is much less time consuming than those involving wave packets. Skeptics are invited to repeat the analysis with the time-dependent theory of Sec. 12.2.

Let us examine the radial component of the current, j_r, associated with the stationary state wave function $\psi_{\mathbf{k}}(\mathbf{r})$. Recalling (18), we see that the asymptotic form of j_r is*

$$j_r \sim \frac{\hbar}{2\mu} (2\pi)^{-3} 2 \,\mathrm{Im} \left\{ \left[e^{-ikr\cos\theta} + \frac{e^{-ikr}}{r} f^*(\theta) \right] \cdot \frac{\partial}{\partial r} \left[e^{ikr\cos\theta} + \frac{e^{ikr}}{r} f(\theta) \right] \right\}. \tag{40}$$

This splits up into the three contributions already mentioned: the incident current $j_{r,\mathrm{inc}}$, the scattered current $j_{r,\mathrm{sc}}$, and the interference term $j_{r,\mathrm{int}}$. Retaining only the leading terms in $1/r$, we easily find that

$$j_{r,\mathrm{inc}} \sim (2\pi)^{-3} v_k \cos\theta, \tag{41}$$

$$j_{r,\mathrm{sc}} \sim (2\pi)^{-3} (v_k/r^2) |f(\theta)|^2, \tag{42}$$

$$j_{r,\mathrm{int}} \sim (2\pi)^{-3} (v_k/r) \,\mathrm{Im}[ie^{ikr(\cos\theta - 1)} f^*(\theta) \cos\theta + ie^{ikr(1-\cos\theta)} f(\theta)]. \tag{43}$$

The interference term has an angular dependence that is strikingly different from that of the other two terms: whereas $j_{r,\mathrm{inc}}$ and $j_{r,\mathrm{sc}}$ are smooth functions of θ as $r \to \infty$, $j_{r,\mathrm{int}}$ oscillates with ever increasing

* Note that j_r has the dimensions of a velocity, instead of (area × time)$^{-1}$. This is due to the fact that $\psi_{\mathbf{k}}$ is not normalized to unity. This difficulty does not arise in the wave packet treatment.

rapidity in this limit unless θ^2 is less than $O(|kr|^{-1})$. Were we to average over a small interval in k-space (i.e., use a wave packet description), $j_{r,\text{int}}$ would disappear everywhere except for $\theta \approx 0$. To see this in detail, let us compute the number of particles passing through the area $r^2\,\delta\Omega$ subtended by the solid angle $\delta\Omega = 2\pi(\delta\theta)^2$. Under the assumption that $f(\theta)$ is well behaved as $\theta \to 0$ (which is true when $\lim_{r\to\infty} rV(r) = 0$), we have

$$\int_{\delta\Omega} r^2\,d\Omega\, j_{r,\text{inc}} \sim (2\pi)^{-3} v_k r^2\,\delta\Omega, \tag{44}$$

$$\int_{\delta\Omega} r^2\,d\Omega\, j_{r,\text{sc}} \sim (2\pi)^{-3} v_k |f(0)|^2\,\delta\Omega, \tag{45}$$

and

$$\int_{\delta\Omega} r^2\,d\Omega\, j_{r,\text{int}} \sim (2\pi)^{-3} v_k r(2\pi)\,\text{Im}\left\{if^*(0)\int_{1-\frac{1}{2}(\delta\theta)^2}^{1} dx\, e^{ikr(x-1)} - \text{c.c.}\right\}.$$

The integral in the last equation equals $[1 - e^{-\frac{1}{2}ikr(\delta\theta)^2}]/ikr$. When $kr(\delta\theta)^2 \gg 1$, which will be the case in essentially all realistic situations, the oscillating term can be dropped if we bear in mind that one should only keep those portions of any result that survive averaging over small intervals $\Delta\mathbf{k}$ of k-space. Therefore

$$\int_{\delta\Omega} r^2\,d\Omega\, j_{r,\text{int}} \sim -(2\pi)^{-3} v_k \frac{4\pi}{k}\,\text{Im}\, f(0). \tag{46}$$

As we shall prove shortly, $\text{Im}\, f(0)$ is positive definite. The interference term therefore contributes a current that flows back towards the scattering center along the incident direction, and thereby decreases the current along $\hat{\mathbf{k}}$ below the value one would have if the scattering center were absent. This is merely a complicated way of saying that the scattering center casts a shadow in the forward direction.

We must still show that $\text{Im}\, f(0) > 0$. Actually we will find the explicit value of $\text{Im}\, f(0)$. As we have just seen, (46) expresses the depletion of the forward-moving beam due to scattering. This leads one to suspect that further information about $\text{Im}\, f(0)$ can be obtained by examining the consequences of probability conservation. As we are dealing with a stationary state, the continuity equation implies

$$\int_S d^3r\, \nabla \cdot \mathbf{j} = r^2 \int d\Omega\, j_r = 0, \tag{47}$$

where S is a sphere of radius r. The various parts of j_r are given by (41)–(43). The first of these vanishes identically when substituted into (47). Using the argument that led to (46), we then find that the conservation requirement (47) is only satisfied if

$$\sigma(k) = \frac{4\pi}{k} \operatorname{Im} f(0). \tag{48}$$

This very important connection between the total cross section and the imaginary part of the forward scattering amplitude is due to Bohr, Peierls and Placzek, and is usually called the *Optical Theorem*. As we have seen, the optical theorem is a succinct way of saying that the depletion of the beam in the forward direction is proportional to everything taken out of the beam by scattering into all angles.

13. The Born and Eikonal Approximations

We must now address ourselves to the question of how the scattering amplitude f is to be calculated. This is a vast subject, and we shall only scratch its surface.

13.1 THE BORN APPROXIMATION

The first thing that comes to mind on looking at the formula for f,

$$f(\mathbf{k}', \mathbf{k}) = -\frac{4\pi^2\mu}{\hbar^2} \int \phi^*_{\mathbf{k}'}(\mathbf{r}) V(r) \psi_{\mathbf{k}}(\mathbf{r})\, d^3r, \tag{1}$$

is the first Born approximation, $\psi_{\mathbf{k}}(\mathbf{r}) \approx \phi_{\mathbf{k}}$. If the potential is weak enough, in some sense that remains to be specified, then this would appear to be a reasonable approximation. It leads to the extremely simple formula

$$f_B(\mathbf{k}', \mathbf{k}) = -\frac{\mu}{2\pi\hbar^2} \int e^{i(\mathbf{k}-\mathbf{k}')\cdot\mathbf{r}}\, V(r)\, d^3r \equiv -\frac{\mu}{2\pi\hbar^2} \tilde{V}(q), \tag{2}$$

where

$$\hbar|\mathbf{q}| \equiv \hbar q = \hbar|\mathbf{k} - \mathbf{k}'| = 2\hbar k \sin \tfrac{1}{2}\theta \tag{3}$$

is the momentum transfer. The amplitude for scattering from a central potential in the first Born approximation is seen to be a function of q only. This is not a general result; it is a special property of the first Born approximation.

Consider the example of a Yukawa potential,

$$V(r) = V_0 \frac{e^{-\alpha r}}{\alpha r}. \tag{4}$$

The Fourier transform of (4) is

$$\tilde{V}(q) = \frac{4\pi V_0}{\alpha^3}\left(\frac{\alpha^2}{q^2 + \alpha^2}\right). \tag{4'}$$

Hence the Born approximation to the differential cross section is

$$\sigma_B(k,\theta) = \left(\frac{2V_0\mu}{\hbar^2\alpha^3}\right)^2 \left[\frac{\alpha^2}{4k^2 \sin^2 \frac{1}{2}\theta + \alpha^2}\right]^2. \tag{5}$$

Note that $\sigma_B(k, \theta)$ is isotropic for wavelengths $\lambdabar = 1/k$ long compared to the range $1/\alpha$ of the potential, and progressively more peaked in the forward direction when $\lambdabar$ becomes small compared to $1/\alpha$. In Sec. 14 we shall show that these features also arise in an accurate treatment. The angular distribution is plotted in Fig. 13.1.

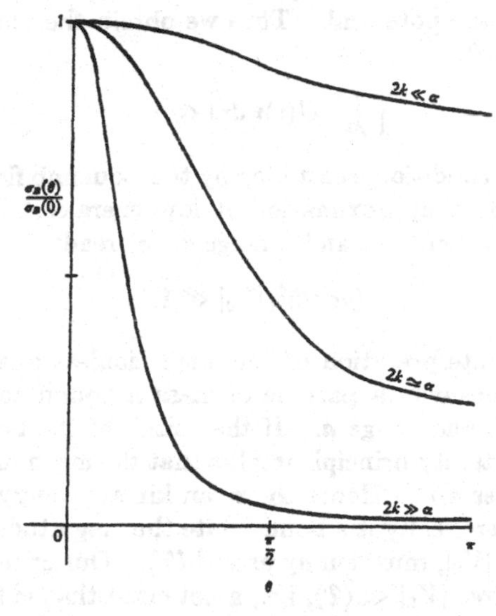

Fig. 13.1. The angular distribution for scattering from a Yukawa potential in the first Born approximation.

The Yukawa potential (4) becomes the Coulomb field in the limit $\alpha \to 0$, provided we set $V_0/\alpha = Z_1Z_2e^2$. The Born approximation to the Coulomb cross section is, therefore,

$$\sigma_B(k, \theta) = \frac{(Z_1Z_2e^2\mu)^2}{4p^4 \sin^4 \frac{1}{2}\theta}, \tag{6}$$

where $p = \hbar k$ is the relative momentum. This is known as the *Rutherford formula*. For obscure reasons, (6) is the exact nonrelativistic result in both classical and quantum mechanics (see Sec. 17).

The Born approximation can only be accurate if the distortion of the wave (i.e., $\psi_{\mathbf{k}} - \phi_{\mathbf{k}}$) by the potential is small in the interaction region. We may obtain a validity criterion by estimating $(\psi_{\mathbf{k}} - \phi_{\mathbf{k}})$ at $r = 0$; using (12.15) we obtain

$$C(k) \equiv \frac{\phi_{\mathbf{k}}(0) - \psi_{\mathbf{k}}(0)}{\phi_{\mathbf{k}}(0)} \approx (2\pi)^{\frac{3}{2}} \frac{1}{4\pi} \int \frac{e^{ikr}}{r} U(r) \phi_{\mathbf{k}}(\mathbf{r})\, d^3r$$

$$= \frac{1}{k} \int_0^\infty e^{ikr} \sin kr\, U(r)\, dr. \qquad (7)$$

The magnitude of this quantity must be small compared to one. The accuracy must depend on energy, because at high energy the wave is distorted less by the potential. Thus we obtain the most stringent requirement at $k = 0$:

$$\left| \int_0^\infty U(r) r\, dr \right| \ll 1. \qquad (8)$$

According to this condition, scattering by the Coulomb field can never be treated by the Born approximation at low energies. For a potential characterized by a depth V_0 and a range a, (8) reads

$$(\mu a^2/\hbar^2)|V_0| \ll 1. \qquad (9)$$

A rather simple interpretation of the dimensionless quantity $\mu V_0 a^2/\hbar^2$ emerges if one considers a particle of mass μ bound to a potential of average depth V_0 and range a. If the "size" of the bound state is of order a, the uncertainty principle implies that the mean momentum must be at least of order $\hbar/a$. Hence the mean kinetic energy $\langle T \rangle$ is of order $(\hbar/a)^2/2\mu$. If there really is a bound state the magnitude of the average potential energy, $|V_0|$, must surely exceed $\langle T \rangle$. Our criterion (9), on the other hand, requires $|V_0| \ll \langle T \rangle$, i.e., a potential that is far too weak to bind a state. Observe that it is $|V_0|$, and not V_0, that appears in (9). Hence a repulsive potential $V(r)$ must be sufficiently weak so that its attractive counterpart, $-V(r)$, is not nearly strong enough to bind a state. These qualitative considerations concerning the relationship between bound states and the accuracy of the Born approximation at $k = 0$ will be supplemented by an examination of special examples in Sec. 15.1 and Prob. 4, and by a more sophisticated analysis in Sec. 52.

If the $k = 0$ criterion (8) is violated, it may still happen that $|C(k)| \ll 1$ at sufficiently high energy. We may explore this point in detail for the

Yukawa potential, because the integral in (7) can be evaluated explicitly in this case*:

$$C(k) = \frac{2\mu V_0}{\hbar^2\alpha^2}\left(\frac{\alpha}{k}\right)\left[\tfrac{1}{2}\tan^{-1}\frac{2k}{\alpha} + \frac{i}{4}\ln\left(1 + \frac{4k^2}{\alpha^2}\right)\right]. \tag{10}$$

At high energies ($k \gg \alpha$) this becomes

$$C(k) \simeq \frac{2\mu V_0}{\hbar^2\alpha^2}\left(\frac{\alpha}{k}\right)\left(\frac{\pi}{4} + \frac{i}{2}\ln\frac{2k}{\alpha}\right).$$

This expression vanishes as $k \to \infty$, and the Born approximation therefore becomes very accurate in this limit. At low energies ($k \ll \alpha$) we have

$$C(k) \simeq \frac{2\mu V_0}{\hbar^2\alpha^2}\left(1 + \frac{ik}{\alpha} - \frac{4k^2}{3\alpha^2} + \cdots\right).$$

As for the square well, the dimensionless quantity $2\mu|V_0|/\hbar^2\alpha^2$ must be small compared to one if the Born approximation is to be valid at all energies.

The requirement that the kinetic energy should be large compared to typical values of the potential energy is not sufficient to insure the accuracy of the Born approximation. Thus if the potential is very weak but extends over a region very large compared to the incident wavelength, the magnitude of the exact wave function in the interaction region will not differ much from the plane wave, but its phase may be very different. If this is the case, the Born approximation breaks down. A familiar example is the scattering of light by a macroscopic target; even if the index of refraction is almost one, the Born approximation will not produce the laws of refraction and reflection.

There are potentials that are too singular to possess a Fourier transform, such as $V_0(r_0/r)^s$, with $s > 2$. The Born approximation cannot be applied to such potentials, no matter how small V_0 may be.

The Born approximation encounters a rather peculiar difficulty in the case of the Coulomb field. If in (10) we replace V_0 by $Z_1Z_2e^2\alpha$, and let $\alpha \to 0$, we obtain $C(k)$ for the Coulomb field:

$$C(k) = \gamma\left(\frac{\pi}{2} + i\lim_{\alpha\to 0}\ln\frac{2k}{\alpha}\right), \tag{10'}$$

where $\gamma = Z_1Z_2e^2/\hbar v$, and $v = \hbar k/\mu$ is the relative velocity. The imaginary part of this expression is infinite for all finite k! As we shall

* For a detailed examination of corrections to the Born approximation in the case of a Yukawa potential see R. H. Dalitz, *Proc. Roy. Soc.* **A206**, 509 (1951).

discover in Sec. 17, the modulus of the *exact* Coulomb scattering amplitude is given by the Born approximation, but the phase of this amplitude is a rather pathological function and gives rise to the divergence in (10′). The cross section does not depend on the phase of f, however, hence the Rutherford formula fails to reveal this dismal failure of the Born approximation. If the Coulomb field is screened at distances large compared to R (by atomic electrons, for example), we can use (10′) to conclude that the Born approximation will provide an accurate scattering amplitude whenever $|\gamma| \sqrt{\frac{1}{4}\pi^2 + (\ln 2kR)^2} \ll 1$.

•All the criteria stated above are necessary conditions for the validity of the Born approximation. The sufficient conditions are not known in general.* Let us note some of the mathematical assumptions that we have tacitly made. Let us replace U by gU everywhere, where g is to be a dimensionless strength parameter; e.g., $2\mu V_0/\hbar^2\alpha^2$ for the Yukawa potential. When $g \to 0$ we obtain the force-free situation. The first Born approximation is then the first term in the iterative solution of (12.15):

$$\psi_{\mathbf{k}}(\mathbf{r}) = \phi_{\mathbf{k}}(\mathbf{r}) + g \int G_k(\mathbf{r} - \mathbf{r}')U(r')\phi_{\mathbf{k}}(\mathbf{r}')\, d^3r' + g^2 \int G_k(\mathbf{r} - \mathbf{r}')U(r')G_k(\mathbf{r}' - \mathbf{r}'')U(r'')\phi_{\mathbf{k}}(\mathbf{r}'')\, d^3r'\, d^3r'' + O(g^3). \quad (11)$$

The scattering amplitude is

$$f(k, \theta, g) = -2\pi^2(\phi_{\mathbf{k}'}, gU\psi_{\mathbf{k}});$$

and if we use (11) we obtain the so-called Born series,

$$-\frac{1}{2\pi^2} f(k, \theta, g) = g(\phi_{\mathbf{k}'}, U\phi_{\mathbf{k}}) + g^2(\phi_{\mathbf{k}'}, UG_kU\phi_{\mathbf{k}}) + g^3(\phi_{\mathbf{k}'}, UG_kUG_kU\phi_{\mathbf{k}}) + \cdots,$$

using a rather obvious notation. We may write this result as

$$f(k, \theta, g) = \sum_{n=1}^{\infty} g^n f_n(k, \theta). \quad (12)$$

Thus the Born approximation is a power series expansion in the strength parameter g, and therefore automatically supposes that f is analytic in some region surrounding the origin of the g-plane. This implies, in particular, that when we treat a repulsive potential (Re $g > 0$, Im $g = 0$) with the Born approximation, we are inevitably concerned with the same potential changed in sign, i.e., attractive (Re $g < 0$, Im $g = 0$). As we shall see, the amplitude $f(g)$ has poles when the problem in question has bound state solutions for Re $g < 0$, Im $g = 0$. The occurrence of such bound states then limits the radius of convergence of the series (12).

* For a detailed discussion of the convergence problem, see W. Kohn, *Rev. Mod. Phys.* **26**, 294 (1954).

Finally we note an important shortcoming of the first Born approximation: $f_1(k, \theta)$ is real, and therefore seems to contradict the optical theorem (12.48), i.e., the Born approximation appears to violate the conservation of probability. This is due to the nonlinearity of the condition (12.48). Upon substituting the series (12) into the optical theorem we obtain

$$\int d\Omega \left| \sum_{n=1}^{\infty} g^n f_n(k, \theta) \right|^2 = \frac{4\pi}{k} \text{Im} \sum_{n=1}^{\infty} g^n f_n(k, 0).$$

Equating powers of g yields

$$\text{Im}\, f_1(k, 0) = 0, \tag{13}$$

$$\int d\Omega |f_1(k, \theta)|^2 = \frac{4\pi}{k} \text{Im}\, f_2(k, 0), \tag{14}$$

and so forth. Hence if f is approximated by any polynomial in g it will violate the optical theorem. In spite of this, the first few terms of the Born series will give an accurate fit to the exact f if the series converges rapidly for the values of k and g in question.•

13.2 THE EIKONAL APPROXIMATION

If the potential is a smooth function of r, the wavelength will become short compared to all distances that characterize $V(r)$ in the high energy limit. Under these conditions the exact wave function in (1) can be replaced by the semi-classical wave function of Sec. 8.3:

$$\psi_{\mathbf{k}}(\mathbf{r}) \simeq e^{i\hat{S}_{\mathbf{k}}(\mathbf{r})/\hbar}. \tag{15}$$

Here $\hat{S}_{\mathbf{k}}(\mathbf{r})$ is the solution of the stationary Hamilton-Jacobi equation,

$$|\nabla \hat{S}_{\mathbf{k}}(\mathbf{r})| = \sqrt{2\mu[E - V(r)]}. \tag{16}$$

The suffix $\mathbf{k}$ signifies that the classical trajectory is characterized by the incident momentum $\hbar\mathbf{k}$.

Ideally we should integrate (16) exactly, i.e., determine the classical trajectory. This is a rather formidable task in itself, however; furthermore, the $\mathbf{r}$-integration that must still be carried out when evaluating the scattering amplitude presents a frightening prospect with this exact form for $\hat{S}_{\mathbf{k}}$. We therefore resort to a further approximation, to wit, we replace the exact classical trajectory by a straight line when evaluating $\hat{S}_{\mathbf{k}}$. Obviously this requires the energy E to be very high, and the scattering angle θ to be very small. The approximation (15) is only permissible when this condition on E is met. The restriction to small θ is also not an important shortcoming because the angular distribution tends to be highly collimated in the forward direction when the energy is high. The curvature of the classical trajectory is therefore a higher-order correction under the circumstances envisaged here.*

Let $\mathbf{b}$ be a vector perpendicular to the trajectory, and having a length equal to the classical impact parameter b. The equation of the orbit is thus $\mathbf{r} = \mathbf{b} + \hat{\mathbf{k}}z$. The partial differential equation (16) is reduced to the ordinary differential equation

$$\left|\frac{d\hat{S}_{\mathbf{k}}}{dz}\right| = \{2\mu[E - V(\sqrt{b^2 + z^2})]\}^{\frac{1}{2}}$$

by our assumption that the classical trajectory is approximately a straight line. On demanding that $\psi \to \phi$ as $z \to -\infty$, we have

$$\hat{S}_{\mathbf{k}}/\hbar = kz + \int_{-\infty}^{z} \{[k^2 - U(\sqrt{b^2 + z'^2})]^{\frac{1}{2}} - k\}\, dz', \tag{17}$$

where we have adjusted the constant of integration so as to isolate the term kz which survives when $U = 0$. By hypothesis, $k^2 \gg |U(r)|$, and it is therefore legitimate to replace (17) by

$$\hat{S}_{\mathbf{k}}/\hbar \simeq kz - \frac{1}{2k}\int_{-\infty}^{z} U(\sqrt{b^2 + z'^2})\, dz'.$$

Our final result for the wave function is thus

$$\psi_{\mathbf{k}}(\mathbf{b} + \hat{\mathbf{k}}z) \simeq \phi_{\mathbf{k}}(\mathbf{b} + \hat{\mathbf{k}}z) \cdot \exp\left\{-\frac{i}{2k}\int_{-\infty}^{z} U(\sqrt{b^2 + z'^2})\, dz'\right\}. \tag{18}$$

In this approximation the exact wave function is replaced by the undistorted wave function throughout all space except in a cylinder of radius $\sim a$ to the "right" of the scattering center, where a is the range of U.

One cannot deduce the scattering amplitude directly from our approximate $\psi_{\mathbf{k}}$. The latter is too crude for that purpose; in particular, it does

* We shall not estimate the accuracy of the eikonal approximation in much detail. Interested readers are referred to D. S. Saxon and L. I. Schiff, *Nuovo Cimento* **6**, 614 (1957); R. J. Glauber, *Lectures in Theoretical Physics*, Vol. 1, 1958, edited by W. E. Brittin and L. G. Dunham, Interscience (New York, 1959). A very elegant and instructive derivation from a completely nonclassical point of view can be found in R. Blankenbecler and M. L. Goldberger, *Phys. Rev.* **126**, 766 (1962). Corrections due to the curvature of the classical trajectory are discussed by D. R. Yennie, F. L. Boos, Jr., and D. G. Ravenhall, *Phys. Rev.* **137**, B882 (1965). The article by Glauber also describes a variety of applications.

not assume the correct asymptotic form (12.18). But we can use (18) in the expression (1) for the scattering amplitude*:

$$f(\mathbf{k}',\mathbf{k}) \simeq -\frac{1}{4\pi}\int d^2b\, dz\, e^{i\mathbf{q}\cdot(\mathbf{b}+\hat{\mathbf{k}}z)}U(\sqrt{b^2+z^2}) \times \exp\left\{-\frac{i}{2k}\int_{-\infty}^{z} U(\sqrt{b^2+z'^2})\,dz'\right\}. \tag{19}$$

Here $\hbar\mathbf{q} = \hbar(\mathbf{k}-\mathbf{k}')$ is the momentum transfer, and $\int d^2b$ signifies an integration over a plane perpendicular to $\mathbf{k}$. Because θ is small, $\mathbf{q}\cdot\mathbf{b} = -\mathbf{k}'\cdot\mathbf{b} \simeq -kb\theta\cos\chi$, where χ is the azimuthal angle that specifies the orientation of $\mathbf{b}$. The term $\mathbf{q}\cdot\hat{\mathbf{k}}z \simeq \frac{1}{2}kz\theta^2$ in the phase factor can be dropped provided

$$\theta^2 \ll \frac{1}{ka}. \tag{20}$$

If we assume that θ is small enough to satisfy (20), we obtain

$$f \simeq \frac{k}{2\pi i}\int_0^\infty b\,db\int_0^{2\pi} d\chi\int_{-\infty}^{\infty} dz\, e^{-ikb\theta\cos\chi} \cdot \frac{\partial}{\partial z}\exp\left\{-\frac{i}{2k}\int_{-\infty}^{z} U(\sqrt{b^2+z'^2})\,dz'\right\}.$$

Now the z-integration is trivial. The integral over χ is merely a familiar integral representation of the Bessel function of order zero:

$$J_0(x) = \frac{1}{2\pi}\int_0^{2\pi} e^{-ix\cos\chi}\,d\chi.$$

Our final result for the scattering amplitude is therefore

$$f(\mathbf{k}',\mathbf{k}) \simeq -ik\int_0^\infty b\,db\, J_0(kb\theta)[e^{2i\Delta(b)}-1], \tag{21}$$

where

$$\Delta(b) = -\frac{1}{4k}\int_{-\infty}^{\infty} U(\sqrt{b^2+z^2})\,dz. \tag{22}$$

* Equation (1) only requires a knowledge of $\psi_{\mathbf{k}}$ in the interaction region. The eikonal wave function (18) is reasonably accurate for such small values of r, and therefore leads to a sensible result for f when it is used in (1). On the other hand, (18) does not display the diffraction phenomena of wave optics, and it therefore provides a very poor approximation to the true wave function in the asymptotic region. By using (1) we circumvent this difficulty.

We shall call (21) the eikonal approximation to the scattering amplitude, because the arguments used in arriving at (18) are borrowed from geometrical optics.

In (21) the factor $e^{2i\Delta} - 1$ tends to zero when b exceeds the range a of U. The integral over b therefore extends over the impact parameters that we would expect on classical grounds. The contribution to the angular distribution of each annulus of radius b is proportional to $J_0(kb\theta)$, a result that also holds in Fraunhofer diffraction by a spherically symmetric object.*

In contrast to the Born approximation, (21) is not linear in U. However, when $|\Delta| \ll 1$, we can make the further approximation $e^{2i\Delta} \simeq 1 + 2i\Delta$, and regain the Born approximation:

$$f_B(\mathbf{k}',\mathbf{k}) \simeq -\frac{1}{2}\int_0^\infty b\,db \int_{-\infty}^{\infty} dz\, J_0(kb\theta)U(\sqrt{b^2+z^2}). \tag{23}$$

One easily verifies that (23) agrees with (2) when (20) holds. The advantage of the eikonal approximation over the Born approximation is that the former applies when the wave function suffers a large change of phase, while the latter requires the change of both phase and amplitude to be small. On the other hand, the Born approximation, when applicable, is not restricted to small scattering angles and short wavelengths.†

The statements of the preceding paragraph become somewhat clearer if one examines a specific example. Consider the Gaussian potential, $U_0 \exp(-r^2/a^2)$. An elementary integration yields

$$\Delta(b) = -\frac{\sqrt{\pi}}{8}\frac{U_0 a}{k}\, e^{-b^2/a^2}. \tag{24}$$

The Born approximation requires $2|\Delta| \ll 1$, i.e.,

$$|U_0|a \ll k. \tag{25}$$

For fixed values of U_0 and k, this condition becomes progressively more stringent as the range a increases. An equivalent way of writing (25) is

$$|U_0| \ll k^2\left(\frac{\lambda\!\!\!^{-}}{a}\right). \tag{25'}$$

* Cf. Landau and Lifshitz, *CTF*, p. 170; Born and Wolf, p. 395.

† As we have already noted, $k\theta \simeq q$. Detailed calculations show that when the exact expression (3) is used for the momentum transfer, i.e., $J_0(|\mathbf{k} - \mathbf{k}'|b)$ instead of $J_0(kb\theta)$, the angular range over which (21) gives accurate results is increased well beyond the limits set by (20). Another virtue of using $|\mathbf{k} - \mathbf{k}'|$ is that it yields a scattering amplitude that satisfies $f(\mathbf{k},\mathbf{k}') = f(-\mathbf{k}',-\mathbf{k})$, a symmetry of f that is required by time reversal invariance (see Sec. 40.5).

This should be contrasted with the condition imposed on the eikonal approximation,

$$|U_0| \ll k^2. \tag{26}$$

At high energy (25′) is always more restrictive than (26).

The eikonal approximation satisfies the optical theorem. To prove this we must compute the total cross section:

$$\sigma = 2\pi k^2 \int_0^\infty b\, db\, b'\, db' \int_0^\pi \sin\theta\, d\theta\, J_0(kb\theta) J_0(kb'\theta) \cdot [e^{2i\Delta(b)} - 1][e^{-2i\Delta(b')} - 1].$$

When the eikonal approximation is justified, the angular distribution is sharply peaked in the forward direction. We may therefore replace $\sin\theta$ by θ, and extend the upper limit on the θ-integral to infinity. The completeness relation for the Bessel functions,

$$\int_0^\infty J_0(x\theta) J_0(x'\theta)\theta\, d\theta = \frac{1}{x}\,\delta(x - x'),$$

then gives

$$\begin{aligned}\sigma &= 2\pi \int_0^\infty b\, db\, |e^{2i\Delta(b)} - 1|^2 \\ &= 8\pi \int_0^\infty b\, db \sin^2 \Delta(b).\end{aligned} \tag{27}$$

Because $J_0(0) = 1$, the imaginary part of the forward scattering amplitude is

$$\operatorname{Im} f(\mathbf{k},\mathbf{k}) = -k \int_0^\infty b\, db\, \operatorname{Re}\, [e^{2i\Delta(b)} - 1].$$

This proves that the optical theorem, Eq. (12.48), is satisfied.

14. Partial Waves

We now turn to a method for solving scattering problems which is exact, at least in principle. In practise it is most useful if the energy is low, or more specifically, if the wavelength is not short compared to the range a of the potential. The technique in question therefore complements the Born and eikonal approximations.

Let the incident momentum be $\hbar\mathbf{k}$. In classical mechanics the largest angular momentum $\hbar l_{\max}$ that can take part in the collision is $\sim\hbar ka$, because a is the largest impact parameter for which there is significant scattering. Thus

$$l_{\max} \simeq ka. \tag{1}$$

We can write the incident wave as (see 11.22)

$$\phi_{\mathbf{k}}(\mathbf{r}) = \sum_{l=0}^{\infty} \sqrt{\frac{2l+1}{2\pi^2}}\, i^l j_l(kr) Y_{l0}(\theta). \tag{2}$$

Because of the spherical symmetry of the potential, the total wave function $\psi_{\mathbf{k}}$ also possesses the quantum number $m = 0$. Therefore it can also be written as a superposition of *partial waves* having the angular momentum quantum numbers l and $m = 0$:

$$\psi_{\mathbf{k}}(\mathbf{r}) = \sum_{l=0}^{\infty} \sqrt{\frac{2l+1}{2\pi^2}}\, i^l A_l(k; r) Y_{l0}(\theta). \tag{3}$$

Our classical argument would lead us to expect that for $l > l_{\max}$, $A_l \simeq j_l(kr)$, i.e., partial waves of $l > l_{\max}$ should not be scattered. If $l_{\max}$ is not too large, we would then have reduced the problem of finding the three-dimensional unknown $\psi_{\mathbf{k}}$ to a small number (of order $l_{\max}$) of one-dimensional problems for the unknown radial functions A_l.

Our task therefore is to (a) derive an equation to determine A_l; (b) express $f(k, \theta)$ in terms of it; and (c) verify the condition (1).

14.1 THE RADIAL INTEGRAL EQUATION

Because $\psi_{\mathbf{k}}$ is a solution of the three-dimensional Schrödinger equation, A_l is a solution of the radial equation (9.16):

$$\left\{\frac{1}{r^2}\frac{d}{dr}r^2\frac{d}{dr} - \frac{l(l+1)}{r^2} - U(r) + k^2\right\} A_l(k; r) = 0. \tag{4}$$

We must now make sure that A_l is the solution of (4) that has the correct normalization and satisfies the boundary conditions. That is to say, we want (3) to satisfy

$$\psi_{\mathbf{k}}(\mathbf{r}) = \phi_{\mathbf{k}}(\mathbf{r}) + \int G_k(|\mathbf{r} - \mathbf{r}'|) U(r') \psi_{\mathbf{k}}(\mathbf{r}')\, d^3r'. \tag{5}$$

For this purpose it is again best to study the integral equation satisfied by A_l, instead of the corresponding differential equation [i.e., Eq. (4)]. In order to construct this radial integral equation we expand G_k in spherical harmonics*:

$$G_k(|\mathbf{r} - \mathbf{r}'|) = \sum_{lm} Y_{lm}(\hat{\mathbf{r}}) Y_{lm}^*(\hat{\mathbf{r}}') G_k^{(l)}(r; r'). \tag{6}$$

* A function of $|\mathbf{r} - \mathbf{r}'|$ is actually a function of r, r', and $\cos\theta$, where θ is the angle between $\mathbf{r}$ and $\mathbf{r}'$. It can therefore be represented as a superposition of Legendre polynomials $P_l(\cos\theta)$. The addition theorem, (10.22), then leads to a representation having the form of (6).

Substituting this expansion, and (2) and (3), into (5), we easily find the desired equation:

$$A_l(k; r) = j_l(kr) + \int_0^\infty G_k^{(l)}(r; r')U(r')A_l(k; r')r'^2\, dr'. \tag{7}$$

Equation (7) is equivalent to (4) plus the boundary conditions implicit in (5).

We must now evaluate $G_k^{(l)}$. In

$$G_k(|\mathbf{r} - \mathbf{r}'|) = \int \frac{d^3q}{(2\pi)^3} \frac{e^{i\mathbf{q}\cdot(\mathbf{r}-\mathbf{r}')}}{k^2 - q^2 + i\epsilon}$$

expand $e^{i\mathbf{q}\cdot\mathbf{r}}$ and $e^{-i\mathbf{q}\cdot\mathbf{r}}$ in spherical waves [see (11.25)], and integrate over $\hat{\mathbf{q}}$. This leads immediately to

$$G_k(|\mathbf{r} - \mathbf{r}'|) = \frac{2}{\pi} \sum_{lm} Y_{lm}(\hat{\mathbf{r}}) Y_{lm}^*(\hat{\mathbf{r}}') \int_0^\infty q^2\, dq \frac{j_l(qr)j_l(qr')}{k^2 - q^2 + i\epsilon}.$$

According to (11.4) the integrand is even, and therefore

$$G_k^{(l)}(r; r') = \frac{1}{\pi} \int_{-\infty}^{+\infty} q^2\, dq \frac{j_l(qr)j_l(qr')}{k^2 - q^2 + i\epsilon}.$$

This expression can be evaluated by contour integration. First we observe that $j_l(z)$ is an entire function of z; hence the denominator $(k^2 - q^2 + i\epsilon)$ provides all the singularities in the finite q-plane. Let $r' > r$, for instance. We then replace $j_l(qr')$ by $\frac{1}{2}[h_l(qr') + h_l^*(qr')]$. As $z \to \infty$, $h_l(z) \to i^{-l-1}e^{iz}/z$, and $h_l(qr')j_l(qr)$ therefore decreases exponentially in the upper half q-plane as $|q| \to \infty$. In this term we can therefore add a large semi-circle to the contour in the upper half-plane, and apply Cauchy's theorem. By the same token we close the contour in the lower half-plane in the term containing $h_l^*(qr')j_l(qr)$. With the help of the identity $h_l^*(-z) = (-1)^l h_l(z)$, we finally obtain *

$$G_k^{(l)}(r; r') = -ikj_l(kr_<)h_l(kr_>). \tag{8}$$

14.2 SCATTERING AMPLITUDE AND PHASE SHIFTS

Consider the asymptotic form of (7). Assuming † that U falls off fast enough at infinity, we can suppose that $r > r'$; using the asymptotic

* Here $r_<$ ($r_>$) is the smaller (larger) of r and r'.

† The conditions on U which must be satisfied for (9) to be valid are discussed in the next subsection, (14.3).

forms (11.5) and (11.11) we obtain

$$A_l(k;r) \sim \frac{1}{kr}\sin(kr - \tfrac{1}{2}l\pi) - \frac{ike^{ikr}}{kri^{l+1}}\int_0^\infty j_l(kr')U(r')A_l(k;r')r'^2\,dr'$$

$$\sim -\frac{1}{2ikr}\left\{e^{-i(kr-\frac{1}{2}l\pi)} - e^{i(kr-\frac{1}{2}l\pi)}\left[1 - 2ik\int_0^\infty j_l(kr')U(r')A_l(k;r')r'^2\,dr'\right]\right\}. \quad (9)$$

The first term in (9) is the incoming spherical wave, and its amplitude and phase are independent of U, as we would expect from the boundary condition built into (7). The outgoing wave (second term) is changed by the potential, however. Because we have assumed a real potential, probability is conserved, and therefore the amplitude of the outgoing wave must equal that of the incoming. Hence we can write

$$1 - 2ik\int_0^\infty j_l(kr')U(r')A_l(k;r')r'^2\,dr' = e^{2i\delta_l(k)}, \quad (10)$$

where $\delta_l(k)$ *is real.* In terms of $\delta_l(k)$, (9) reads

$$A_l(k;r) \sim -\frac{1}{2ikr}\{e^{-i(kr-\frac{1}{2}l\pi)} - e^{i(kr-\frac{1}{2}l\pi)}e^{2i\delta_l(k)}\}$$

$$\sim \frac{e^{i\delta_l}}{kr}\sin(kr - \tfrac{1}{2}l\pi + \delta_l). \quad (11)$$

Thus the interaction merely shifts the phase of the asymptotic radial wave function. The phase shifts $\delta_l(k)$ play a central role in scattering theory, as we shall see.

We shall now express the scattering amplitude in terms of the phase shifts. According to (12.18), (2) and (3),

$$f(k,\theta) = (2\pi)^{\frac{3}{2}}\lim_{r\to\infty}[(\psi_{\mathbf{k}} - \phi_{\mathbf{k}})re^{-ikr}]$$

$$= \sum_{l=0}^{\infty} i^l\sqrt{4\pi(2l+1)}\,Y_{l0}(\theta)\lim_{r\to\infty}[(A_l - j_l)re^{-ikr}].$$

But from (11) and (11.5)

$$(A_l - j_l) \sim \frac{e^{ikr}}{ri^l}\frac{e^{2i\delta_l} - 1}{2ik},$$

and therefore

$$f(k, \theta) = \frac{1}{2ik} \sum_{l=0}^{\infty} \sqrt{4\pi(2l+1)}\, (e^{2i\delta_l} - 1) Y_{l0}(\theta) \tag{12}$$

$$= \frac{1}{k} \sum_{l=0}^{\infty} \sqrt{4\pi(2l+1)}\, e^{i\delta_l} \sin \delta_l Y_{l0}(\theta). \tag{13}$$

We should also note that (10) leads to the important formula

$$\frac{1}{k} e^{i\delta_l} \sin \delta_l = -\int_0^\infty j_l(kr) U(r) A_l(k; r) r^2\, dr. \tag{14}$$

We have now succeeded in expressing explicitly the contribution of each angular momentum (partial wave) to the scattering amplitude. The various partial waves can be treated separately, as we have seen throughout, because L^2 is a constant of the motion. If the target were only cylindrically symmetric about the direction $\hat{\mathbf{k}}$ we could still expand $\psi_{\mathbf{k}}$ in L_z-eigenfunctions with $m = 0$ as in (3), but the differential (or integral) equations determining the expansion coefficients would mix various A_l's together, and the scattering amplitude could not be expressed in the simple forms (12) and (13).

The only partial waves that contribute significantly to the scattering amplitude are those having phase shifts that differ appreciably from zero.* Assuming the argument at the beginning of this section is correct, we expect $\delta_l \to 0$ for all $l \gg l_{max}$. At low energies (i.e., such that $k \ll a^{-1}$) only the $l = 0$ or s-wave is scattered, and the angular distribution of scattered particles is isotropic. As the energy is raised the phase shifts for small but nonzero l will become appreciable, and the differential cross section will vary slowly with scattering angle (see Prob. 7). We already found a qualitatively similar energy dependence of the angular distribution with the Born approximation (see Fig. 13.1).

When we form the differential cross section $|f(k, \theta)|^2$ from (13) the various l-waves interfere. This is not true of the total cross section, however, because of the orthogonality of the spherical harmonics:

$$\sigma(k) = \frac{4\pi}{k^2} \sum_{l=0}^{\infty} (2l+1) \sin^2 \delta_l \equiv \sum_{l=0}^{\infty} \sigma_l(k). \tag{15}$$

Comparing (15) and (13), and recalling that $Y_{l0}(0) = \sqrt{(2l+1)/4\pi}$, we see that the optical theorem is satisfied, which is to be expected

* Equation (10) only determines the phase shift modulo π. We always choose the branch with the property $\delta_l \to 0$ as $U(r) \to 0$.

because no approximation has been made. Note, incidentally, that the optical theorem also requires $\text{Im } \delta_l(k) = 0$.

The partial wave expansion (13) and (15) have played a very important role in nuclear physics. Scattering experiments serve to determine the phase shifts, and they in turn help to determine the unknown potential $V(r)$. (The question of whether $\{\delta_l(k)\}$ uniquely determines $V(r)$ is complicated and involved, and we shall not enter into it here.) From the series (13) and (15) we may draw some general conclusions that are of importance in the analysis of experiments. Since the angular distribution contains interference terms, the relative signs of the phase shifts can be measured. On the other hand, the substitution $\delta_l \to -\delta_l$ for all l, which merely replaces f by $-f^*$, does not change the differential cross section. Hence the angular distribution leaves the overall sign of the set $\{\delta_l\}$ indeterminate. This sign has physical significance, however, as one may see from (14) in the case where U is sufficiently weak to permit the approximations * $A_l \sim j_l$, $\delta_l \ll 1$:

$$\delta_l(k) \approx -k \int_0^\infty [j_l(kr)]^2 U(r) r^2 \, dr. \tag{14'}$$

Hence if U is everywhere repulsive $\delta_l < 0$, and vice versa for the purely attractive case. These results are actually exact if U never changes its sign as r is varied (see Sec. 49.1). This overall sign of $\{\delta_l\}$ can only be determined by further interference effects; we shall consider the most important example of this sort when we come to Coulomb scattering (see Sec. 17.3).

Note that the contribution of any one term to the total cross section is bounded by

$$\sigma_l \leq 4\pi(2l + 1)\lambda\!\!\!^{-2}. \tag{16}$$

Even if the differential cross section is isotropic we therefore know that the scattering occurs also in the $l \neq 0$ states if $\sigma(k) > 4\pi\lambda\!\!\!^{-2}$. This situation actually arises in proton-proton scattering in the 100–250 Mev region (i.e., an isotropic angular distribution and a total cross section in excess of $4\pi\lambda\!\!\!^{-2}$), and therefore one can conclude unambiguously that the higher partial waves somehow interfere in such a way as to give a θ-independent cross section.

We must still show that $\delta_l \to 0$ for $l \gg l_{max}$. To do this it will suffice to estimate δ_l on the basis of intuitive considerations. The argument is

* Equation (14′) is called the Born approximation to the phase shift. If one substitutes (14′) into (13) and carries out the sum on l, one does not regain the three-dimensional Born approximation (13.2) however. This is clear because (13) will satisfy the optical theorem with any set of phases, while (13.2) does not conserve probability. To retrieve the Born approximation one must make the further approximation $e^{i\delta_l} \sin \delta_l \approx \delta_l$ in (13).

probably best understood if one thinks of a one-dimensional problem that is completely equivalent to the radial Schrödinger equation. In this one-dimensional problem the wave function is $xA_l(k;x)$, and the potential is $\hbar^2 W_l(x)/2\mu$, where

$$W_l(x) = U(x) + \frac{l(l+1)}{x^2} \qquad (x > 0),$$
$$= \infty \qquad (x < 0);$$

the hard wall at the origin insures that the one-dimensional solution corresponds to a function $A_l(k;r)$ that is regular at $r = 0$. $W_l(x)$ is

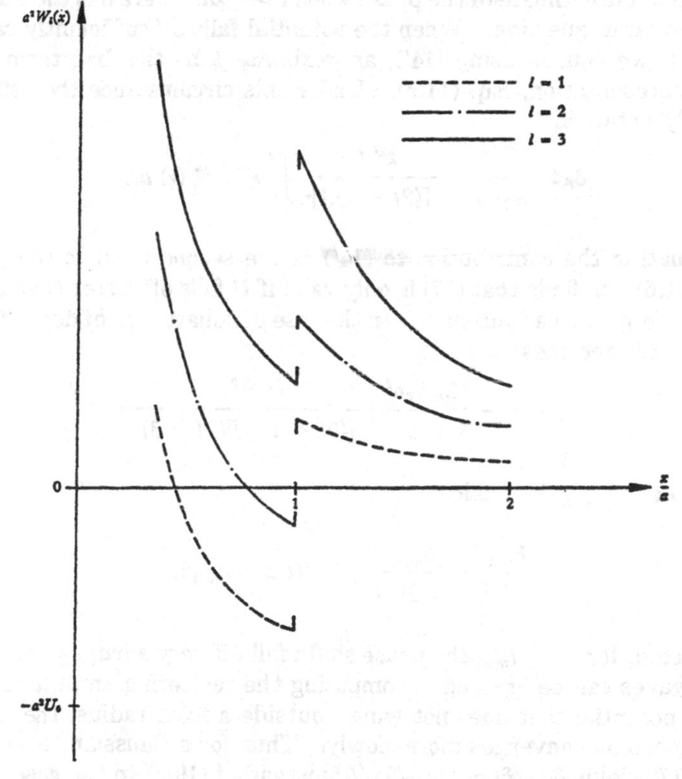

Fig. 14.1. The equivalent one-dimensional potentials $W_l(x)$ when $U(r)$ is a square well of radius a and depth U_0.

sketched in Fig. 14.1 when $U(r)$ is a square well of radius a and depth U_0. We observe that if $k^2a^2 < l(l+1)$ there is insufficient energy to pass the centrifugal barrier classically, and the solution decays (though only algebraically, not exponentially) just to the left of $x = a$. Furthermore, for any fixed set of values of the dimensionless parameters U_0a^2 and ka one can, by simply increasing l, always make the barrier as formidable as one pleases, thereby reducing the influence of the true potential $U(r)$ on $A_l(k;r)$ correspondingly. Thus for every U_0a^2 and ka there exists a number $l_{\max}(U_0a^2, ka)$ such that if $l > l_{\max}$, $j_l(kr)$ and δ_l approximate A_l and $e^{i\delta_l} \sin \delta_l$ to some desired accuracy. For most purposes this complicated condition can be replaced by the rougher and intuitively more obvious inequality $l \gg ka$. More elaborate, but in essence similar, considerations must be used for other potentials. Except for special circumstances which we shall not dwell on here, the final conclusion is that the Born approximation (14′) gives a reliable estimate of the phase when $l \gg ka$, where a is the range of the potential in question. When the potential falls off sufficiently rapidly as $r \to \infty$, we can, in using (14′), approximate j_l by the first term in its series representation, Eq. (11.6). Under this circumstance the estimate we finally obtain is *

$$\delta_l(k) \underset{l \gg l_{\max}}{\sim} -\frac{k^{2l+1}}{[(2l+1)!!]^2} \int_0^\infty r^{2l+2} U(r)\, dr. \tag{17}$$

By evaluating the contribution to (14′) of the second term in the power series (11.6) one finds that (17) is only valid if U falls off faster than an exponential (e.g., like a Gaussian). In the case of square well of depth V_0 and range a, (17) becomes

$$\delta_l(k) \sim -\left(\frac{2\mu V_0 a^2}{\hbar^2}\right) \frac{(ka)^{2l+1}}{[(2l+1)!!]^2(2l+3)}, \tag{18}$$

Writing $ka \equiv l_{\max}$, this yields

$$\frac{\delta_{l+1}}{\delta_l} \sim \left(\frac{\delta_{\max}}{2l}\right)^2, \qquad (l \gg l_{\max}). \tag{19}$$

As expected, for $l \gg l_{\max}$ the phase shifts fall off very abruptly, and such partial waves can be ignored in computing the scattering amplitude.

For a potential that does not vanish outside a fixed radius, the partial wave expansion converges more slowly. Thus for a Gaussian, $U(r) \sim r^{-1} e^{-(r/a)^2}$, (17) yields $\delta_{l+1}/\delta_l \sim (l_{\max}/2\sqrt{l})^2$ instead of (19). In the case of the Yukawa potential, $U \sim r^{-1} e^{-r/a}$, one must return to (14′), which leads to $\delta_l \sim (l\, l_{\max})^{-\frac{1}{2}} e^{-l/l_{\max}}$ when $l \gg l_{\max}$ and $l_{\max} \gg 1$.

* For further discussion of the phase shifts near threshold (i.e., as $k \to 0$) see Sec. 49.3.

14.3 ASYMPTOTIC FORM OF RADIAL WAVE FUNCTIONS

•Up to now we have failed to ascertain the validity of the asymptotic expansion (9), and the corresponding one in three dimensions, Eq. (12.17). The most convenient way to approach this question is with the differential equation (4). We begin with $l \neq 0$. If $\lim_{r\to\infty} r^2 U = 0$ the centrifugal term $l(l+1)/r^2$ dominates, and hence the interacting problem reduces to the free ($U = 0$) differential equation in the asymptotic region. Thus

$$A_l(k;r) \sim a_l(k) j_l(kr) + b_l(k) n_l(kr);$$

the asymptotic forms for j_l and n_l then yield

$$A_l(k;r) \sim \frac{\gamma_l(k)}{kr} \sin(kr - \tfrac{1}{2}l\pi + \delta_l).$$

This agrees with (11). For a bound state, and under the same conditions on U and l, we have the asymptotic form ($k \to i\alpha$, $\alpha > 0$)

$$A_l(\alpha;r) \sim a_l h_l(i\alpha r) + b_l h_l^*(i\alpha r), \tag{20}$$

where $-\hbar^2\alpha^2/2\mu$ is the binding energy. Because exponentially growing solutions are unacceptable $b_l = 0$, and hence the asymptotic form of a bound state is

$$A_l \sim \text{const.} \frac{e^{-\alpha r}}{r}. \tag{21}$$

Observe that this asymptotic form is completely determined by the binding energy.

Consider now a potential U that has the asymptotic form

$$U(r) \sim \frac{U_0}{r^s}. \tag{22}$$

When $1 \leq s \leq 2$ this will take care of the $l \neq 0$ cases not treated above, and it will also dispose of the general case when $l = 0$. If we substitute $\chi = rA$ into (4), we find that in the asymptotic region

$$\left(\frac{d^2}{dr^2} - \frac{U_0}{r^s} + k^2\right)\chi(r) = 0. \tag{23}$$

As $r \to \infty$, the potential changes very slowly and the semiclassical approximation is valid. We therefore put

$$\chi(r) = e^{\pm i(kr + u(r))}$$

into (23), and drop u'' and $(u')^2$ in comparison to u':

$$\frac{du}{dr} + \frac{U_0}{2kr^s} = 0. \tag{24}$$

Hence if $s > 1$

$$\chi \sim \exp\left[\pm ikr\left(1 - \frac{U_0}{2k^2(1-s)r^s}\right)\right] \sim e^{\pm ikr}.$$

Therefore the asymptotic radial forms (11) and (21) are valid for all values of l for any potential that obeys

$$\lim_{r\to\infty} rU(r) = 0. \tag{25}$$

The situation is different for the Coulomb field ($s = 1$), however. Integrating (24) we have

$$\chi \sim \exp\left[\pm i\left(kr - \frac{U_0}{2k}\ln kr\right)\right]. \tag{26}$$

Although kr eventually dominates the phase factor as $r \to \infty$, the asymptotic form (12.18) breaks down in this case. This can be seen more clearly for bound states, for which the semiclassical method yields the asymptotic form

$$\chi \sim e^{-\left(\alpha r + \frac{U_0}{2\alpha}\ln \alpha r\right)} = (\alpha r)^{-U_0/2\alpha}e^{-\alpha r}. \tag{27}$$

Because $U_0 < 0$ if bound states are to exist, this shows that the bound-state eigenfunctions for a Coulomb potential fall off much more slowly than (21). It is clear from this whole discussion that the Coulomb field must be treated by specially adapted methods. This is not tragic, of course, because the partial wave expansion surely fails to converge for $1/r$ potentials.●

14.4 BOUND STATES

One usually finds the radial wave functions for the bound states directly from the differential equation (4), with $k^2 < 0$; the quantization arises from the normalization requirement. Nevertheless, for certain purposes the radial integral equation arising from (12.20) is more useful. Calling the bound-state eigenfunction $R_l(\alpha; r)$, we obtain the desired equation by deleting the incident wave and by setting $k = i\alpha$ in (7):

$$R_l(\alpha; r) = \alpha \int_0^\infty j_l(i\alpha r_<)h_l(i\,\alpha r_>)U(r')R_l(\alpha; r')r'^2\,dr'. \tag{28}$$

(If $\alpha < 0$, h_l^* must appear in the kernel. We shall *always* use the convention $\alpha > 0$.) When (25) is satisfied, this will yield the correct asymptotic form (21).

Note that (28) arises from (7) if we substitute $k = i\alpha$, and *drop* the incident wave $j_l(i\alpha r)$. This close relationship between (7) and (28) motivates the following line of reasoning. If we replace k by a complex variable z in (4) or (7), we define a function $A_l(z; r)$ throughout the z-plane. As soon as $\operatorname{Im} z \neq 0$, however, A_l is not square integrable because $j_l(zr)$ diverges exponentially as $|z| \to \infty$ unless $\operatorname{Im} z = 0$. The scattering term in (7), on the other hand, converges as $|zr| \to \infty$ if

$\text{Im}\, z > 0$ because $h_l(z)$ decays exponentially in the upper half-plane. Consider the integral equation

$$(z - z_0)A_l(z;r) = (z - z_0)j_l(zr) \\ - iz \int_0^\infty j_l(zr_<)h_l(zr_>)U(r')(z - z_0)A_l(z;r')r'^2\, dr' \quad (29)$$

obtained from (7) after multiplying by $(z - z_0)$. In general (29) merely reads $0 = 0$ at $z = z_0$. If, however, $A_l(z;r)$ has a simple pole at z_0, i.e., if in the neighborhood of z_0

$$A_l(z;r) = \frac{R_l(z_0;r)}{z - z_0} + \text{reg.}, \quad (30)$$

where "reg." stands for a function that is analytic at z_0, then (29) becomes

$$R_l(z_0;r) = -iz_0 \int_0^\infty j_l(z_0r_<)h_l(z_0r_>)U(r')R_l(z_0;r')r'^2\, dr' \quad (31)$$

at z_0 (recall that $j_l(z)$ is an entire function). But (31) is just (28) with $i\alpha$ replaced by z_0. Hence if $\text{Im}\, z_0 > 0$, $R_l(z_0;r)$ is a square integrable solution of the Schrödinger equation with eigenvalue $\hbar^2 z_0{}^2/2\mu$. Because the square integrable solutions all have negative, real, energy eigenvalues, z_0 must lie on the positive imaginary axis, i.e., at the points $z = i\alpha$, where $-\hbar^2\alpha^2/2\mu$ are energy levels of angular momentum l.

Our conclusion is therefore that $A_l(k;r)$ has poles on the positive imaginary k-axis at the points $i\sqrt{2\mu|E|/\hbar^2}$, where E is one of the bound-state energy eigenvalues of angular momentum l, the residues being the bound state radial wave functions. It is also true that these poles are simple, but we have not shown this as yet. Furthermore, it is clear from what we have said that $A_l(k;r)$ cannot have any other poles in the upper-half k-plane. (Our argument tells us nothing about the lower-half k-plane, however.) We must expect the partial wave amplitude $e^{i\delta_l} \sin \delta_l/k$ to have poles at the points $k = i\alpha$ in virtue of (14).* On the other hand, the partial wave amplitude can have singularities that bear no relation to the spectrum of the Hamiltonian. This fact is demonstrated in terms of an example in Prob. 4.

14.5 SCATTERING BY COMPLEX SYSTEMS

The formulation of collision theory that we have presented is based on the one-body Schrödinger equation with a Hermitian interaction, and one could be led to suppose that none of our results apply to scattering by complex systems. To see whether this is actually so, let us examine

* The analytic properties of radial wave functions are discussed by R. Newton, *Jour. Math. Phys.* **1**, 319 (1960).

the derivation of formula (12) for the scattering amplitude once more. The basic assumptions that we shall now invoke are: (a) the interaction between the projectile and the target is negligible outside a sphere of radius R; (b) this interaction is spherically symmetric; (c) the ground state of the target has zero angular momentum and is nondegenerate; and (d) the target is initially in this state. In contrast to our earlier discussion, we do not assume that the target is structureless; on the contrary, it may be a complicated system with numerous excited states. If the projectile is sufficiently energetic, these states can be excited. It may even be that the target can absorb the projectile and emit other types of particles (as in $\pi^- + O^{16} \rightarrow N^{14} + n + n$). Nevertheless, assumption (a) tells us the form of the projectile's wave function when $r > R$, and this shall suffice for our modest purposes.

Because of assumption (b), the incident wave of angular momentum l is dynamically uncoupled from all final configurations having a different total angular momentum. Let us focus our attention on elastic scattering in the lth partial wave. When $r > R$ the interaction is inoperative, and the portion of the total wave function describing the elastic process must therefore be the ground state wave function of the target multiplied by a linear combination of free-particle wave functions:

$$\psi_{\text{el}}^{(l)} = C\chi_0[h_l^*(kr) + \eta_l h_l(kr)]Y_{l0}(\theta). \tag{32}$$

Here χ_0 is the target's ground state, C is a normalization constant, and η_l gives the fraction of the incident wave to be found in the outgoing wave. If the energy is such that elastic scattering is the only process allowed, probability conservation demands $|\eta_l| = 1$. Comparing with (11) we note that in this case $\eta_l = e^{2i\delta_l}$. When inelastic processes are also energetically possible, the elastically scattered wave must suffer a decrease of amplitude, and therefore $|\eta_l| < 1$ (equivalently, one can take the phase shift to be complex). A repetition of the argument that led from (11) to (12) then shows that the elastic amplitude is given by

$$f_{\text{el}}(k, \theta) = \frac{1}{2ik} \sum_{l=0}^{\infty} \sqrt{4\pi(2l + 1)}\,(\eta_l - 1)Y_{l0}(\theta). \tag{33}$$

The total elastic cross section, $\int |f_{\text{el}}(k, \theta)|^2\, d\Omega$, is therefore

$$\sigma_{\text{el}}(k) = \pi\lambda^2 \sum_{l=0}^{\infty} (2l + 1)|\eta_l - 1|^2. \tag{34}$$

The total cross section for *all* inelastic processes $\sigma_{\text{inel}}(k)$ (i.e., regardless of what the final particles may be, or at what angles they emerge) can also be expressed in terms of the amplitudes η_l. The sought-after expression follows from the optical theorem, provided the latter is generalized to include inelastic processes. We recall from (12.46) that the number of incident particles taken out of the beam per second by the scattering center is $(2\pi)^{-3}v_k(4\pi/k)\,\mathrm{Im}\,f_{el}(k, 0) \equiv \dot{N}$, where v_k is the relative velocity in the initial state. By definition, the total cross section,

$$\sigma_{tot}(k) = \sigma_{el}(k) + \sigma_{inel}(k), \tag{35}$$

is given by $\dot{N}/F$, where $F = (2\pi)^3/v_k$ is the incident flux. Hence

$$\sigma_{tot}(k) = \frac{4\pi}{k}\,\mathrm{Im}\,f_{el}(k, 0); \tag{36}$$

this is the desired generalization of the optical theorem. The total and inelastic cross sections are therefore

$$\sigma_{tot}(k) = 2\pi\lambda\!\!\!^{-2} \sum_{l=0}^{\infty} (2l + 1)(1 - \mathrm{Re}\,\eta_l), \tag{37}$$

$$\sigma_{inel}(k) = \pi\lambda\!\!\!^{-2} \sum_{l=0}^{\infty} (2l + 1)(1 - |\eta_l|^2). \tag{38}$$

The inelastic cross section is a measure of the depletion of the outgoing elastic wave, and is therefore determined by the quantities $1 - |\eta_l|^2$.

Because a single set of (complex) parameters determines both the elastic and inelastic cross sections, there are certain relations between them.* Of these one deserves special mention: Elastic scattering always accompanies inelastic scattering (i.e., when $|\eta_l| < 1$, $|\eta_l - 1| > 0$). This is intuitively obvious, because the inelastic processes necessarily lead to the formation of a shadow.

It should be noted that we have only obtained a parametrization of the elastic and inelastic cross sections. In the case of potential scattering we were able to go beyond this and to determine the amplitudes η_l in terms of the interactions (i.e., Eq. (14)). A detailed dynamical theory of the target and its interaction with the projectile would be required if one wished to compute η_l in the case of a complex target.

14.6 RELATIONSHIP TO THE EIKONAL APPROXIMATION

The partial wave method decomposes the collision process into various sub-processes having definite angular momenta, whereas the eikonal approximation involves an analogous decomposition with respect to impact parameters. These methods are simply related when we can form

* See Blatt and Weisskopf, p. 322.

a packet of fairly definite angular momentum and impact parameter, because we then have

$$kb \simeq l. \tag{39}$$

A packet of the type just described can only be constructed when many angular momentum states participate, i.e., when many terms in the partial wave series contribute to the scattering amplitude. If many terms in the partial wave series must be summed, it is natural to approximate (33) by an integral over l. (As (12) is a special case of (33), we shall use the latter here.)

If (33) is to be replaced by an integral, Y_{l0} and η_l must be reasonably smooth functions of l. The l-dependence of the spherical harmonic is given by

$$P_l(\cos\theta) \simeq J_0(2l \sin \tfrac{1}{2}\theta) \tag{40}$$

when $\sin^2 \frac{1}{2}\theta \ll 1$ and $l \gg 1$. The fact that (40) is not very accurate for small l is presumably not too important if many partial waves contribute significantly. We therefore replace (33) by

$$f_{el}(k,\theta) \simeq \frac{1}{ik}\int_0^\infty l\,dl\,J_0(2l\sin\tfrac{1}{2}\theta)(\eta_l - 1).$$

The impact parameter can then be introduced via (39):

$$f_{el}(k,\theta) \simeq -ik\int_0^\infty b\,db\,J_0(qb)[\eta(b) - 1], \tag{41}$$

where $q = 2k\sin\frac{1}{2}\theta$, and $\eta(b)$ is a smooth function that satisfies $\eta(l/k) = \eta_l$. This expression for f becomes the eikonal approximation when $\theta \ll 1$ and $\eta(b) = e^{2i\Delta(b)}$. Hence the phase shifts and $\Delta(b)$ are related by $\Delta(l/k) \simeq \delta_l(k)$ when the conditions set out in Sec. 13.2 are satisfied.

A crude evaluation of the formulas of the preceding subsection can be carried out with the help of (41). Consider an interaction where any collision with an impact parameter $b \lesssim R$ leads to an inelastic process. Furthermore, assume that $kR \gg 1$. These conditions actually apply in the scattering of energetic pions and K-mesons by complex nuclei. To describe such a situation we make the very crude approximation

$$\begin{aligned} \eta(b) &= 0 \qquad (b < R) \\ &= 1 \qquad (b > R). \end{aligned} \tag{42}$$

Equation (41) can then be evaluated by using

$$\int_0^R b\,db\,J_0(xb) = \frac{R}{x}J_1(xR),$$

i.e.,

$$f_{el} = iR^2k\frac{J_1(2kR\sin\frac{1}{2}\theta)}{2kR\sin\frac{1}{2}\theta}. \tag{43}$$

As one might expect, this angular distribution also applies to the diffraction of a short-wavelength electromagnetic wave by a conducting sphere.* The total cross section follows immediately from (36), because $J_1(x) \to \frac{1}{2}x$ as $x \to 0$:

$$\sigma_{tot} = 2\pi R^2. \tag{44}$$

The total elastic cross section is evaluated from (34):

$$\sigma_{el} \simeq 2\pi\lambda\!\!\!^{-2}\int_0^{kR} l\,dl = \pi R^2. \tag{45}$$

Hence σ_{inel} is also πR^2.

Observe that σ_{inel} equals the geometrical cross section of the target, whereas σ_{tot} is *twice* as large. This factor of two occurs because the diffraction (or shadow) scattering described by (43) also contributes an amount πR^2 to the total cross section.

15. The Delta-Shell Potential

In order to familiarize ourselves with the general theory developed in the previous sections, we shall study a somewhat artificial example in considerable detail. Our example has the virtue of great mathematical simplicity. In spite of this it will enable us to study the following subjects: (a) bound states, (b) scattering resonances, (c) the convergence of the Born series, and (d) the analytic properties of partial wave amplitudes in the complex momentum and energy planes, and the physical significance of the singularities. The investigation of analytic properties of scattering amplitudes has become very fashionable in recent years, and the following pages are intended, in part, as a first introduction to this subject.

The potential we consider is

$$U(r) = -\lambda\,\delta(r - a); \tag{1}$$

i.e., a force field that vanishes everywhere except on a sphere of radius a. The strength parameter λ has the dimension $(\text{length})^{-1}$. One can look upon (1) as a crude model of the interaction experienced by a neutron

* Jackson, Sec. 9.10.

when it interacts with a nucleus of radius a. The resonances we shall find actually bear some resemblance to the fine structure resonances observed in low energy neutron scattering. But the mechanism causing these resonances involves many particle interactions in a very fundamental way, and therefore (1) cannot really do justice to the situation found in nature. Nevertheless, a careful study of this potential will provide a first orientation to the theory of nuclear reactions.

When the potential is given by (1) the radial integral equation (14.7) reduces to the *algebraic* equation

$$A_l(k; r) = j_l(kr) + ik\lambda a^2 A_l(k; a) \times \begin{cases} j_l(kr)h_l(ka) & (a > r) \\ j_l(ka)h_l(kr) & (r > a) \end{cases}. \tag{2}$$

This can be solved immediately for the only unknown, namely

$$A_l(k; a) = \frac{j_l(ka)}{1 - ik\lambda a^2 j_l(ka)h_l(ka)}. \tag{3}$$

When we substitute (3) into (2) we have the complete solution.

We can construct the scattering amplitude from (14.14):

$$\begin{aligned} e^{i\delta_l} \sin \delta_l &= k\lambda a^2 j_l(ka) A_l(k; a) \\ &= \frac{k\lambda a^2 [j_l(ka)]^2}{1 - ik\lambda a^2 j_l(ka)h_l(ka)}. \end{aligned} \tag{4}$$

The tangent of δ_l is also a convenient quantity for some purposes. Because $h_l(z) = j_l(z) + in_l(z)$, we easily find

$$\tan \delta_l = \frac{k\lambda a^2 [j_l(ka)]^2}{1 + k\lambda a^2 j_l(ka)n_l(ka)}. \tag{5}$$

It is natural to express all lengths in units of a, and all wave numbers in units of $1/a$. Hence define the dimensionless variables

$$\rho = \frac{r}{a}, \qquad \xi = ka, \qquad g = \lambda a, \tag{6}$$

in terms of which, for example,

$$e^{i\delta_l} \sin \delta_l = \frac{g\xi[j_l(\xi)]^2}{1 - i\xi g j_l(\xi)h_l(\xi)}. \tag{7}$$

15.1 BOUND STATES

In Sec. 14.4 we concluded that the existence of bound states requires the occurrence of poles in the radial continuum functions when the latter are treated as functions of the complex variable k. Furthermore, it was shown that these poles must lie on the positive imaginary axis in the k-plane. We therefore seek the poles of (2). These poles must arise from zeros of the denominator of $A_l(k; a)$, because j_l is entire and the singularity in $h_l(kr)$ at $kr = 0$ does not appear in (2). Hence we must determine the location of the pure imaginary zeros of the function

$$D_l(g; \zeta) = 1 - i\zeta g j_l(\zeta) h_l(\zeta), \tag{8}$$

where $\zeta = \xi + i\eta$. For general values of g and l, this is a messy transcendental equation. We may, however, ask for the least value of g that can bind a state of angular momentum l_0. Since this state will, by hypothesis, have zero binding energy $E = -\hbar^2\eta^2/2\mu a^2$, we can expand (8) about $\zeta = 0$ in finding it. Because

$$\lim_{\zeta \to 0} \zeta j_{l_0}(\zeta) h_{l_0}(\zeta) = \frac{-i}{2l_0 + 1}, \tag{9}$$

we have

$$D_{l_0}(g, 0) = 1 - \frac{g}{2l_0 + 1}. \tag{10}$$

Therefore if $g = 2l_0 + 1$ there is a bound state of angular momentum l_0 with energy zero. As g increases beyond $2l_0 + 1$, the zero of (8) presumably moves up the imaginary ζ-axis, and, by the same token, it would appear that there are bound states for all values of $l < l_0$. Rather than verifying these guesses for general l, we shall confine ourselves to a detailed examination of the s-wave.

•Before embarking on this, we can make some remarks about the Born series for partial wave amplitudes that should help to clarify the discussion at the end of Sec. 13.1. By definition we shall call the iterative solution of (14.7) the Born series for A_l, and we call the corresponding expansion,

$$\begin{aligned} \frac{1}{k} e^{i\delta_l} \sin \delta_l = &- \int_0^\infty [j_l(kr)]^2 U(r) r^2 \, dr \\ &- \int_0^\infty j_l(kr) U(r) G_k^{(l)}(r, r') U(r') j_l(kr') r^2 r'^2 \, dr \, dr' + \cdots, \end{aligned} \tag{11}$$

the Born series for the partial wave amplitude. According to (7), this series in our present example is

$$e^{i\delta_l} \sin \delta_l = g\xi[j_l(\xi)]^2 \sum_{n=0}^{\infty} g^n [i\xi j_l(\xi) h_l(\xi)]^n, \tag{12}$$

and its radius of convergence (note: we are now in the g-plane!) is

$$\frac{1}{\xi |j_l(\xi) h_l(\xi)|} \equiv g_l^{(B)}(\xi). \tag{13}$$

As we would expect from our earlier work, this radius depends on the energy, and it is smallest at low energy. In particular, (9) shows that

$$g_l^{(B)}(0) = 2l + 1,$$

which simply says that the Born series for the lth partial wave converges at zero energy if the potential cannot bind a state of angular momentum l. This result actually holds for a wide class of potentials, but it must be remembered here that we are talking about a *radius* of convergence, and hence the Born series will not converge for a repulsive potential if its attractive counterpart does in fact bind a state. As we raise the energy, the radius of convergence increases and attains the asymptotic value $g_l^{(B)} \sim \xi$.•

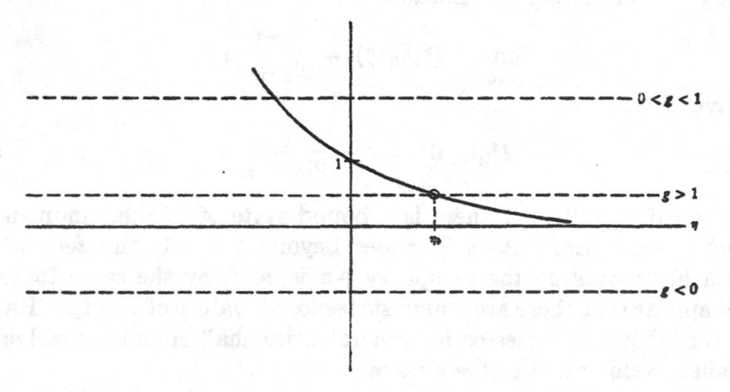

Fig. 15.1. Determination of the binding energy in the s-wave.

Let us now look for the $l = 0$ bound states in detail. We have

$$D_0(g; \zeta) = 1 - g\zeta^{-1}e^{i\zeta}\sin\zeta, \tag{14}$$

and the eigenvalue equation $D_0(g; i\eta) = 0$ is therefore

$$\frac{1}{g} = \frac{1 - e^{-2\eta}}{2\eta}. \tag{15}$$

In Fig. 15.1, $1/g$ is shown as a dashed line, the right side of (15) as a solid curve, and the bound state energy is given by $-\hbar^2\eta_b^2/2\mu a^2$. In agreement with the discussion for general l, Fig. 15.1 shows that there is only an s-wave bound state if $g > 1$. (Recall that η must be positive.) As we would expect, the binding energy increases with g. Because the right side of (15) is monotonic, there is at most one bound state.

Needless to say, the same results can be obtained by integrating the differential equation by more elementary methods. Thus we know that the bound state solution $\rho^{-1}\chi_0(\eta; \rho)$ must be of the form $Aj_0(i\eta\rho)$ for $\rho < 1$ and $Bh_0(i\eta\rho)$ for $\rho > 1$. Continuity at $\rho = 1$ requires $A/B = h_0(i\eta)/j_0(i\eta)$. The discontinuity in the potential then implies that

$$\lim_{\epsilon\to 0}\left\{\frac{d\chi_0}{d\rho}\bigg|_{\rho=1+\epsilon} - \frac{d\chi_0}{d\rho}\bigg|_{\rho=1-\epsilon}\right\} = -g\chi_0(\eta; 1),$$

which immediately yields the condition (15). This discontinuity condition applies to all the states (i.e., scattering and bound states of every l), and one can verify that it is met by our solutions (2). One might worry that such discontinuous derivatives could lead to violations of the continuity equation. This is not the case, however; a direct calculation shows that the current constructed from any linear combination of our solutions (2) is a continuous function of r.

The actual form of the radial wave function can also be obtained by the circuitous route of evaluating the residue of (2) at the bound state pole. If this pole is at $\zeta = i\eta_b^{(l)}$, say, we have

$$R_l(\rho) = \begin{cases} Nj_l(i\eta_b^{(l)}\rho)h_l(i\eta_b^{(l)}) & (\rho < 1), \\ Nh_l(i\eta_b^{(l)}\rho)j_l(i\eta_b^{(l)}) & (\rho > 1), \end{cases}$$

where N is the normalization constant.

15.2 SCATTERING STATES

Having disposed of the bound states, we now examine the scattering solutions more carefully. Consider the radial wave function in the inside region. From (2) and (3) we have

$$A_l(\xi; \rho) = \frac{j_l(\xi\rho)}{D_l(g; \xi)} \qquad (\rho < 1). \tag{16}$$

Because $D_l(g; \xi)$ also occurs in $e^{i\delta_l}\sin\delta_l$ (see (7)), there cannot be any zero of $D_l(g; \xi)$ for real ξ. Hence by making $|g|$ large it appears as though we can make $A_l(\xi; \rho)$ as small as we please inside ($\rho < 1$). In this limit the potential acts like an impenetrable sphere. (Note that this is true for both signs of g, which is due to the very poor impedance match that must be made at $\rho = 1$.) The similarity to hard sphere scattering is also clear from (5), because as $|g| \to \infty$

$$\tan\delta_l(\xi) \to \frac{j_l(\xi)}{n_l(\xi)}. \tag{17}$$

This is precisely the phase shift that results from the requirement that the free *outside* wave function $\alpha_l j_l(\xi\rho) + \beta_l n_l(\xi\rho)$ vanish at $\rho = 1$.

Of course our potential (1) is not really an impenetrable sphere, because $U = 0$ for $\rho < 1$; it is just the surface $\rho = 1$ that becomes impermeable as $|g| \to \infty$. In other words, in this limit we would also expect solutions for $\rho < 1$ provided $j_l(\xi) \simeq 0$, i.e., solutions that almost vanish at $\rho = 1$. These are just the solutions *inside* an enclosure of radius a. In the $|g| \to \infty$ limit the potential (1) decouples the inside and outside regions completely, and prevents a particle that is inside from ever reaching the outside, and vice versa. Moreover, in this limit $D_l(g; \zeta) \to -i\zeta g j_l(\zeta) h_l(\zeta)$, and the zeros of D_l therefore tend to the roots of $j_l(\zeta) = 0$. These roots, which lie on the real axis, correspond precisely to the energy eigenvalues for motion inside an impermeable enclosure of radius a.

It is now natural to ask: What happens for finite, but large, g? ("Large" remains to be defined.) The zeros of $D_l(g; \zeta)$ cannot lie on the real axis any longer, because the wave functions extend over all space and are not normalizable. Because they cannot be in the upper half-plane, they must lie near the solutions of $j_l(\xi) = 0$, but in the lower half-plane. If ξ is not near such a zero, A_l will be very small for $\rho < 1$.

Let us then look for these other zeros of D_l (or poles of $e^{i\delta_l} \sin \delta_l$). We again confine ourselves to $l = 0$. It will prove convenient to write $D_0 = 0$ as $2i\zeta = g(e^{2i\zeta} - 1)$; the real and imaginary parts of this equation are then

$$\begin{aligned} 2\eta &= g(1 - e^{-2\eta} \cos 2\xi), \\ 2\xi &= g e^{-2\eta} \sin 2\xi, \end{aligned} \tag{18}$$

whence $\eta = \frac{1}{2}g - \xi \cot 2\xi$, or

$$e^{1-2\xi \cot 2\xi} = g e^{1-g} \frac{\sin 2\xi}{2\xi}. \tag{19}$$

In studying the solutions of (19) it should be observed that $ge^{1-g} \lll 1$ if $g \gg 1$. The location of the roots is again most conveniently studied by graphical means, and the relevant curves are sketched in Fig. 15.2. Because both sides of (19) are even functions, we only show $\xi > 0$. The solutions for an attractive ($g > 0$) case are also shown as dots in the figure. For $g \gg 1$ these roots approach $\xi = n\pi$, the zeros of $j_0(\xi)$, in agreement with the remark made previously.

We shall now solve for these roots $\zeta_n = \xi_n + i\eta_n$, under the assumption that $|\zeta_n| \simeq n\pi$, i.e., large $|g|$. Putting $\xi_n = n\pi + \nu$ in (19) gives

$$e^{-2n\pi[1+(\nu/n\pi)] \cot 2\nu} = g e^{-g} \frac{\sin 2\nu}{2n\pi[1 + (\nu/n\pi)]},$$

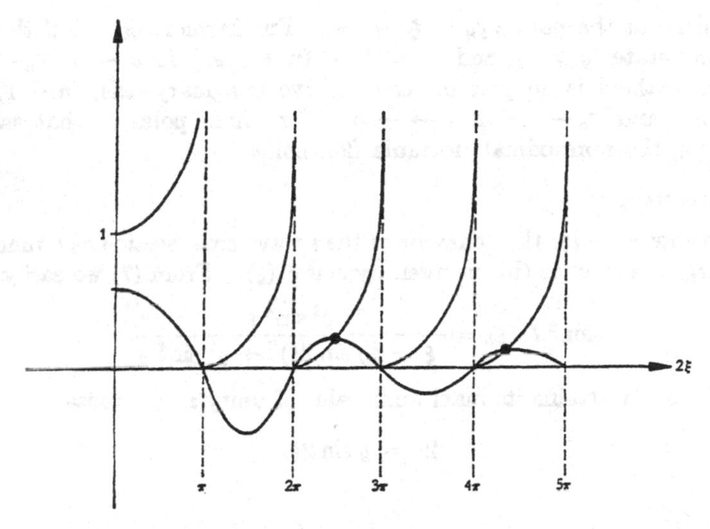

Fig. 15.2. The continuous curve shows the right-hand side of Eq. (19), whereas the curve with singularities is the left-hand side.

which for $\nu \to 0$ yields $\nu / n\pi = 1/g$. Therefore the real part of the root is at

$$\xi_n \simeq n\pi \left(1 + \frac{1}{g}\right). \tag{20}$$

We should note that these approximations require that the shift ν be small compared to π, i.e., $g \gg n$. Next we compute $\eta_n = \operatorname{Im} \zeta_n$. Presumably $|\eta_n| \ll 1$, and so we can expand (18) about $\eta = 0$ to obtain

$$\eta_n \simeq \frac{g(1 - \cos 2\xi_n)}{2(1 - g \cos 2\xi_n)} \simeq -\frac{1 - \cos 2\xi_n}{2 \cos 2\xi_n}, \tag{21}$$

since $|g \cos 2\xi_n| \gg 1$. Using (20) we find

$$\eta_n \simeq -\left(\frac{n\pi}{g}\right)^2. \tag{22}$$

Finally, therefore,

$$\zeta_n \simeq n\pi \left(1 + \frac{1}{g}\right) - i\left(\frac{n\pi}{g}\right)^2, \qquad n = \pm 1, \pm 2, \ldots \tag{23}$$

As promised, all these poles lie in the lower half plane for both signs of g.

To summarize: If $g > 0$ the s-wave scattering amplitude has one pole on the imaginary axis at $i\eta_b$, and an infinite number of poles in the lower half plane at the points $\zeta_n = \xi_n + i\eta_n$. Furthermore $\eta_0 > 0$ if there is a bound state ($g > 1$), and $n\pi < \xi_n < (n+1)\pi$. As $g \to \infty$, $\zeta_n \to n\pi$. If $g < 0$, there is no pole on the positive imaginary axis, $(n-1)\pi < \xi_n < n\pi$, and $\zeta_n \to n\pi$ as $g \to -\infty$. For those poles ζ_n that satisfy $|n| \ll |g|$, the approximate formula (23) holds.

15.3 RESONANCES

Let us now consider the behavior of the s-wave cross section as a function of energy; in essence this is given by $\sin^2 \delta_0(\xi)$. From (7) we easily find

$$\sin^2 \delta_0(\xi) = \frac{g^2 \sin^4 \xi}{(\xi - \frac{1}{2}g \sin 2\xi)^2 + g^2 \sin^4 \xi}. \tag{24}$$

This function attains its maximum value of unity at the roots of

$$2\xi = g \sin 2\xi. \tag{25}$$

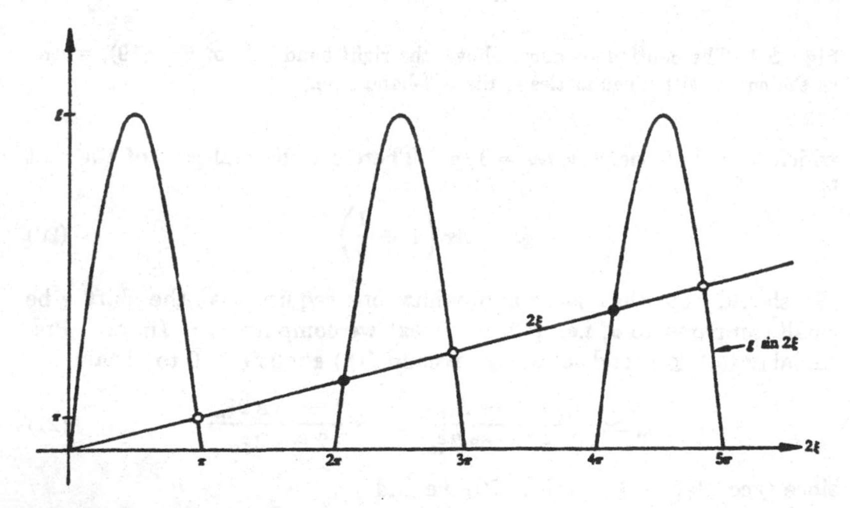

Fig. 15.3. The solutions of Eq. (25) when $g = 10\pi$. The roots ξ_b are shown as open circles, and the roots ξ_a as solid dots.

The solution of (25) is sketched in Fig. 15.3. When the cross section of the lth partial wave passes through the maximum value $4\pi(2l+1)\lambda\!\!\!^{-2}$, one says that that wave has a *resonance* at that energy. Thus the roots $\xi^{(n)}$ of (25) give us the s-wave resonances for our problem. We note that there are a finite number n_0 of roots of (25), with $n_0 \approx g/\pi$. Assume $g \gg \pi$, i.e., $n_0 \gg 1$, and consider the roots below $n\pi$, with $n \ll n_0$. It is clear from the sketch that $\xi^{(n)}$ will lie near the points $n\pi/2$. This is somewhat surprising, because the poles ζ_n of $e^{i\delta_0} \sin \delta_0$ lie near $n\pi$; therefore there are more resonances (at least for $\xi < g/2$) than there are poles. But if we look at (24) we immediately note that these resonances fall into two very different classes, $\xi_b^{(n)}$ and $\xi_s^{(n)}$: (a) $\xi_b^{(n)}$ near $\xi = \pi/2, 3\pi/2, 5\pi/2, \ldots$, which are broad because $\sin \xi^4 \approx 1$ at these points, and (b) $\xi_s^{(n)}$ near $\xi = \pi, 2\pi, 3\pi, \ldots$, which are very sharp because $\sin^4 \xi \approx 0$ there.* To verify these statements we expand about $\xi^{(n)}$. From the definition of $\xi^{(n)}$

$$\xi - \tfrac{1}{2}g \sin 2\xi \simeq -(\xi - \xi^{(n)})g \cos 2\xi^{(n)};$$

in the immediate vicinity of $\xi^{(n)}$ Eq. (24) therefore assumes the form

$$\sin^2 \delta_0(\xi) \approx \frac{\sin^4 \xi^{(n)}}{(\xi - \xi^{(n)})^2 + \sin^4 \xi^{(n)}}, \tag{26}$$

because $|\cos 2\xi^{(n)}| \simeq 1$. Thus when $\xi^{(n)} \simeq \pi/2, 3\pi/2, \ldots$, the resonances are indeed very broad. In fact, for all $\xi \ll \frac{1}{2}g$ and not near the sharp resonances $\xi_s^{(n)}$, the argument pertaining to (17) shows that the cross section is approximately the same as for a hard sphere, i.e.,

$$\sin^2 \delta_0 \simeq (\sin^2 \delta_0)_{\text{hard sphere}} = \sin^2 \xi. \tag{27}$$

The location of the sharp resonances follows immediately from (25); that is,

$$\xi_s^{(n)} \simeq \pi n \left(1 + \frac{1}{g}\right) = \operatorname{Re} \zeta_n$$

if $g \gg n$. Furthermore $\sin^2 \xi_s^{(n)} \simeq (n\pi/g)^2$. Using our earlier notation (23) we can therefore write Eq. (26) as †

$$\sin^2 \delta_0(\xi) \simeq \frac{|\operatorname{Im} \zeta_n|^2}{(\xi - \operatorname{Re} \zeta_n)^2 + |\operatorname{Im} \zeta_n|^2} \tag{28}$$

in the immediate vicinity of the sharp resonances. We have now established the connection (which is actually much more general than our specific example) between the poles ζ_n of the partial wave amplitude just below the real axis of the ζ-plane and the sharp resonances in the partial wave cross section: $\operatorname{Re} \zeta_n$ gives the location of the resonances and $\operatorname{Im} \zeta_n$ their width. This is actually somewhat oversimplified, as we already know, because there are only a finite number of resonances and an infinite number of poles. We note that as n increases the resonances become

* The distinction between the broad and sharp resonances is also discussed in Sec. 49.4.

† In nuclear physics this would be called a Breit-Wigner formula.

broader, and it follows from the preceding drawing that the whole discussion culminating in (28) and the subsequent statement are only valid if the width $|\text{Im}\,\zeta_n|$ is small compared to the distance between neighboring resonances, i.e.,

$$|\text{Im}\,\zeta_n| \ll \text{Re}\,\zeta_{n+1} - \text{Re}\,\zeta_n. \tag{29}$$

Once (29) is violated, the resonances soon disappear. When the energy reaches values such that $\xi \geq \frac{1}{2}g$ there are no resonances whatsoever. At these higher energies the potential barrier is too permeable to support quasi-stationary states in the interior region, and by the same token, it no longer acts like a hard sphere for energies between the roots of $j_0(\xi) = 0$.

We can now complete our study of the interior wave function. In the discussion following (16) we concluded that $|A_0(\xi;\rho)| \ll |j_0(\xi\rho)|$ if ξ is *not* near ξ_n, and $g \gg \pi$, but we did not know what the interior amplitude is near ξ_n, because we still did not know the value of ξ_n. Now we do, and so we can easily compute that

$$D_0(g;\xi_n) \simeq -\frac{in\pi}{g}.$$

Hence $|A_0(\xi_n;\rho)| \gg |j_0(\xi_n\rho)|$. If we recall that our incident wave is precisely $j_0(\xi\rho)$, we see that the interior amplitude is much *smaller* than the incident amplitude away from the sharp resonances (hard-sphere scattering), but far *larger* at resonance. In view of this let us, for the moment, change normalization to a new function $\bar{A}_0(\xi;\rho)$ normalized to unity inside the sphere of radius a. Then $\lim_{g\to\infty} \bar{A}_0(\xi_n;\rho)$ is the interior eigenfunction of an enclosure of radius a.

In Fig. 15.4 $\sin^2\delta_0$ is plotted for $g \simeq 15$, as well as the hard-sphere phase shift, which is shown as a dashed line. We see that $|\sin\delta_0(\xi)|^2$ agrees very well with the hard-sphere result except in the vicinity of $\xi = \pi$, $2\pi, \ldots$. Although it is not apparent from the figure, there is also an important difference from the hard-sphere result as $\xi \to 0$. According to (24)

$$\sin^2\delta_0(\xi) \underset{\xi\to 0}{\longrightarrow} \xi^2\left(\frac{g}{1-g}\right)^2;$$

in this limit only s-waves scatter, and so the total cross section approaches the value

$$\sigma_t(\xi) \underset{\xi\to 0}{\longrightarrow} \frac{4\pi}{k^2}\sin^2\delta_0 = 4\pi\left(\frac{ag}{1-g}\right)^2. \tag{30}$$

It is customary to write the zero energy cross section as

$$\sigma_t(0) = 4\pi a_s^2, \tag{31}$$

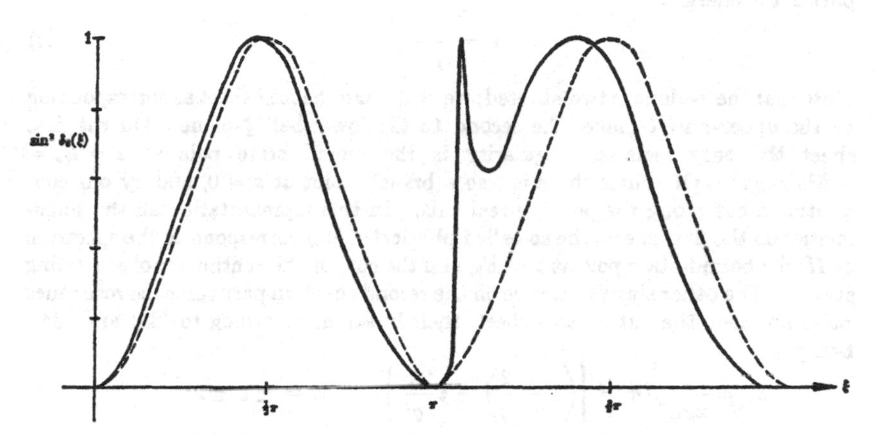

Fig. 15.4. The s-wave cross section (divided by $\lambda\!\!\!^{-2}$) when $g = 15$.

where a_s is called the *scattering length*. In our example

$$a_s = a\frac{g}{g-1}. \tag{32}$$

Note that a_s is not always a length characterizing the size of the system. In particular, a_s can be very much larger than a if $g \simeq 1$, i.e., if we either have a weakly bound state, or not quite enough strength to bind a state. This is a general result. The most famous example is neutron-proton scattering in the singlet spin state, where the scattering length is -24×10^{-13} cm, whereas the range of nuclear forces is about 1.4×10^{-13} cm. There is no bound state of the system in this partial wave (the deuteron is a spin triplet), but the force is almost strong enough to bind a state.

Let us briefly examine the low energy limit of the higher partial waves. From (5) or (7) we have

$$\delta_l \xrightarrow[\xi\to 0]{} \frac{\xi^{2l+1}}{[(2l+1)!!]^2}\, g\left(1 - \frac{g}{2l+1}\right)^{-1} \tag{33}$$

This agrees with our earlier estimate (14.17). On the other hand, if g is near the value for binding a state of angular momentum l the cross section is anomalously large, and the usual ξ^{2l+1} estimate is incorrect.

15.4 ANALYTIC PROPERTIES IN THE COMPLEX ENERGY PLANE

•Our discussion of the analytic properties of the s-wave amplitude has taken place in the complex k-plane (or ζ-plane). Frequently one studies these analytic properties in the energy plane. Let us define a complex variable z whose real part is the energy:

$$z = \frac{\hbar^2}{2\mu a^2}\zeta^2. \tag{34}$$

Note that the z-plane is two-sheeted; we shall take the first sheet as corresponding to the upper-half ζ-plane, the second to the lower-half ζ-plane. On the first sheet the only isolated singularity is the bound state pole at $z = E_b = -\hbar^2\eta_b^2/2\mu a^2$. Of course there is also a branch point at $z = 0$, and by our convention a cut along the positive real axis. In this representation all the singularities on the first sheet (the so-called physical sheet) correspond to the spectrum of H: the bound-state pole at $z = E_b$, and the cut for the continuum of scattering states. The other singularities lie on the second sheet; in particular the resonance poles are near the cut on this sheet, their location, according to (23) and (34), being

$$z_n = \frac{\hbar^2}{2\mu a^2}(n\pi)^2\left\{\left(1 + \frac{2}{g}\right) - i\frac{2n\pi}{g^2}\right\}, \qquad n = \pm 1, \pm 2, \ldots, \tag{35}$$

when $g \gg n\pi$ (see Fig. 15.5).

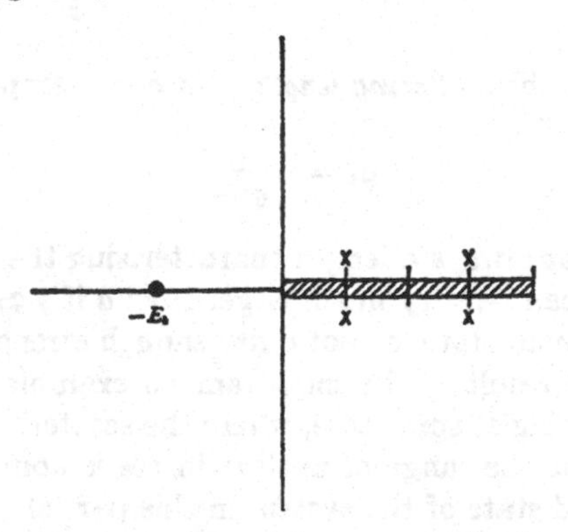

Fig. 15.5. A partial wave amplitude in the z-plane showing one bound-state. The resonances are on the second Riemann sheet and are marked as crosses.

It is also of some interest to compute the discontinuity of the partial wave amplitude,

$$f_l(\xi) = \frac{1}{\xi}e^{i\delta_l(\xi)}\sin\delta_l(\xi),$$

across the cut. By our definitions of the variable z, the boundary value above the cut is $f_l(\xi)$, and below the cut $f_l(-\xi)$, with $\xi > 0$. According to (7)

$$f_l(-\xi) = \frac{g[j_l(\xi)]^2}{1 + i\xi g j_l(\xi) h_l^*(\xi)},$$

where we used $j_l(\xi) = (-1)^l j_l(-\xi)$, and $h_l(-\xi) = (-1)^l h_l^*(\xi)$. Hence

$$f_l(-\xi) = f_l^*(\xi). \tag{36}$$

Therefore the discontinuity is

$$f_l(\xi) - f_l(-\xi) = 2i \operatorname{Im} f_l(\xi) = \frac{2i}{\xi} \sin^2 \delta_l(\xi), \tag{37}$$

i.e., it is proportional to the partial cross section. The resonances therefore show up as sharp maxima in the discontinuity, and thereby reflect the proximity of the resonance poles.

As a curiosity we may note that as g is decreased to unity the bound-state pole proceeds towards the branch point. If g is lowered still further it ducks through the cut and runs back out the negative real axis on the "unphysical" sheet.

The simple analytic properties that we have found here are partly due to the vanishing of the potential when $r > a$. Potentials that fall off exponentially as $r \rightarrow \infty$ lead to somewhat more complicated analytic properties *; for example, the partial wave amplitudes have cuts along the negative energy axis in addition to the singularities shown in Fig. 15.5. Nevertheless, it is generally true that the spectrum of the Hamiltonian makes its appearance as singularities of the scattering amplitudes. This remark applies not only to potential scattering,† but also to inelastic scattering and reactions.*

16. Resonance Scattering and Exponential Decay

We shall now study some time-dependent properties of resonance scattering from the delta-shell potential. When we constructed wave packets for the time-dependent description of scattering, we always assumed the spread in k-space, Δk, small enough so that the scattering amplitude could be taken as slowly varying in the interval Δk. If we are near a resonance ‡ the previous discussion will, therefore, still apply provided the spread of the packet is small compared to the resonance width, i.e., $a\, \Delta k \ll |\eta_n|$. Frequently one cannot satisfy this criterion and then the argument must be amended. A very important situation occurs when the spread of the incident packet is large compared to the width of the resonance. In terms of the characteristics of the delta-shell potential,

* Cf., e.g., R. Blankenbecler, M. L. Goldberger, N. Khuri and S. Treiman, *Ann. Phys.* 10, 62 (1960).

† A somewhat more systematic treatment of analytic properties in potential scattering will be found in Sec. 49.

‡ We only consider the sharp resonances, whose widths are small compared to the distance between resonances. The energies where the phase shift goes slowly through $\frac{1}{2}\pi$ are usually not called resonances.

this happens when

$$a\,\Delta k \gg |\eta_n|. \tag{1}$$

When this inequality is satisfied, time-dependent phenomena of considerable interest arise. For reasons which will soon be clear, we shall, in addition to (1), assume that the spread Δk is small compared to the distance between resonances:

$$a\,\Delta k \ll \pi. \tag{2}$$

Equations (1) and (2) are only possible simultaneously if the coupling is very strong, $(n\pi/g)^2 \ll \pi$. Under these circumstances there are many sharp resonances. Our discussion shall only deal with the strong coupling regime. For simplicity we assume that the incident wave packet is centered on the nth resonance. The mean wave number is therefore

$$k_n = \xi_n/a, \tag{3}$$

and the group velocity is $\hbar k_n/\mu \equiv v_n$.

Because the roots for different values of l of $j_l(\xi) = 0$ do not coincide, we can assume that there is no $l \neq 0$ resonance in the interval $|k - k_n| \lesssim \Delta k$, and hence the $l \neq 0$ terms in the scattering amplitude all vary slowly in Δk.

As we saw in the preceding section, the s-wave amplitude has a simple pole due to the bound state, and an infinite number of simple poles in the lower half of the k-plane. Examination of (15.7) also reveals that there is an essential singularity at infinity. Hence we may represent the scattering amplitude by

$$e^{i\delta_0(\zeta)}\sin\delta_0(\zeta) = E(\zeta) + \frac{R_b}{\zeta - i\eta_b} + \sum_n \frac{R_n}{\zeta - \zeta_n}, \tag{4}$$

where R_b is the residue at the bound state pole, R_n the residue at the pole in the lower half-plane which tends to $n\pi$ as $|g| \to \infty$, and $E(\zeta)$ is an entire function. In the vicinity of ζ_n only the nth term of the sum in (4) will be singular, and so we can write

$$e^{i\delta_0(\xi)}\sin\delta_0(\xi) = \frac{R_n}{\xi - \zeta_n} + \mathcal{R}_0(\xi) \tag{5}$$

when ζ is near ξ_n, where $\mathcal{R}_0(\xi)$ is a regular function in this region. $\mathcal{R}_0$ is slowly varying in Δk also, of course. Therefore we can write the complete scattering amplitude as the sum of two terms, one containing the higher partial waves and the $\mathcal{R}_0$ part of the s-wave amplitude, and the other the resonant s-wave term. The first part will lead to an essentially undistorted scattered wave packet of precisely the type (12.32) which we call Ψ_{nr}, where "nr" stands for nonresonant. The remainder of the scattered wave will be quite different, however.

Let us compute this resonant part of the scattered packet, which we call Ψ_{res}. According to (12.31)

$$\Psi_{res}(\mathbf{r}, t) \sim \frac{1}{(2\pi)^{\frac{3}{2}} r} \int \frac{\chi(\mathbf{q})}{q} \frac{R_n/a}{q - k_n + i\kappa_n} e^{i(qr - \omega_q t)} \, d^3q, \tag{6}$$

where

$$a\kappa_n = -\eta_n, \tag{7}$$

and

$$R_n = \frac{g\zeta_n[j_0(\zeta_n)]^2}{\partial D_0/\partial \zeta|_{\zeta=\zeta_n}}. \tag{8}$$

Using (15.23) and (15.14) we easily find

$$R_n \simeq -\kappa_n a. \tag{9}$$

Taking advantage of the narrow spread Δk about k_n of $\chi(\mathbf{q})$, and carrying out the angular integration over $\hat{\mathbf{q}}$, we find

$$\Psi_{res}(\mathbf{r}, t) \sim -e^{i\omega_n t}\left(\frac{1}{2\pi}\right)^{\frac{3}{2}} \frac{1}{r} \int_{-\infty}^{\infty} \bar{\chi}(q) \frac{\kappa_n}{q - k_n + i\kappa_n} e^{iq(r - v_n t)} \, q \, dq, \tag{10}$$

where

$$\bar{\chi}(q) = \int \chi(\mathbf{q}) \, d\hat{\mathbf{q}}.$$

In (10) we have taken advantage of the narrow spread in $\bar{\chi}(q)$ to extend the lower limit of integration to $-\infty$.

By approximating the time-dependent phase in (10) we have discarded those terms that prevent the construction of sharp wave fronts (see end of Sec. 4.4). This approximation is permissible, because the wave front passes a point in a time small compared to the time $t_p = 1/v_n \Delta k$ required for the free packet to pass a point. Recalling the discussion pertaining to (4.43), we can construct a $\bar{\chi}(q)$ that leads to a sharp front and a spread Δk by writing

$$\bar{\chi}(q) = \frac{A}{q - k_n + i\,\Delta k}, \tag{11}$$

where $\Delta k > 0$, and according to (1), $\Delta k \gg \kappa_n$. Two remarks concerning (11) are necessary: (a) more general packets with sharp fronts are easily built by writing a number of poles all in the lower half plane with imaginary parts exceeding $|\kappa_n|$; (b) $\bar{\chi}$ corresponds to an incident beam that strikes the target at $t \simeq 0$, not at $-z_0/v_n$ as in Sec. 12. The main virtue of (11) is that it allows us to evaluate (10) by contour integration:

$$\Psi_{res}(\mathbf{r}, t) = 0 \qquad (r > v_n t) \tag{12}$$

and if $r < v_n t$

$$\Psi_{res}(\mathbf{r}, t) \sim -\left(\frac{1}{2\pi}\right)^{\frac{1}{2}} \frac{2\pi i}{r} \left\{ \bar{\chi}(k_n - i\kappa_n)\kappa_n(k_n - i\kappa_n)e^{i(k_n - i\kappa_n)(r - v_n t)} + A\,\frac{\kappa_n(k_n - i\,\Delta k)}{i(\kappa_n - \Delta k)}\, e^{i(k_n - i\Delta k)(r - v_n t)} \right\} \tag{13}$$

Suppose an observer is to be stationed at the point $\mathbf{r}$. For early times ($t < t_r$, where $t_r = r/v_n$) he won't see anything. At $t = t_r$ the probability density will suddenly jump to a finite value due to the arrival of the two packets Ψ_{res} and Ψ_{nr} arising from the resonant and nonresonant parts of the scattering amplitude. Ψ_{nr} passes the observer in a time t_p of order $1/v_n\,\Delta k$. Consider next the probability density due to Ψ_{res} for $t > t_r$—we shall deal with the interference term $|\Psi_{nr}\Psi_{res}^*|$ in a moment. According to (13)

$$|\Psi_{res}(\mathbf{r}, t)|^2 \sim \frac{\kappa_n^2 k_n^2}{2\pi r^2} \left| \bar{\chi}(k_n)e^{-\kappa_n v_n(t - t_r)} + \frac{iA}{\Delta k}\, e^{-(t - t_r)/t_p} \right|^2, \tag{14}$$

where we took advantage of $k_n \gg \Delta k$, $k_n \gg \kappa_n$, $\Delta k \gg \kappa_n$. The second term in (14) becomes negligible when $t - t_r$ is large compared to the passage time t_p, and so for $t - t_r \gg t_p$, only the first term survives. The interference term mentioned above is also negligible unless $t - t_r \lesssim t_p$, while the diffuseness of the wave front that we have ignored all along is only visible when $|t - t_r| \ll t_p$. At sufficiently late times (i.e., $t - t_r \gg t_p$) the observer therefore sees a spherically symmetric * probability density that decays exponentially,

$$|\Psi_{res}(\mathbf{r}, t)|^2 \xrightarrow[t - t_r \gg t_p]{} \frac{\text{const.}}{r^2}\, e^{-(t - t_r)/\tau_n}, \tag{15}$$

where

$$\tau_n = \frac{1}{2\kappa_n v_n} \tag{16}$$

is the so-called *lifetime* of the nth resonance. It is simply related to the

* If the incident energy is near a resonance of angular momentum l, the angular distribution becomes $|Y_{l0}(\theta)|^2$ for late times.

width in energy Γ_n of the resonance. If we call

$$\Gamma_n = \frac{\hbar^2}{2u}[(k_n + \kappa_n)^2 - (k_n - \kappa_n)^2] \simeq 2\hbar\kappa_n v_n,$$

we obtain

$$\tau_n \Gamma_n \simeq \hbar, \tag{17}$$

which is reminiscent of the energy-time uncertainty relationship (1.6). The present example actually affords a striking illustration of the uncertainty principle. We have seen that if we prepare a packet of ill-defined energy ($\Delta E \gg \Gamma_n$) we can (eventually) measure a time τ_n that uniquely characterizes the system. By "unique" we mean that τ_n does not depend on the detailed form of the initial state: In the asymptotic form (15) the properties of the initial packet are completely buried in the multiplicative constant. On the other hand, an incident packet with a well-defined energy ($\Delta E \ll \Gamma_n$) does not suffer an appreciable change of shape upon scattering, and the only time revealed by it is t_p, which does not characterize the system in any intrinsic fashion, but rather depends on the form of the particular packet.

The example of resonant scattering discussed here can serve as a very schematic model of radioactive states. In fact, the combination of nuclear and Coulomb interaction experienced by an α-particle bears some resemblance to the delta-shell potential. Consider the bombardment of the stable nucleus Pb^{206} by α-particles of energy 5.4 Mev. We can try to get a very monochromatic α-beam, with a spread of say 1 kev, in which case the passage time t_p is of order 10^{-19} sec. When the cyclotron is shut off, the prompt packets will, therefore, pass an observer almost instantly. The observer will, however, continue to see an exponentially decaying number of α-particles, with a time constant of 138 days! Clearly there is an incredibly sharp resonance in the cross section for $\alpha + Pb^{206}$ scattering whose width is of order 10^{-18} ev. When a resonance is so ridiculously narrow, we can never observe it directly, and we usually call it a particle—in our example, the α-unstable nucleus Po^{210}. This is a reasonable description because for $t \lesssim 138$ days there will be some nuclei in the target that have all the chemical and physical properties associated with $Z = 84$, $A = 210$. It is therefore natural to say that Pb^{206} has "captured" the α-particle, i.e., that the latter is "inside" the nucleus. The probability of finding the α "inside" is not stationary however, and decreases with a characteristic decay time of 138 days. For times short compared to this (i.e., all times of importance in most of nuclear physics) Po^{210} is to all intents and purposes a stationary state.

17. The Coulomb Field

The Kepler problem plays a central role in atomic physics, and phenomena involving scattering by a Coulomb field have also been of great importance in nuclear physics. We shall therefore study the solutions of the wave equation for motion in a Coulomb field in some detail. We begin our discussion by constructing the continuum solutions in closed form, and then we will extract the bound-state spectrum and eigenfunctions from the partial waves. Finally we shall take up the situation most relevant to nuclear physics, where scattering is due to a combination of Coulomb and short range potentials.

17.1 THREE-DIMENSIONAL CONTINUUM SOLUTIONS

The Schrödinger equation in question is

$$\left(\nabla^2 + k^2 + \frac{2\gamma k}{r}\right)\psi(\mathbf{r}) = 0, \tag{1}$$

where $\gamma = Z_1 Z_2 e^2/\hbar v$, v being the incident velocity, and $\hbar k = \mu v$. The dimensionless parameter γ can also be expressed as $(Z_1 Z_2 c/v)/137.04$, where c is the speed of light. The field is *attractive* when $\gamma > 0$. Loosely speaking the interaction is weak when $|\gamma| \ll 1$.

The solution of (1) is greatly facilitated by the following considerations due to Gordon, who first gave the complete solution of the Kepler problem in 1928. At large distances (i.e., compared to $1/k$) the Coulomb field varies slowly compared to the wavelength and the semiclassical method applies—we already took advantage of this in Sec. 14.3. Gordon showed that the surfaces orthogonal to a family of hyperbolic orbits with parallel asymptotes and fixed energy are

$$kz + \gamma \ln kr(1 - \cos\theta) = \text{const.}$$

The asymptotic wave fronts are therefore given by this family of surfaces. This suggests the introduction of a new unknown function χ by means of

$$\psi_{\mathbf{k}}(\mathbf{r}) = e^{i\mathbf{k}\cdot\mathbf{r}}\chi(u), \tag{2}$$

where the new variable is

$$u = ikr(1 - \cos\theta) = i(kr - \mathbf{k}\cdot\mathbf{r}) = 2ikr\sin^2\tfrac{1}{2}\theta. \tag{3}$$

Substituting into (1) we have

$$\left\{u\frac{d^2}{du^2} + (1 - u)\frac{d}{du} - i\gamma\right\}\chi(u) = 0. \tag{4}$$

We shall be concerned with the properties of χ as a function of the complex variable u. These properties are determined by the singularities of the functions that multiply the various terms in the differential equation.* If we rewrite (4) as

$$\chi'' + \frac{1-u}{u}\chi' - \frac{i\gamma}{u}\chi = 0,$$

we see that the coefficients of χ' and χ both have simple poles at $u = 0$, and are analytic everywhere else in the finite u-plane. It then follows that χ can have a pole or branch point at $u = 0$, but nowhere else in the finite u-plane. According to (3), we can only accept a solution that is regular at $u = 0$, and so our solution will be a polynomial in u or an entire function of u. The behavior at infinity can be deduced by the substitution $u = 1/w$, which yields

$$\left\{\frac{d^2}{dw^2} + \frac{1}{w}\left(1 + \frac{1}{w}\right)\frac{d}{dw} - \frac{i\gamma}{\omega^3}\right\}\chi = 0.$$

Now the coefficient of $d\chi/dw$ and χ have higher-order poles at $w = 0$, which implies that χ has an essential singularity at $u = \infty$, and it is therefore an entire function.

We can reduce (4) to a first order equation by expressing χ as a Laplace transform,

$$\chi(u) = \int_{t_1}^{t_2} e^{ut} f(t)\, dt, \tag{5}$$

where both $f(t)$ and the integration path in the t-plane remain to be specified. Applying (4) gives

$$\begin{aligned}
&\int_{t_1}^{t_2} [ut^2 + (1-u)t - i\gamma]e^{ut} f(t)\, dt \\
&\quad = \int_{t_1}^{t_2} f(t)\left[(t - i\gamma) + t(t-1)\frac{d}{dt}\right] e^{ut}\, dt \\
&\quad = t(t-1)f(t)e^{ut}\Big|_{t_1}^{t_2} + \int_{t_1}^{t_2} e^{ut}\left[(t-i\gamma)f(t) - \frac{d}{dt}t(t-1)f(t)\right] dt
\end{aligned}$$

after an integration by parts. Hence if $f(t)$ satisfies

$$(t - i\gamma)f(t) - \frac{d}{dt}[t(t-1)f(t)] = 0, \tag{6}$$

* An account of the theory of ordinary differential equations that is well adapted to our present purpose will be found in Morse and Feshbach, Chap. 5.

and t_1 and t_2 are chosen so as to make the integrated-by-parts term vanish, a solution of the original problem has been achieved. The first-order equation (6) has the solution

$$f(t) = At^{i\gamma-1}(1-t)^{-i\gamma}.$$

Thus

$$\chi(u) = A\int_C e^{ut}t^{i\gamma-1}(1-t)^{-i\gamma}\,dt, \tag{7}$$

provided C is closed, or if not, provided

$$t^{i\gamma}(1-t)^{1-i\gamma}e^{ut}$$

takes on the same value at both ends of C. Note that the integrand in (7) has branch points at $t = 1$ and $t = 0$. The cut can be taken to join these points; equivalently, one can draw two cuts extending to the point at infinity.

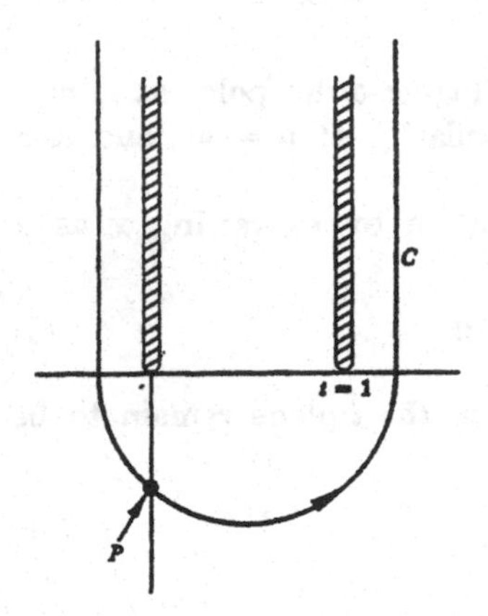

Fig. 17.1. Integration path in the t-plane.

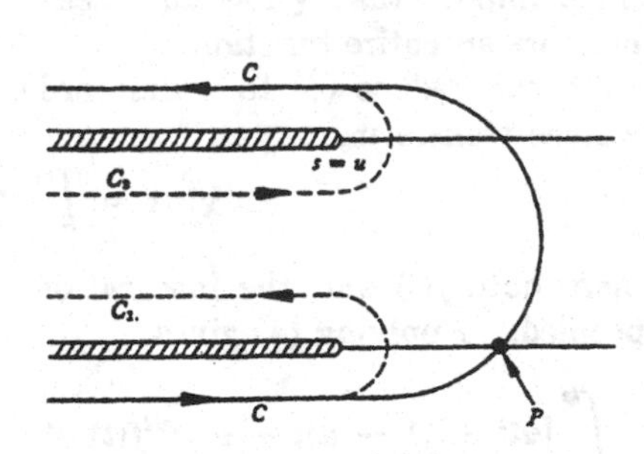

Fig. 17.2. Integration paths in the s-plane.

There are a number of different possibilities for choosing C. Any C that is open, and on which $e^{ut} \to 0$ at the end points, will do. Because the physical value of u is on the positive imaginary axis (see (3)), one could take C to start at $i\infty$, enclose either of the two branch lines, and then bring it back to $i\infty$. These two choices actually yield linearly independent solutions of (4) known as Whittaker functions. It will turn out, however, that the choice that satisfies the scattering boundary conditions encloses both cuts, as shown in Fig. 17.1.

The asymptotic expansion, which we shall take up in a moment, is most easily carried out after the substitution $s = ut$ is made. This changes (7) to

$$\chi(u) = A \int_C e^s s^{i\gamma-1}(u - s)^{-i\gamma} \, ds. \tag{8}$$

We note that this function is regular at $u = 0$, as it should be:

$$\chi(0) = (-1)^{-i\gamma} A \int_C e^s s^{-1} \, ds = (-1)^{-i\gamma} 2\pi i A. \tag{9}$$

We are now ready to evaluate the behavior of χ for large u. This will clearly be essential in determining the scattering amplitude. It is important to note that large u is not guaranteed by large values of r; in the forward direction u is always zero. From Fig. 17.2 we see that as $u \to \infty$, s/u is small on the lower parts of C, and for that part of the integration we could expand in inverse powers of u. This is not so for the upper leg of the circuit, however. We therefore deform C into C_1 and C_2 as shown, and in the C_2 integral make the change of variables $s \to s + u$. C_2 is then translated into C_1, and we therefore obtain

$$\begin{aligned}
\chi(u) &= A \int_{C_1} \{e^s s^{i\gamma-1}(u - s)^{-i\gamma} + e^{s+u}(s + u)^{i\gamma-1}(-s)^{-i\gamma}\} \, ds \\
&= A u^{-i\gamma} \int_{C_1} e^s s^{i\gamma-1}\left(1 - \frac{s}{u}\right)^{-i\gamma} ds \\
&\qquad - A(u)^{i\gamma-1} e^u \int_{C_1} e^s s^{-i\gamma}\left(1 + \frac{s}{u}\right)^{i\gamma-1} ds.
\end{aligned} \tag{10}$$

Finally we expand the integrands in powers of (s/u), and evaluate the coefficients with the help of Hankel's formula,

$$\int_{C_1} e^s s^{-z} \, ds = \frac{2\pi i}{\Gamma(z)}, \tag{11}$$

and the functional equation

$$\Gamma(z + 1) = z\Gamma(z). \tag{11'}$$

In detail, (10) becomes

$$\begin{aligned}
\chi(u) \xrightarrow[u\to\infty]{} &\frac{2\pi i A}{\Gamma(1 - i\gamma)} u^{-i\gamma} \left\{1 - \frac{\gamma^2}{u} + O(u^{-2})\right\} \\
&- \frac{2\pi i A}{\Gamma(i\gamma)} e^u (-u)^{i\gamma-1} \left\{1 + \frac{(i\gamma - 1)^2}{u} + O(u^{-2})\right\}.
\end{aligned} \tag{12}$$

This representation of the function has the somewhat distressing property of being multivalued, which appears to contradict the earlier assertion that χ is entire. The point is that (12) is an asymptotic expansion in the vicinity of an essential singularity, and does not converge to χ in the same way as a power series. The difference between this asymptotic representation and the function is negligible only if $|u| \to \infty$, and $0 \leqslant \arg u \leqslant \pi$. In our case $\arg u = \frac{1}{2}\pi$ if θ is real, and therefore (12) is the appropriate expansion.* In order to evaluate (12) we write $u = e^{\ln|u|}e^{i\omega}$, and therefore $u^{-i\gamma} = e^{\omega\gamma}e^{-i\gamma\ln|u|}$, etc. Since $|u| = kr - \mathbf{k}\cdot\mathbf{r}$, $\omega = \frac{1}{2}\pi$, (12) leads to

$$\psi_{\mathbf{k}}(\mathbf{r}) \xrightarrow[u\to\infty]{} \frac{2\pi i A e^{\frac{1}{2}\pi\gamma}}{\Gamma(1-i\gamma)} e^{i[\mathbf{k}\cdot\mathbf{r} - \gamma\ln(kr-\mathbf{k}\cdot\mathbf{r})]} + \frac{2\pi A e^{\frac{1}{2}\pi\gamma}}{\Gamma(i\gamma)} \frac{e^{2i\gamma\ln\sin\frac{1}{2}\theta}}{2k\sin^2\frac{1}{2}\theta} \frac{e^{i(kr+\gamma\ln 2kr)}}{r}. \tag{13}$$

That the asymptotic Coulomb wave functions contain a logarithmic term in the phase was already proven in Eq. (14.26).

Let us compute the probability current at large u. Here we ignore the interference term as usual, the justification being the same as in our earlier discussion. From the first term of (13) we obtain a current proportional to $\mathbf{k}$, and independent of $\mathbf{r}$, as well as a term proportional to $1/u$ coming from the gradient of the logarithmic phase. As $u \to \infty$ the first term therefore provides a current incident in the direction of $\mathbf{k}$. The second term in (13) is dominated by a current proportional to $\mathbf{r}/r^3$, and corrections of order $1/r^3$ arising from the logarithmic phases. This term therefore represents the outgoing scattered wave (note the factor e^{ikr}/r). In view of these remarks we choose

$$A = \frac{e^{-\frac{1}{2}\pi\gamma}\Gamma(1-i\gamma)}{2\pi i\,(2\pi)^{\frac{3}{2}}}. \tag{14}$$

Then †

$$\psi_{\mathbf{k}}(\mathbf{r}) \xrightarrow[u\to\infty]{} (2\pi)^{-\frac{3}{2}} \left\{ e^{i[\mathbf{k}\cdot\mathbf{r} - \gamma\ln(kr-\mathbf{k}\cdot\mathbf{r})]} + f_C(k,\theta)\frac{e^{i(kr+\gamma\ln 2kr)}}{r} \right\}, \tag{15}$$

* The multivalued nature of asymptotic expansions is treated in Morse and Feshbach, p. 609.

† This formula is not valid at $\theta = 0$. A related remark is that Im f_C diverges as $\theta \to 0$, and the total cross section therefore fails to exist. These are essential differences between the Coulomb field and those that fall off more rapidly at infinity. On the other hand, just because of this long range character, the Coulomb field is screened by compensating charges under almost all circumstances. Let R be the radius outside of which the target appears as a neutral object. Then (16) is the correct scattering amplitude for all momentum transfers $2\hbar k \sin\frac{1}{2}\theta$ large compared to $\hbar/R$.

where $f_C(k, \theta)$ is called the Coulomb scattering amplitude. According to (13) it is given by

$$f_C(k, \theta) = \frac{\Gamma(1 - i\gamma)}{i\Gamma(i\gamma)} \frac{e^{2i\gamma \ln \sin \frac{1}{2}\theta}}{2k \sin^2 \frac{1}{2}\theta},$$

which can be rewritten as

$$f_C(k, \theta) = \gamma \frac{\Gamma(1 - i\gamma)}{\Gamma(1 + i\gamma)} \frac{e^{2i\gamma \ln \sin \frac{1}{2}\theta}}{2k \sin^2 \frac{1}{2}\theta} \tag{16}$$

when (11′) is used. The differential cross section is therefore

$$\sigma_C(k, \theta) = |f_C(k, \theta)|^2 = \frac{\gamma^2}{4k^2 \sin^4 \frac{1}{2}\theta} = \left(\frac{Z_1 Z_2 e^2}{4E \sin^2 \frac{1}{2}\theta}\right)^2, \tag{17}$$

which is precisely *the Rutherford formula.* We note that $\hbar$ does not appear in this result; on the other hand, this is not true of f_C.

17.2 PARTIAL WAVE DECOMPOSITION; BOUND STATES

Our next task is the partial wave decomposition of $\psi_{\mathbf{k}}(\mathbf{r})$. This decomposition is not of any use in Coulomb scattering *per se*, but it is essential if one wishes to determine the wave functions in the discrete spectrum by analytic continuation, and also for the study of scattering by a potential that differs from the Coulomb field only at small distances (e.g., proton-proton scattering). Let us first write ψ explicitly:

$$\psi_{\mathbf{k}}(\mathbf{r}) = \frac{e^{-\frac{1}{2}\pi\gamma}\Gamma(1 - i\gamma)}{2\pi i(2\pi)^{\frac{3}{2}}} \int_C e^{ikrt} e^{i\mathbf{k}\cdot\mathbf{r}(1-t)} t^{i\gamma - 1}(1 - t)^{-i\gamma}\, dt. \tag{18}$$

Then we expand $e^{i\mathbf{k}\cdot\mathbf{r}(1-t)}$ in terms of spherical harmonics:

$$\psi_{\mathbf{k}}(\mathbf{r}) = \sum_{l=0}^{\infty} \left(\frac{2l + 1}{2\pi^2}\right)^{\frac{1}{2}} i^l C_l(k; r)\, Y_{l0}(\theta), \tag{19}$$

where

$$C_l(k; r) = \frac{e^{-\frac{1}{2}\pi\gamma}\Gamma(1 - i\gamma)}{2\pi i} \int_C e^{ikrt} j_l(kr(1 - t)) t^{i\gamma - 1}(1 - t)^{-i\gamma}\, dt. \tag{20}$$

This function is a solution of the radial equation corresponding to (1):

$$\left\{\frac{1}{r^2}\frac{d}{dr} r^2 \frac{d}{dr} - \frac{l(l + 1)}{r^2} + k^2 + \frac{2\gamma k}{r}\right\} C_l(k; r) = 0. \tag{21}$$

In fact, the boundary conditions imposed on (18) imply that $C_l(k; r)$ is just the function $A_l(k; r)$ that we used in our earlier discussion of scattering (see Sec. 14). It therefore satisfies the integral equation

$$C_l(k, r) = j_l(kr) + 2i\gamma k^2 \int_0^\infty j_l(kr_<)h_l(kr_>)C_l(k; r')r'\,dr'. \qquad (22)$$

As in Secs. 14 and 15, we shall determine the bound-state wave function and energies by analytic continuation of $C_l(k; r)$ to the positive imaginary k-axis. This can only be done if we have a representation of the function that reveals its analytic properties in the variable k. Clearly (20) is useless in this respect, because γ is inversely proportional to k. To get a transparent representation we must carry out the t-integration. This can be done by expanding $e^{ikrt}j_l[kr(1 - t)]$ in a power series; as this is an entire function, the series converges everywhere except at infinity. The most convenient way of doing the expansion is to put $j_l(z) = e^{iz}z^l f_l(z)$ into Bessel's equation. Then f_l satisfies

$$zf_l'' + 2(iz + l + 1)f_l' + 2i(l + 1)f_l = 0.$$

Of course f_l is also an entire function, and we may solve for it as a power series. In this way we obtain the formula

$$j_l(z) = 2^l z^l e^{iz} \sum_{\nu=0}^{\infty} \frac{(l + \nu)!}{(2l + 1 + \nu)!} \frac{(-2iz)^\nu}{\nu!}, \qquad (23)$$

which is valid throughout the finite z-plane. When (23) is substituted into (20) the exponential conveniently drops out, and we are left with integrals of the type

$$\int_C t^{a-1}(1 - t)^{b-1}\,dt,$$

where $a + b =$ integer. We may therefore deform the branch cuts and contour C of Fig. 17.1 into those shown in Fig. 17.3. In evaluating the

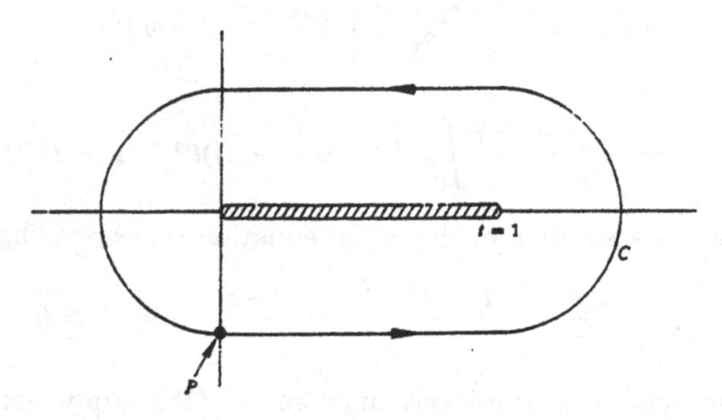

Fig. 17.3. Contour used in evaluation of Eq. 20.

integrals we must be very careful to choose the phase of the multivalued integrand correctly. If we refer back to our asymptotic expansion we note that in (11) we assumed the phase of s to be 0 at the point P in Fig. 17.2. In the t-plane this means that the phase of t is zero as shown in Figs. 17.1 and 17.3. Hence

$$\int_C t^{a-1}(1-t)^{b-1}\,dt = (1-e^{2\pi ib})\int_0^1 t^{a-1}(1-t)^{b-1}\,dt = (1-e^{2\pi ib})\frac{\Gamma(a)\Gamma(b)}{\Gamma(a+b)},$$

and therefore

$$C_l(k;r) = \frac{e^{-\frac{1}{2}\pi\gamma}\Gamma(1-i\gamma)}{2\pi i}\,2^l(kr)^l e^{ikr}\sum_{\nu=0}^{\infty}\frac{(l+\nu)!}{(2l+1+\nu)!}\frac{(-2ikr)^\nu}{\nu!} \times (1-e^{2\pi\gamma})\frac{\Gamma(i\gamma)\Gamma(l+\nu+1-i\gamma)}{\Gamma(l+\nu+1)}.$$

Using (11′) and

$$\Gamma(1-i\gamma)\Gamma(i\gamma) = \frac{\pi}{\sin i\pi\gamma}$$

we therefore have

$$C_l(k;r) = \frac{(2kr)^l e^{ikr}e^{\frac{1}{2}\pi\gamma}}{(2l+1)!}\Gamma(l+1-i\gamma)\;{}_1F_1(l+1-i\gamma|2l+2|-2ikr), \tag{24}$$

where

$${}_1F_1(l+1-i\gamma|2l+2|-2ikr) = 1+\frac{l+1-i\gamma}{2l+2}\frac{(-2ikr)}{1!} + \frac{(l+1-i\gamma)(l+2-i\gamma)}{(2l+2)(2l+3)}\frac{(-2ikr)^2}{2!}+\cdots \tag{25}$$

is a confluent hypergeometric function. As one readily verifies, this power series in kr converges in the whole finite kr-plane for all values of γ. Hence the only singularities of $C_l(k;r)$ are due to

$$e^{\frac{1}{2}\pi\gamma}\Gamma(l+1-i\gamma). \tag{26}$$

The function $e^{\frac{1}{2}\pi\gamma}$ has an essential singularity at $k=0$; we shall simply ignore this somewhat mysterious fact. The gamma function has simple poles whenever its argument is a negative integer or zero. Since $i\gamma$ is proportional to i/k, we have the expected result that only an attractive field ($\gamma > 0$) will bind states. Let us put

$$l + 1 - i\gamma = -n' \qquad (n' = 0, 1, 2, 3, \ldots). \tag{27}$$

The energies of the bound states are therefore given by

$$E_{ln'} = -\left(\frac{Z_1 Z_2 e^2}{\hbar c}\right)^2 \frac{\mu c^2}{2(l + n' + 1)^2}. \tag{28}$$

This is the celebrated *Bohr formula* for the energy levels of hydrogenic atoms. The principal quantum number is defined as $n = l + 1 + n' = 1, 2, \ldots$, and the energy eigenvalues depend only on it. As a consequence, levels with different angular momenta may have precisely the same energy. This remarkable degeneracy of the spectrum comes about because the Schrödinger equation (1) is invariant under a group that is larger than the three-dimensional rotation group.* As we shall see in Sec. 46, these unexpected degeneracies are removed by relativistic corrections.

The dimensionless quantity

$$\frac{e^2}{\hbar c} = \frac{1}{137.04} \equiv \alpha \tag{29}$$

plays an essential role in all electromagnetic phenomena, and is called *the fine structure constant*. As we shall see later on, the fact that $\alpha \ll 1$ implies that transverse electromagnetic interactions are weak enough to permit the application of the Born approximation method. We note that if α were not small, the binding energies (28) would be comparable to the rest mass of the electron, and the nonrelativistic theory we are using would not be applicable. Thus in the case of hydrogen the ground state energy is -13.60 ev, which is to be compared with the rest mass of the electron, 511 kev.

According to our previous discussion, the residue of (24) at the point (27) is the (un-normalized) radial wave function. For the case of hydrogen ($Z_1 = Z_2 = 1$) we must therefore make the change of variable $-2ikr \to 2re^2 m_e/\hbar^2 i\gamma \equiv 2r/a_0 n$, where

$$a_0 = \frac{\hbar^2}{m_e e^2}.$$

The length $a_0 = 0.529 \times 10^{-8}$ cm is *the Bohr radius of hydrogen;* it is the characteristic length of atomic structure. Except for a multiplicative

* See V. Fock, *Z. Phys.* **98**, 145 (1935); V. Bargmann, *ibid.* **99**, 576 (1936).

constant the bound state wave function is therefore

$$(2r/a_0n)^l e^{-r/a_0n}\, {}_1F_1(-n'|2l+2|2r/a_0n)$$
$$= \left(\frac{2r}{a_0n}\right)^l e^{-r/a_0n} \left\{1 + \frac{(-n')}{2l+2}\left(\frac{2r}{a_0n}\right) + \frac{(-n')(1-n')}{(2l+2)(2l+3)}\left(\frac{2r}{a_0n}\right)^2 \frac{1}{2!} + \frac{(-n')(-n'+1)(-n'+2)}{(2l+2)(2l+3)(2l+4)}\left(\frac{2r}{a_0n}\right)^3 \frac{1}{3!} + \cdots\right\}. \tag{30}$$

If the nuclear charge is Z instead of unity, one merely replaces a_0 by (a_0/Z) in (30). The function in braces is obviously a polynomial of degree n', the so-called Laguerre polynomial. The asymptotic form of the bound-state wave functions is therefore

$$\sim \left(\frac{2r}{a_0n}\right)^{n-1} e^{-r/a_0n}, \tag{31}$$

which decreases much more slowly than the e^{-r}/r fall-off characteristic of bound states for potentials having a finite range. This is the counterpart of the logarithmic phase in the scattering states. (Note that (31) agrees with (14.27).)

The normalized radial wave functions for the low lying bound states are often used in applications, and we shall therefore give the explicit formulas for them here.* Our notation is $R_{nl}(r)$, where n is the principal and l the angular momentum quantum number; the conventional spectroscopic notation for the states is also given below.

K-shell ($n = 1$)

$$1s: \quad R_{10} = \left(\frac{Z}{a_0}\right)^{\frac{3}{2}} 2e^{-\frac{1}{2}\rho}$$

L-shell ($n = 2$)

$$2s: \quad R_{20} = \left(\frac{Z}{a_0}\right)^{\frac{3}{2}} \frac{1}{2\sqrt{2}} (2-\rho)e^{-\frac{1}{2}\rho}$$

$$2p: \quad R_{21} = \left(\frac{Z}{a_0}\right)^{\frac{3}{2}} \frac{1}{2\sqrt{6}} \rho e^{-\frac{1}{2}\rho}$$

* A great deal of information concerning hydrogenic wave functions can be found in Condon and Shortley, Chapt. V; Pauling and Wilson, Chapt. V; Bethe and Salpeter, Secs. 2 and 3.

M-shell $(n = 3)$

$$3s: \quad R_{30} = \left(\frac{Z}{a_0}\right)^{\frac{3}{2}} \frac{1}{9\sqrt{3}} (6 - 6\rho + \rho^2) e^{-\frac{1}{3}\rho}$$

$$3p: \quad R_{31} = \left(\frac{Z}{a_0}\right)^{\frac{3}{2}} \frac{1}{9\sqrt{6}} (4 - \rho)\rho e^{-\frac{1}{3}\rho}$$

$$3d: \quad R_{32} = \left(\frac{Z}{a_0}\right)^{\frac{3}{2}} \frac{1}{9\sqrt{30}} \rho^2 e^{-\frac{1}{3}\rho}$$

In these formulas $\rho = 2Zr/na_0$.

•Let us return to the scattering problem in the partial wave formulation. Because we shall require the asymptotic form of $C_l(k;r)$ we must determine the asymptotic expansion of (25). For this purpose we need a closed expression, which is provided by the contour integral *

$$_1F_1(l + 1 - i\gamma|2l + 2| - 2ikr) = \frac{(2l+1)!}{2\pi i} \int_C e^s (s + 2ikr)^{-l-1+i\gamma} s^{-l-1-i\gamma}\, ds, \tag{32}$$

where C is the same contour as in Fig. 17.2. The manipulations that led from (8) to (12) can be used once more to give

$$C_l(k;r) = \frac{(2kr)^l e^{\frac{1}{2}\pi\gamma}\Gamma(l + i - i\gamma)}{2\pi i} \left\{ e^{ikr} \int_{C_1} e^s (s + 2ikr)^{-l-1+i\gamma} s^{-l-1-i\gamma}\, ds \right.$$
$$\left. + e^{-ikr} \int_{C_1} e^s (s - 2ikr)^{-l-1-i\gamma} s^{-l-1+i\gamma}\, ds \right\}$$

$$\xrightarrow[2ikr \to \infty]{} (2kr)^l e^{\frac{1}{2}\pi\gamma}\Gamma(l + 1 - i\gamma) \left\{ e^{ikr} \frac{(2ikr)^{-l-1+i\gamma}}{\Gamma(l+1+i\gamma)} \right.$$
$$\left. + e^{-ikr} \frac{(-2ikr)^{-l-1-i\gamma}}{\Gamma(l+1-i\gamma)} \right\}$$

$$\sim -\frac{1}{2ikr} \left\{ e^{-i(kr - \frac{1}{2}l\pi + \gamma \ln 2kr)} - \frac{\Gamma(l+1-i\gamma)}{\Gamma(l+1+i\gamma)} e^{i(kr - \frac{1}{2}l\pi + \gamma \ln 2kr)} \right\}. \tag{33}$$

This relation bears a striking resemblance to the asymptotic form (14.11) of $A_l(k;r)$ that we found for potentials of finite range:

$$A_l(k;r) \sim -\frac{1}{2ikr} \{ e^{-i(kr - \frac{1}{2}l\pi)} - e^{2i\delta_l(k)} e^{i(kr - \frac{1}{2}l\pi)} \} \tag{34}$$

* Expansion in powers of $2ikr$ will yield (25). Equation (32) can be derived by the method that led to (7).

where $\delta_l(k)$ is the phase shift. It is therefore customary to call the (real) quantity $\eta_l(k)$, defined by

$$e^{2i\eta_l(k)} = \frac{\Gamma(l + 1 - i\gamma)}{\Gamma(l + 1 + i\gamma)}, \tag{35}$$

the Coulomb phase shift. The analogy between the Coulomb wave functions and the various Bessel functions is now clear too. Thus $C_l(k; r)$ is the counterpart of j_l, while the functions

$$H_l^{(\pm)}(k; r) = \frac{(2kr)^l}{\pi i} e^{\frac{1}{2}\pi\gamma}\Gamma(l + 1 - i\gamma)e^{\pm ikr} \int_{C_1} e^s(s \pm 2ikr)^{-l-1\pm i\gamma}s^{-l-1\mp i\gamma}\, ds \tag{36}$$

are the analogues of h_l and h_l^*.•

17.3 COMBINED NUCLEAR AND COULOMB SCATTERING

•Consider finally the situation where we have a short range potential $\hbar^2 U(r)/2\mu$, as well as the Coulomb field. The pertinent radial wave equation is

$$\left\{\frac{1}{r^2}\frac{d}{dr}\left(r^2\frac{d}{dr}\right) - \frac{l(l+1)}{r^2} + k^2 + \frac{2\gamma k}{r} + U(r)\right\} \chi_l(k; r) = 0. \tag{37}$$

As $r \to \infty$, the Coulomb field dominates both $U(r)$ and $l(l + 1)/r^2$, and therefore χ_l has the asymptotic form

$$\chi_l \sim \frac{a}{r}\{e^{-i(kr - \frac{1}{2}l\pi + \gamma \ln 2kr)} - e^{2i\Delta_l(k)}e^{i(kr - \frac{1}{2}l\pi + \gamma\ln 2kr)}\},$$

where Δ_l is real because of probability conservation. If $U = 0$, $\Delta_l = \eta_l$; we therefore put $\Delta_l = \eta_l + \delta_l'$, and choose $a = -i/2k$ as in (34) and (33). Thus

$$\chi_l \sim -\frac{1}{2ikr}\{e^{-i(kr - \frac{1}{2}l\pi + \gamma\ln 2kr)} - e^{2i(\eta_l + \delta_l')}e^{i(kr - \frac{1}{2}l\pi + \gamma\ln 2kr)}\} \tag{38}$$

where δ_l' is the additional phase shift due to U. The prime serves to emphasize that δ_l' is not the same phase as U would yield in the absence of the Coulomb interaction. Nevertheless these phase shifts δ_l' still give a convenient way of parametrizing the scattering amplitude, because they are small under the same conditions as those that make the phases δ_l belonging to the purely short range potential U small.

With the help of (38) we can now construct the asymptotic form of the total wave function

$$\begin{aligned}\Psi_{\mathbf{k}}(\mathbf{r}) &= \sum_{l=0}^{\infty}\left(\frac{2l+1}{2\pi^2}\right)^{\frac{1}{2}} i^l\chi_l(k; r)Y_{l0}(\theta) \\ &\sim \psi_{\mathbf{k}}(\mathbf{r}) + \frac{1}{2ikr}e^{i(kr + \gamma\ln 2kr)}\sum_{l=0}^{\infty}\left(\frac{2l+1}{2\pi^2}\right)^{\frac{1}{2}} e^{2i\eta_l}(e^{2i\delta_l'} - 1)Y_{l0}(\theta). \qquad (39)\end{aligned}$$

Using (15) this becomes

$$\Psi_{\mathbf{k}}(\mathbf{r}) \sim (2\pi)^{-\frac{3}{2}} \left\{ e^{i[\mathbf{k}\cdot\mathbf{r} - \gamma \ln(kr - \mathbf{k}\cdot\mathbf{r})]} + \frac{e^{i(kr + \gamma \ln 2kr)}}{r} [f_C(\theta) + f'(\theta)] \right\}, \tag{40}$$

where

$$f'(\theta) = \frac{1}{k} \sum_{l=0}^{\infty} \sqrt{4\pi(2l+1)}\, e^{2i\eta_l} e^{i\delta_l'} \sin \delta_l' Y_{l0}(\theta). \tag{41}$$

The differential cross section is therefore

$$\sigma(\theta) = \sigma_C(\theta) + |f'(\theta)|^2 + 2\,\mathrm{Re}\{f_C^*(\theta) f'(\theta)\}. \tag{42}$$

The interference between Coulomb and "nuclear" scattering displayed here is an important tool in nuclear physics. As an illustration, consider proton-proton scattering at 10 Mev. In this case the coupling parameter γ is approximately $\frac{1}{20}$. Furthermore, only nuclear s-wave scattering is important at this energy. When $|\gamma| \ll 1$ the Coulomb phase shifts are small * (cf. (35)), and we may replace $e^{2i\eta_0}$ by one and δ_0' by δ_0. Hence f' assumes the approximate form $\lambda e^{i\delta_0} \sin \delta_0$. The appropriate formula for the cross section is therefore

$$\sigma(\theta) = \sigma_C(\theta) + \lambda^2 \sin^2 \delta_0 + \frac{\gamma \lambda \sin \delta_0 \cos[\delta_0 - 2\gamma \ln \sin \frac{1}{2}\theta]}{k \sin^2 \frac{1}{2}\theta}. \tag{43}$$

Observe that the nuclear cross section, $\lambda^2 \sin^2 \delta_0$, is independent of the sign of δ_0, while the interference term is not, and therefore enables one to determine sgn δ_0. Because sgn δ_0 is related to the sgn U ($\delta_0 < 0$ for repulsion, $\delta_0 > 0$ for attraction), this information is very valuable if, as is frequently the case in particle physics, one does not know the sign of the interaction. (In this connection, recall the remarks pertaining to (14.15).)•

REFERENCES

1. Applications of elastic scattering theory to problems in nuclear physics can be found in the following:

   de Benedetti, Chapt. 3;

   Bethe and Morrison, Chapts. 9, 10, 13, and 16.

2. A very detailed account of the continuum wave functions for the Coulomb field is contained in the article by M. H. H. Hull and G. Breit in the *Encyclopedia of Physics*, vol. XLI/1, Springer-Verlag (Berlin, 1959).
3. Group theoretic aspects of that Coulomb problem are reviewed in M. Bander and C. Itzykson, *Rev. Mod. Phys.* **38**, 330, 346 (1966).

PROBLEMS

1. (a) Show that the "spherical" components of the vector $\mathbf{r}$, namely $x \pm iy$ and z, are proportional to the functions $Y_{1m}(\theta, \phi)$.

(b) Rotate the coordinate system through 90° about the y-axis so that the new coordinates x', y', z' are related to the old by $x' = -z$, $y' = y$, $z' = x$. Let

* This does not mean that the partial wave expansion for Coulomb scattering converges rapidly at high energy; on the contrary, at all energies the whole series must be summed to give the Rutherford formula.

Y'_{1m} be eigenfunctions of L^2 and $L_{z'}$, the latter being the angular momentum about the new (z') axis. Construct the linear relationships between the functions Y'_{1m} and the functions Y_{1m} belonging to the original coordinate system.

(c) Consider the functions $Y_{\frac{1}{2},\pm\frac{1}{2}} = \text{const}(\sin\theta\, e^{\pm i\phi})^{\frac{1}{2}}$. Show that under the rotation of part (b) these do not transform linearly among themselves.

2. (a) Verify the representation of the δ-function in spherical coordinates given in (11.18).

(b) Recall that the Green's function for Laplace's equation is $|\mathbf{r} - \mathbf{r}'|^{-1}$. Use this fact, the multipole formula

$$\frac{1}{|\mathbf{r} - \mathbf{r}'|} = \sum_{l=0}^{\infty} \frac{r_<^l}{r_>^{l+1}} P_l(\hat{\mathbf{r}} \cdot \hat{\mathbf{r}}'),$$

and the addition theorem to derive the completeness relation (10.21). (Hint: Rewrite the expansion of $|\mathbf{r} - \mathbf{r}'|^{-1}$ in terms of step functions, and note that the derivative of a step function is a δ-function.)

3. A monoenergetic beam of particles is incident on a set of N identical static potentials. Show that the differential cross section in the lowest Born approximation is given by

$$\sigma(\mathbf{q}) = \sigma_0(\mathbf{q}) \left| \sum_{n=1}^{N} e^{i\mathbf{q}\cdot\mathbf{r}_n} \right|^2,$$

where $\hbar\mathbf{q}$ is the momentum transfer in the collision, $\sigma_0(\mathbf{q})$ the cross section for scattering off one potential, and $\mathbf{r}_n$ the center of the nth potential. Assume now that these scatterers are equally spaced on a line. Discuss the angular distribution.

4. Consider the Schrödinger equation

$$-\frac{\hbar^2}{2\mu}\nabla^2\psi(\mathbf{r}) + \int V(\mathbf{r}, \mathbf{r}')\psi(\mathbf{r}')\, d^3r' = E\psi(\mathbf{r}), \tag{1}$$

where the nonlocal potential has the form

$$V(\mathbf{r}, \mathbf{r}') = -\frac{\hbar^2}{2\mu}\lambda u(|\mathbf{r}|)u(|\mathbf{r}'|). \tag{2}$$

(a) Show that only the s-wave is affected by the interaction.

(b) Establish the integral equation that is equivalent to (1) and the scattering boundary conditions.

(c) Note that because of the special form of (2) this integral equation can be reduced to an algebraic equation, and therefore is solvable. Show that the scattering amplitude for incident momentum $\hbar\mathbf{k}$ is

$$f(k) = \frac{4\pi\lambda|v(k)|^2}{1 + \dfrac{2\lambda}{\pi}\displaystyle\int d^3q \cdot \frac{|v(q)|^2}{k^2 - q^2 + i\epsilon}} \tag{3}$$

where $v(k) = \int_0^\infty (\sin kr/kr)u(r)r^2\, dr$.

(d) Compare (3) to the Born series. What is the condition on λ if the Born series is to converge for wave number k?

(e) Consider the special case

$$u(r) = \frac{e^{-r/b}}{r}$$

henceforth. Show that

$$k \cot \delta(k) = \frac{(k^2b^2 + 1)^2 + \xi(k^2b^2 - 1)}{2\xi b}$$

where $\xi = 2\pi\lambda b^3$.

(f) Determine the condition on ξ necessary for the existence of a bound state. Can there be more than one bound state for this potential?

(g) For every value of k determine the largest value of ξ, ξ_k, for which the Born series converges. In the light of Prob. 4(f), what is the significance of ξ_0?

(h) For low energies, $k \to 0$, one can expand

$$k \cot \delta = -\frac{1}{a} + O(k^2)$$

where a is called the scattering length. Evaluate a. Express the zero-energy limit of the cross section in terms of a.

(i) Discuss the behavior of a as a function of ξ. How is the occurrence of a bound state reflected in a and in the cross section?

5. A particle of mass m moves in the one-dimensional potential shown below. Discuss the solutions of Schrödinger's equation in a qualitative fashion. Does the barrier B lead to any resonance phenomena?

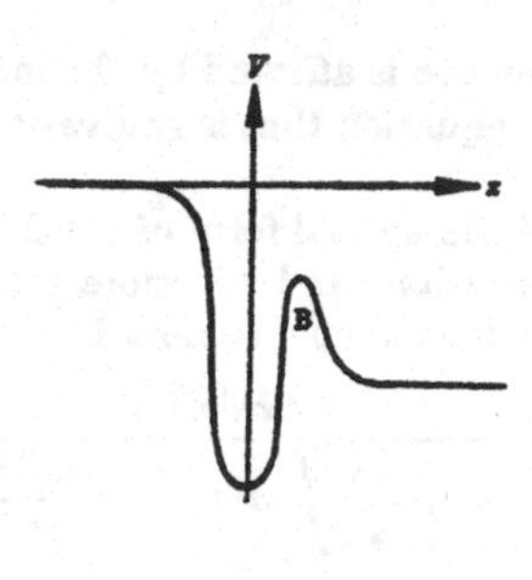

6. Consider the one-dimensional Schrödinger equation (7.8):

$$\left(\frac{d^2}{dx^2} + k^2 + \gamma\,\delta(x)\right)\psi_k(x) = 0. \tag{1}$$

Let $u_k(q)$ be the momentum space wave function, i.e.,

$$u_k(q) = \int_{-\infty}^{\infty} \frac{dx}{\sqrt{2\pi}}\, e^{-iqx}\psi_k(x). \tag{2}$$

(a) Convert (1) into an integral equation in momentum space and show that the solution that satisfies the outgoing wave boundary conditions for *both* signs of k is given by

$$u_k(q) = \delta(k - q) - \frac{\gamma}{2\pi}\frac{1}{D(k)}\frac{1}{k^2 - q^2 + i\epsilon}, \tag{3}$$

where

$$D(k) = 1 + \frac{\gamma}{2\pi}\int_{-\infty}^{\infty}\frac{dq}{k^2 - q^2 + i\epsilon}. \tag{4}$$

Verify that the transmission and reflection coefficients are given by (7.13).

(b) Derive the orthogonality relation

$$\int_{-\infty}^{\infty} dq\, u_k^*(q)u_{k'}(q) = \delta(k - k').$$

(c) Define

$$\Delta(q, q') = \int_{-\infty}^{\infty} dk\, u_k(q)u_k^*(q').$$

By direct calculation show that $\Delta(q, q') = \delta(q - q')$ in the case of repulsion ($\gamma < 0$), and that

$$\Delta(q, q') = \delta(q - q') - \frac{\gamma^3}{4\pi}\frac{1}{(q^2 + \frac{1}{4}\gamma^2)(q'^2 + \frac{1}{4}\gamma^2)} \tag{5}$$

in the case of attraction. Note that you have proven (7.34) and the statements subsequent to it in Sec. 7.3.

(d) Use (5) to evaluate the bound-state wave function (7.10).

(e) Formulate the three-dimensional delta-shell problem of Sec. 15 in momentum space. Derive completeness and orthogonality relations for the wave functions belonging to a single value of l, and determine the maximum number of bound states in each partial wave.

7. A particle of mass m is scattered by the potential

$$\begin{aligned} V(r) &= 0 \qquad (r > a) \\ &= V_0 \qquad (r < a), \end{aligned}$$

with $V_0 > 0$.

(a) Estimate the differential scattering cross section if $(\hbar k)^2 \gg 2mV_0$, and discuss the validity of the approximation used.

(b) In the limit $ka \ll 1$, show that the cross section takes on the form

$$\frac{d\sigma}{d\Omega} = A + B\cos\theta,$$

with $B \ll A$.

(c) Now assume that the parameter

$$x = \frac{\hbar^2}{mV_0a^2}$$

is small compared to unity. What is the physical significance of this assumption when $ka \ll 1$? Evaluate the leading nonvanishing contributions to A and B in the limit $x \to 0$, $ka \to 0$.

(d) Calculate the s-wave phase shift exactly for this potential, and compare it to the Born formula (14.15).

8. Show that the closed formula (12.19) for the scattering amplitude can be obtained by summing the partial wave series (14.13).

9. Verify the qualitative features of Fig. 49.2. In doing so, bear in mind that the phase shift is only defined *modulo* π (Why?). To remove this ambiguity, adopt the requirement that $\delta \to 0$ as $\xi \to \infty$ when the interaction is no longer effective. Once this definition is adopted, show that $\delta_0 > 0$ if $g > 0$ (attraction), and $\delta_0 < 0$ if $g < 0$ (repulsion). Also show that $\delta_0(0) = \pi$ if $g > 1$ (one bound state), whereas $\delta_0(0) = 0$ if $g < 1$ (no bound states). These results are special cases of theorems proven in Sec. 49.

IV

The Measurement Process and the Statistical Interpretation of Quantum Mechanics

In this chapter we shall apply the formalism of quantum mechanics to the description of measurements. The object whose properties we wish to measure and the apparatus used for this purpose shall be treated as an interacting dynamical system. Our discussion will expose the basic concepts, but it will be highly stylized and not do justice to the enormous complexity of an actual laboratory experiment. Here, as nowhere else, it is crucial to keep one's eye firmly focused on the ultimate objective: *To verify the consistency of the mathematical formalism with the interpretation of the scalar product as a probability amplitude.* A somewhat secondary objective is to show that the result of a measurement of an observable is one of the eigenvalues of the associated mathematical operator. In carrying out this program we shall attempt to define what is meant by measurement, by the preparation of a system in a specified state, and other concepts that we have been using rather loosely.

The analysis of measurements will serve to motivate the reformulation of quantum mechanics on which the balance of this book will be based. This reformulation of the theory will be carried out in the subsequent two chapters.

As a prototype of a measurement we shall study the Stern-Gerlach experiment. In this experiment an atomic or molecular beam is passed through an inhomogeneous magnetic field. The field causes a spatial separation of systems having different values of the magnetic moment and allows one to determine the moment by measuring the deflection of the emergent beams. In all experiments of this type the externally applied fields vary negligibly over distances of order the atom's (or molecule's) dimension. We shall therefore begin with a short detour, and describe how an atomic system behaves in a slowly varying applied field.

18. *The Two-Body System in a Slowly Varying External Field*

As a simple but fairly realistic model of an atom or ion we consider a two-body system with constituent masses m_1 and m_2, having electric charges e_1 and e_2. Let $\mathbf{A}(\mathbf{x})$ and $V(\mathbf{x})$ be the applied electromagnetic potentials; whenever $\mathbf{A}_i$ or V_i appears in the following equations we mean the potential evaluated at the location of the ith particle. The Hamiltonian is then

$$H = \sum_{i=1}^{2} \left\{ \frac{1}{2m_i}\left(\mathbf{p}_i - \frac{e_i}{c}\mathbf{A}_i\right)^2 + e_i V_i \right\} + U_{12}, \tag{1}$$

where U_{12} is the interaction potential between the two particles.

Because it is rather simpler, we shall first discuss the electrostatic case, $\mathbf{A} = 0$. Upon introducing the center-of-mass ($\mathbf{R}$) and relative ($\mathbf{r}$) coordinates, and their conjugate momenta $\mathbf{P} = -i\hbar\,\partial/\partial\mathbf{R}$, $\mathbf{p} = -i\hbar\,\partial/\partial\mathbf{r}$, (1) becomes

$$H = \frac{1}{2M}P^2 + QV(\mathbf{R}) + \frac{p^2}{2\mu} + U_{12} + \sum_i e_i[V_i - V(\mathbf{R})],$$

where $Q = e_1 + e_2$ is the total charge. As we assume a slowly varying potential V, this Hamiltonian can be approximated by $H = H_0 + H_1$, where

$$\begin{aligned} H_0 &= \frac{1}{2M}P^2 + QV(\mathbf{R}), \\ H_1(\mathbf{R}) &= \frac{1}{2\mu}p^2 + U_{12} - \mathbf{d}\cdot\boldsymbol{\mathcal{E}}(\mathbf{R}), \end{aligned} \tag{2}$$

where $\mathcal{E}$ is the electric field, and

$$\mathbf{d} = \frac{1}{M}(e_1m_2 - e_2m_1)\mathbf{r} \tag{3}$$

is the dipole moment in the c.o.m. system. For a neutral atom, $\mathbf{d} = e\mathbf{r}$. The interaction $\mathbf{d} \cdot \mathcal{E}$ is the cause of the *Stark effect* (see Prob. 2, Chapt. VII).

Consider the Schrödinger equation

$$i\hbar \frac{\partial}{\partial t} \Psi(\mathbf{r}, \mathbf{R}, t) = H\Psi(\mathbf{r}, \mathbf{R}, t). \tag{4}$$

Note that if $\mathcal{E} = 0$, we may construct a solution Ψ as the product of an eigenfunction $\Phi_n(\mathbf{r})$ of $(p^2/2\mu) + U_{12}$ and a wave packet in the coordinate $\mathbf{R}$ describing the rectilinear motion of the center of mass. For slowly varying fields, we expect that this c.o.m. motion will not vary drastically from the $\mathcal{E} = 0$ situation and, more important, that one will still be able to construct Ψ out of a single eigenfunction of the internal motion. With this in mind, consider the solutions $\Phi_n(\mathbf{r}, \mathbf{R})$ of

$$[H_1(\mathbf{R}) - E_n(\mathbf{R})]\Phi_n(\mathbf{r}, \mathbf{R}) = 0. \tag{5}$$

Here the eigenvalue and wave function depend on the parameter $\mathbf{R}$, because the electrostatic field appearing in H_1 depends on the location of the center of mass. Instead of the solution

$$\Phi_n(\mathbf{r}) \int f(\mathbf{P})\, e^{i\mathbf{P}\cdot\mathbf{R}/\hbar}\, e^{-iP^2t/2M\hbar}\, d^3P,$$

which we would have in the absence of $\mathcal{E}$, we now attempt a solution

$$\Psi \simeq u_n(\mathbf{R}, t)\Phi_n(\mathbf{r}, \mathbf{R}), \tag{6}$$

where u_n is determined from (4). When the operator H_0 is applied to (6), it will also work on $\Phi_n(\mathbf{r}, \mathbf{R})$, but because $\partial\Phi/\partial\mathbf{R}$ is, by hypothesis, exceedingly small, (i.e., Φ divided by a macroscopic length) we can neglect these derivatives. Therefore

$$i\hbar \frac{\partial}{\partial t} u_n(\mathbf{R}, t) \simeq \left\{\frac{1}{2M} P^2 + QV(\mathbf{R}) + E_n(\mathbf{R})\right\} u_n(\mathbf{R}, t), \tag{7}$$

which is just the equation for a particle of mass M in the effective potential $QV(\mathbf{R}) + E_n(\mathbf{R})$. This potential varies negligibly in the de Broglie wavelength characterizing the c.o.m. motion, and the problem (7) can therefore be solved semiclassically. We can therefore construct wave packets in **R** that move along the classical trajectory dictated by the force

$$\mathbf{F} = -\frac{\partial}{\partial \mathbf{R}} \{QV(\mathbf{R}) + E_n(\mathbf{R})\}. \tag{8}$$

The appearance of $E_n(\mathbf{R})$ in **F** is an essential feature, because it permits a classical analysis to determine a "quantum mechanical property" of the system—in this instance the electric dipole moment associated with some particular internal motion Φ_n.

The approximation (6), where u_n is determined by (7), is called the *adiabatic approximation*. Although we shall not discuss its limitations and precision,* it is clear that it is applicable under the conditions stated here.

For reasons that will be discussed in Sec. 38, essentially all atomic systems have a vanishing electric dipole moment, and therefore the above discussion is rather academic. A far more important case is the motion of atoms possessing a magnetic dipole moment through an inhomogeneous field. This constitutes the *Stern-Gerlach experiment*. The analysis of this case is based on precisely the same approximations, but it is somewhat more complicated in detail than the electric-dipole experiment. The final result for the case where $e_1 = -e_2 = e$ is

$$H_1(R) = \frac{1}{2\mu} p^2 + U_{12} - \mathcal{H}(\mathbf{R}) \cdot \mathfrak{M}, \tag{9}$$

where the magnetic moment is

$$\mathfrak{M} = \frac{e\hbar}{2Mc} \left(\frac{m_2}{m_1} - \frac{m_1}{m_2} \right) \mathbf{L}, \tag{10}$$

and $\hbar\mathbf{L} = \mathbf{r} \times \mathbf{p}$ is the relative angular momentum. If one of the masses (say m_1) is much lighter than the other, as is true of the electron's mass in comparison to that of the nucleus, (10) reduces to the familiar expression

$$\mathfrak{M} = \frac{e\hbar}{2m_1c} \mathbf{L}. \tag{11}$$

For the electron $e\hbar/2mc = 0.58 \times 10^{-8}$ ev/gauss; this is called the Bohr magneton. Thus even in a field of 10^3 gauss, the magnetic interaction energy is smaller by a factor of 10^6 than the level spacing in hydrogen.

The remainder of the discussion is quite the same as in the electrostatic case. We again make the approximation (6), with Φ_n now being the

* See Pauli, Sec. 11, and Messiah, Chapt. 17, Sec. 12.

eigenfunction of (9) belonging to the eigenvalue $E_n(\mathbf{R})$. The functions $u_n(\mathbf{R}, t)$ are wave packets that follow the trajectory determined by

$$M\ddot{\mathbf{R}} = -\frac{\partial}{\partial \mathbf{R}} E_n(\mathbf{R}). \tag{12}$$

Finally we must solve the Schrödinger equation pertaining to (9). Consider the case of an electron moving about a nucleus, so that $m_e \simeq \mu$. Then

$$\left(\frac{1}{2m_e} p^2 + U_{12} - \mu_0 \mathbf{L} \cdot \mathcal{H}(\mathbf{R})\right) \Phi_n = E_n(\mathbf{R}) \Phi_n \tag{13}$$

where $\mu_0 = e\hbar/2m_e c$. Now we note that

$$\left[\mathbf{L} \cdot \mathcal{H}, \frac{1}{2m_e} p^2 + U_{12}\right] = 0$$

because of the rotational invariance of $(p^2/2m_e) + U_{12}$, and therefore the perturbing term $-\mu_0 \mathbf{L} \cdot \mathcal{H}$ is a constant of the motion.* We may therefore construct simultaneous eigenfunctions of H_1, L^2 and $L_\parallel$, where $L_\parallel$ is the projection of $\mathbf{L}$ onto $\hat{\mathcal{H}}$. These eigenfunctions have the form $R_{nl}(r) Y_{lm}(\theta\phi)$, the angles being measured from the axis $\hat{\mathcal{H}}$. These functions are already eigenfunctions of $(p^2/2m_e) + U_{12}$, the eigenvalue being the m-independent energy E_{nl}^0. Having chosen the quantization axis properly,† we can easily evaluate the energy in the presence of the field because

$$\mu_0 \mathcal{H} L_\parallel R_{nl} Y_{lm} = \mu_0 \mathcal{H} m R_{nl} Y_{lm}.$$

The required energy eigenvalues are therefore

$$E_{nlm}(\mathbf{R}) = E_{nl}^0 - \mu_0 m \mathcal{H}(\mathbf{R}). \tag{14}$$

The degeneracy with respect to m has been lifted by the applied magnetic field; this is known as the *Zeeman effect.*

* The fact that the cylindrically symmetric Hamiltonian (9) commutes with L^2 is a very special property of the magnetic interaction. That this is not a general property can be seen from the electrostatic Hamiltonian (2), which is also symmetric about $\mathcal{E}$, but for which L^2 is not a constant of the motion.

† If we were to use spherical harmonics referred to a different axis, they would not be eigenfunctions of $H_1(\mathbf{R})$. Because the wave functions of a given l transform linearly among themselves, the eigenfunctions of $H_1(\mathbf{R})$ would then be a linear combination of these inappropriate Y's. The energy eigenvalues are independent of the choice of coordinate system; the number m appearing in (14) is the eigenvalue of the component of $\mathbf{L}$ along $\mathcal{H}$, which is an invariant statement.

19. The Stern-Gerlach Experiment

A beam of atoms (Ag in the original experiment) is passed through a magnet whose cross section is shown in Fig. 19.1. Because of the inhomogeneous field, the atoms are subjected to the force $\mu_0 m\, \partial\mathcal{H}/\partial Z$ (cf. (18.12)), and the beams will be separated according to the spectrum of m values. By measuring the deflection $\mu_0 m$ is determined.

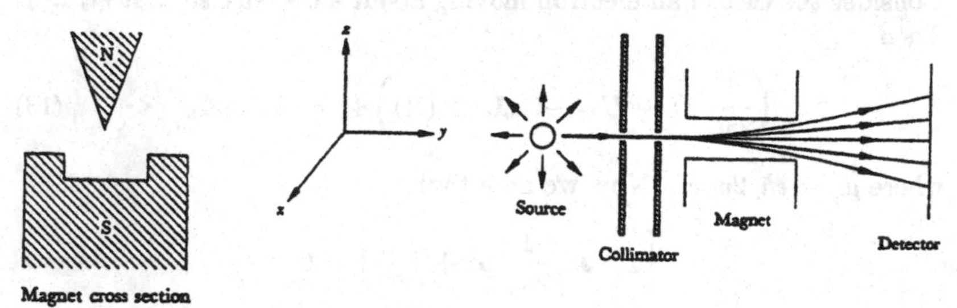

Fig. 19.1. The Stern-Gerlach experiment.

At early times atoms in the state m are described by the wave function $u^0(\mathbf{R}t)\Phi_m(\mathbf{r})$, where u^0 is a free wave packet for the center-of-mass motion collimated as shown and moving towards the magnet system; Φ is now independent of $\mathbf{R}$ because the system is not in the field as yet. After the system is in the field, $u^0(\mathbf{R}t)\Phi_m(\mathbf{r}) \rightarrow u_m(\mathbf{R}t)\Phi_m(\mathbf{r}, \mathbf{R})$. For large values of t the wave function is $u_m(\mathbf{R}t)\Phi_m(\mathbf{r})$, where $u_m(\mathbf{R}t)$ is nonzero in some spatial region $V_m(t)$. These regions have the property that there is *no overlap* between $V_m(t)$ and $V_{m'}(t)$ if $m \neq m'$.

Let $u^0(\mathbf{R}t)$ be normalized to one. We assume that Φ_m is normalized:

$$\int d^3r\, \Phi_m^*(\mathbf{r}, \mathbf{R})\Phi_{m'}(\mathbf{r}, \mathbf{R}) = \delta_{mm'}, \tag{1}$$

hence

$$\int d^3R |u_m(\mathbf{R}t)|^2 = 1. \tag{2}$$

Consider as an initial wave packet on the left of $\mathcal{H}$ the linear combination

$$\Psi_0(t) = u^0(\mathbf{R}t) \sum_m c_m \Phi_m(\mathbf{r}), \tag{3}$$

with

$$\sum_m |c_m|^2 = 1. \tag{4}$$

For later times the linearity of the wave equation implies that

$$\Psi_0(t) \rightarrow \Psi(t) = \sum_m c_m u_m(\mathbf{R}t)\Phi_m(\mathbf{r}, \mathbf{R}). \tag{5}$$

When the beam has emerged into the field-free region on the right, the wave function is

$$\Psi(t) = \sum_m c_m u_m(\mathbf{R}t)\Phi_m(\mathbf{r}). \tag{6}$$

The essential characteristic of this wave function is that it uniquely correlates the center-of-mass wave function of the atom with its internal state of motion: if $\mathbf{R}$ *is in the volume* $V_m(t)$*, then the internal state of the system is represented by* Φ_m. This correlation enables us to determine the internal state of motion by merely measuring the coordinate $\mathbf{R}$. The measurement of $\mathbf{R}$ can be carried out with a photographic emulsion or a γ-ray microscope. Alternatively, one could erect a screen on the right of the apparatus with one hole that permits only the beam having the internal state m to pass on. In the latter arrangement, those systems that have passed through the screen to the right are then represented by the wave function

$$\Psi_m(t) = u_m(\mathbf{R}t)\Phi_m(\mathbf{r}); \tag{7}$$

as always, we normalize the wave function to unity. The fraction of the initial ensemble (beam) that passes through this filter arrangement is given by

$$\int_{V_m(t)} d^3R \int d^3r |\Psi(\mathbf{r}\mathbf{R}t)|^2 = \sum_{m'} |c_{m'}|^2 \int_{V_m(t)} d^3R |u_{m'}(\mathbf{R}t)|^2 = |c_m|^2. \tag{8}$$

Here the first equality is due to the orthogonality of the functions Φ_m, and the second follows from the fact that $u_{m'}(\mathbf{R}t)$ vanishes in $V_m(t)$ unless $m = m'$. Therefore $|c_m|^2$ is the probability of finding that the value of the magnetic moment is $\mu_0 m$ when an ensemble of systems is in the state represented by the wave function $\Sigma_m c_m \Phi_m$.

One should bear in mind that we have now derived the fact that c_m in (4) is a probability amplitude. The basic elements in the derivation were

1. *the linearity of the Schrödinger equation,*
2. *the interpretation of* $|u_m(\mathbf{R})|^2$ *as a probability distribution for the coordinate* $\mathbf{R}$, and
3. *the physical requirement that a measurement must produce a situation where the occurrence of an event constitutes an unambiguous determination of a physical attribute, in our case "atom is in the volume* V_m*, hence its magnetic moment is* $\mu_0 m$*."*

The argument of Sec. 7.6 that first led us to the interpretation of c_m as a probability amplitude was rather formal and hence less satisfactory.

The same interpretation of c_m will also appear if we consider the probability for finding the relative coordinate $\mathbf{r}$, with $\mathbf{R}$ going undetected. This probability enters into the expression $\int W(\mathbf{r}, t)F(\mathbf{r})\, d^3r$ that determines the expectation value of all observables $F(\mathbf{r})$ that are functions of the internal coordinates of the atom. From the rules of wave mechanics the probability density in question is

$$W(\mathbf{r}, t) = \int d^3R \left| \sum_m c_m u_m(\mathbf{R}t)\Phi_m(\mathbf{r},\mathbf{R}) \right|^2. \tag{9}$$

For sufficiently late times the packet has traversed the field, and then the different u_m do not overlap. For such values of t, Φ_m no longer depends on $\mathbf{R}$, and (9) therefore reduces to

$$W(\mathbf{r}, t) \to \sum_m |c_m|^2 |\Phi_m(\mathbf{r})|^2. \tag{10}$$

Thus c_m again appears as the probability that the system is in the state m.

Now contrast (10) with the same probability distribution before the systems entered the apparatus; according to (3) this is given by

$$W_0(\mathbf{r}, t) = \left| \sum_m c_m \Phi_m(\mathbf{r}) \right|^2. \tag{11}$$

In the original state (3) *the relative phases of the coefficients c_m enter into the probability distribution, but this is no longer true of the same probability distribution subsequent to the measurement.*

We shall explore the significance of this conclusion in considerable detail, but before doing so we should make a few remarks concerning the length of time the beam must stay in the Stern-Gerlach field. Let the x- and z-directions be those shown in Fig. 19.1, and let $R_x = X, R_z = Z$. The collimators between the source and the magnet produce a beam whose width (in the z-x plane) is a. This beam will spread with time: If $t \simeq 0$ is the time at which the packet leaves the collimators, its spread in the z-direction at later times, as given by the uncertainty principle, is

$$\Delta Z \simeq \frac{\hbar}{a}\frac{t}{M}. \tag{12}$$

The deflection caused by $\mathcal{H}$ must exceed this. Let $Z_m(t)$ be the location of the mth beam referred to the axis of the magnet. From (18.12) we have

$$Z_m(t) \simeq \frac{1}{2M}\frac{\partial E_m}{\partial Z}t^2. \tag{13}$$

The condition that two beams with different m's be sufficiently separated is therefore $|Z_m(t) - Z_{m'}(t)| \gg \Delta Z$, or

$$\tfrac{1}{2}at \left| \frac{\partial}{\partial Z}(E_m - E_{m'}) \right| \gg \hbar. \tag{14}$$

Thus t cannot be as small as we please: *One cannot ascertain whether a system is in one of two stationary states (m and m') in an arbitrarily short time.* For fixed experimental conditions (i.e., collimator opening a and field gradient $\partial\mathcal{H}/\partial Z$) the time required for this decision is roughly inversely proportional to $|E_m - E_{m'}|$. One can make this somewhat more precise by noting that $\partial(E_m - E_{m'})/\partial Z \approx (E_m - E_{m'})/L$, where L is the distance in which $\mathcal{H}$ varies appreciably. However $a/L < 1$ in general,* and therefore the minimum time t required is †

$$t \gtrsim \frac{\hbar}{|E_m - E_{m'}|}. \tag{15}$$

This inequality is frequently referred to as *the energy-time uncertainty principle.* Nevertheless, one should *not* think of (15) as being on an equal footing with $\Delta p\, \Delta x \gtrsim \hbar$. Time is only a parameter in the theory, and is not canonically conjugate to the energy. In particular, it is not possible to construct an operator for t, as we shall see eventually.

20. Changes of State Resulting from Measurement ‡

The analysis of the measurement process outlined in the preceding section raises a number of questions. Among these we may mention the following:

* If $a/L > 1$, different parts of the beam are subjected to different forces, and then the detailed nature of the incident wave packet becomes relevant to the analysis of the experiment. Therefore an arrangement wherein $a/L > 1$ is useless as a measurement device.

† In an actual Stern-Gerlach experiment the passage time greatly exceeds the lower limit specified by (15).

‡ This section is meant to serve a dual purpose, and to some extent it therefore falls between two chairs. Primarily, it is intended to motivate the formulation of quantum mechanics that will be developed in Chapt. V. But it also is an introduction to some of the difficult aspects of measurement theory. Unfortunately, it is well-nigh impossible to treat the latter properly without the formal machinery that will be developed in the subsequent chapters. In their first reading of this book, students are therefore advised to ignore the mysterious footnotes, and to forge ahead, even when lack of total comprehension would appear to counsel further meditation. The rather superficial understanding thus gained should suffice to make Chapt. V approachable. Once Chapts. V and VI have been mastered, a second and more systematic study of this section is recommended.

1. The apparatus necessary to determine the center-of-mass coordinate **R** was not incorporated into the description; what are the consequences of enlarging the dynamical description of the measurement process?
2. Born's interpretation of the wave function as a probability amplitude was used in the derivation of Eqs. (19.8) and (19.9). To what extent does the conclusion that c_m is itself a probability amplitude depend on this assumption?

This section is devoted to these problems and related matters. In Sec. 20.2 we shall embellish the Stern-Gerlach apparatus with devices capable of measuring **R**, and investigate question 1 above. In Sec. 20.3 we shall examine the validity of the interpretation of $|c_m|^2$ as a probability in some detail. The beginning of this last subsection also contains a concise definition of what we mean by a measurement. From a logical viewpoint this definition should be stated at the outset, but it would be rather incomprehensible without a preparatory discussion.

20.1 THE DENSITY MATRIX. PURE STATES AND MIXTURES

In order to carry the analysis further we shall have to introduce a formal quantity that will play a central role in the remainder of this chapter, and in numerous other connections. This quantity is the density matrix of Landau and von Neumann. We shall study the density matrix in detail in Sec. 26, and therefore confine ourselves to its most essential properties here.

Consider a system in a state described by the wave function $\psi(x, t)$, where x can stand for any number of variables. Let A be a linear Hermitian operator. The most general form such a linear operator can take is that of an integral operator:

$$A[\psi(x, t)] = \int \langle x|A|x'\rangle\psi(x', t)\, dx'. \tag{1}$$

Instead of using the notation of (7.47) for the integral operator A, we have written its kernel as $\langle x|A|x'\rangle$. The reason behind this notation will emerge in a moment, and even more clearly in the next chapter.

The expectation value of A in the state ψ is

$$(\psi, A\psi) \equiv \int \psi^*(x, t)\langle x|A|x'\rangle\psi(x', t)\, dx\, dx'. \tag{2}$$

The most familiar operators are the coordinate operator, $\langle \mathbf{x}|\bar{\mathbf{x}}|\mathbf{x}'\rangle = \mathbf{x}\,\delta(\mathbf{x} - \mathbf{x}')$, and the momentum operator

$$\langle \mathbf{x}|\bar{\mathbf{p}}|\mathbf{x}'\rangle = \frac{\hbar}{i}\,\delta^{(1)}(\mathbf{x} - \mathbf{x}').$$

These are *local operators*, because they only involve the wave function (or its derivative) at one point. *Nonlocal operators* can, in principle, also exist. We have already encountered them in (4.34′), and shall employ them again in many-particle problems (see the Hartree-Fock exchange potential in Chapt. XI). Essentially all quantities actually known to occur in nature are represented by operators having a nonlocality in configuration space that is at most of microscopic dimensions.* We shall reserve the name "observable" for such *physically significant* Hermitian operators.

The density matrix associated with the state ψ is defined as

$$\langle x|\rho(t)|x'\rangle \equiv \psi(x, t)\psi^*(x', t). \tag{3}$$

For the moment let us suppress the variable t. If we think of x as labeling the rows, and x' the columns, then ρ is seen to be a Hermitian matrix, because

$$\langle x|\rho|x'\rangle^* = \langle x'|\rho|x\rangle. \tag{4}$$

The diagonal elements of ρ are the probabilities for finding the system at the coordinate x. The expectation value of A can then be written as

$$\langle A\rangle = \int \langle x|A|x'\rangle\langle x'|\rho|x\rangle\, dx\, dx'. \tag{5}$$

For our present purpose, the most important property of this particular ρ is

$$\rho^2 = \rho. \tag{6}$$

This follows immediately from the definition (3) and the fact that ψ is normalized to one:

$$\int \langle x|\rho|x''\rangle\langle x''|\rho|x'\rangle\, dx'' = \psi(x)\psi^*(x') = \langle x|\rho|x'\rangle.$$

Thus we conclude that for an arbitrary wave function $\psi(x, t)$ the density matrix as defined by (3) is an idempotent matrix.

Not all states can be described by a single wave function, however. Consider, for example, an ensemble of atoms prepared by two *different* Stern-Gerlach apparatuses using *different* sources. Each Stern-Gerlach setup is provided with a screen that permits only atoms with some particular value of the magnetic quantum number (m_1 and m_2, say) to pass on. Each of these arrangements acts as a filter, because no matter what state is fed in, a system with prescribed properties emerges. These two beams

* We have used the qualification "essentially" here because superfluids and superconductors possess observables that are macroscopically nonlocal in coordinate space. Further remarks related to this will be made at the appropriate juncture.

are then combined by suitable fields into one beam. If p_1 and p_2 ($p_1 + p_2 = 1$) are the fractional intensities of the two sub-beams, then it is clear that in the ensemble thus prepared the expectation value of an operator relating to the internal variables of the atom will be given by

$$\langle A \rangle = \sum_{i=1}^{2} p_i \int \Phi^*_{m_i}(\mathbf{r}) \langle \mathbf{r} | A | \mathbf{r}' \rangle \Phi_{m_i}(\mathbf{r}')\, d^3r\, d^3r'.$$

For the state just described, the density matrix is defined as

$$\langle \mathbf{r} | \rho | \mathbf{r}' \rangle = \sum_{i=1}^{2} p_i \Phi_{m_i}(\mathbf{r}) \Phi^*_{m_i}(\mathbf{r}'). \tag{7}$$

When we compute ρ^2 with this expression, we obtain

$$\langle \mathbf{r} | \rho^2 | \mathbf{r}' \rangle = \sum_{i=1}^{2} p_i^2 \Phi_{m_i}(\mathbf{r}) \Phi^*_{m_i}(\mathbf{r}'), \tag{8}$$

which is *not* equal to ρ.

We can make this somewhat more precise by computing the trace of the matrix ρ^2. By definition the trace of any matrix in the present notation is

$$\operatorname{tr} M \equiv \int \langle x | M | x \rangle\, dx.$$

When a state can be described by a single wave function as in (3), $\operatorname{tr} \rho^2 = 1$. On the other hand, (8) shows that for states that must be described by more than one wave function, $\operatorname{tr} \rho^2 = \Sigma_i p_i^2 < 1$. A state described by a density matrix for which $\operatorname{tr} \rho^2 = 1$ is said to be a *pure state*, whereas one for which $\operatorname{tr} \rho^2 < 1$ is said to be a *mixture*. The only restriction imposed on these statements is that the various wave functions ψ_n that enter into the mixture $\langle x | \rho | x' \rangle = \Sigma_n p_n \psi_n(x) \psi_n^*(x')$ must be normalized to one and linearly independent. They need not be energy eigenfunctions.

The expectation value of an observable A in an ensemble described by the density matrix ρ can be expressed succinctly in terms of the trace:

$$\langle A \rangle = \int \langle x | A | x' \rangle \langle x' | \rho | x \rangle\, dx\, dx' = \operatorname{tr} A\rho. \tag{9}$$

The experimental significance of $\langle A \rangle$, it will be remembered, is claimed to be the following: Consider N identical systems prepared in the identical fashion, and let a_i be the value of A measured on the ith member of the ensemble; then

$$\langle A \rangle = \lim_{N \to \infty} N^{-1} \sum_{i=1}^{N} a_i.$$

It should be noted that *we always deal with states of an ensemble, whether the ensemble in question is pure or mixed.*

If the density matrix is given by

$$\langle x|\rho|x'\rangle = \sum_n p_n \psi_n(x)\psi_n^*(x')$$

at $t = t_0$, it evolves into

$$\langle x|\rho(t)|x'\rangle = \sum_n p_n \Psi_n(x, t)\Psi_n^*(x', t)$$

at later times, where $\Psi_n(x, t)$ satisfies the Schrödinger equation $i\hbar\, \partial\Psi_n/\partial t = H\Psi_n$, and the initial condition $\Psi_n(x, t_0) = \psi_n(x)$. The time dependence of tr ρ^2 is easily determined. We observe that the wave functions $\Psi_n(x, t)$ enter tr ρ^2 in the form of scalar products. But (Ψ_n, Ψ_m) is time independent when H is Hermitian, and we therefore conclude that tr ρ^2 is itself time independent. As a consequence, *a pure state cannot evolve into a mixture if the time evolution is governed by a Schrödinger equation.*

20.2 COHERENCE PROPERTIES OF STATES FOLLOWING MEASUREMENT

We are now in a position to extend our analysis of the Stern-Gerlach (S-G) experiment. We recall from (19.5) that the S-G magnet forces the wave function to assume a form wherein the various trajectories of the atoms are *uniquely correlated* with *distinct* internal states Φ_m:

$$\Psi(\mathbf{r}\mathbf{R}t) = \sum_m c_m u_m(\mathbf{R}t)\Phi_m(\mathbf{r}). \tag{10}$$

The essential piece of "apparatus" in the experiment is really not the magnet, but the center of mass of the atom! It plays the same role as the pointer on a voltmeter: If $\mathbf{R}$ points to $V_m(t)$, the internal state of the system is Φ_m. Instead of measuring a property related to Φ_m, we can make measurements on $\mathbf{R}$. This leads to an unambiguous determination of the state Φ_m of the "object" because the different states of the "apparatus" $u_m(\mathbf{R}t)$ are *macroscopically* distinguishable. In any useful measurement apparatus the pointer positions, in our generalized sense of the term, must be uniquely correlated to the states of the object. Furthermore, because we must eventually use an actual piece of laboratory apparatus to determine the position of the pointer, the different positions must be macroscopically distinguishable.

The density matrix associated with the pure state (10) is

$$\langle \mathbf{rR}|\rho(t)|\mathbf{r'R'}\rangle = \sum_{mm'} c_m c^*_{m'} u_m(\mathbf{R}t)\Phi_m(\mathbf{r})u^*_{m'}(\mathbf{R'}t)\Phi^*_{m'}(\mathbf{r'}). \tag{11}$$

The expectation value in the state Ψ of an observable A is

$$\begin{aligned}\langle A\rangle &= \int d^3r\, d^3r'\, d^3R\, d^3R' \langle \mathbf{rR}|\rho(t)|\mathbf{r'R'}\rangle\langle \mathbf{r'R'}|A|\mathbf{rR}\rangle \\ &\equiv \operatorname{tr} \rho A.\end{aligned} \tag{12}$$

If t is such that the different regions $V_m(t)$ are macroscopically separated, then the terms $m \neq m'$ in (12) will drop out because, for all known observables, $\langle \mathbf{r'R'}|A|\mathbf{rR}\rangle$ vanishes when $|\mathbf{R} - \mathbf{R'}|$ is macroscopic. *It is therefore impossible to distinguish ρ from the density matrix $\hat{\rho}$ defined by*

$$\langle \mathbf{rR}|\hat{\rho}(t)|\mathbf{r'R'}\rangle = \sum_m |c_m|^2 u_m(\mathbf{R}t)\Phi_m(\mathbf{r})u^*_m(\mathbf{R'}t)\Phi^*_m(\mathbf{r'}) \tag{13}$$

for all times during which $V_m(t)$ and $V_{m'}(t)$ are macroscopically separated. Note that $\hat{\rho}$ is not the density matrix of a pure state, because

$$\operatorname{tr} \hat{\rho}^2 = \sum_m |c_m|^4 < 1. \tag{14}$$

Nevertheless, it is impossible by any known experiment to display the coherence effects that distinguish between the pure state ρ from the mixture $\hat{\rho}$ for the indicated values of t. Because the phases of the coefficients c_m do not enter into $\hat{\rho}$, one frequently hears the somewhat mysterious statement that the "phases are destroyed in the measurement act."

Let us explore the role of the phases of the c_m somewhat further. It is now abundantly clear that these phases can only be displayed by an experiment if the various sub-beams produced by the S-G apparatus are first recombined. Let us therefore provide a set of fields that perform this function. For sufficiently large values of t, the wave function of the recombined ensemble will then be

$$\Psi(\mathbf{rR}t) = u^0(\mathbf{R}t)\sum_m c_m e^{i\nu_m}\Phi_m(\mathbf{r}), \tag{15}$$

where the phases ν_m depend on path lengths traversed by the different beams. Whereas the density matrix ρ could not be distinguished from the mutilated matrix $\hat{\rho}$ when Ψ had the form (10), this is no longer true once Ψ has assumed the form (15). On the contrary, the pure state (15) is easily distinguished from the mixture

$$\langle \mathbf{rR}|\hat{\rho}(t)|\mathbf{r}'\mathbf{R}'\rangle = u^0(\mathbf{R}t)u^0(\mathbf{R}'t)^* \sum_m |c_m|^2 \Phi_m(\mathbf{r})\Phi_m^*(\mathbf{r}') \tag{16}$$

by a further S-G experiment. This is best understood in terms of an explicit example. Let us assume that the input, i.e., the wave function of Eq. (19.3), is an eigenfunction of $\mathbf{L} \cdot \hat{\mathbf{n}}$ with eigenvalue zero, where $\hat{\mathbf{n}}$ is a unit vector in the x-z plane that makes an angle of Θ with the z-axis of the S-G arrangement (see Fig. 19.1). The initial state is therefore * $u^0(\mathbf{R}t)Y_{l0}(\vartheta_n)$, where ϑ_n is the polar angle referred to the direction $\hat{\mathbf{n}}$. The coefficients c_m are then determined by expanding $Y_{l0}(\vartheta_n)$ in terms of eigenfunctions appropriate to our particular S-G magnet. These are the function $Y_{lm}(\vartheta\varphi)$, where the angles are referred to the coordinate system of Fig. 19.1. The addition theorem,

$$Y_{l0}(\vartheta_n) = \left(\frac{4\pi}{2l+1}\right)^{\frac{1}{2}} \sum_{m=-l}^{l} Y_{lm}(\vartheta\varphi)Y_{lm}(\Theta, 0),$$

then informs us that the desired coefficients are

$$c_m = \left(\frac{4\pi}{2l+1}\right)^{\frac{1}{2}} Y_{lm}(\Theta, 0). \tag{17}$$

In order to simplify the discussion, let us assume that the paths of the various beams are chosen so as to make all the ν_m equal (modulo 2π). In this case the internal wave function of the atom is again $Y_{l0}(\vartheta_n)$ after the beam has been split and recombined. That this state is pure is easily demonstrated by the fact that the beam will not be split if it is allowed to pass through a second S-G apparatus whose field points along $\hat{\mathbf{n}}$, i.e., the second S-G arrangement acts as a *pass filter*. On the other hand, the mixture (16) will be split by a S-G apparatus, no matter what the latter's orientation may be. This is due to the fact that there does not exist a rotation of the coordinate system that transforms the expression

$$\sum_m |Y_{lm}(\vartheta\varphi)|^2 |Y_{lm}(\Theta, 0)|^2$$

into the square of a single spherical harmonic. The interference terms

* We shall not bother to indicate the dependence of the wave function on $|\mathbf{r}|$; as we have seen in Sec. 18, this variable does not play a role in the S-G experiment.

(i.e., those in $\Psi\Psi^*$ with $m \neq m'$) must be present if the probability is to be expressible as the square of a single spherical harmonic.*

The experiment which recombines the output from the S-G apparatus to form the state (15) does not provide us with any information. The S-G magnets and the recombining fields merely force the wave function to undergo a sequence of contortions, which are then carefully undone; the experiment is vacuous, and is not a measurement. If we wish to obtain some information, we shall have to insert devices between the S-G apparatus and the recombining fields that count the number of atoms (i.e., measure the numbers $|c_m|^2$) that traverse the various distinct trajectories. We shall soon show that once this is done, the output of this whole contraption is the mixture $\hat{\rho}$ of Eq. (16), and *not* the pure state (15).

By introducing the counting devices just referred to, we have enlarged the dynamical system with which we are dealing, and so we shall have to increase the number of variables that appear in the Schrödinger equation. For simplicity we assume that each of the $(2l + 1)$ counters is a system possessing only † the degree of freedom z_m. This degree of freedom is

* When $\Theta \neq 0$ it is impossible to construct a pass filter of any kind for the beam described by $\hat{\rho}$. According to (16), the eigenvalues of $\hat{\rho}$ are $|c_m(\Theta)|^2$, which lie between zero and one when $\Theta \neq 0$. These eigenvalues are invariant under all unitary transformations on the spherical harmonics. On the other hand, a pass filter can only be constructed if the density matrix can be cast into diagonal form with all elements but one vanishing, i.e., for a pure state.

The representation which diagonalizes ρ is constructed from the simultaneous eigenfunctions of a certain set of observables. By definition, a pure state pass filter is an arrangement that measures all of these observables and, subsequent to measurement, rejects all systems that do not have the eigenvalues of these observables corresponding to the sole nonvanishing element of the density matrix.

Unfortunately these elegant mathematical statements can only be translated into feasible experimental measuring apparatus under very special circumstances. In general it is not really possible to construct instruments that measure the density matrix completely. Thus for a beam of spin $\frac{1}{2}$ atoms the situation is fairly straightforward, but for larger spins it quickly becomes forbiddingly complex. The $j = \frac{1}{2}$ density matrix is completely specified by the polarization $\mathbf{P}$ (see Sec. 40). The S-G experiment can measure $\mathbf{P}_\perp$, the component of $\mathbf{P}$ perpendicular to the momentum vector of the atoms. One can use a field along the x-direction of Fig. 19.1 to force the missing component $\mathbf{P}\text{-}\mathbf{P}_\perp$ into the z-direction and thereby render it susceptible to measurement. For $j = 1$ atoms ρ is characterized by eight real numbers. Three of these are related to the magnetic moments, and the remaining five to the quadrupole moments. In order to measure ρ completely for a beam of spin-one atoms one therefore needs instruments capable of measuring the quadrupole tensor in addition to the S-G apparatus. Another method for determining ρ in the $j = \frac{1}{2}$ case is taken up in Prob. 10, Chapt. VI. Above and beyond these difficulties of an essentially practical nature, there is our earlier observation that one cannot distinguish ρ from $\hat{\rho}$ by any experiment when the sub-beams are spatially separated.

† Actual counters possess an immense number of degrees of freedom. The consequences of this fact are discussed in Sec. 20.3.

coupled to the center of mass of the atom by an interaction of finite range $U(\mathbf{z}_m - \mathbf{R})$. Before the atoms come within range of the counters, each of the latter is in the ground state described by the wave function $\chi_0(\mathbf{z}_m t)$. As U does not depend on the internal coordinate $\mathbf{r}$, the atom's internal state will still be Φ_m after passage through the counter. Furthermore, if U is a slowly varying function of $|\mathbf{z}_m - \mathbf{R}|$, the atoms will follow a classical trajectory through the counter. This trajectory will differ somewhat from the $U = 0$ trajectory, but the effect can be compensated by the recombining fields. One can therefore produce a single beam, as in the apparatus that did not include the counters. Our final and most important assumption concerning U is that it forces the counter to assume the state $\chi_1(\mathbf{z}_m t)$ after passage of the atom, where χ_1 is macroscopically distinguishable from the counter's initial state χ_0. (Here again we have in mind a situation where χ_0 and χ_1 are nonvanishing in macroscopically separated domains of the variable $\mathbf{z}$.) We shall not attempt to describe a device with these properties here. Suffice it to say that the unadorned S-G arrangement has the same characteristics: it forces the apparatus to assume the state $u_m(\mathbf{R}t)$ subsequent to traversal of the atom, where u_m is macroscopically distinguishable from the initial state $u^0(\mathbf{R}t)$.

We can now write down the explicit wave functions that describe the atomic beam at various important points on the journey through our array of instruments. Consider first the case where the initial internal state of the atoms is one of the Φ_m. Before the beam enters the S-G apparatus the wave function is

$$\Psi_m(\mathbf{rRZ}t) = u^0(\mathbf{R}t)\Phi_m(\mathbf{r}) \prod_n \chi_0(\mathbf{z}_n t). \tag{18}$$

Here we used the abbreviation $\mathbf{Z} \equiv (\mathbf{z}_1, \mathbf{z}_2, \ldots, \mathbf{z}_{2l+1})$. After traversing the S-G magnet, but before interacting with the counters, Ψ_m becomes

$$\Psi_m(t) = u_m(\mathbf{R}t)\Phi_m(\mathbf{r}) \prod_n \chi_0(\mathbf{z}_n t). \tag{19}$$

Subsequent to passage through the counters, all counters but the mth one remain in their ground state, and the wave function assumes the form

$$\Psi_m(t) = u_m(\mathbf{R}t)\Phi_m(\mathbf{r})\chi_1(\mathbf{z}_m t) \prod_{n \neq m} \chi_0(\mathbf{z}_n t). \tag{20}$$

The only further change resulting from the recombining fields is $u_m(\mathbf{R}t) \rightarrow u^0(\mathbf{R}t)e^{i\gamma_m}$. If the initial internal state is $\Sigma\, c_m\Phi_m$, the linearity of the

Schrödinger equation implies that the final state will be

$$\Psi(t) = u^0(\mathbf{R}t) \sum_m e^{i\gamma_m} c_m \Phi_m(\mathbf{r}) \chi_1(\mathbf{z}_m t) \prod_{n \neq m} \chi_0(\mathbf{z}_n t). \tag{21}$$

We hope not to insult the reader by reminding him that this is a pure state.

Consider the dependence of the density matrix $\Psi(\mathbf{rRZ}t)\Psi^*(\mathbf{r'R'Z'}t)$ on the counter coordinates. A typical interference term, associated with $c_1 c_2^*$, say, will have the form

$$\chi_1(\mathbf{z}_1 t)\chi_0^*(\mathbf{z}_1' t)\chi_0(\mathbf{z}_2 t)\chi_1^*(\mathbf{z}_2' t) \prod_{n>2} \chi_0(\mathbf{z}_n t)\chi_0^*(\mathbf{z}_n' t).$$

Because of our assumptions concerning χ_0 and χ_1, it would take a macroscopically nonlocal observable involving the counter variables to detect such an interference term in the density matrix associated with (21). Hence we may drop such terms from the density matrix, and work with the equivalent mixture

$$\langle \mathbf{rRZ}|\hat{\rho}(t)|\mathbf{r'R'Z'}\rangle = u^0(\mathbf{R}t)u^0(\mathbf{R'}t)^* \sum_m |c_m|^2 \Phi_m(\mathbf{r})\Phi_m^*(\mathbf{r'}) \cdot \chi_1(\mathbf{z}_m t)\chi_1^*(\mathbf{z}_m' t) \prod_{n \neq m} \chi_0(\mathbf{z}_n t)\chi_0^*(\mathbf{z}_n' t). \tag{22}$$

Here the phases of the coefficients c_m no longer appear.

It should be stressed that (22) gives a complete * description of the system after the beam has run through the apparatus whether we "look at" the counters or not. To emphasize this point, let us compute the density matrix that describes the atomic beam irrespective of the counter variables. This density matrix must have the property that it gives the correct expectation value of all observables that depend only on the atomic variables. In the evaluation of such an expectation value one merely integrates over the variables in the probability distribution that do not appear in the operator. Hence the density matrix we are seeking is obtained from $\Psi\Psi^*$ by setting all $\mathbf{z}_n = \mathbf{z}_n'$, and integrating over the entire range of these variables. Because χ_0 and χ_1 do not overlap, the interfer-

* The adjective "complete" is only correct if we refrain from subsequently manipulating the counter system so as to reveal interference effects between the states that evolve out of χ_0 and χ_1. Further important remarks concerning this point will be found in Sec. 20.3.

ence terms drop out once more, and we find that *

$$\langle \mathbf{r}\mathbf{R}|\hat{\rho}(t)|\mathbf{r}'\mathbf{R}'\rangle = u^0(\mathbf{R}t)u^0(\mathbf{R}t)^* \sum_m |c_m|^2 \Phi_m(\mathbf{r})\Phi_m^*(\mathbf{r}') \qquad (23)$$

is the density matrix associated with the beam under the stated circumstances. Needless to say, this same expression emerges if we perform the contraction of variables on the mixture (22) instead of on $\Psi\Psi^*$. Equation (23) is precisely the density matrix (16) discussed earlier. Whether we examine the counters or not is therefore irrelevant. The distinction between the experiments in which the mixture $\hat{\rho}$ and the pure state $u^0(\mathbf{R}t)\Sigma e^{i\gamma_m}c_m\Phi_m$ emerge is that in the former case the counters interact with the atoms, whereas in the latter they do not, i.e., in the first case the interaction operator U appears in the Hamiltonian, and in the second it does not. Quantum mechanics may have some rather bizarre and peculiar features, but it is not a subjective and mystical creed.

When the counters are switched on, the experiment may be used to measure the probabilities $|c_m|^2$. To determine these, we could employ a beam that is sufficiently dilute so as to assure that only one atom is in the apparatus at any one time. After each passage we then determine which one of the counters is in the state χ_1 by a measurement of all the z-coordinates. Subsequently we force this counter back into the null position χ_0. By repeating this experiment indefinitely we would find that the fraction of atoms that excited the mth counter is $|c_m|^2$. Note that in each of the separate trials just described one ascertains the complete trajectory of the atom. It is therefore quite natural that the mixture $\hat{\rho}$ of Eq. (23) is precisely the same as would be prepared by combining the output of $2l + 1$ *different* S-G filters, each of which prepares one of the states Φ_m, with each S-G apparatus employing its own source of atoms of relative strength $|c_m|^2$.

We shall briefly examine one other experiment in the hope of elucidating further the coherence properties of states following measurement. Consider again an S-G apparatus backed up by a screen with two holes placed symmetrically about the axis of the field, but with adjustable separation between the holes. Beyond the pierced screen we place a huge detector that can only tell whether an atom was transmitted or not, but not from which hole it comes. By changing the distance between the holes we can

* As Landau observed long ago, this expression also results when we trace over the counter variables under the weaker condition that χ_0 and χ_1 be orthogonal. We prefer to restrict the appellation "measurement" to processes where the states of the object are uniquely correlated to macroscopically distinguishable states of the apparatus. Bohr first emphasized that the measurement process must, in the last analysis, be of this nature if it is to lead to interpretable results.

measure how many beams there are, as well as their deflections. It would seem that we have not "destroyed" any phases, and yet we have measured the angular momentum l ($2l$ equals number of beams) and the elementary magnetic moment μ_0. To analyze this situation it will be sufficient to assume that the source produces a species of atoms that can exist in a ground state $\Phi_{l_1m_1}$, and an excited state $\Phi_{l_2m_2}$ ($l_1 \neq l_2$). Further, assume that the magnetic moments are $\pm\mu_1 l_1$, $\pm\mu_1(l_1 - 1)$, . . . , and $\pm\mu_2 l_2$, $\pm\mu_2(l_2 - 1)$, . . . , respectively. By moving the holes about we shall find that there are l_1 separations that are multiples of a distance d_1 (proportional to μ_1) for which the detector triggers, and another set of l_2 separations, multiples of another length d_2. If we introduce a packet

$$\Psi = \sum_{lm} c_{lm}\Phi_{lm}$$

into the apparatus and have the holes adjusted to multiples of d_1, we shall not "lose" all the phases of the c_{lm}'s. In particular, the phase relations between the appropriate pair of c_{l_1m}'s will be preserved. However, all phase relations between the part of Ψ belonging to $l = l_1$, and the part with $l = l_2$, will be "destroyed." Because this set-up only serves to distinguish states with different values of l and μ, it only "destroys" phase relations between states which differ in these properties. It does not select states with a definite value of m.

•If one wishes to demonstrate the remaining interference terms in

$$\sum_l \left|\sum_m c_{lm}\Phi_{lm}\right|^2, \tag{24}$$

some care must be exercised. By making the holes in the screen that separates out l_1, say, small enough, one can obtain a diffraction pattern in the c.o.m. motion's wave function $u_l(\mathbf{R})$ beyond the screen. The resultant overlap between waves of different m cannot be observed by a photographic plate because this only determines $\mathbf{R}$, and averages over the relative coordinate; because of the orthogonality relation (19.1) the interference term then vanishes. A more appropriate set-up for revealing the interferences in Eq. (24) would be a further S-G apparatus whose collimating slits select out different parts of the diffraction pattern.•

It should be remembered that the profound difference in the coherence properties of states before and after measurement that we have discussed here in the context of the Stern-Gerlach experiment is not really new to us. At the very outset, in Sec. 1, we discovered a similar disappearance of interference patterns when we tried to ascertain the "trajectory" of photons through a grating.

20.3 THE STATISTICAL INTERPRETATION OF QUANTUM MECHANICS

In the preceding pages we have been rather cavalier in deducing the notion that c_m is a probability amplitude. We shall close our discussion of measurement theory by examining the plausibility of this interpretation of c_m. Before doing so we shall review and summarize the salient features of our rather lengthy analysis of the Stern-Gerlach experiment.

Let Φ_m, with $m = 1, 2, \ldots, k$, be a set of orthonormal "object" eigenfunctions belonging to the object observable Λ. Furthermore, let $\Xi(z_1 \cdots z_N t)$ be the initial wave function of the "apparatus" whose variables are $(z_1 \cdots z_N)$. Before measurement the system is in the pure state *

$$\Psi(t) = \Xi^0 \sum_m c_m \Phi_m. \tag{25}$$

An experimental arrangement capable of measuring † the observable Λ forces the wave function to evolve from (25) into the correlated form

$$\Psi(t) = \sum_m c_m \Phi_m \Xi_m. \tag{26}$$

We recall that in the simple Stern-Gerlach experiment (cf. (10)), and also in the arrangement with the counter (cf. (21)), the final wave functions had the form (26).

Let ρ be the density matrix belonging to the pure state (26), i.e., $\rho = \Psi\Psi^*$. If we write $c_m \equiv |c_m|e^{i\alpha_m}$, we may define another density matrix $\hat{\rho}$ by the purely formal operation

$$\hat{\rho} = \int_0^{2\pi} \frac{d\alpha_1}{2\pi} \cdots \frac{d\alpha_k}{2\pi} \rho. \tag{27}$$

* It would certainly be more realistic to treat the initial state of the apparatus as a mixture. We have refrained from incorporating such complications because von Neumann (Chapt. VI) has shown that the argument and conclusions are not essentially altered thereby.

† In our discussion the word "measurement" has been applied to an experimental arrangement that can also be used to prepare ensembles described by eigenfunctions of the observable in question. This is well suited to observables that are (more or less) exact constants of the motion. There is also the important class of measurement devices that change the state of the system, but still allow one to conclude unambiguously what this was before the measurement act. This is exemplified by a Stern-Gerlach apparatus that sprays the various sub-beams onto a photographic emulsion. It can be shown (see Pauli, Sec. 9) that no new and fundamental principles are involved here, and we shall therefore forego an analysis of this type of experiment. In the last analysis all information obtained in the laboratory depends, either directly or indirectly, on measurements of one of these two types.

Averaging over phases has removed the interference terms from ρ, and consequently tr $\hat{\rho}^2 < 1$, whereas tr $\rho^2 = 1$.

An experimental arrangement is a measuring device if and only if the different Ξ_m are macroscopically distinguishable. When this is the case, tr $A\hat{\rho}$ = tr $A\rho$ *for all observables A known to occur in nature.** *Consequently the pure state ρ and the mixture $\hat{\rho}$ are indistinguishable. In this sense it is permissible to say that the measurement process turns a pure state into a mixture, in spite of the fact that* tr ρ^2 *is a rigorous constant of the motion.*

In the simple S-G experiment, where the atom's center of mass is considered to be the apparatus, the indistinguishability of the pure state and the corresponding mixture $\hat{\rho}$ was easily removed by recombining the beams. It is true, of course, that once this is done, the *total* arrangement no longer constitutes a measurement device. Nevertheless, in this example, one can *choose* to forego the possibility of measuring and, after a more or less reasonable period of time, demonstrate that $\hat{\rho} \neq \rho$.

In a realistic experimental arrangement, on the other hand, it is essentially impossible to demonstrate the mathematical fact that $\hat{\rho} \neq \rho$, no matter how much time one has at one's disposal. This statement is based on the following reasoning. In the approximation where one treats an object as a one-body system, the time T required for overlap to occur between two wave packets of dimension a originally separated by a distance R is given by $T \simeq MaR/\hbar$, where M is the total mass of the object. When this object is macroscopic (e.g., a single grain in an emulsion) the large values of M and a imply that T will be very large. For example, for grains of Ag of linear dimension $a \simeq 10^{-4}$ cm, T is of order 10^3 years when $R \simeq 10^{-2}$ cm. This calculation of the time required to establish coherence is actually a gross underestimate, because it does not take the complexity of the grains into account. The apparatus is described by wave functions Ξ_m containing a vast number of variables $(z_1 \cdots z_N)$. Consequently the interference term $\Xi_m^* \Xi_{m'}$ vanishes unless there is overlap in the N-dimensional configuration space, and not only in three-dimensional space. The overlap in the individual variables must be

* As we remarked in Sec. 20.1, this statement requires some qualification. The one particle density matrix of superfluid He, $\langle \mathbf{r}|\rho_1|\mathbf{r}'\rangle$, apparently tends to a constant as $|\mathbf{r} - \mathbf{r}'| \to \infty$ (see Chapt. XII). Superfluids, and also superconductors, therefore provide us with observables that are macroscopically nonlocal in coordinate space. However, it would seem that any contraption that takes advantage of these long-range coherence properties of superfluids would not comply with our definition of a "measuring device," because it would not act as a generalized pointer. For example, if a superfluid could be used to display interference between the separated beams emerging from an S-G apparatus, we could not use the complete arrangement to measure the magnetic moment.

exceedingly close to unity because, loosely speaking, it is raised to the Nth power in $\Xi_m^*\Xi_{m'}$. Clearly the time T_c required to establish coherence is fantastically large.

If we take advantage of the indistinguishability of ρ and $\hat{\rho}$ to say that $\hat{\rho}$ is the state of the system subsequent to measurement, the intuitive interpretation of c_m as a probability amplitude emerges without further ado. This is so because c_m enters $\hat{\rho}$ only via $|c_m|^2$, and the latter quantity appears in $\hat{\rho}$ in precisely the same manner as probabilities do in classical statistical physics. Moreover, *once the experimental arrangement is such as to make the replacement of ρ by $\hat{\rho}$ permissible, the addition of further observational apparatus does not alter the nature of the state of the system in a fundamental way: the c_m continue to appear only in the form of $|c_m|^2$.*

In contrast to Sec. 19, we have not used the probability amplitude interpretation of the coordinate space wave functions explicitly in arriving at these conclusions. Nevertheless, the notion of macroscopically distinguishable states of the apparatus assumes a relationship between the wave function and the spatial location of a system. This relationship can be quite crude, however. For our purposes it would more than suffice to assume that the system is not to be found in a region where the coordinate space wave function vanishes.

In von Neumann's formulation of the measurement process, the practical requirement that Ξ_m and $\Xi_{m'}$ should be macroscopically distinguishable is not taken advantage of, and the mathematically incontestable fact that $\text{tr}\,\rho^2$ is a constant of the motion occupies the center of the stage. Within the ground rules laid down by von Neumann, the interpretation of c_m as a probability amplitude cannot be inferred from the Schrödinger equation. To the extent that nonclassical interference terms (such as $c_m c_{m'}^*$) are present in the mathematical expression for ρ no matter what the interaction between "object" and apparatus may be, the numbers c_m are intuitively uninterpretable, and the theory is an empty mathematical formalism. If one does not wish to surrender the conventional interpretation of c_m, the only alternative is to insist on the validity of this interpretation, and then to make it logically consistent with the mathematical theory. This objective is achieved by von Neumann by enlarging the body of basic laws with the assertion that a state evolves in one of *two* distinct ways: (1) between measurement acts, the wave function develops in a causal fashion according to a linear (Schrödinger) equation, and (2) *if* a measurement *happens* to occur at time t_0, the state changes abruptly according to the recipe $\rho(t_0) \rightarrow \hat{\rho}(t_0)$.*

* $\hat{\rho}(t_0)$ then serves as the initial condition for the further evolution of the state, which is again governed by the linear equation $i\hbar\, \partial\hat{\rho}/\partial t = [H, \hat{\rho}]$.

This second type of change is called *the reduction of the wave function.**

Let us focus our attention once more on the fact that we do not observe ρ but only tr $A\rho$, and that for all *observables* A (though *not* all Hermitian operators!) tr $A\rho$ = tr $A\hat{\rho}$ once the measurement apparatus has been brought into play. Consequently the question of whether the pure state ρ or the mixture $\hat{\rho}$ emerges can only be answered if we are willing to wait for coherence to be re-established between the various wave functions Ξ_m. As we have seen, under realistic conditions this time T_c is monstrously large, and therefore plays a role analogous to that of the Poincaré recurrence time T_P in statistical mechanics. Just as the second law of thermodynamics is only correct over times small compared to T_P, so is the replacement of ρ by $\hat{\rho}$ only valid for times small compared to T_c.† Because our knowledge of the microcosmos is, in any case, restricted to times that are exceedingly short compared to T_c, a discussion of precisely what form the laws of microscopic physics have over times of order T_c takes us outside the realm of purely scientific speculation.

To recapitulate: We are free to replace ρ by $\hat{\rho}$ after the measurement, safe in the knowledge that the error will never be found. Put another way, *only* $|c_m|^2$ appears in the results of *all* observations carried out on the state of the entire system (apparatus + object), and therefore the conventional statistical interpretation of quantum mechanics follows by employing concepts familiar to us on the macroscopic (or classical) level

* The reduction postulate is an independent axiom. One might think that the reduction comes about when one traces over the variables beyond those of (object + apparatus). This is erroneous, however. Let us make the assumption (which is common to almost all theories of physics) that it is possible to abstract a part of the world, and ignore its interaction with the remainder. In terms of our discussion, we enlarge the "apparatus" until the interaction between (object + apparatus) and the rest of the world is negligible. If $\Theta(\xi_1, \ldots)$ is the wave function of the rest of the world, the total wave function after measurement is $\Psi = \Theta(\xi_1, \ldots)\Sigma_m c_m \Phi_m \Xi_m$. By tracing $\Psi\Psi^*$ over the variables ξ_i we obtain ρ and not $\hat{\rho}$.

If one accepts von Neumann's formulation of quantum mechanics, one is led to far reaching and not entirely palatable conclusions. We shall not give the arguments here, and instead refer the interested reader to the articles by Wigner referred to at the end of this chapter. The outcome of these considerations is that quantum mechanics cannot give a complete description of the physical world because there must exist systems (called "conscious" by Wigner) that are beyond the theory's powers of description, i.e., that cannot be incorporated into the part of the world that we treat with the Schrödinger equation.

† In this connection one should note that in approximating ρ by $\hat{\rho}$ one introduces irreversibility, because the time reversed Schrödinger equation cannot retrieve ρ from $\hat{\rho}$.

of perception.* To be sure, a reduction does occur in the statistical distributions that arise from $\hat{\rho}$, but there is nothing novel to quantum mechanics in this. In classical probability theory the state of a coin following the toss is, say, "heads," whereas before the toss it was 50% "tails" and 50% "heads." In the same way the state of any member of the ensemble is $\hat{\rho}$ after the experimental arrangement has done its work, but immediately after an observation ascertains that, say, the mth possibility has actually occurred, that particular member is in the ensemble described by $\Phi_m \Xi_m$. *The essentially new feature in quantum mechanics is that the wave function is asserted to provide the most complete description conceivable, and that the statistical nature of the theory cannot (as in the case of coin tossing) be removed by a more detailed theory involving further variables and/or a more precise specification of initial conditions.*

In closing we emphasize that our discussion has merely consisted of several demonstrations of internal consistency. All that we have shown is that (a) the eigenvalues of certain operators agree with the macroscopic definitions of the associated observables, (b) that it is possible to design experiments that amplify differences between microscopic states into macroscopically distinguishable states of the apparatus, and (c) that the theory involves statistical concepts already familiar to us in classical physics. We have not attempted to reconcile the fact that the theory only makes statistical predictions with our observation of individual, solitary, events.

REFERENCES

1. The classic discussions of measurement theory are in von Neumann, Chapt. 6, Pauli, Sec. 9, and F. London et E. Bauer, *La Théorie de l'Observation en Méchanique Quantique* (Hermann et Cie., Paris 1939). See also Bohm, Part VI.
2. For more recent developments, especially concerning the question of "conscious" systems, see E. P. Wigner, *Am. J. Phys.* **31,** 1 (1963), and an article by the same author in *The Scientist Speculates*, edited by I. J. Good (Heinemann, London 1962).

* It should be noted that the operators that represent observables in the mathematical formalism are also linked to our macroscopic level of perception. In the Stern-Gerlach experiment we deduce the value of the magnetic moment by analyzing the trajectories of the atoms with classical mechanics and electromagnetic theory. The magnetic moment of the atom is thereby *defined* in a purely macroscopic fashion. The mathematical formalism of quantum mechanics also allows us to compute the beam trajectory. A comparison of the quantum and classical calculations of the trajectory then yields a relationship between a mathematical operator (i.e., $e\hbar\mathbf{L}/2mc$) and a quantity that we are used to calling "the magnetic moment."

3. A very ingenious experiment that sheds some light on measurement theory is discussed by Y. Aharonov and D. Bohm, *Phys. Rev.* **115**, 485 (1959), and by W. H. Furry and N. F. Ramsay, *ibid.* **118**, 623 (1960).
4. A similar point of view to the one expressed in Sec. 20 is found in A. Peres and N. Rosen, *Phys. Rev.* **135B**, 1486 (1964). See also J. M. Jauch, *Helv. Phys. Acta* **37**, 293 (1964).
5. The problem of hidden variables is discussed in J. S. Bell, *Rev. Mod. Phys.* **38**, 447 (1966).

PROBLEM

By using a mass spectograph and a Stern-Gerlach apparatus demonstrate that L_z and the Hamiltonian are compatible observables for an isolated atom.

V

States and Observables. Transformation Theory

This and the following chapter are devoted to a formulation of quantum mechanics that is based rather directly on the analysis of measurements. In essence, this was also the guiding philosophy in Heisenberg's first paper. We shall see that the correspondence between the elements of a linear vector space and the pure states of an ensemble emerges quite naturally from such an analysis. In the subsequent chapter we shall show that the form of Schrödinger's equation, and the wave-particle dualism, can be inferred from rather general symmetry considerations. Although the formulation at which we shall arrive is rather abstract, it is considerably more powerful than wave mechanics. As a matter of fact, the equations of wave mechanics are merely one particular concrete realization of the laws of quantum mechanics when these are stated in the abstract language. One might think that this is only a formal convenience, but this is not so. In many practical electromagnetic calculations one uses the methods of vector analysis instead of explicit coordinate systems because the laws of electrodynamics are form-invariant in the vector language. The abstract formulation enjoys similar advantages over wave mechanics, and appears to capture the essence of quantum mechanics. The abstract approach is also well suited to the relativistic domain, but wave mechanics is not. Perhaps it is not surprising that Dirac invented not only the abstract formalism, but also quantum electrodynamics.

21. Measurement Symbols and Transformation Functions *

Let $\{a'\} \equiv (a', a'', \ldots)$ be the set of real values that some particular physical quantity pertaining to the system of interest can assume.† We call such a physical quantity an observable, designate it by A, and call the set of numbers $\{a'\}$ the spectrum of A. In general the system will possess an infinite number of different observables, which we denote by $A, B, C, \ldots$, with spectra $\{a'\}$, $\{b'\}$, $\{c'\}$, ... For the moment these spectra are assumed to be discrete. The generalization to continuous spectra will be discussed eventually.

The square of the angular momentum is a familiar example of an observable. In this case the numbers a' are 0, 2, 6, 12, . . . , etc. (i.e., $l(l+1)$). The arguments that follow are actually best understood in terms of one component of the angular momentum vector, say L_i, and the associated magnetic moment $\mathfrak{M}_i$. In the sequel it is very important to understand that components of $\mathbf{L}$ in different directions are distinct observables, and are therefore not designated by the same symbol A.

Assume that we are presented with an ensemble of systems, such as a beam of particles. From this ensemble we can, by filtration, separate out a sub-ensemble in which all members have the definite value a' of the observable A. The phrase "has the definite value a'" is a frequently used shorthand for the statement that if we measure the observable A in the filtered sub-ensemble, we are sure to obtain the result a'. In a similar vein the phrase "a system is in the *state* a'" is an abbreviation for the statement that the system in question is a member of a sub-ensemble prepared by the filtration described above. For verification we can use a second identical filter and see if the entire sub-ensemble manages to run

* As we have already said, the development of the theory given here is originally due to Dirac. It is expounded at length in his celebrated book, which should be read by all serious students of the subject. Our treatment is based on a "linear" combination of Dirac's work and the papers by J. Schwinger, *Proc. Nat. Acad. Sc.* **45**, 1542 (1959); **46**, 257, 570 (1960).

† The numbers a' shall sometimes be referred to as quantum numbers. They should not be confused with the q-numbers in the nomenclature originally used by Dirac: q-numbers are operators and c-numbers are ordinary numbers!

the gauntlet. The Stern-Gerlach (S-G) experiment followed by a screen with only one hole constitutes such a filter, the ensemble that emerges from it being characterized by a definite value of $\mathfrak{M}$ in some particular direction.

Because there are usually an infinite number of different observables, there are also an infinite number of distinct filtration experiments. In the case of a beam of atoms of angular momentum unity, there are three possible different filtrations *for every orientation* of the vector $\mathfrak{M}$.

We shall now set up a formal scheme—called the measurement algebra—which expresses the essential properties of such filtration experiments in a mathematically fruitful way.* With each filtration measurement we associate an abstract symbol. These *measurement symbols* are labeled by the quantum numbers selected by the associated experiment; thus $M(a')$ corresponds to the process that selects systems that have the value a' of A. The algebraic laws governing the operations on the M's shall be inferred by requiring that the results of these operations are again symbols that may be uniquely associated with measurements. Thus the laws of addition are defined by requiring that $M(a') + M(a'')$ corresponds to a filter that accepts systems that have *either* the value a' *or* a''. This is the case of the S-G experiment followed by a screen with two holes. Because such an experiment is symmetric in a' and a'', we require

$$M(a') + M(a'') = M(a'') + M(a'),$$

and more generally we may also conclude that

$$(M(a') + M(a'')) + M(a''') = M(a') + (M(a'') + M(a''')).$$

The filter that passes everything will be designated by **1**. According to the definition of addition, we see that

$$\sum_{a'} M(a') = \mathbf{1}. \tag{1}$$

At the opposite extreme from **1** we introduce **0** to designate the filter that rejects everything, such as the S-G experiment backed up by a screen without any holes.

Two successive filtrations (such as two S-G experiments in series) will be associated with the product of two M's, the symbol for the first filtra-

* Students with a good knowledge of modern mathematics should consult the discussions of this question due to J. von Neumann and G. Birkhoff, *Annals of Math.* **37**, 823 (1936), D. Finkelstein, J. M. Jauch, S. Schiminovich and D. Speiser, *J. Math. Phys.* **3**, 207 (1962), and Mackey.

tion appearing on the right. According to the definition of **0**, we have

$$M(a')M(a'') = \mathbf{0} \qquad (a' \neq a''), \tag{2}$$

because nothing is transmitted by this double filter. On the other hand, if $a' = a''$ the second filter simply passes everything impinging on it, and therefore

$$M(a')M(a') = M(a'). \tag{3}$$

Hence $M(a')$ is idempotent. If we introduce the notation

$$\delta(a', a'') = \begin{cases} 1 & (a' = a''), \\ 0 & (a' \neq a'') \end{cases}$$

we can combine (2) and (3) into

$$M(a')M(a'') = \delta(a', a'')M(a'). \tag{4}$$

From their physical significance it also follows that

$$\begin{aligned} \mathbf{1} \cdot \mathbf{1} &= \mathbf{1}, \\ \mathbf{0} \cdot \mathbf{0} &= \mathbf{0}, \\ \mathbf{1} \cdot \mathbf{0} &= \mathbf{0}, \\ \mathbf{1}M(a') &= M(a')\mathbf{1} = M(a'), \\ \mathbf{0}M(a') &= M(a')\mathbf{0} = \mathbf{0}, \\ M(a') + \mathbf{0} &= M(a'). \end{aligned}$$

Because **1** and **0** play exactly the same role as 1 and 0 do in arithmetic, we shall cease to designate them by special symbols, and we shall ignore 1 in multiplication.

Let A_1 and A_2 be two *compatible observables*. The adjective *compatible* has the same meaning as in the previous chapters: The filtration associated with A_2 on an ensemble that has previously been prepared in the state a_1' will not destroy the purity of the ensemble with respect to this value of the observable A_1. Therefore the sequence of filtrations $M(a_1')$ and $M(a_2')$ will select an ensemble in which A_1 and A_2 have the values a_1' and a_2', respectively. Let us introduce a symbol for this sequence,

$$M(a_1'a_2') = M(a_1')M(a_2') = M(a_2')M(a_1'). \tag{5}$$

The multiplication is commutative here just because the observables A_1 and A_2 are compatible. Let $A_1, \ldots, A_f$ be a complete set of compatible observables. Then

$$M(a_1' \cdots a_f') = \prod_{i=1}^{f} M(a_i') \tag{6}$$

stands for the sequence of filtrations that prepares an ensemble concerning which we have optimal knowledge. By definition, the measurement of any observable that is not a function of the set $\{A_i\}$ will produce an ensemble in which the observables A_i no longer have definite values. It is now a simple matter to show that the symbols $M(a_1' \cdots a_f')$ have the same properties as the $M(a')$, provided

$$\delta(a_1' \cdots a_f', a_1'' \cdots a_f'') = \prod_{i=1}^{f} \delta(a_i', a_i''). \tag{7}$$

Because the algebra of M's belonging to a (complete) compatible set is the same as that of the simple $\{M(a')\}$, we shall work with one operator A most of the time. It will be understood that A stands for a complete set, and that the quantity a' actually stands for f real numbers.

Up to now we have only considered filtrations concerning the spectrum of compatible observables. The essence of quantum mechanics, of course, is that there exist incompatible observables. That is to say, we could equally well describe our system in terms of a property B with spectrum $(b', b'', \ldots)$, and the associated measurement symbols $M(b')$. We may now ask: What is $M(a')M(b')$? This sequence of filtrations accepts systems in b' and manufactures systems in a'. Clearly no single symbol introduced so far can be associated with this process, and hence our algebra must be augmented with further objects of a more complicated type.

Before embarking on this general problem, we begin by considering something only one step more complicated, namely an operation on an ensemble that selects systems in the state a' and issues them forth in the state a''. This could, for example, be an S-G apparatus that selects m', followed by a radio-frequency field that induces transitions between the various Zeeman states, and then another S-G set-up that selects out some other magnetic quantum number m''. The symbol to be defined momentarily actually stands for an idealization of such an experiment, i.e., an apparatus that accepts the state m', and then puts *all* accepted systems into m''. In the case of an arbitrary observable A, we shall associate the symbol* $M(a', a'')$ with a device that only accepts systems in the state a'', and then issues forth every accepted system in the state a'. Clearly

$$M(a', a') = M(a'). \tag{8}$$

A sequence of such measurements, $M(a^{\text{iv}}, a''')M(a'', a')$, will not, by definition, pass anything if $a'' \neq a'''$. On the other hand, if $a'' = a'''$,

* This symbol is not to be confused with the one appearing in (5).

the middle selector is irrelevant. Therefore

$$M(a^{iv}, a''')M(a'', a') = \delta(a''', a'')M(a^{iv}, a'). \tag{9}$$

Because

$$M(a'', a')M(a^{iv}, a''') = \delta(a', a^{iv})M(a'', a'''), \tag{10}$$

multiplication is not commutative. From the definition it is also clear that

$$M(a', a'') \neq M(a'', a'). \tag{11}$$

The next step in the argument is more difficult, but also more incisive, because it concerns incompatible observables. Let B designate a set of mutually compatible observables that are *incompatible* with A. Thus A could stand for the set L^2 and L_z, B for L^2 and L_y. Let $M(a', b')$ stand for the measurement that accepts only systems in state b', and ejects them in a state a'. Note that the admitted and ejected states refer to incompatible observables. A Stern-Gerlach apparatus set up to accept a particular m, followed by another apparatus with a twisted magnetic field that rotates the orientation of the accepted magnetic moment, is an example of this kind of device. Consider $M(a', b')$ $M(c', d')$; clearly this performs the same selection and preparation function as $M(a', d')$. It is not true, however, that $M(a', b')M(c', d') = M(a', d')$, as we can see from a succession of two of our twisting S-G's. Let m_z', m_x', etc., stand for magnetic quantum numbers referred to the z-, x-, etc. axes. Then $M(m_x', m_z')$ symbolizes two S-G's; the first one selects m_z', and the second has a twisted field that rotates the moment into the x-direction and then forces it into m_x'. A second experiment $M(m_y', m_z'')$ will not accept everything that has come out of $M(m_x', m_z')$, no matter what the values of m_z'' and m_x' are, because they refer to different axes. (The first measurement produces a wave function that is a linear combination of eigenfunctions appropriate to the accepting filter of the second apparatus.) We therefore put

$$M(a', b')M(c', d') = \langle b'|c'\rangle M(a', d'), \tag{12}$$

where $\langle b'|c'\rangle$ is related to the fraction of systems in the state c' that can be accepted by a B-filter when it is set to admit state b'. We already know from (9) and (10) that

$$\langle a'|a''\rangle = \delta(a', a''). \tag{13}$$

*We shall assume that in the general case $\langle b'|c'\rangle$ is also a number, i.e., an object that commutes with the M-symbols.** The deviation of $\langle b'|c'\rangle$ from 1

* It is also possible to construct a mathematically consistent theory in which the quantities $\langle b'|c'\rangle$ are quaternions. See von Neumann and Birkhoff, *loc. cit.*, and D. Finkelstein *et al.*, *loc. cit.*

or 0 is, in some as yet unspecified sense, a measure of the incompatibility of B and C.

Equations (8) and (12) imply that

$$M(a')M(b', c') = \langle a'|b'\rangle M(a', c'),$$
$$M(a', b')M(c') = M(a', c')\langle b'|c'\rangle.$$

Because of the completeness relation (1), we then have

$$M(b', c') = \sum_{a'} \langle a'|b'\rangle M(a', c'), \tag{14}$$

$$M(a', b') = \sum_{c'} M(a', c')\langle b'|c'\rangle, \tag{15}$$

and by combining these relations we obtain

$$\begin{aligned} M(a', b') &= \sum_{c',d'} M(c')M(a', b')M(d') \\ &= \sum_{c',d'} \langle c'|a'\rangle\langle b'|d'\rangle M(c', d'). \end{aligned} \tag{16}$$

This identity shows that the measurement symbols of the A-B type are *linearly* related to those of any other type (C-D, for example). The array of numbers $\langle a'|b'\rangle$, etc., effect the transformations between the various complete (and therefore equivalent) descriptions, and are called *transformation functions.* Equation (16) also shows that the set of symbols $\{M(a', b')\}$ is closed, and the deficiency found earlier for the simpler symbols $M(a')$, etc., is therefore removed.

A very important relation between the transformation functions themselves can be derived from the preceding equations:

$$\begin{aligned} M(a')M(c') &= \langle a'|c'\rangle M(a', c') \\ &= \sum_{b'} M(a')M(b')M(c') = \sum_{b'} \langle a'|b'\rangle\langle b'|c'\rangle M(a', c'), \end{aligned}$$

or

$$\sum_{b'} \langle a'|b'\rangle\langle b'|c'\rangle = \langle a'|c'\rangle. \tag{17}$$

By setting $c' = a''$, and using (13), we also obtain

$$\sum_{b'} \langle a'|b'\rangle\langle b'|a''\rangle = \delta(a', a''). \tag{18}$$

22. Probabilities

Let $\{\lambda(a')\}$ and $\{\lambda(b')\}$ be arbitrary sets of nonzero numbers. We observe that the transformation

$$M(a', b') \rightarrow \lambda(a')M(a', b')[\lambda(b')]^{-1}, \qquad (1)$$
$$\langle a'|b'\rangle \rightarrow \frac{1}{\lambda(a')}\langle a'|b'\rangle\lambda(b'),$$

leaves the measurement algebra invariant. Hence the numbers $\langle b'|c'\rangle$ cannot themselves be physically significant, even though the argument preceding (21.12) might have led one to suppose that $\langle b'|c'\rangle$ is the probability of finding the state b' when a measurement is performed on a system known to be in the state c'.

Consider the filter sequence $M(a')M(b')M(a')$, which differs from $M(a')$ only in that a b' measurement is carried out between the two a'-determinations:

$$M(a')M(b')M(a') = \langle a'|b'\rangle\langle b'|a'\rangle M(a'). \qquad (2)$$

The coefficient of $M(a')$,

$$p(a', b') = \langle a'|b'\rangle\langle b'|a'\rangle, \qquad (3)$$

is invariant under the transformation (1). According to (21.18), furthermore, $p(a', b')$ is normalized to unity:

$$\sum_{a'} p(a', b') = \sum_{b'} p(a', b') = 1. \qquad (4)$$

We shall therefore attempt to interpret $p(a', b') = \langle a'|b'\rangle\langle b'|a'\rangle$ *as the probability for finding a system in the state* b' *when a B-filtration is performed on an ensemble known to be in the state* a'. Note that (3) is symmetric in a', b', and therefore the previous statement also holds if the roles of A and B are interchanged. This is a necessary symmetry because the probability that a system be found in the state b' when it is prepared to be in a' must surely equal the probability that a system be in the state a' when it was originally prepared in the state b'.

If $p(a', b')$ is to be a probability, it must be real and satisfy

$$0 \leqslant p(a', b') \leqslant 1. \qquad (5)$$

An especially trivial probability is

$$p(a', a'') = \delta(a', a''). \qquad (6)$$

22. Probabilities

Let us consider successive measurements once more. Take an ensemble in the state c', perform the B-filtration b', and then the A-filtration a'. The probability for surviving these discriminations is

$$p(a', b', c') = p(a', b')p(b', c'). \tag{7}$$

When the B-filter operates (i.e., records the value of b' each time a member of the ensemble goes through) but does not select, the relevant probability is

$$\begin{aligned} p(a', B, c') &\equiv \sum_{b'} p(a', b', c') = \sum_{b'} \langle a'|b'\rangle\langle b'|a'\rangle\langle b'|c'\rangle\langle c'|b'\rangle \\ &= \sum_{b'} \langle a'|b'\rangle\langle b'|c'\rangle\langle c'|b'\rangle\langle b'|a'\rangle. \end{aligned} \tag{8}$$

If, on the other hand, the B-filter was inoperative (i.e., either not there, or not working, which is the same thing) the probability for a member of the c'-ensemble to appear in the state a' is $p(a', c') = \langle a'|c'\rangle\langle c'|a'\rangle$. With the help of (21.17) this probability can be written as

$$p(a', c) = \sum_{b'b''} \langle a'|b'\rangle\langle b'|c'\rangle\langle c'|b''\rangle\langle b''|a'\rangle. \tag{9}$$

Observe that this expression does *not* equal $p(a', B, c')$. Although this result is completely unexpected from the classical point of view, it agrees with the conclusions reached in Sec. 20.2 from the analysis of the Stern-Gerlach apparatus with recombined beams. It is a virtue of the present formulation that this very important aspect of measurements appears in the theory at so rudimentary a stage. If we look back over the preceding pages, we note that the essential step that leads inevitably to the difference between (8) and (9) is the recognition that all physical quantities are *not* compatible with each other in microscopic physics.* From the point of view of the present formulation this is an *empirical* fact which we express by (21.12). The notion of incompatible observables is completely foreign to classical physics, on the other hand. Classically L_x and L_y can simultaneously have precise values, i.e., one can prepare a classical system with some value of L_x, and then measure L_y without disturbing the previously assigned value of L_x. Although filtration measurements are conceivable classically, we could say that classical measurement algebra is solely concerned with symbols like those of (21.6), and never with A-B symbols. Hence the only transformation functions

* Planck's constant scales the magnitude of the incompatibility between various observables. As long as $\hbar \neq 0$, we will be lead to some essential modification of the classical measurement concept. Thus $\hbar$ plays a role rather similar to that of c in relativity theory, as we had already observed in Sec. 2.

that would occur are $\delta(a', a'')$, and the "interference of probabilities" expressed by (9) would never appear.*

The final statistical notion that we introduce at this time is that of *the expectation value of an observable.* By definition, an ensemble in the state a' will, upon measurement, be sure to reveal the value a' of observable A. What about an ensemble prepared by $M(b')$? Members of this ensemble will not with certainty reveal any particular value of A. In fact, the probability that such an ensemble will be found in a' is precisely $p(b', a')$. The mean value of A in this distribution is called the expectation value of A in the state b', $\langle A\rangle_{b'}$:

$$\langle A\rangle_{b'} = \sum_{a'} a' p(a', b') = \sum_{a'} \langle b'|a'\rangle a' \langle a'|b'\rangle. \tag{10}$$

For any function of the observable A, $f(A)$, we define the expectation value as

$$\langle f(A)\rangle_{b'} = \sum_{a'} f(a') p(a', b'). \tag{11}$$

In particular, the dispersion or uncertainty of A in b', $\Delta A_{b'}$, is given by the second cumulant of the distribution,

$$\langle(\Delta A_{b'})^2\rangle = \langle(A - \langle A\rangle_{b'})^2\rangle = \sum_{a'} [a'^2 - \langle A\rangle_{b'}^2] p(a', b'). \tag{12}$$

23. State Vectors and Operators; Unitary Transformations

It is now clear that an essential task of quantum theory is the evaluation of the transformation functions. In fact, the time dependence of these quantities determines the dynamical properties of the system. (At the moment time has not entered as yet; we are dealing with "quantum statics.") A further task is the determination of the spectra of observables. Traditionally this program has been carried out with the help of a multidimensional (usually ∞) vector space, from which the measurement algebra can subsequently be derived. Whether it is actually possi-

* It is of course true that (9) holds only if we make the identification $p(a', b') = \langle a'|b'\rangle\langle b'|a'\rangle$. Because of the symmetry (1), the probabilities must be bilinear, or quadrilinear, etc., in the transformation function. Equation (3) is merely the simplest choice. Clearly a more complicated definition of p would also differ from the classical result.

ble to construct a complete theory without introducing the vector space is a question that we shall not dwell on. In any case, the "geometric" language has the virtue of being somewhat more "visualizable."

23.1 INTRODUCTION OF A VECTOR SPACE

The possibility of a geometric realization is contained in the basic properties of transformation functions

$$\sum_{b'} \langle a'|b'\rangle\langle b'|c'\rangle = \langle a'|c'\rangle, \tag{1}$$

$$\langle a'|a''\rangle = \delta(a', a''), \tag{2}$$

and the related linearity of the M-algebra

$$M(a', b') = \sum_{c'd'} \langle c'|a'\rangle\langle b'|d'\rangle M(c', d'). \tag{3}$$

The probability interpretation also requires $\langle a'|b'\rangle\langle b'|a'\rangle$ to be real and between 0 and 1.

Now we must choose the kind of numbers that are to be assigned to the transformation functions. By restricting oneself to real numbers one eventually constructs a rather clumsy formalism, and so we shall use the wisdom of hindsight and choose complex numbers. Since $\langle a'|b'\rangle\langle b'|a'\rangle$ is to be real and positive for all a' and b', it is natural to assume that

$$\langle a'|b'\rangle = \langle b'|a'\rangle^*, \qquad p(a', b') = |\langle a'|b'\rangle|^2. \tag{4}$$

The basic relations (1) and (2) now read

$$\sum_{b'} \langle a'|b'\rangle\langle c'|b'\rangle^* = \langle a'|c'\rangle, \tag{5}$$

and

$$\sum_{b'} \langle a'|b'\rangle\langle a''|b'\rangle^* = \delta(a', a''). \tag{6}$$

Consider now a complex N-dimensional vector space. A vector $\mathbf{U}$ is the N-tuple of complex numbers $(U_1, \ldots, U_N)$. The scalar product between two vectors $\mathbf{U}$ and $\mathbf{V}$ is defined as

$$\mathbf{U}^* \cdot \mathbf{V} = \sum_{i=1}^{N} U_i^* V_i. \tag{7}$$

A set of N linearly independent vectors $\mathbf{u}^i$ $(i = 1, \ldots, N)$ is said to form an orthonormal basis if

$$\mathbf{u}^{i*} \cdot \mathbf{u}^j = \delta_{ij}. \tag{8}$$

An example of a basis are the vectors $\mathbf{e}^1 = (1, 0, 0, \ldots)$, $\mathbf{e}^2 = (0, 1, 0, \ldots)$, etc. In terms of these

$$\mathbf{U}^* \cdot \mathbf{V} = \sum (\mathbf{U}^* \cdot \mathbf{e}^i)(\mathbf{e}^i \cdot \mathbf{V}). \tag{7'}$$

The connection between (5) and (6) on the one hand, and (7) and (8) on the other, is now apparent. Let $\mathbf{a}'$ and $\mathbf{c}'$ be the vectors having the following components:

$$\mathbf{a}' = (\langle a'|b'\rangle, \langle a'|b''\rangle, \ldots),$$
$$\mathbf{c}' = (\langle c'|b'\rangle, \langle c'|b''\rangle, \ldots).$$

Then

$$\mathbf{a}' \cdot \mathbf{c}'^* = \sum_{i=1}^{N} \langle a'|b^i\rangle\langle c'|b^i\rangle^* = \langle a'|c'\rangle.$$

Comparing (8) and (6) we observe that the set of vectors $(\mathbf{a}', \mathbf{a}'', \ldots)$ forms an orthonormal basis, as do $(\mathbf{b}', \mathbf{b}'', \ldots)$ and $(\mathbf{c}', \mathbf{c}'', \ldots)$.

23.2 BRAS, KETS, AND LINEAR OPERATORS

We shall not use the conventional notation of vector algebra any further, and shall follow Dirac instead. To every state of an ensemble specified by a complete set of quantum numbers a' we associate an abstract vector $|a'\rangle$. This set of vectors $\{|a'\rangle\}$ spans an abstract vector space (or Hilbert space when the number N of distinct values that the observable A can assume is infinite). By "span" we mean that if $|\beta\rangle$ is a vector associated with a state specified by the values of any other set of observables, there exists a unique expansion $|\beta\rangle = \Sigma_{a'} C_{a'}|a'\rangle$, where $C_{a'}$ is a complex number. We also define a dual vector space spanned by the dual vectors $\langle a'|$. These dual vectors stand in one-to-one correspondence with the vectors $|a'\rangle$: If $|\beta\rangle$ is the vector given above, then $\langle\beta| = \Sigma_{a'} C_a^*\langle a'|$. To every pair of vectors $|a'\rangle$ and $\langle b'|$ we assign a scalar product $\langle b'|a'\rangle = \langle a'|b'\rangle^*$. As a consequence, if $|\alpha\rangle = \lambda'|a'\rangle + \lambda''|a''\rangle$, and $|\beta\rangle = \mu'|b'\rangle$, then

$$\langle\beta|\alpha\rangle = \mu'^*\lambda'\langle b'|a'\rangle + \mu'^*\lambda''\langle b'|a''\rangle. \tag{9}$$

A vector $|\alpha\rangle$ is said to be normalized to unity if $\langle\alpha|\alpha\rangle = 1$, and $|\alpha\rangle$ and $\langle\beta|$ are orthogonal if $\langle\beta|\alpha\rangle = 0$. Although we shall frequently say that $|\alpha\rangle$ and $|\beta\rangle$ are orthogonal if $\langle\beta|\alpha\rangle = 0$, it should be remembered that the scalar product is only defined between a vector and a dual vector. The sets $\{|a'\rangle\}$, $\{|b'\rangle\}$, etc., are all orthonormal. By hypothesis they span the space, and each of these sets therefore constitutes a basis.

A variety of names are associated with the vectors that we have just defined, and we shall use them all rather indiscriminately. The objects $|a'\rangle$, etc., are frequently referred to as *state vectors*, or simply vectors, and occasionally they are even called states. We shall also use the Dirac terminology: $|a'\rangle$ is called a *ket*, and $\langle a'|$ a *bra*.

By hypothesis an arbitrary ket $|\beta\rangle$ can be written as $|\beta\rangle = \Sigma_{a'} C_{a'}|a'\rangle$. Because of the orthonormality properties of the set $\{|a'\rangle\}$, we immediately obtain $C_{a'} = \langle a'|\beta\rangle$, whence

$$|\beta\rangle = \sum_{a'} |a'\rangle\langle a'|\beta\rangle. \tag{10}$$

By the same token

$$\langle\alpha| = \sum_{a'} \langle\alpha|a'\rangle\langle a'|, \tag{11}$$

and therefore

$$\langle\alpha|\beta\rangle = \sum_{a'} \langle\alpha|a'\rangle\langle a'|\beta\rangle. \tag{12}$$

This agrees with (5), as it must.

Finally we construct operators that map vectors into other vectors. With but one exception (see Sec. 27) we shall only be concerned with *linear operators*. By definition, an operator $\mathcal{O}$ is linear if for every pair of vectors $|\alpha\rangle$ and $|\beta\rangle$ it satisfies $\mathcal{O}[\lambda_\alpha|\alpha\rangle + \lambda_\beta|\beta\rangle] = \lambda_\alpha\mathcal{O}|\alpha\rangle + \lambda_\beta\mathcal{O}|\beta\rangle$, where λ_α and λ_β are complex numbers. The most elementary operator is one that maps a basis vector ($|a'\rangle$, say) belonging to one set of observables into a basis vector belonging to some other set (say $|b'\rangle$). We may write such an operator as $|b'\rangle\langle a'|$, with the understanding that when it multiplies a ket $|\alpha\rangle$, it stands for the operations

$$(|b'\rangle\langle a'|) \cdot |\alpha\rangle \equiv |b'\rangle\langle a'|\alpha\rangle. \tag{13}$$

Thus $|b'\rangle\langle a'|$ is an operator that takes any ket and "rotates" it into the "direction" $|b'\rangle$, the coefficient being the scalar product of $|a'\rangle$ with the ket in question. We also define operation to the left onto bras by

$$\langle\alpha| \cdot (|b'\rangle\langle a'|) \equiv \langle\alpha|b'\rangle\langle a'|. \tag{14}$$

The multiplication law of such operators is obviously

$$(|a'\rangle\langle b'|) \cdot (|c'\rangle\langle d'|) = \langle b'|c'\rangle \cdot (|a'\rangle\langle d'|). \tag{15}$$

Usually we shall write relations such as (15) in the more convenient and completely equivalent form

$$|a'\rangle\langle b'| \cdot |c'\rangle\langle d'| = |a'\rangle\langle b'|c'\rangle\langle d'|. \tag{15'}$$

The action of this operator onto a ket or bra can then be seen by inspection:

$$(|a'\rangle\langle b'| \cdot |c'\rangle\langle d'|)|\alpha\rangle = |a'\rangle \cdot \langle b'|c'\rangle\langle d'|\alpha\rangle,$$
$$\langle\alpha|(|a'\rangle\langle b'| \cdot |c'\rangle\langle d'|) = \langle\alpha|a'\rangle\langle b'|c'\rangle \cdot \langle d'|.$$

If we compare (15) with (21.12), we observe that $|a'\rangle\langle b'|$ has exactly the same algebraic properties as $M(a', b')$. We therefore identify the operator $|a'\rangle\langle b'|$ with the measurement symbol $M(a', b')$:

$$|a'\rangle\langle b'| = M(a', b'). \tag{16}$$

Note that $M(a') = |a'\rangle\langle a'|$ is *a projection operator*, because for any vector $|\alpha\rangle = \Sigma_{a'}|a'\rangle\langle a'|\alpha\rangle$ it projects out the portion of the vector in the "direction" $|a'\rangle$:

$$M(a')|\alpha\rangle = |a'\rangle\langle a'| \cdot \sum_{a''} |a''\rangle\langle a''|\alpha\rangle$$
$$= |a'\rangle\langle a'|\alpha\rangle.$$

The relation $\Sigma_{a'}M(a') = 1$ now takes on the form

$$\sum_{a'} |a'\rangle\langle a'| = 1, \tag{17}$$

and will be called the *completeness relation*.

Equation (17) is, from a purely arithmetic point of view, one of Dirac's most useful notational innovations. Thus, with its help, (10) may be "computed" as follows:

$$|\beta\rangle = 1|\beta\rangle = \left(\sum_{a'} |a'\rangle\langle a'|\right) |\beta\rangle = \sum_{a'} |a'\rangle\langle a'|\beta\rangle.$$

Similarly (12) is "evaluated" by the steps

$$\langle\alpha|\beta\rangle = (\langle\alpha|)1(|\beta\rangle) = (\langle\alpha|) \left(\sum_{a'} |a'\rangle\langle a'|\right) (|\beta\rangle)$$
$$= \sum_{a'} \langle\alpha|a'\rangle\langle a'|\beta\rangle.$$

We may represent any linear operator X in a fashion similar to (16). Let X be defined by its action on the A-basis,

$$X|a'\rangle = \sum_{a''} |a''\rangle\langle a''|X|a'\rangle,$$

where $\langle a''|X|a'\rangle$ is a complex number. Then we can write

$$X = \sum_{a'a''} |a''\rangle\langle a''|X|a'\rangle\langle a'|, \tag{18}$$

where one should note how (17) has been "used" again. The array of numbers $\langle a''|X|a'\rangle$ is called the *matrix* of the operator X in the A-representation. We may also construct the representation of the product of two operators X and Y:

$$XY = \sum_{a'a''} |a'\rangle\langle a'|XY|a''\rangle\langle a''|$$
$$= \sum_{a'a''a'''} |a'\rangle\langle a'|X|a''\rangle\langle a''|Y|a'''\rangle\langle a'''|. \tag{19}$$

Thus

$$\langle a'|XY|a''\rangle = \sum_{a'''} \langle a'|X|a'''\rangle\langle a'''|Y|a''\rangle,$$

which shows that the nomenclature of matrix element for $\langle a'|X|a''\rangle$ is indeed appropriate. The *trace* of an operator is defined as

$$\operatorname{tr} X = \sum_{a'} \langle a'|X|a'\rangle, \tag{20}$$

and the *determinant* as

$$\det X = \det\{\langle a'|X|a''\rangle\}. \tag{21}$$

Thus, for example,

$$\operatorname{tr} M(a',b') = \sum_{a''} \langle a''|a'\rangle\langle b'|a''\rangle = \langle b'|a'\rangle. \tag{22}$$

We also define the transpose, complex conjugate, and adjoint operators as

$$X^T = \sum_{a'a''} |a'\rangle\langle a''|X|a'\rangle\langle a''|, \tag{23}$$
$$X^* = \sum_{a'a''} |a'\rangle\langle a'|X|a''\rangle^*\langle a''|, \tag{24}$$
$$X^\dagger = X^{*T} = \sum_{a'a''} |a'\rangle\langle a''|X|a'\rangle^*\langle a''|, \tag{25}$$

respectively. From (25) one readily shows

$$M(a',b')^\dagger = (|a'\rangle\langle b'|)^\dagger = |b'\rangle\langle a'| = M(b',a'), \tag{26}$$

and also

$$(XY)^\dagger = Y^\dagger X^\dagger, \qquad (XY)^T = Y^T X^T, \tag{27}$$
$$(\lambda X)^\dagger = \lambda^* X^\dagger, \tag{28}$$

where λ is a number. If $X = X^\dagger$, X is said to be *Hermitian;* if $X = X^T$, it is called *symmetric;* and if $X = X^*$, it is called *real.*

23.3 UNITARY OPERATORS

Another very important class of objects is the *unitary operators*. Let $|a'\rangle, \ldots, |a^k\rangle, \ldots$, be the A-basis ordered in some fixed way, $|b'\rangle, \ldots, |b^k\rangle, \ldots$, a uniquely ordered B-basis, and so forth. For example, $\{|a'\rangle\}$ could be the various projections of a magnetic moment along the z-axis ordered from the largest to the smallest value of m, whereas the $\{|b'\rangle\}$ could be the projections along the y-axis ordered from smallest to largest value of m. It is the task of the unitary operators to "turn" the basis vectors of one representation into those of another. Thus we require the unitary operator U_{ab} to perform as follows:

$$\begin{aligned} U_{ab}|b^k\rangle &= |a^k\rangle \\ \langle a^k|U_{ab} &= \langle b^k| \end{aligned} \qquad (k = 1, 2, \ldots, N). \tag{29}$$

Note that these defining relations assume a definite pairing of the sets $\{|a'\rangle\}$ and $\{|b'\rangle\}$. Because $\langle a^k|a^q\rangle = \delta_{kq}$,

$$U_{ab} = \sum_k |a^k\rangle\langle b^k|, \tag{30}$$

where we again note that the sum is over ordered pairs.* According to (26) the adjoint of U_{ab} is

$$U_{ab}^\dagger = \sum_k |b^k\rangle\langle a^k| = U_{ba}. \tag{31}$$

Moreover

$$U_{ab}U_{bc} = \sum_{k,q} |a^k\rangle\langle b^k| \cdot |b^q\rangle\langle c^q| = \sum_k |a^k\rangle\langle c^k|,$$

or

$$U_{ab}U_{bc} = U_{ac}, \tag{32}$$

and

$$U_{ab}U_{ba} = 1. \tag{33}$$

Hence

$$U_{ab} = (U_{ba})^{-1} = U_{ba}^\dagger, \tag{34}$$

which means that

$$U_{ab}U_{ab}^\dagger = U_{ab}^\dagger U_{ab} = 1. \tag{35}$$

This is the usual definition of a unitary operator.

An essential property of the unitary operators is the following. Let $|\alpha\rangle$ and $|\beta\rangle$ be two arbitrary kets, and let $|\alpha'\rangle$ and $|\beta'\rangle$ be their images under

* Aside from the nontrivial operators (30), there are also very trivial unitary operators that merely reshuffle the conventional order of vectors within a fixed representation.

the transformation U, i.e., $U|\alpha\rangle = |\alpha'\rangle$, and $U|\beta\rangle = |\beta'\rangle$; then $\langle\alpha|\beta\rangle = \langle\alpha'|\beta'\rangle$. This is summarized by the statement that *the scalar product is invariant under a unitary transformation of the product vectors.*

Let U_1 and U_2 be unitary operators. Because of (27)

$$U_1U_2(U_1U_2)^\dagger = 1. \tag{36}$$

Hence the product of any two unitary operators is again unitary. Equation (34) shows that the inverse of a unitary operator is also a unitary operator. As the unit operator is obviously unitary, we therefore conclude that *the set of all the unitary operators on the vector space forms a group.* We shall explore the consequences of this fact in Chapt. VI.

We shall now study the transformation of operators under a change of representation. Here the basic question is the following: Let $|\alpha\rangle$ be an arbitrary ket, and $|\alpha_X\rangle$ its image under the operator X, i.e., $|\alpha_X\rangle = X|\alpha\rangle$; what is the operator X' that performs the mapping $U|\alpha\rangle \rightarrow U|\alpha_X\rangle$? To answer this we note that $|\alpha\rangle$ is arbitrary, and hence $X'U|\alpha\rangle = UX|\alpha\rangle$ implies $X'U = UX$. Therefore (35) yields

$$X' = UXU^\dagger, \qquad X = U^\dagger X'U. \tag{37}$$

The transformation $X \rightarrow X'$ is called *a unitary transformation* on the operator X, and two operators connected by a unitary transformation are said to be *unitary equivalents.* In passing we may note some simple properties of the transformation (37). If the transformation is generated by U_{ba}, the matrix elements of X and X' are connected by

$$\langle a^k|X|a^q\rangle = \langle b^k|X'|b^q\rangle, \tag{38}$$

and therefore *

$$X' = \sum_{kq} |b^k\rangle\langle a^k|X|a^q\rangle\langle b^q|. \tag{39}$$

It is also easy to prove the following useful identities:

$$\begin{aligned} |\det U| &= 1, \\ \operatorname{tr} UXU^\dagger &= \operatorname{tr} X, \\ \det UXU^\dagger &= \det X, \\ (UXU^\dagger)^\dagger &= UX^\dagger U^\dagger. \end{aligned} \tag{40}$$

* One should not confuse this formula with

$$X = \sum_{kqk'q'} |b^k\rangle\langle b^k|a^q\rangle\langle a^q|X|a^{q'}\rangle\langle a^{q'}|b^{k'}\rangle\langle b^{k'}|.$$

23.4 OBSERVABLES

Next we consider the question of constructing operators for the observables themselves. The information at our disposal is that A is completely specified by the (real) numbers a', a'', . . . in the representation $|a'\rangle$, $|a''\rangle$, ... Because a measurement of A on any ensemble in a state $|a'\rangle$ cannot disturb this state, we must have $A = \Sigma_{a'} f(a')M(a')$, where $f(a')$ is a numerical function of a'. It is most convenient (though not necessary) to simply put

$$A \equiv \sum_{a'} a'M(a') = \sum_{a'} |a'\rangle a'\langle a'|. \tag{41}$$

When we recall that the expectation value of A in $|b'\rangle$ is, by definition, $\langle A\rangle_{b'} = \Sigma_{a'} a'|\langle b'|a'\rangle|^2$, we see that (41) implies the formula

$$\langle A\rangle_{b'} = \langle b'|A|b'\rangle. \tag{42}$$

Because $\{a'\}$ is real, $A = A^\dagger$: *observables are represented by Hermitian operators.* Furthermore, (41) implies

$$\begin{aligned} (A - a')|a'\rangle &= 0, \\ \langle a'|(A - a') &= 0, \end{aligned} \tag{43}$$

or $|a'\rangle$ *is an eigenket of A with eigenvalue* a'. If N is finite the spectrum is found from the Cayley-Hamilton theorem

$$\det(A - \lambda) = 0, \tag{44}$$

where the roots λ are then the numbers $\{a'\}$. In the case where N is infinite one must usually resort to the methods of analysis to find the eigenvalues.

Let us apply a unitary transformation to (43), but for once let us make the fact that A really stands for a complete set of f compatible observables explicit. Thus (43) stands for the f equations $(A_i - a_i')|a_1' \cdots a_f'\rangle = 0$, with $i = 1, \ldots, f$. Applying U_{ba} to this last equation we immediately obtain

$$(U_{ba}A_iU_{ba}^\dagger)|b_1^k \cdots b_f^k\rangle = a_i^k|b_1^k \cdots b_f^k\rangle, \tag{45}$$

i.e., $|b_1^k \cdots b_f^k\rangle$ is a simultaneous eigenket of the f observables $U_{ba}A_iU_{ba}^\dagger$, and we therefore have the important theorem: *Unitary equivalent observables have the same spectra.* On the other hand, $|b_1^k \cdots b_f^k\rangle$ is also an eigenket of the complete set $\{B_i\}$, and therefore for all i and j

$$[B_i, U_{ba}A_jU_{ba}^\dagger] = 0. \tag{46}$$

By hypothesis, the set $\{B_i\}$ is complete. Hence any operator that commutes with all members of this set is a function of the B_i. Therefore the transformed observables $U_{ba}A_iU_{ba}^\dagger$ are functions of $\{B_i\}$. Usually, in fact, $U_{ba}A_iU_{ba}^\dagger$ will actually be *equal* to one of the observables B_i; this is the case when $\{A_i\} = (L^2, L_z)$, and $\{B_i\} = (L^2, L_x)$, and the statement that $U_{ba}A_iU_{ba}^\dagger$ is a function of the B_i will only apply to such bizarre choices as $\{B_i\} = (L^2, \cosh L_x/(L^2)^{\frac{1}{2}})$. In terms of the first example the theorem following Eq. (45) is also understandable because we already know that all components of the vector $\mathbf{L}$ have the same spectrum. When the number of states N is finite it is therefore possible (and usually most convenient) to assume that the observables that define *all* the different representations are unitary equivalents. Then, to repeat, the spectra will be the same in *all* representations. Furthermore, *all the algebraic laws relating the various observables—whether they are compatible or not—are invariant under unitary transformations.* It is clear from all that has been said that *the unitary transformations play the same role in quantum mechanics as do the canonical transformations in classical mechanics.*

23.5 CONTINUOUS SPECTRA

Although our arguments have really only applied to the case where N is finite, essentially all the results obtained so far also hold true if $N \to \infty$, as long as the spectra remain discrete. Continua are much more subtle, however, and we shall be forced to take a rather cavalier attitude towards their mathematical formulation. That some of our results do not apply to the case of continuous spectra is obvious from our earlier experience with the free particle in three dimensions. As we know, in the momentum representation, where the observables are p_x, p_y, p_z, the spectra are continuous, $-\infty < p_i' < \infty$. In the (p^2, L^2, L_z) representation, on the other hand, the spectra are $0 \leqslant (p^2)' < \infty$, $(L^2)' = l(l+1)$, $-l \leqslant L_z' \leqslant l$. This peculiar but familiar situation is due to the fact that the transformation $(x, y, z) \to (r, \theta, \phi)$ is singular at $r = 0$ and is therefore not strictly unitary.* The situation is quite the same in classical mechanics, because this transformation does not leave the Poisson brackets invariant.

Let ξ be an observable with a continuous spectrum, and let us continue to use the notation ξ' for the numerical values that this observable can display. The eigenkets will be written as $|\xi'\rangle$, and by definition

$$\xi|\xi'\rangle = \xi'|\xi'\rangle, \qquad \langle\xi'|\xi = \langle\xi'|\xi'. \tag{47}$$

* A detailed discussion of this question can be found in L. C. Biedenharn and P. J. Brussard, *Ann. Phys.* **16**, 1 (1961).

The normalization is taken to be

$$\langle\xi'|\xi''\rangle = \delta(\xi' - \xi''), \tag{48}$$

it being understood that a product of δ-functions is to appear in (48) if ξ stands for a set of observables with continuous spectra. The basic measurement symbols will again be

$$M(\xi') = |\xi'\rangle\langle\xi'|, \tag{49}$$

but $M(\xi')$ clearly cannot correspond to a physically realizable filtration process. For example, we may think of a momentum filter for a beam of charged particles made of a magnet followed by a set of slits. Such a beam has a nonzero momentum spread for every finite slit separation. Thus a symbol like

$$M_{\Delta\xi}(\xi_0) = \int_{\xi_0'-\Delta\xi}^{\xi_0'+\Delta\xi} d\xi' |\xi'\rangle\langle\xi'| \tag{50}$$

is really more closely related to filtration and preparation in the case of a continuous spectrum. In fact, our belief in the existence of continuous spectra is based on the observation that we keep finding states as we decrease $\Delta\xi$, as long as $\Delta\xi > 0$. Returning to (50), we note that $M_{\Delta\xi}(\xi_0')$ is a projection operator onto the subspace $\xi_0' - \Delta\xi' < \xi' < \xi_0' + \Delta\xi$. If this subspace does not overlap $\xi_1' - \Delta\xi < \xi' < \xi_1' + \Delta\xi$, then

$$M_{\Delta\xi}(\xi_0')M_{\Delta\xi}(\xi_1') = 0.$$

Moreover, the sum of all $M_{\Delta\xi}(\xi_i')$ that do not overlap but cover the spectrum equals unity. Although one can work with the more realistic operators $M_{\Delta\xi}(\xi_i')$ if $\Delta\xi$ is small enough, the idealized objects (49) are far easier to use.

We shall now record some obvious relationships involving the various mathematical entities pertaining to continuous spectra. First of all, the observable ξ can be expressed as *

$$\xi = \int d\xi' |\xi'\rangle\xi'\langle\xi'|; \tag{51}$$

i.e.,

$$\langle\xi'|\xi|\xi''\rangle = \xi'\,\delta(\xi' - \xi''). \tag{52}$$

The transformation functions satisfy

$$\int d\eta' \langle\xi'|\eta'\rangle\langle\eta'|\zeta'\rangle = \langle\xi'|\zeta'\rangle, \tag{53}$$

* If the spectrum of ξ is in part discrete, then the integration signs in the following equations should be read as sums over the discrete and integrals over the continuous parts of the spectrum.

and

$$\int d\eta' \langle \xi' | \eta' \rangle \langle \eta' | \xi'' \rangle = \delta(\xi' - \xi''). \tag{54}$$

When one transforms from representations with continuous spectra to those with discrete spectra one uses

$$\sum_{a'} \langle \xi' | a' \rangle \langle a' | \eta' \rangle = \langle \xi' | \eta' \rangle \tag{55}$$

and

$$\int d\xi' \langle a' | \xi' \rangle \langle \xi' | b' \rangle = \langle a' | b' \rangle \tag{56}$$

instead of (53).*

The most familiar observables with continuous spectra are the momentum and position, denoted by p_i ($i = 1, 2, 3$) and x_i; their spectra cover the entire real line. Since $|\langle x' | a' \rangle|^2\, dx'$ is the probability for finding the position of the system in the interval $(x', x' + dx')$, when it is in the state $|a'\rangle$, we see that *the Schrödinger wave function* $\psi_{a'}(x')$ *is equal to the transformation function* $\langle x' | a' \rangle$.

23.6 COMPOSITE SYSTEMS

A rather trivial extension of the notation is necessary for problems concerning composite systems such as the hydrogen atom. Let A_I, $\{|a'_I\rangle\}$ and A_{II}, $\{|a'_{II}\rangle\}$, be the observables and kets of systems I and II (e.g., electron and proton). In the case where these systems are not coupled by an interaction, there must exist states in which there are no statistical correlations.† For such states, the probability for finding properties b'_I and b'_{II} in a state specified by the quantum numbers a'_I and a'_{II} has the form

$$p(a'_I a'_{II}, b'_I b'_{II}) = p(a'_I, b'_I)\, p(a'_{II}, b'_{II}). \tag{57}$$

We easily achieve this property by extending the space of state vectors as follows. Let $\mathfrak{H}_I$ and $\mathfrak{H}_{II}$ be the spaces spanned by the sets $\{|a'_I\rangle\}$ and $\{|a'_{II}\rangle\}$, respectively. We then construct the product space $\mathfrak{H} = \mathfrak{H}_I \otimes \mathfrak{H}_{II}$ spanned by the vectors $|a'_I\rangle|a'_{II}\rangle$, which we write as $|a'_I a'_{II}\rangle$.

* In the remainder of this book we shall frequently use the notation developed for discrete spectra even though some of the observables may possess partially continuous spectra. When the latter is the case summation signs are to be read as integrals, and Kronecker symbols as Dirac δ-functions.

† If the constituents of the composite system are indistinguishable there are correlations even in the absence of any interaction; this matter will be discussed in Secs. 41 and 48.2, and in greater detail in Chapt. XII.

The observable A for the whole system is $A = A_I + A_{II}$, and

$$\begin{aligned} A|a'_I a'_{II}\rangle &= (a'_I + a'_{II})|a'_I a'_{II}\rangle, \\ \langle b'_I b'_{II}|a'_I c'_{II}\rangle &= \langle b'_I|a'_I\rangle\langle b'_{II}|c'_{II}\rangle, \end{aligned} \tag{58}$$

and so forth. If X_I operates only on I, X_{II} on II, the matrix element of $X_I X_{II}$ factorizes:

$$\langle a'_I b'_{II}|X_I X_{II}|a''_I b''_{II}\rangle = \langle a'_I|X_I|a''_I\rangle\langle b'_{II}|X_{II}|b''_{II}\rangle. \tag{59}$$

If the systems interact, there then exist operators $V_{I,II}$ whose matrix elements do not factor. Thus

$$\langle a'_I b'_{II}|A_I + A_{II}|c'_I d'_{II}\rangle = \langle a'_I|A_I|c'_I\rangle\langle b'_{II}|d'_{II}\rangle + \langle a'_I|c'_I\rangle\langle b'_{II}|A_{II}|d'_{II}\rangle, \tag{60}$$

while, on the other hand,

$$\langle a'_I b'_{II}|V_{I,II}|c'_I d'_{II}\rangle$$

cannot be factorized. The Coulomb interaction in the hydrogen atom is an operator of the type $V_{I,II}$.

23.7 SUMMARY

Because of the length and central importance of the present section, we shall now summarize our essential conclusions:

(1) *The composition properties of the transformation functions and their interpretation as probability amplitudes imply that the measurement algebra can be realized as a set of linear operators on a complex vector space with Hermitian metric.*

(2) *With every maximally filtered (i.e., pure) state characterized by the quantum numbers a' we can associate a vector (or ket) $|a'\rangle$, and a dual vector (or bra) $\langle a'|$; $\{|a'\rangle\}$ is an orthonormal basis in the vector space, and $\{\langle a'|\}$ is an orthonormal basis in the dual space.*

(3) *The measurement symbol $M(a') = |a'\rangle\langle a'|$ is a projection operator onto $|a'\rangle$, and $M(a') + M(a'')$ projects onto the subspace spanned by $|a'\rangle$ and $|a''\rangle$. In terms of $M(a')$ a linear Hermitian operator corresponding to the observable A can be constructed: $A = \sum_{a'}|a'\rangle a'\langle a'|$.*

(4) *$|a'\rangle$ is an eigenket of A with eigenvalue a'.*

(5) *Since the whole theory depends on the existence of the completeness relation $\sum_{a'} M(a') = 1$, only Hermitian operators with complete sets of eigenkets are candidates for observables.*

(6) *The algebra of observables and the scalar product between states (and ipso facto, the probabilities) are invariant under unitary transformations. Unitary equivalent observables have the same spectra.*

(7) *The unitary operators form a group.*

(8) *The formulation of the theory in terms of operators and vectors in an abstract space has the distinct advantage that the equations involving these abstract quantities are form-invariant under the group of unitary transformations. This is not true of wave mechanics. The relation between the abstract and wave mechanical formalisms is analogous to the relation between the equations* **curl F** $= 0$, *and* $\partial F_x/\partial y - \partial F_y/\partial x = 0$, *etc.; the latter depend on the coordinate system, and the former does not.*

(9) *Because any ket can be brought into another by a linear combination of measurement symbols, all kets must have physical significance, i.e., correspond to "states."**

Concerning this last point, which is called the *Superposition Principle*, some further explanation is in order. Let $|\alpha\rangle$ be some vector. When we say that we are dealing with a system in a state represented by $|\alpha\rangle$, we always choose $|\langle\alpha|\alpha\rangle| = 1$, because the probability of being in the state is, by hypothesis, one. Now we observe that the replacement $|\alpha\rangle \rightarrow e^{i\varphi}|\alpha\rangle$, which is a special case of (22.1), leaves the physically significant numbers such as $\langle\alpha|A|\alpha\rangle$ and $|\langle\alpha|\beta\rangle|^2$ unchanged. Hence if $|\alpha\rangle$ represents a physical state of the system, the ket $e^{i\varphi}|\alpha\rangle$ represents the *same* physical state. The totality of vectors $e^{i\varphi}|\alpha\rangle$ is called a ray; *all vectors belonging to a ray represent the same physical state.* Strictly speaking, we are really concerned with a ray space, and not a vector (or Hilbert) space. This distinction has surprisingly few consequences,† however, and we shall be able to ignore the difference between vector and ray spaces in almost all our work.

24. The Uncertainty Relations for Arbitrary Observables

The uncertainty relation for the canonical momenta and coordinates has already been studied in Sec. 4.3. We now extend this discussion to arbitrary pairs of incompatible observables. The essential tool in the argument is the *Schwarz inequality*, which states that for any pair of kets $|\alpha\rangle$ and $|\beta\rangle$

$$\langle\alpha|\alpha\rangle\langle\beta|\beta\rangle \geqslant |\langle\alpha|\beta\rangle|^2. \tag{1}$$

This inequality is familiar from three-dimensional vector algebra: For any pair of vectors **a** and **b**, $|ab| \geqslant |\mathbf{a} \cdot \mathbf{b}|$, and the equality only applies if **a** and **b** are collinear. The proof of (1) runs as follows. Consider

* This statement requires some qualification. See the discussion of superselection rules in Sec. 36.3.

† See Sec. 27.

$|\gamma\rangle = |\alpha\rangle + \lambda|\beta\rangle$; then

$$\langle\gamma|\gamma\rangle = \langle\alpha|\alpha\rangle + \lambda\lambda^*\langle\beta|\beta\rangle + \lambda\langle\alpha|\beta\rangle + \lambda^*\langle\beta|\alpha\rangle \geqslant 0$$

for all λ. To obtain the minimum value of $\langle\gamma|\gamma\rangle$ we differentiate with respect to λ (or λ^*), obtain $\lambda^*_{\min} = -\langle\alpha|\beta\rangle/\langle\beta|\beta\rangle$, and substitute this value of λ into $\langle\gamma|\gamma\rangle$ to obtain (1). As in the three-dimensional case, the equality only holds if $|\beta\rangle = c|\alpha\rangle$.

Now let $|g'\rangle$ be an arbitrary normalized ket. By definition (see (22.12))

$$(\Delta A_{g'}\,\Delta B_{g'})^2 = \langle g'|(A - \langle A\rangle_{g'})^2|g'\rangle\langle g'|(B - \langle B\rangle_{g'})^2|g'\rangle. \tag{2}$$

If we define the kets $|g'_{A,B}\rangle$ by

$$(A - \langle A\rangle_{g'})|g'\rangle = |g'_A\rangle,$$
$$(B - \langle B\rangle_{g'})|g'\rangle = |g'_B\rangle,$$

we can write (2) as

$$(\Delta A_{g'}\,\Delta B_{g'})^2 = \langle g'_A|g'_A\rangle\langle g'_B|g'_B\rangle.$$

The Schwarz inequality then yields

$$(\Delta A_{g'}\,\Delta B_{g'})^2 \geqslant |\langle g'|(A - \langle A\rangle_{g'})(B - \langle B\rangle_{g'})|g'\rangle|^2. \tag{3}$$

Let $A - \langle A\rangle_{g'} = \bar{A}$. We can write $\bar{A}\bar{B}$ as

$$\bar{A}\bar{B} = \tfrac{1}{2}(\bar{A}\bar{B} + \bar{B}\bar{A}) + \tfrac{1}{2}(\bar{A}\bar{B} - \bar{B}\bar{A})$$
$$= \tfrac{1}{2}\{\bar{A}, \bar{B}\} + \tfrac{1}{2}[A, B],$$

where $\{\bar{A}, \bar{B}\} = (\bar{A}\bar{B} + \bar{B}\bar{A})$ is called the anticommutator. Note that $\{\bar{A}, \bar{B}\}^\dagger = \{\bar{A}, \bar{B}\}$, $[A, B]^\dagger = -[A, B]$. Put

$$[A, B] = iC, \tag{4}$$

where C is Hermitian. Hence

$$(\Delta A_{g'}\,\Delta B_{g'})^2 \geqslant \tfrac{1}{4}|\langle\{\bar{A}, \bar{B}\}\rangle_{g'} + i\langle C\rangle_{g'}|^2.$$

Because both expectation values are real we have the final inequality

$$(\Delta A_{g'})^2(\Delta B_{g'})^2 \geqslant \tfrac{1}{4}\langle\{\bar{A}, \bar{B}\}\rangle^2_{g'} + \tfrac{1}{4}\langle C\rangle^2_{g'}. \tag{5}$$

By a judicious choice of $|g'\rangle$ one can usually make the expectation value $\langle\{\bar{A}, \bar{B}\}\rangle = \langle\{A, B\}\rangle - 2\langle A\rangle\langle B\rangle$ vanish. It is therefore customary to write (5) as

$$\Delta A\,\Delta B \geqslant \tfrac{1}{2}|\langle[A, B]\rangle|. \tag{6}$$

This is the generalized form of Heisenberg's uncertainty principle.

•The equality sign in (6) only applies to a special set of kets. To be precise, $\bar{A}|g'\rangle$ and $\bar{B}|g'\rangle$ must be collinear,

$$\bar{A}|g'\rangle = \lambda\bar{B}|g'\rangle, \tag{7}$$

and $\langle\{\bar{A}, \bar{B}\}\rangle_{g'}$ must vanish. With the help of (7) we can write this last condition as

$$\langle\{\bar{A}, \bar{B}\}\rangle_{g'} = (\lambda + \lambda^*)(\Delta B)^2_{g'} = \left(\frac{1}{\lambda} + \frac{1}{\lambda^*}\right)(\Delta A)^2_{g'} = 0.$$

These expressions will vanish if λ is pure imaginary. The actual value of λ can then be determined from

$$\langle[\bar{A}, \bar{B}]\rangle_{g'} = (\lambda^* - \lambda)(\Delta B)^2 = (\Delta A)^2\left(\frac{1}{\lambda} - \frac{1}{\lambda^*}\right) = i\langle C\rangle_{g'},$$

i.e.,

$$\lambda = -\frac{\langle[A, B]\rangle_{g'}}{2(\Delta B_{g'})^2} = \frac{2(\Delta A_{g'})^2}{\langle[A, B]\rangle_{g'}}.\bullet$$

25. Addition of Angular Momentum

In classical mechanics the addition of two dynamically independent angular momenta is carried out by applying the elementary laws of vector algebra. The analogous quantum mechanical problem is considerably more involved, and the machinery of transformation theory as developed in Sec. 23 is ideally suited to its solution.

Before studying the question of addition, let us review what we already learned about angular momentum in Sec. 10. Let J_x, J_y, J_z be three observables that have the following algebraic properties (they are *not* to be thought of as differential operators):

$$[J_x, J_y] = iJ_z \qquad \text{(and cyclic permutations).} \tag{1}$$

As we know,

$$J^2 = J_x^2 + J_y^2 + J_z^2$$

commutes with J_i. Hence we may take (J_z, J^2) as our compatible set, and * $|jm\rangle$ as the eigenkets of J^2, J_z:

$$\begin{aligned} J^2|jm\rangle &= j(j+1)|jm\rangle, \\ J_z|jm\rangle &= m|jm\rangle. \end{aligned} \tag{2}$$

* Instead of being systematic and using the notation $|J^{2\prime}J_z'\rangle$, we shall use the conventional symbols for the eigenvalues.

The arguments used in discussing the spectrum of the orbital angular momentum in Sec. 10 can be taken over *in toto*, and we find that $j = 0, \frac{1}{2}, 1, \ldots, m = -j, -j+1, \ldots, j-1, j$. Here we cannot eliminate $j = \frac{1}{2}, \frac{3}{2}, \ldots$, because we are not trying to construct functions of (θ, ϕ). We again define the non-Hermitian operators

$$J_{\pm} = J_x \pm iJ_y. \tag{3}$$

We choose our phases such that

$$J_{\pm}|jm\rangle = \sqrt{j(j+1) - m(m \pm 1)}\,|jm \pm 1\rangle, \tag{4}$$

and normalize the kets in the conventional manner:

$$\langle jm|j'm'\rangle = \delta_{jj'}\,\delta_{mm'}. \tag{5}$$

Now we come to something new. Let us consider two systems, with independent angular momentum operators $\mathbf{J}_1, \mathbf{J}_2$; by hypothesis $[\mathbf{J}_1, \mathbf{J}_2] = 0$. The problem then is to construct the eigenkets and determine the eigenvalues pertaining to the total angular momentum

$$\mathbf{J} = \mathbf{J}_1 + \mathbf{J}_2. \tag{6}$$

As always, only one component of $\mathbf{J}$, and J^2, can be specified simultaneously. Because J_1^2 and J_2^2 are scalars, they also commute with J^2. On the other hand,

$$[J^2, J_{1z}] = 2[\mathbf{J}_1 \cdot \mathbf{J}_2, J_{1z}] \neq 0,$$
$$[J^2, J_{2z}] \neq 0,$$

and therefore J_{1z} and J_{2z} cannot be specified simultaneously with the square of the total angular momentum operator. We therefore have two different complete sets of compatible angular momentum observables *: (a) the set we already know in detail is associated with the individual angular momenta, i.e., $J_1^2, J_{1z}, J_2^2, J_{2z}$; and (b) the sought-after set J^2, J_z, J_1^2, J_2^2. The kets associated with set (a) are already known. They are merely products of kets of the type specified by (2). They will be written as $|j_1m_1j_2m_2\rangle$, and they satisfy ($\alpha = 1$ or 2)

$$\begin{aligned}[J_\alpha^2 - j_\alpha(j_\alpha + 1)]|j_1m_1j_2m_2\rangle &= 0,\\ [J_{\alpha z} - m_\alpha]|j_1m_1j_2m_2\rangle &= 0.\end{aligned} \tag{7}$$

* A complete description of a system cannot be given in terms of the angular momentum variables alone. In what follows we shall not concern ourselves with these further observables (e.g., the Hamiltonian) because they are supposed to commute with the angular momentum operators. Although all kets should be specified by further quantum numbers, we shall suppress these as they never change throughout the equations which will be derived.

Our task is the evaluation of the eigenkets belonging to the observables (b), and the determination of the spectrum of J^2 for fixed values of j_1 and j_2.

The eigenkets belonging to set (b) above are defined by

$$\begin{aligned}[J^2 - j(j+1)]|j_1j_2jm\rangle &= 0,\\ (J_z - m)|j_1j_2jm\rangle &= 0,\\ [J_\alpha^2 - j_\alpha(j_\alpha+1)]|j_1j_2jm\rangle &= 0,\end{aligned} \tag{8}$$

and

$$\langle j_1j_2jm|j_1'j_2'j'm'\rangle = \delta_{jj'}\,\delta_{mm'}\,\delta_{j_1j_1'}\,\delta_{j_2j_2'}, \tag{9}$$

apart from some arbitrary phases.

Important simplifications occur because the operators J_1^2 and J_2^2 appear in both compatible sets. In particular, $\langle j_1j_2jm|j_1'm_1j_2'm_2\rangle = 0$ unless $j_1 = j_1'$ and $j_2 = j_2'$. To take advantage of this fact we decompose the Hilbert space into orthogonal $(2j_1+1)(2j_2+1)$-dimensional subspaces $\mathfrak{H}_{j_1j_2}$. Each such subspace is spanned by the kets $|j_1m_1j_2m_2\rangle$, i.e.,

$$\sum_{m_1m_2} |j_1m_1j_2m_2\rangle\langle j_1m_1j_2m_2| = 1_{j_1j_2},$$

where $1_{j_1j_2}$ is the unit operator in $\mathfrak{H}_{j_1j_2}$, and zero outside. The basis defined by (8) also spans $\mathfrak{H}_{j_1j_2}$:

$$\sum_{jm} |j_1j_2jm\rangle\langle j_1j_2jm| = 1_{j_1j_2}.$$

We may therefore write the unknown kets $|j_1j_2jm\rangle$ in terms of the known kets as

$$|j_1j_2jm\rangle = \sum_{m_1m_2} |j_1m_1j_2m_2\rangle\langle j_1m_1j_2m_2|j_1j_2jm\rangle. \tag{10}$$

The unknown transformation functions that appear in this linear relationship are called *Clebsch-Gordan or vector addition coefficients*, and we shall write them more compactly as $\langle j_1m_1j_2m_2|jm\rangle$. The inverse relation to (10) is

$$|j_1m_1j_2m_2\rangle = \sum_{jm} |j_1j_2jm\rangle\langle jm|j_1m_1j_2m_2\rangle. \tag{11}$$

The completeness relations in $\mathfrak{H}_{j_1j_2}$ then imply

$$\sum_{jm} \langle j_1m_1j_2m_2|jm\rangle\langle jm|j_1j_2m_1'm_2'\rangle = \delta_{m_1m_1'}\,\delta_{m_2m_2'}, \tag{12}$$

$$\sum_{m_1m_2} \langle jm|j_1m_1j_2m_2\rangle\langle j_1m_1j_2m_2|j'm'\rangle = \delta_{jj'}\,\delta_{mm'}. \tag{13}$$

Let us now determine the spectrum of J^2 and J_z. We already have a number of helpful facts at our disposal: (a) $(2j_1 + 1)(2j_2 + 1)$ kets span $\mathfrak{H}_{j_1j_2}$; (b) to every value of j there correspond $(2j + 1)$ kets having the eigenvalues $m = j, j - 1, \ldots, -j$; and (c) $|j_1m_1j_2m_2\rangle$ is already an eigenket of J_z with eigenvalue $m_1 + m_2$.

From (c) we conclude that the largest value of m is $j_1 + j_2$. Hence the largest value of j is also $j_1 + j_2$. As there is only one state with $m = j_1 + j_2$, it must also have the eigenvalue $j_1 + j_2$ of j. We are free to choose the phase so that

$$|j_1j_2\, j_1 + j_2\, j_1 + j_2\rangle = |j_1j_1j_2j_2\rangle. \tag{14}$$

There are two kets $|j_1m_1j_2m_2\rangle$ with $m_1 + m_2 = j_1 + j_2 - 1$, i.e., $m_1 = j_1, m_2 = j_2 - 1$ and $m_1 = j_1 - 1, m_2 = j_2$. Hence there are two linearly independent $|j_1j_2jm\rangle$-states with $m = j_1 + j_2 - 1$. According to (b) one of these must have the eigenvalue $j = j_1 + j_2$, hence the other has $j = j_1 + j_2 - 1$. There are three kets $|j_1m_1j_2m_2\rangle$ with $m_1 + m_2 = j_1 + j_2 - 2$, and therefore three linearly independent $|jm\rangle$ states with $m = j_1 + j_2 - 2$; one belongs to $j = j_1 + j_2$, one to $j = j_1 + j_2 - 1$, and hence the last one must belong to $j = j_1 + j_2 - 2$. We keep repeating this argument, but then there comes a point where the number of linearly independent kets for a given value of m no longer increases. That this must be so is obvious since the degeneracies for $m = -(j_1 + j_2) + 2$, $-(j_1 + j_2) + 1$, $-(j_1 + j_2)$, are also 3, 2, 1, respectively. After some reflection we realize that below $m = |j_1 - j_2|$ the degeneracy stops growing or actually decreases, and therefore the smallest value of j is $|j_1 - j_2|$. It is already clear that there is only one state for each value of j and m, and this is verified by

$$\sum_{j'=|j_1-j_2|}^{j_1+j_2} (2j' + 1) = (2j_1 + 1)(2j_2 + 1).$$

The summary of our results is then:

(a) the spectrum of J^2 is $j = j_1 + j_2, j_1 + j_2 - 1, \ldots, |j_1 - j_2|$;
(b) there is only *one* sequence of $(2j + 1)$ kets to each value of j.

Instead of (a), one frequently sees the somewhat misleading relationship

$$j_1 + j_2 \geqslant j \geqslant |j_1 - j_2|. \tag{15}$$

This is known as *the triangular inequality*. It has an obvious classical interpretation in terms of vector addition.

Finally we must compute the Clebsch-Gordan (CG) coefficients themselves. We already know that $\langle j_1m_1j_2m_2|jm\rangle = 0$ unless j satisfies (15), $m = m_1 + m_2$, and $m = j, j-1, \ldots, -j$. Because j_1, j_2 are fixed we suppress these quantum numbers for the moment. The computations take advantage of various recursion relations between the CG-coefficients. The most important of these follows from the application of $J_\mp$ to (10):

$$\begin{aligned} J_\mp|jm\rangle &= \sqrt{j(j+1) - m(m \mp 1)} \sum_{m_1m_2} |m_1m_2\rangle\langle m_1m_2|j\, m \mp 1\rangle \\ &= \sum_{m_1m_2} \Big\{\sqrt{j_1(j_1+1) - m_1(m_1 \mp 1)}\, |m_1 \mp 1\, m_2\rangle \\ &\quad + \sqrt{j_2(j_2+1) - m_2(m_2 \mp 1)}\, |m_1\, m_2 \mp 1\rangle\Big\}\langle m_1m_2|jm\rangle, \end{aligned} \tag{16}$$

or

$$\begin{aligned} &\sqrt{j(j+1) - m(m \mp 1)}\, \langle m_1m_2|j\, m \mp 1\rangle \\ &\qquad = \sqrt{j_1(j_1+1) - m_1(m_1 \pm 1)}\, \langle m_1 \pm 1\, m_2|jm\rangle \\ &\qquad\quad + \sqrt{j_2(j_2+1) - m_2(m_2 \pm 1)}\, \langle m_1\, m_2 \pm 1|jm\rangle. \end{aligned} \tag{17$_\mp$}$$

We shall now show that these two recursion formulas, together with the normalization condition (13), suffice to determine the CG-coefficients to within a phase. Let us first determine the coefficients for $m = j$. Because $J_+|jj\rangle = 0$, we should be able to obtain the answer to this first step by setting $m = j$ in (17_+). If we do this, and take into account that $m = m_1 + m_2$ in any CG-coefficient, we find

$$\begin{aligned} &\sqrt{j_1(j_1+1) - m_1(m_1-1)}\, \langle m_1 - 1\, j + 1 - m_1|jj\rangle \\ &\quad + \sqrt{j_2(j_2+1) - (j - m_1)(j - m_1 + 1)}\, \langle m_1\, j - m_1|jj\rangle = 0. \end{aligned} \tag{18}$$

Let us call $\langle j_1\, j - j_1|jj\rangle = C(j)$; then it is clear that (18) determines all the coefficients $\langle m_1\, j - m_1|jj\rangle$ in terms of $C(j)$. This last unknown is determined by

$$\sum_{m_1m_2} |\langle m_1m_2|jj\rangle|^2 = 1,$$

and the phase is such that $C(j)$ is real and positive; all the CG-coefficients will then be real. Having determined the CG-coefficients for $m = j$, we now turn to those with smaller values of m. It is clear that these must be determined by applying J_- to the ket $|jj\rangle$ the appropriate number of times, and in fact we see that (17_-) determines $\langle m_1m_2|\, j\, j-1\rangle$ in terms of $\langle m_1m_2|jj\rangle$. Therefore repeated application of (17_-) will generate all the coefficients for a fixed value of j once the $m = j$ coefficients are determined.*

One can check one's understanding of the procedure just outlined by verifying the entries in the table given here.†

$\langle j_1 m_1 \tfrac{1}{2} m_2 | jm \rangle$

	$m_2 = \frac{1}{2}$	$m_2 = -\frac{1}{2}$
$j = j_1 + \frac{1}{2}$	$\sqrt{\dfrac{j_1 + m + \frac{1}{2}}{2j_1 + 1}}$	$\sqrt{\dfrac{j_1 - m + \frac{1}{2}}{2j_1 + 1}}$
$j = j_1 - \frac{1}{2}$	$-\sqrt{\dfrac{j_1 - m + \frac{1}{2}}{2j_1 + 1}}$	$\sqrt{\dfrac{j_1 + m + \frac{1}{2}}{2j_1 + 1}}$

$\langle j_1 m_1 1 m_2 | jm \rangle$

$j =$	$m_2 = 1$	$m_2 = 0$	$m_2 = -1$
$j_1 + 1$	$\sqrt{\dfrac{(j_1 + m)(j_1 + m + 1)}{(2j_1 + 1)(2j_1 + 2)}}$	$\sqrt{\dfrac{(j_1 - m + 1)(j_1 + m + 1)}{(2j_1 + 1)(j_1 + 1)}}$	$\sqrt{\dfrac{(j_1 - m)(j_1 - m + 1)}{(2j_1 + 1)(2j_1 + 2)}}$
j_1	$-\sqrt{\dfrac{(j_1 + m)(j_1 - m + 1)}{2j_1(j_1 + 1)}}$	$\dfrac{m}{\sqrt{j_1(j_1 + 1)}}$	$\sqrt{\dfrac{(j_1 - m)(j_1 + m + 1)}{2j_1(j_1 + 1)}}$
$j_1 - 1$	$\sqrt{\dfrac{(j_1 - m)(j_1 - m + 1)}{2j_1(2j_1 + 1)}}$	$-\sqrt{\dfrac{(j_1 - m)(j_1 + m)}{j_1(2j_1 + 1)}}$	$\sqrt{\dfrac{(j_1 + m + 1)(j_1 + m)}{2j_1(2j_1 + 1)}}$

• The CG-coefficients are actually rather symmetric functions of their arguments. This is not too apparent from the CG-coefficient itself, and can best be seen by introducing a more convenient set of definitions due to Wigner. The idea is that one adds together three angular momenta to a vanishing resultant, $\mathbf{J}_1 + \mathbf{J}_2 + \mathbf{J}_3 = 0$, instead of the less symmetric addition (6). For this purpose we introduce a new coefficient, called a *3-j symbol or Wigner coefficient,* by

$$\begin{pmatrix} j_1 & j_2 & j_3 \\ m_1 & m_2 & m_3 \end{pmatrix} = \frac{(-1)^{j_1 - j_2 - m_3}}{\sqrt{2j_3 + 1}} \langle j_1 m_1 j_2 m_2 | j_3 - m_3 \rangle. \tag{19}$$

We shall list a number of properties of the 3-j symbols, but not give any proofs

* Condon and Shortley (p. 74, Eq. 4) give a more complicated recursion formula which relates CG-coefficients of *different* j. These identities permit the evaluation of all the coefficients from $\langle j_1 j_1 j_2 j_2 | j_1 + j_2\ j_1 + j_2 \rangle = 1$; the phase chosen here is the same as in Condon and Shortley.

† Further tables of formulas can be found in Condon and Shortley, pp. 76–77. Extensive numerical tables are now available: M. Rotenberg, R. Bivins, N. Metropolis and J. K. Wootens, *The 3-j and 6-j Symbols* (Technology Press, Cambridge 1959).

because they are not of great interest to the nonspecialist.* The 3-j symbols are symmetric under a cyclic permutation of the columns:

$$\begin{pmatrix} j_1 & j_2 & j_3 \\ m_1 & m_2 & m_3 \end{pmatrix} = \begin{pmatrix} j_3 & j_1 & j_2 \\ m_3 & m_1 & m_2 \end{pmatrix} = \begin{pmatrix} j_2 & j_3 & j_1 \\ m_2 & m_3 & m_1 \end{pmatrix}. \tag{20}$$

For noncyclic permutations it is multiplied by a phase:

$$\begin{pmatrix} j_1 & j_2 & j_3 \\ m_1 & m_2 & m_3 \end{pmatrix} = (-1)^{j_1+j_2+j_3} \begin{pmatrix} j_2 & j_1 & j_3 \\ m_2 & m_1 & m_3 \end{pmatrix}; \tag{21}$$

the same holds true if all the magnetic quantum numbers are inverted, viz.

$$\begin{pmatrix} j_1 & j_2 & j_3 \\ -m_1 & -m_2 & -m_3 \end{pmatrix} = (-1)^{j_1+j_2+j_3} \begin{pmatrix} j_1 & j_2 & j_3 \\ m_1 & m_2 & m_3 \end{pmatrix}. \tag{22}$$

In terms of the new coefficients the completeness and orthogonality relations become

$$\sum_{m_1 m_2} \begin{pmatrix} j_1 & j_2 & j_3 \\ m_1 & m_2 & m_3 \end{pmatrix} \begin{pmatrix} j_1 & j_2 & j_3' \\ m_1 & m_2 & m_3' \end{pmatrix} = \frac{\delta_{j_3 j_3'}\, \delta_{m_3 m_3'}}{2j_3 + 1}, \tag{23}$$

$$\sum_{j_3 m_3} (2j_3 + 1) \begin{pmatrix} j_1 & j_2 & j_3 \\ m_1 & m_2 & m_3 \end{pmatrix} \begin{pmatrix} j_1 & j_2 & j_3 \\ m_1' & m_2' & m_3 \end{pmatrix} = \delta_{m_1 m_1'}\, \delta_{m_2 m_2'}. \quad \bullet \tag{24}$$

26. Mixtures and the Density Matrix

To every ray there corresponds a physical state, and we may represent such a state by the measurement symbol $M(a') = |a'\rangle\langle a'|$. The converse of this statement is not true. There are physical states that do not correspond to any one ray in the Hilbert space. We already encountered this important fact in Sec. 20, and introduced the density matrix for the purpose of describing such states. We recall that we introduced the term *pure state* to describe the situation where one wave function (i.e., one ray) represented a state, and when this was not true we called the state in question a *mixture*.

When an ensemble is in the pure state $|a'\rangle$, the probability of finding the eigenvalue a' is unity, and the expectation value of an arbitrary observable B is $\langle a'|B|a'\rangle$, or

$$\langle B\rangle_{a'} = \operatorname{tr} BM(a'). \tag{1}$$

Consider the ensemble formed by combining sub-ensembles prepared by

* A thorough discussion will be found in de Shalit and Talmi, Sec. 13, and Edmonds, Sec. 3.7.

the filtrations $\{M(a')\}$. Let the fractional populations in these subensembles be given by $p(a')$, with

$$\sum_{a'} p(a') = 1. \tag{2}$$

The expectation value of B in this ensemble is

$$\langle B\rangle = \sum_{a'} p(a')\langle a'|B|a'\rangle. \tag{3}$$

If we define the operator ρ as $\Sigma p(a')M(a')$, or

$$\rho = \sum_{a'} |a'\rangle p(a')\langle a'|, \tag{4}$$

we can write (3) in a fashion similar to (1):

$$\langle B\rangle = \operatorname{tr} B\rho. \tag{5}$$

We shall call ρ the density matrix, even though it is an operator. (For the sake of consistency it should perhaps be called the statistical operator, but our usage has become traditional.) In Sec. 20.1 we derived some important properties of the density matrix, and we shall rederive these here. Some new properties of this important operator will also emerge.

In spite of the defining relation (4), one should not think that ρ is always a diagonal operator. The definition and (2) merely show that *the eigenvalues of ρ lie between zero and one and add up to unity.* In another representation (B, say) its matrix elements are $\Sigma_{a'}\langle b'|a'\rangle p(a')\langle a'|b''\rangle$. There is one case, however, in which ρ is diagonal in all representations. Consider the situation where the population of the states a' is completely random, i.e., $p(a') = 1/N$, where N is the total number of states. In this case ρ itself is merely the number $1/N$, and is therefore diagonal in all representations. The fractional populations of the pure states in such a completely random ensemble are therefore $1/N$ in *all* representations.

An important property of the measurement symbol is that it is a projection operator onto a one-dimensional subspace. In general ρ is not a projection operator, however. By direct calculation we have

$$\rho^2 = \sum_{a'} |a'\rangle[p(a')]^2\langle a'|, \tag{6}$$

and this only equals ρ if one of the $p(a')$ is one—the others must then vanish because of (2). In the latter instance ρ is just a measurement symbol. We can obtain an invariant characterization of ρ by taking the trace of (6). We then obtain

$$\operatorname{tr} \rho^2 \leqslant 1. \tag{7}$$

The equality only holds if we are dealing with a pure state. An equivalent statement to (7) is that $\rho - \rho^2$ is a positive-definite Hermitian operator unless the state in question is pure, in which case $\rho = \rho^2$. To repeat, Eq. (7) provides us with a criterion for deciding whether a state is pure or not that is invariant under all unitary transformations.

•Consider a system made up of N *independent* constituents, whose observables are $A_1, A_2, \ldots, A_N$ (which of course commute, $[A_i, A_j] = 0$). The density matrix is

$$\rho = \sum_{a_1' \cdots a_N'} |a_1' \cdots a_N'\rangle p(a_1' \cdots a_N') \langle a_1' \cdots a_N'|$$

quite generally. The assumption of statistical independence means that $p(a_1' \cdots a_n') = p_1(a_1')p_2(a_2') \cdots p_N(a_N')$.* Because

$$|a_1' \cdots a_N'\rangle \equiv |a_1'\rangle|a_2'\rangle \cdots |a_N'\rangle,$$

we have

$$\rho = \prod_{i=1}^{N} \rho_i, \qquad \rho_i = \sum_{a_i'} |a_i'\rangle p(a_i') \langle a_i'|.$$

We can now define a new operator which instead of being multiplicative is additive, namely $\ln \rho$. This operator has the advantage of being extensive for a large system, i.e., proportional to the number of constituent systems. For a uniform thermodynamic system, this implies proportionality to the volume. Consider the average value of $\ln \rho$, $\operatorname{tr} \rho \ln \rho$. This quantity, aside from a constant, is called the *entropy* S:

$$S = -k \operatorname{tr} \rho \ln \rho, \tag{8}$$

where k is Boltzmann's constant. By definition, S is additive:

$$S = -k \operatorname{tr} \rho \sum_i \ln \rho_i = -k \sum_i \operatorname{tr}_i \rho_i \ln \rho_i = \sum_i S_i,$$

where tr_i means a trace over the Hilbert space of the ith system.

Consider S in more detail for one subsystem (we drop the label i). Assuming that there are a finite number $\bar{n}$ of states, we have

$$S = -k \sum_{\alpha=1}^{\bar{n}} p(a^\alpha) \ln p(a^\alpha). \tag{9}$$

* Here we assume that the separate systems are distinguishable. For systems of identical particles special considerations are necessary; see Sec. 41 and Chapt. XII.

For a pure state, where only one $p(a^\alpha)$ equals one and all others vanish, $S = 0$. Since $0 \leqslant p \leqslant 1$, we have $S \geqslant 0$. We may now ask for the maximum value of S. Since $\delta\Sigma p(a') = 0$, we introduce the Lagrange multiplier λ, and ask for

$$\delta \sum_{\alpha=1}^{\bar{n}} p(a^\alpha)[\ln p(a^\alpha) + \lambda] = 0,$$

or

$$\sum_{\alpha} \delta p(a^\alpha)[\ln p(a^\alpha) + \lambda + 1] = 0,$$

where the δp are now independent. Hence $\ln p(a^\alpha) + \lambda + 1 = 0$, and therefore $p(a^\alpha)$ is independent of α. Since $\Sigma p(a^\alpha) = 1$, the maximum value of S is therefore attained when $p(a^\alpha) = 1/\bar{n}$. From (9) we then see that

$$0 \leqslant S \leqslant k \ln \bar{n}. \tag{10}$$

The operator

$$\rho = \frac{1}{\bar{n}} \sum_{a'} |a'\rangle\langle a'|$$

represents the completely random situation, where we know as little as possible concerning the probabilities of finding the properties a'. Hence $S_{\max}$ corresponds to the least information, $S = 0$ to complete knowledge (pure state). When $S = S_{\max}$ the probabilities equal $1/\bar{n}$ in all representations, i.e., $\rho = 1/\bar{n}$.

The most important density matrix in physics is the one describing a system in thermal equilibrium. Let H be the Hamiltonian of the system. Then the ρ which maximizes S for a fixed value of the internal energy

$$U = \operatorname{tr} \rho H,$$

is found from,

$$\delta\{\operatorname{tr} \rho[\ln \rho - \beta H]\} = 0.$$

The solution of this variational problem is

$$\rho = \frac{e^{-\beta H}}{Z}, \tag{11}$$

where

$$Z = \operatorname{tr} e^{-\beta H}.$$

The Lagrange multiplier β is related to the absolute temperature T by $\beta = 1/kT$. The ensemble described by (11) is called the canonical ensemble, and Z is called the partition function. The usual thermodynamic identity

$$F = U - TS$$

follows if we identify the free energy F as

$$F = -kT \ln Z.$$

The density matrix (11) is frequently written in the representation in which the Hamiltonian is diagonal. We write this set of kets as $\{|Ea'\rangle\}$, where E is the energy eigenvalue, and a' designates the remaining quantum numbers. Then

$$\langle Ea'|\rho|E'a''\rangle = \delta(E, E')\, \delta(a', a'')Z^{-1}e^{-\beta E}.$$

Note the appearance of the Boltzmann factor $e^{-E/kT}$. The canonical partition function can also be written in a compact fashion in this representation:

$$Z = \sum_{Ea'} e^{-\beta E}. \quad \bullet$$

27. Equivalent Descriptions

Consider two experimenters O and O' using differently oriented coordinate frames with a common origin. Let R specify their relative orientation—as given by the Euler angles, for example. Let $|jm\rangle$ and $|jm; R\rangle$ be the angular momentum eigenkets prepared by O and O', respectively. Thus $|jm\rangle$ is a state with the eigenvalues $j(j+1)$ and m of the operators J^2 and J_z, while $|jm; R\rangle$ is an eigenstate of the z'-component of $\mathbf{J}$ with the *same* eigenvalue m. Furthermore, let O prepare a one-electron state of momentum p traveling along the z-axis, and let O' prepare a one electron state of the same energy traveling along the z'-axis. We shall designate these states by $|u\rangle$ and $|u; R\rangle$, respectively. If O carries out an experiment to determine the probability that the state $|u\rangle$ be found to have definite eigenvalues of J^2 and J_z, he will find that this probability is $|\langle u|jm\rangle|^2$. But O can also carry out a similar experiment on the states prepared by O', and measure the probability $|\langle u; R|jm; R\rangle|^2$. If we assume that the relationship between rotations and translations is the same in the two frames, we must require $|\langle u|jm\rangle| = |\langle u;R|jm;R\rangle|$. This relation must also hold for an arbitrary state $|u\rangle$, provided $|u; R\rangle$ has precisely the same properties when observed by O' as $|u\rangle$ has when observed by O. Note that absolute value signs appear because the rays, and not the kets, correspond uniquely to physical states. It is also important to realize that we have *not* assumed rotational invariance of the equations of motion; in fact, we have not even mentioned these equations.

The considerations of the preceding paragraph can be generalized to other relationships between pairs of observers. Other important examples are pairs of observers whose coordinate systems are translated with respect to each other, and pairs of observers who use a different convention for the origin of time. In relativistic quantum mechanics important consequences also emerge when one considers observers who use different conventions for the sign of the electric charge.

27.1 WIGNER'S THEOREM

We now inquire into the mathematical nature of the relationship between the two complete, though equivalent, descriptions used by O and O'. In the case of the example used in the preceding paragraph, we should like to determine the properties of the operator that performs the mapping $|u\rangle \to |u; R\rangle$ on the Hilbert space. If the vectors, and scalar products, were themselves of physical significance, we would immediately conclude that the correspondence between the two equivalent descriptions is given by a unitary transformation. Because there need only be a one-to-one correspondence between physical states prepared by O and O', however, only a one-to-one correspondence between rays is required. We must therefore be prepared for more complicated transformations than the unitary ones. Fortunately Wigner has proven the remarkably simple result that we need only concern ourselves with one kind of transformation beyond the unitary type. Moreover, we shall see that non-unitary operators can only be associated with a very restricted class of transformations.

Henceforth we shall use the following notation: O uses the kets $|\alpha\rangle$, $|\beta\rangle$, to describe a pair of states with certain properties, O' associates $|\alpha'\rangle$ and $|\beta'\rangle$ with states which, in his description, have identical properties; if, moreover, $c_1|\alpha_1\rangle + c_2|\alpha_2\rangle$ is a ket describing some state prepared by O, then $(c_1|\alpha_1\rangle + c_2|\alpha_2\rangle)'$ is the corresponding O'-ket. The ray correspondence that we are about to investigate is then *

$$|\langle\alpha|\beta\rangle| = |\langle\alpha'|\beta'\rangle|. \tag{1}$$

Let $\{|a_n\rangle\}$ be an orthonormal basis prepared by O, and let $\{|a_n'\rangle\}$ be the equivalent basis prepared by O'. By hypothesis

$$\sum_n |a_n\rangle\langle a_n| = \sum_n |a_n'\rangle\langle a_n'| = 1. \tag{2}$$

The phases of these sets can be chosen at our convenience. Furthermore, if $|\alpha\rangle = \Sigma_n|a_n\rangle\langle a_n|\alpha\rangle$, and $|\alpha'\rangle = \Sigma_n|a_n'\rangle\langle a_n'|\alpha'\rangle$, we are also free to choose the arbitrary overall phases of $|\alpha\rangle$ and $|\alpha'\rangle$; only the ray containing $|\alpha\rangle$ must be mapped into the ray containing $|\alpha'\rangle$ in the transformation from the O- to the O'-description.

* In this connection, see Prob. 7.

The content of this section is then expressed by Wigner's theorem: We may choose the arbitrary phases so that *either*

$$\text{(I)} \qquad \begin{aligned} \langle\alpha|\beta\rangle &= \langle\alpha'|\beta'\rangle \\ (c_1|\alpha_1\rangle + c_2|\alpha_2\rangle)' &= c_1|\alpha_1'\rangle + c_2|\alpha_2'\rangle, \end{aligned} \tag{3}$$

or

$$\text{(II)} \qquad \begin{aligned} \langle\alpha|\beta\rangle &= \langle\beta'|\alpha'\rangle = \langle\alpha'|\beta'\rangle^* \\ (c_1|\alpha_1\rangle + c_2|\alpha_2\rangle)' &= c_1^*|\alpha_1'\rangle + c_2^*|\alpha_2'\rangle. \end{aligned} \tag{4}$$

In case (I) the transformation to the primed system is simply unitary, but in (II) it is not; it is then said to be *anti-unitary*. (II is *not* a linear transformation!) In a theory employing a complex vector space where the phases of scalar products have physical significance, one would be obliged to use (I). In quantum mechanics (II) is a possibility that cannot be dismissed out of hand.

•To prove the theorem, we begin with the ket $|\varphi_n\rangle = |a_1\rangle + |a_n\rangle$, and show that we can arrange the phases of $\{|a_n'\rangle\}$ so that $|\varphi_n'\rangle = |a_1'\rangle + |a_n'\rangle$. From (2) we have

$$|\varphi_n'\rangle = \sum_m |a_m'\rangle\langle a_m'|(|a_1\rangle + |a_n\rangle)',$$

and from (1)

$$\begin{aligned} \langle a_m'|(|a_1\rangle + |a_n\rangle)' &= e^{i\gamma_m}\{\langle a_m|a_1\rangle + \langle a_m|a_n\rangle\} \\ &= e^{i\gamma_m}(\delta_{m1} + \delta_{mn}), \end{aligned}$$

or

$$|\varphi_n'\rangle = e^{i\gamma_1}|a_1'\rangle + e^{i\gamma_n}|a_n'\rangle.$$

We are also free to redefine the phases of $\{|a_n'\rangle\}$ so as to absorb the phase factors $e^{i\gamma_n}$. Then we have

$$|\varphi_n'\rangle = |a_1'\rangle + |a_n'\rangle. \tag{5}$$

Consider the ket

$$|\alpha\rangle = \sum_n c_n|a_n\rangle,$$

and its image

$$|\alpha'\rangle = \sum_n c_n'|a_n'\rangle.$$

Of course, $c_n = \langle a_n|\alpha\rangle$, $c_n' = \langle a_n'|\alpha'\rangle$, and therefore

$$|c_n| = |c_n'|. \tag{6}$$

But

$$\begin{aligned} |\langle\varphi_n|\alpha\rangle| &= |c_1 + c_n|, \\ |\langle\varphi_n'|\alpha'\rangle| &= |c_1' + c_n'|, \end{aligned}$$

and therefore

$$|c_1 + c_n| = |c_1' + c_n'|.$$

We are still free to choose the phase of $|\alpha'\rangle$, and we do this so as to make $c_1 = c_1'$. Thus we must solve

$$|c_1 + c_n| = |c_1 + c_n'|,$$

which gives us the quadratic equation

$$c_1 c_n^* + c_1^* c_n = c_1 c_n'^* + c_1^* c_n'.$$

Multiplying by c_n' gives

$$c_1^* c_n'^2 - c_n'(c_1 c_n^* + c_1^* c_n) + c_1 |c_n|^2 = 0.$$

The roots of this equation are

$$c_n' = \begin{cases} c_n & (7) \\ \dfrac{c_1}{c_1^*} c_n^*. & (8) \end{cases}$$

We can still redefine the phase of $|\alpha\rangle$ itself, and make c_1 real. Thus we finally have

$$|\alpha'\rangle = \sum_n c_n^* |a_n'\rangle \tag{9}$$

in case (8). Hence (7) corresponds to the unitary transformation, (8) to the anti-unitary.

Consider another ket $|\beta\rangle = \Sigma_n d_n |a_n\rangle$. Its transform is either $\Sigma_n d_n |a_n'\rangle$ or $\Sigma_n d_n^* |a_n\rangle$. Assume $|\alpha'\rangle = \Sigma_n c_n |a_n'\rangle$ (unitary case). Then we see that if $|\beta\rangle$ undergoes the unitary transformation, $\langle\beta'|\alpha'\rangle = \Sigma_n d_n^* c_n = \langle\beta|\alpha\rangle$, while if $|\beta\rangle$ undergoes an anti-unitary transformation, $|\langle\beta'|\alpha'\rangle| = |\Sigma_n d_n c_n| \neq |\langle\beta|\alpha\rangle|$. Hence in a particular transformation $O \to O'$, *all* vectors are transformed according to a unitary or anti-unitary transformation. In detail, if

$$|\alpha'\rangle = \sum_n c_n^* |a_n'\rangle, \qquad |\beta'\rangle = \sum_n d_n^* |a_n'\rangle,$$

then

$$\langle\beta'|\alpha'\rangle = \sum_n d_n c_n^* = \langle\alpha|\beta\rangle. \tag{10}$$

This completes the proof of the theorem.•

27.2 UNITARY AND ANTI-UNITARY TRANSFORMATION

For some purposes it is convenient to introduce an operator Λ that transforms $\{|a_n\rangle\} \to \{|a_n'\rangle\}$ according to the phase convention fixed by (5):

$$\Lambda|a_n\rangle = |a_n'\rangle, \qquad \Lambda^{-1}|a_n'\rangle = |a_n\rangle. \tag{11}$$

In the unitary case the action of Λ on an arbitrary ket is defined by

$$|\alpha'\rangle = \Lambda|\alpha\rangle = \sum_n (\Lambda|a_n\rangle)\langle a_n|\alpha\rangle = \sum_n |a_n'\rangle\langle a_n|\alpha\rangle. \tag{12}$$

In the anti-unitary case the defining relation is

$$|\alpha'\rangle = \Lambda|\alpha\rangle = \sum_n (\Lambda|a_n\rangle)\langle a_n|\alpha\rangle^*$$
$$= \sum_n |a_n'\rangle\langle\alpha|a_n\rangle. \tag{13}$$

Note that in the anti-unitary case Λ is only defined when a representation is given. Its operation on the basis can always be represented by a unitary matrix, but for an arbitrary ket it takes the complex conjugate of the expansion coefficients. Therefore an anti-unitary operator is frequently written as

$$\Lambda = UK, \tag{14}$$

where U is a unitary transformation on the chosen basis, and K the operation of complex conjugation of all expansion coefficients; i.e., if c is a number, $Kc = c^*K$. If the basis is changed, the work of U and K has to be reapportioned. A detailed discussion of this matter will be given in Sec. 39.

The foregoing discussion does not tell us whether a particular correspondence $|\alpha\rangle \rightarrow |\alpha'\rangle$ is mediated by a unitary or anti-unitary operator Λ. The answer to this question is a matter of physics and/or geometry; it depends *uniquely* on the actual connection between the equivalent descriptions, i.e., whether they differ by a rotation, a reflection, etc. An important observation almost eliminates the anti-unitary possibility completely, however. Consider the case where the connection between the equivalent descriptions depends on a continuous set of parameters. This is true of the example quoted at the very beginning of this section, since one may go from O to O' by a continuous rotation R. To each R there corresponds an operator on the Hilbert space Λ_R. Furthermore, to two successive rotations R_1 and R_2 that connect a chain of three observers according to $O \underset{R_1}{\rightarrow} O' \underset{R_2}{\rightarrow} O''$ there corresponds the operator $\Lambda_{R_2}\Lambda_{R_1}$. One can, however, go directly from O to O'' by a single rotation that we designate by R_2R_1, and to which there corresponds the operator $\Lambda_{R_2R_1}$. We therefore have *

$$\Lambda_{R_2R_1} = \Lambda_{R_2}\Lambda_{R_1}. \tag{15}$$

But for every conceivable rotation R there exists another rotation $R_{\frac{1}{2}}$

* The argument is actually slightly more complicated because we could introduce an arbitrary factor of modulus one into the right-hand side of (15). This does not change the conclusions and has therefore been omitted. For full details see Hamermesh, Sec. 12-3.

such that $R_{\frac{1}{2}}R_{\frac{1}{2}} = R$, and therefore for every operator Λ_R we can write

$$\Lambda_R = \Lambda_{R_{\frac{1}{2}}}\Lambda_{R_{\frac{1}{2}}}.$$

On the other hand, (4) implies that the square of an anti-unitary operator is unitary. Therefore we would obtain a contradiction unless all the Λ_R are unitary. This argument obviously generalizes to the following statement: *If the equivalent descriptions are related to each other by a continuous group, every element of which is continuously connected to the identity,* then the operators that carry out the corresponding mappings in the Hilbert space are unitary.* Therefore we need not worry about anti-unitary operators when studying translations or spatial rotations. The anti-unitary operators only come into play for discrete groups like space inversion or time reversal, and we shall see that only the latter transformation actually involves these novel objects.

PROBLEMS

1. Let A be an arbitrary 2×2 matrix. What are the restrictions on the elements if A is to be a density matrix?

2. The Hamiltonian of a system is given by

$$H = \sum_{n,m=1}^{k} a_n^\dagger A_{nm} a_m,$$

where the numbers A_{nm} are the elements of a Hermitian matrix with eigenvalues $E_1, E_2, \ldots, E_k$. The a's are operators and satisfy

$$\begin{aligned} a_n^\dagger a_m + a_m a_n^\dagger &= \delta_{n,m}, \\ a_n a_m + a_m a_n &= 0. \end{aligned} \tag{1}$$

(a) Show that it is possible to introduce a new set of operators

$$\alpha_n = \sum_{m=1}^{k} U_{nm} a_m$$

* This phrase has the following meaning: Consider the group of transformations that leave $x^2 + y^2 + z^2$ invariant; the elements fall into two classes: (I) those that can be reached by a succession of small, in fact infinitesimal, rotations, and (II) those that involve a reflection, e.g. $x \to -x$. The former have a transformation matrix whose determinant is $+1$, the latter have determinant -1. The elements of (II) are *not* continuously connected to the identity, and the theorem does not apply to them.

(the U_{nm} being complex numbers) that also satisfy the relations (1), but in terms of which

$$H = \sum_{n=1}^{k} E_n \alpha_n^\dagger \alpha_n.$$

(b) Show that the operators $N_n = \alpha_n^\dagger \alpha_n$, $n = 1, 2 \ldots, k$, are a complete compatible set, and that their eigenvalues are 0 and 1.

(c) Let $|N_1' N_2' \cdots N_k'\rangle$ be an eigenket of the set $\{N_n\}$. Show that

$$\alpha_1^\dagger |0 N_2' \cdots N_k'\rangle = e^{i\gamma} |1 N_2' \cdots N_k'\rangle,$$
$$\alpha_1^\dagger |1 N_2' \cdots N_k'\rangle = 0,$$

where γ is real.

3. Let $J_x \pm iJ_y = J_\pm$ and J_z be the angular momentum operators, and consider the operator V_+ that satisfies

$$[J_+, V_+] = 0, \qquad [J_z, V_+] = V_+.$$

Let $|jm\rangle$ be a simultaneous eigenket of J^2 and J_z. Show that

$$V_+|jj\rangle = \text{const.}|j{+}1\; j{+}1\rangle.$$

4. Let $|j_1 m_1 j_2 m_2 j_3 m_3\rangle$ be a simultaneous eigenket of J_i^2, J_{iz}, where $i = 1, 2, 3$. If $|0\rangle$ is an eigenket of the total angular momentum with eigenvalue zero, show that

$$|0\rangle = \sum_{m_1 m_2 m_3} \begin{pmatrix} j_1 & j_2 & j_3 \\ m_1 & m_2 & m_3 \end{pmatrix} |j_1 m_1 j_2 m_2 j_3 m_3\rangle.$$

5. Consider the problem of constructing eigenstates of the total angular momentum $\mathbf{J} = \mathbf{J}_1 + \mathbf{J}_2 + \mathbf{J}_3$, where the $\mathbf{J}_i$ commute with each other and each have length one (i.e., $j_i(j_i + 1) = 2$). Let $j(j + 1)$ and m be the eigenvalues of the operators $(\mathbf{J})^2$ and $\mathbf{J}_z$, respectively.

(a) What are the possible values of j?

(b) How many linearly independent eigenstates are there for each value of j and m?

(c) Construct the $j = 0$ state explicitly. Let **a**, **b**, and **c** be ordinary vectors. The only scalar formed by these vectors which is linear in all of them is $\mathbf{a} \times \mathbf{b} \cdot \mathbf{c}$. Establish the connection between this fact and your formula for the $j = 0$ state.

6. Let $H(s)$ be Hermitian, and let $U(s, s_0)$ be the solution of

$$i\hbar \frac{d}{ds} U(s, s_0) = H(s) U(s, s_0), \tag{1}$$

with the initial condition $U(s_0, s_0) = 1$.

(a) Show that the solution that satisfies both the differential equation and the initial condition is

$$U(s, s_0) = 1 - \frac{i}{\hbar}\int_{s_0}^{s} H(s')U(s', s_0)\, ds'.$$

(b) Show that

$$U(s_3, s_2)U(s_2, s_1) = U(s_3, s_1).$$

(Hint: first prove this for $s_3 = s_2 + \Delta s$.)

(c) Show that $U^{\dagger}(s, s_0)U(s, s_0) = 1$.

(d) Show that $U(s, s_0)$ is unitary. (Hint: use the result of (b)).

(e) Finally show that the solution of (1) is unitary if, for $s = s_0$, U equals an arbitrary unitary operator.

7. In the language of Sec. 27, let A and A' be the operators in the Hilbert space used for the same observable by observers O and O'; for example, A could be the operator J_y in the coordinate system used by O. A' would then be the y'-component of $\mathbf{J}$, where y' is the y-coordinate used by O'. Show that the requirement that

$$\langle\alpha|A|\alpha\rangle = \langle\alpha'|A|\alpha'\rangle$$

for an arbitrary ket $|\alpha\rangle$ and all observables A already implies the basic axiom of Sec. 27:

$$|\langle\alpha_1|\alpha_2\rangle| = |\langle\alpha_1'|\alpha_2'\rangle|.$$

VI

Symmetries

The formalism erected in the last chapter is not yet a complete physical theory. Given the algebraic laws obeyed by the observables, it provides us with the means for computing their eigenvalues and the relevant transformation functions. On the other hand, we do not have a way of finding the observables and their algebraic relationships. Nor do we have a dynamical theory, since we have not even introduced the time. The formulation of these notions, and the construction of a dynamical theory, is on a rather different footing from the arguments leading to the static, formal skeleton now in our possession. The latter depends on little more than the concept of incompatible observables. The whole machinery of state vectors, operators, and the like, appears to rest on a very solid foundation. In contrast to this, the dynamical theory will be seen to depend on a host of assumptions of a far more specialized nature (see the remarks of Dirac, p. 84). In nonrelativistic problems the dynamical theory has been so successful that one must almost be a heretic to doubt it. In the relativistic domain the construction of a dynamical theory is still far from complete, however, and is therefore a very controversial question.

The development we shall present in this chapter is based almost entirely on the exploitation of symmetry principles. From this study we will be able to deduce essentially all the properties of those observables that are connected with symmetry operations. In the case of physically isolated systems these observables are just the conserved quantities, but even systems that are not isolated must possess such observables. The shortcoming of this type of approach is that it does not suffice for the systematic construction of a complete dynamical theory. It will only tell us of the existence of operators that generate time translations or spatial rotations (i.e., the Hamiltonian H and the angular momentum $\mathbf{J}$), for example, but not how these operators are to be constructed from the basic dynamical variables. This state of affairs already prevails in classical physics, where symmetry considerations also tell us of the existence of H and $\mathbf{J}$, but do not say how these quantities are to be expressed in terms of the canonical p's and q's. This is clear if one thinks of two systems as different as the free electromagnetic field and two particles interacting through instantaneous forces; in both cases H and $\mathbf{J}$ exist and have similar properties (e.g., their various Poisson brackets have precisely the same form), but in the former example H and $\mathbf{J}$ are constructed from a continuous infinity of dynamical variables—the field strengths, while in the latter example only six p's and six q's are involved. The treatment that follows will not really do justice to the question of how the symmetry generators are to be constructed in a consistent and systematic fashion from the basic dynamical variables. Instead we shall lean rather heavily on the classical analogy—the Correspondence Principle. For an account of the various systematic formulations of quantum dynamics—especially the theories of Feynman and Schwinger—the reader is referred to the book by Mandelstam and Yourgrau.

Before entering into the detailed study of symmetry, we should really make two general remarks in the hope of removing some frequently entertained misconceptions. First of all, as long as we are not talking about the whole universe, it is presumably always possible to translate (or rotate, etc.,) a physical system, and therefore observables like the total momentum should be contained in any realistic (as compared to model) theory. But the existence of such observables and the associated transformations does not, by itself, imply that the dynamical equations governing the behavior of an arbitrary system are covariant under these transformations. On the other hand, only when the equations possess this invariance do symmetry arguments attain their full power. Moreover, the experimental definitions of quantities like the linear or angular momentum depend on the existence of conservation laws. The second remark concerns the fact that many discussions of invariance principles make it appear that they are self-evident. This is hardly the case, because these principles are only simple when they refer to the laws of nature themselves, and not to particular physical situations. A simple and familiar example will illustrate this. Consider Laplace's equation $\nabla^2\phi = 0$; this is a highly symmetric statement whose form does not change under a very wide class of transformations on the coordinate system. Nevertheless, Laplace's equation admits solutions that appear to have no vestige of these invariance properties. Experiments always observe solutions, never basic equations. Therefore the existence of symmetry principles is not always obvious. It is the task of theoretical physics to extract the symmetry principles from data on asymmetric phenomena, and to predict further consequences that can be tested by observation.* Like all the other laws of nature the symmetry principles are not *a priori* self-evident.

28. Displacements in Time and Equations of Motion

In addition to specifying the physical values actually obtained in a measurement, we should also specify the time at which the observation is made. We make the assumption that successive measurements can be carried out with arbitrarily small time delays, i.e., that there is a continuous, real variable t, $-\infty < t < \infty$, which labels the sequence of measurements. Note that *t will only appear as a parameter in the theory*, not as the eigenvalue of an Hermitian operator. In Sec. 30 we shall see why it is not possible to construct an operator for t that has acceptable properties.

We make the further important assumption that at any instant t a complete observation (i.e., the specification of a complete set of observables) is possible. This is an unpleasant idealization, because it seems so unrealistic, but it appears to be essential if one wants to arrive at the conventional theory. Needless to say, the theory will be internally consistent, and one will be able to construct *Gedanken* experiments that perform such an instantaneous complete observation.

28.1 THE TIME EVOLUTION OPERATOR AND SCHRÖDINGER'S EQUATION

We begin by considering *isolated systems*, i.e., systems that do not interact with anything unless a measurement occurs. This restriction will be removed shortly. At $t = t_0$ an observer prepares a variety of states $|a't_0\rangle$, $|a''t_0\rangle$, etc., by the appropriate complete observations. For $t > t_0$ these systems evolve undisturbed, and if we start with a pure state,

* As an illustration we shall (in Sec. 40) see how a scattering experiment can be used to test reflection invariance.

we remain in a pure state until the next observation. In general, of course, this pure state will not correspond to the original vector $|a't_0\rangle$, but it will correspond to some vector. In principle it is possible to check this statement by performing measurements at $t > t_0$ and verifying that it is possible to construct a pass filter for the evolved state (recall that this characterizes a pure state uniquely). Let us then designate the moving ket by $|a't_0;\, t\rangle$; this is an eigenket of some other set of observables corresponding to the mentioned pass filter. Now we have

$$(A - a')|a't_0\rangle = 0, \tag{1}$$

but when $t > t_0$

$$(A - a')|a't_0;\, t\rangle \neq 0 \tag{2}$$

in general. An observer who measures the expectation value of the observable A at time t will in fact obtain

$$\langle A\rangle_t = \langle a't_0;\, t|A|a't_0;\, t\rangle \neq a'. \tag{3}$$

Nevertheless, the actual result of any A-measurement at $t > t_0$ must still be *one* of the values $(a', a'', \ldots)$, even though the probability of obtaining this particular result will no longer be 0 or 1 as it was at $t = t_0$. This is clear from an example: The possible result of a measurement of J^2 is $j(j + 1)$, where $j = 0, \frac{1}{2}, 1, \ldots$. We may prepare a state with $j = 1$, and later on find that it is no no longer an eigenstate of J^2 with $j = 1$, but any J^2-observation must still give a result that belongs to the spectrum. (These remarks obviously do not apply to explicitly time-dependent observables like $\cos(J^2t/\tau)$.) In other words, if we write the operator A in terms of its matrix in the $|a't_0;\, t\rangle$-representation, and carry out the diagonalization of this matrix, we must again find the eigenvalues $(a', a'', \ldots)$. It is therefore natural to assert that *$|a't_0;\, t\rangle$ is obtained from $|a't_0\rangle$ by a unitary transformation*, because the theorem following (23.45) shows that such a transformation does not alter the spectrum.

Let U then be the unitary operator in question. Because the system is isolated, U can only depend on $t - t_0$:

$$|a't_0;\, t\rangle = U(t - t_0)|a't_0\rangle. \tag{4}$$

Consider a time t' such that $t > t' > t_0$. We could permit the system to evolve from $|a't_0\rangle$ until time t', i.e., into the state represented by $U(t' - t_0)|a't_0\rangle$, and then perform a selection of this state. This state $|a't_0;\, t'\rangle$ could then be allowed to evolve until t, at which time it is $U(t - t')|a't_0;\, t'\rangle$. The latter state is clearly identical with (4), and we therefore obtain the *group property* of the unitary operators:

$$U(t - t')U(t' - t_0) = U(t - t_0). \tag{5}$$

Setting $t_0 = t$, and noting that $U(0) = 1$, yields

$$U^{-1}(t - t') = U(t' - t) = U^{\dagger}(t - t'). \tag{6}$$

We are therefore dealing with a continuous one-parameter group of operators; the statement in italics on p. 230 therefore allows us to dismiss the possibility that U is anti-unitary.

Consider the case where $t - t_0$ is an infinitesimal, and put $t_0 = 0$. In accordance with our earlier assumptions about the continuous nature of the evolution in time, we expect that $U(\delta t)$ will differ from 1 by an infinitesimal operator:

$$U(\delta t) \xrightarrow[\delta t \to 0]{} 1 + \delta U.$$

Unitarity then implies

$$1 = UU^{\dagger} = (1 + \delta U)(1 + \delta U^{\dagger}) \simeq 1 + (\delta U + \delta U^{\dagger}) + O[(\delta U)^2].$$

Hence we require $\delta U + \delta U^{\dagger} = 0$, which implies that δU is anti-Hermitian, or $\delta U = i \times$ (Hermitian operator). We make the further observation that the group property,

$$U(\delta t_1)U(\delta t_2) = U(\delta t_1 + \delta t_2),$$

requires $\delta U(\delta t)$ to be proportional to δt. Thus in the case of an isolated system, δU equals $i\ \delta t$ multiplied by a time-independent Hermitian operator. This operator has the dimension of frequency, or energy/$\hbar$. Conventionally one writes

$$\delta U = -iH\ \delta t/\hbar, \tag{7}$$

where H is the *Hamiltonian operator* of the system. It is also customary to call H the *generator* of infinitesimal time displacements.

Using the group property we have

$$U(t + dt) - U(t) = [U(dt) - 1]U(t) = \frac{-i}{\hbar} H\ dt\ U(t),$$

or

$$i\hbar \frac{\partial}{\partial t} U(t) = HU(t). \tag{8}$$

Because H commutes with itself, this equation can be integrated immediately:

$$U(t) = e^{-iHt/\hbar}, \tag{9}$$

since $U(0) = 1$. In the same way we have

$$|a't_0; t + dt\rangle - |a't_0; t\rangle = [U(dt) - 1]|a't_0; t\rangle$$
$$= \frac{-i}{\hbar} H\, dt|a't_0; t\rangle,$$

and therefore $|a't_0; t\rangle$ satisfies the *Schrödinger equation*,

$$\left(i\hbar \frac{\partial}{\partial t} - H\right)|a't_0; t\rangle = 0 \tag{10}$$

with the initial condition $|a't_0; t_0\rangle = |a't_0\rangle$. From (4) and (9), (10) has the formal solution

$$|a't_0; t\rangle = e^{-iH(t-t_0)/\hbar}|a't_0\rangle. \tag{11}$$

Equation (11) assumes a very simple form when the Hamiltonian is itself a member of the complete set of observables that specifies the initial state. When this is the case we shall write $|Ea't_0\rangle$ instead of $|a't_0\rangle$, where E is the eigenvalue of H, and a' designates the eigenvalues of the other constants of the motion. Equation (11) then reads

$$|Ea't_0; t\rangle = e^{-iE(t-t_0)/\hbar}|Ea't_0\rangle.$$

These eigenkets of H therefore belong to the same ray for all values of t. They represent the same physical states at different times and are therefore stationary. The matrix elements of a time-independent observable B between eigenkets of H also have a very simple time dependence:

$$\langle E'a't_0; t|B|E''a''t_0; t\rangle = e^{-i(E''-E')(t-t_0)/\hbar}\langle E'a't_0|B|E''a''t_0\rangle.$$

This equation was one of the basic assumptions in Heisenberg's very first paper of 1925.

We now turn to systems that are not isolated between observations. As a concrete example of such a situation we may think of an atomic or molecular beam passing through an alternating electromagnetic field. Under these circumstances the spectrum of an observable that does not depend explicitly on t must still be time independent, and we therefore require the mapping $|a't_0\rangle \to |a't_0; t\rangle$ to be unitary. The time t_0 at which the system was prepared in the state specified by the quantum numbers a' is not a matter of mere convention any longer, because the nature of the interactions experienced by the system depends on the time. Hence the unitary operator must depend on both t_0 and t,

$$|a't_0; t\rangle = U(t, t_0)|a't_0\rangle. \tag{12}$$

The group property still holds, of course:

$$|a't_0; t'\rangle = U(t', t)|a't_0; t\rangle = U(t', t)U(t, t_0)|a't_0\rangle$$
$$\equiv U(t', t_0)|a't_0\rangle,$$

or

$$U(t_1, t_2) = U(t_1, t_3)U(t_3, t_2), \tag{13}$$

and

$$U(t_1, t_2) = U^{-1}(t_2, t_1). \tag{14}$$

For an infinitesimal time interval we have

$$U(t + dt, t) = 1 + \delta U(t),$$

and so $\delta U(t)$ must again be anti-Hermitian. Equation (13) implies once more that $\delta U(t)$ must be proportional to δt. Now the coefficient of $i\,\delta t$ in $\delta U(t)$ *can* depend on t, however. This operator is then the t-dependent Hamiltonian $H(t)$, and the unitary operator is therefore given by

$$U(t + \delta t, t) = 1 - \frac{i}{\hbar} H(t)\,\delta t. \tag{15}$$

Needless to say, $H(t) = H(t)^\dagger$. Just as before U satisfies the equation

$$i\hbar \frac{\partial}{\partial t} U(t, t_0) = H(t)U(t, t_0) \tag{16}$$

with the initial condition

$$U(t_0, t_0) = 1, \tag{17}$$

and the kets still satisfy the Schrödinger equation

$$i\hbar \frac{\partial}{\partial t} |a't\rangle = H(t)|a't\rangle. \tag{18}$$

In our earlier work (Chap. II) the linearity of the Schrödinger equation appeared as a more or less *ad hoc* assumption. In the present formulation this linearity is an immediate consequence of the unitary and continuous nature of time displacements.

Once the Hamiltonian depends on t, it is no longer possible to obtain a formal solution such as (9) is to (8). In order to investigate the structure of $U(t_1, t_0)$, we replace (16) and (17) by the equivalent integral equation

$$U(t, t_0) = 1 - \frac{i}{\hbar} \int_{t_0}^{t} H(t')U(t', t_0)\,dt'. \tag{19}$$

If we iterate this equation, i.e., substitute the right-hand side into the integrand, we obtain

$$U(t, t_0) = 1 - \frac{i}{\hbar}\int_{t_0}^{t} H(t')\,dt' + \left(\frac{-i}{\hbar}\right)^2 \int_{t_0}^{t}\int_{t_0}^{t'} H(t')H(t'')U(t'', t_0)\,dt'\,dt''.$$

Assuming that this process converges, we eventually obtain the Dyson series

$$U(t, t_0) = 1 + \sum_{n=1}^{\infty}\left(\frac{-i}{\hbar}\right)^n \int_{t_0}^{t} dt_1 \int_{t_0}^{t_1} dt_2 \cdots \int_{t_0}^{t_{n-1}} dt_n\, H(t_1) \cdots H(t_n). \quad (20)$$

Note that the integrand $H(t_1)H(t_2) \cdots H(t_n)$ is time ordered: The H-operators, reading from right to left, have increasing time arguments. This is a consequence of the "causal" nature of the time evolution, because one can also obtain $U(t, t_0)$ by repeated use of the group property and (15):

$$U(t, t_0) = \lim_{\substack{n\to\infty \\ \Delta t\to 0 \\ n\Delta t = t - t_0}} U(t_0 + n\,\Delta t, t_0 + (n-1)\,\Delta t) \times U(t_0 + (n-1)\,\Delta t, t_0 + (n-2)\,\Delta t) \cdots U(t_0 + \Delta t, t_0).$$

When $[H(t), H(t')] = 0$, the order of the operators in (20) is not important, and a bit of work then shows that a significant simplification is possible:

$$\begin{aligned} U(t, t_0) &= 1 + \sum_{n=1}^{\infty}\left(-\frac{i}{\hbar}\right)^n \frac{1}{n!}\int_{t_0}^{t} dt_1 \cdots \int_{t_0}^{t} dt_n\, H(t_1) \cdots H(t_n) \\ &= \exp\left\{-\frac{i}{\hbar}\int_{t_0}^{t} H(t')\,dt'\right\}. \end{aligned} \quad (21)$$

Because $[H(t), H(t')]$ is often nonzero unless $t = t'$, (20) must frequently be used in its full complexity. The most important application of the Dyson series occurs in quantum electrodynamics where most calculations involve its use. (Note that (21) reduces to (9) if H is t-independent.)

28.2 THE SCHRÖDINGER AND HEISENBERG PICTURES

Our discussion has been based on a description where the kets move about in the Hilbert space as time goes on, but the operators stay fixed. This is only one of many possible and completely equivalent modes of description. Because it was first used by Schrödinger, Dirac has called it the *Schrödinger picture.* Another very important way of discussing time development is with the *Heisenberg picture.* Here we fix the kets, and allow the operators to move, i.e., the function (or mappings) they perform depends on t.* The existence of this possibility can be seen by writing an arbitrary matrix element of A at time t:

$$\langle b't_0; t|A|c't_0; t\rangle = \langle b't_0|U^\dagger(t, t_0)AU(t, t_0)|c't_0\rangle. \tag{22}$$

Clearly we can calculate equally well with the stationary kets $|b't_0\rangle$ and the time-dependent operator

$$A(t, t_0) = U^\dagger(t, t_0)AU(t, t_0). \tag{23}$$

In order not to confuse matters with awkward notation, consider first the case where the system is isolated, and choose $t_0 = 0$. Because of (9), (22) then reads

$$\langle b't|A|c't\rangle = \langle b'|e^{iHt/\hbar}Ae^{-iHt/\hbar}|c'\rangle. \tag{24}$$

The Heisenberg picture observable is then defined as

$$A(t) = e^{iHt/\hbar}Ae^{-iHt/\hbar} \tag{25}$$

with $A(0) = A$. Since (25) [and (23)] are unitary transformations they of course leave the spectra unchanged.

Let us briefly summarize the differences between these two pictures. In the S-picture a system evolves in time according to the Schrödinger equation, or more succinctly, according to the moving ket

$$|a't\rangle_S = U(t)|a'\rangle_H, \tag{26}$$

where the subscript H on the right stands for Heisenberg. The ket $|a'\rangle_H$ is just the initial condition which must, in any case, be stipulated in addition to the Schrödinger equation. In the H-picture, the system is described by the time-dependent observable $A(t)$ and the *fixed ket* $|a'\rangle_H$. The equation of motion must now determine $A(t)$; the equation that does this is obtained by differentiating (25):

$$i\hbar \frac{d}{dt} A(t) = [A(t), H]. \tag{27}$$

This is *the Heisenberg equation of motion*, and it replaces the S-equation in the H-picture.† It should be noted that a ket $|\alpha\rangle_S$ which was taken to be fixed in the S-picture, moves in the H-picture, because the inverse transformation must be applied to go from S-kets to H-kets, i.e.,

$$|\alpha\rangle_H = U^\dagger(t)|\alpha\rangle_S. \tag{28}$$

* Pictures in which both the observables and the state vectors move are also of importance. See Prob. 1 and Sec. 58.

† Recall that the expectation value of (27) already appeared in Sec. 6.1.

Only if $|\alpha\rangle_S$ moves according to (26) does the H-ket remain fixed. In fact, if $|\alpha\rangle_S$ in (28) is fixed, then $|\alpha\rangle_H$ "rotates in the opposite sense" to the moving ket in (26). One may therefore characterize the H-picture as the one where the eigenkets of the dynamical observables move, whereas the kets that represent the instantaneous state of the evolving system are fixed.

To understand the two pictures somewhat better, consider the important class of transformation functions that involve the scalar product of the ket representing the evolving system, $|a't_0; t\rangle$, and another fixed ket, say $|b'\rangle$. The quantity $|\langle b'|a't_0; t\rangle|^2$ is then the probability of finding a system in the state $|b'\rangle$ at time t if it was prepared in $|a'\rangle$ at t_0, and is therefore called the *transition probability*. Note that $\langle b'|$ is a fixed bra, and the transformation function in question therefore depends on t. In the H-picture, $|a't_0\rangle$ becomes fixed, but $|b'\rangle$ moves, i.e., $|b'\rangle \to U^\dagger(t, t_0)|b'\rangle$ as in (28). But the transition probability does *not* depend on the picture in which it is evaluated.

•A few embellishments may now be taken up. If the operator in question is explicitly time dependent, it depends on t even in the S-picture and should be written $A_S(t)$. (Our previous example of $\cos J^2t/\tau$ is a case in point.) Then

$$A_H(t) = e^{iHt/\hbar} A_S(t) e^{-iHt/\hbar},$$

still holds, but (27) is replaced by

$$i\hbar \frac{d}{dt} A_H(t) = i\hbar \frac{\partial}{\partial t} A_H(t) + [A_H(t), H]. \tag{29}$$

If the Hamiltonian is explicitly t-dependent, life becomes more complicated. From the basic definition (23), and from

$$i\hbar \frac{\partial}{\partial t} U(t, t_0) = H_S(t) U(t, t_0)$$

and its adjoint

$$i\hbar \frac{\partial}{\partial t} U^\dagger(t, t_0) = -U^\dagger(t, t_0) H_S(t),$$

we easily obtain

$$i\hbar \frac{\partial}{\partial t} A(t, t_0) = [A(t, t_0), H_H(t, t_0)]. \tag{30}$$

(In the simpler case treated previously, $H_H = H_S$).•

From a practical point of view, the S-picture is often (though not always, by any means) the better one to use. For theoretical discussions the H-picture is frequently superior. For example, (27) tells us immediately that $\langle A\rangle$ is t-independent in all states if $[A, H] = 0$. We shall also see that the analogy with classical mechanics is strongest in the H-picture, and when we come to treat the electromagnetic field, we shall find the H-picture almost indispensable.

Up to now we have concerned ourselves with pure states. Let us briefly consider a mixture. For simplicity, we study a system whose Hamiltonian is t-independent. At $t = 0$, say, the density matrix is given by

$$\rho = \sum_{a'} |a'\rangle p(a')\langle a'|. \tag{31}$$

In the S-picture each ket goes into the moving ket $|a't\rangle$ when $t > 0$, and the density matrix evolves into

$$\rho(t) = \sum_{a'} |a't\rangle p(a')\langle a't|. \tag{32}$$

The probabilities $p(a')$ do *not* change because the system is not disturbed by measurement. From $(i\hbar\,\partial_t - H)|a't\rangle = 0$ and $\langle a't|(i\hbar\,\partial_t + H) = 0$ we have

$$i\hbar \frac{\partial}{\partial t}\rho(t) = [H, \rho(t)]. \tag{33}$$

This is the equation of motion of the density matrix. One should not confuse (33) with the Heisenberg equation (27)—note the sign difference, in fact. In the H-picture ρ is *fixed*, and $A \to A(t)$. Thus in an ensemble prepared to be (31) at $t = 0$, $\langle A\rangle = \operatorname{tr} A\rho$ becomes $\operatorname{tr} A\rho(t)$ for $t > 0$ in the S-picture, and $\operatorname{tr} A(t)\rho$ in the H-picture. These numbers are the same, of course. Equation (33) is due to von Neumann and is frequently called the quantal Liouville equation because it assumes the same form as the equation of motion for the phase space probability distribution in classical statistics if $[H, \rho](i\hbar)^{-1}$ is replaced by the corresponding Poisson bracket.

29. Spatial Translations

Consider two observers whose coordinate frames are rigidly displaced with respect to each other. The first observer has a complete arsenal of measurement devices with which he prepares the states $|a'\rangle$, $|b'\rangle$, etc. The second observer's origin is at $\mathbf{a}$ from the first one's point of view.

The second observer also supplies himself with a complete set of states $|a'; \mathbf{a}\rangle$ that, as far as he is concerned, have the same properties as the $|a'\rangle$ of the first observer. Thus $|\langle a'|b'\rangle| = |\langle a'; \mathbf{a}|b'; \mathbf{a}\rangle|$, and by Wigner's theorem we may choose the arbitrary phases in $|a'\rangle$, $|a'; \mathbf{a}\rangle$ so that these two sets of kets are connected by a unitary operator.*

The translations of the preparing and measuring apparatus in ordinary three-dimensional space form a group parametrized by the vectors which specify the origin of the coordinate frame before and after displacement. A group element is written as $\mathcal{G}(\mathbf{a}_1, \mathbf{a}_2)$, and it satisfies the composition law

$$\mathcal{G}(\mathbf{a}_1, \mathbf{a}_2)\mathcal{G}(\mathbf{a}_2, \mathbf{a}_3) = \mathcal{G}(\mathbf{a}_1, \mathbf{a}_3).$$

Because we assume the three-dimensional space to be Euclidean, only the displacements are of significance, and $\mathcal{G}(\mathbf{a}_1, \mathbf{a}_2)$ can be written as $\mathcal{G}(\mathbf{a}_1 - \mathbf{a}_2)$. The group relations are then

$$\mathcal{G}(\mathbf{a})\mathcal{G}(\mathbf{a}') = \mathcal{G}(\mathbf{a} + \mathbf{a}'),$$
$$\mathcal{G}(\mathbf{a}) = \mathcal{G}^{-1}(-\mathbf{a}).$$

The operations $\mathcal{G}(\mathbf{a})$ discussed up to now work on the apparatus and take place in ordinary 3-space. To each of these there must correspond a unitary operator in the Hilbert space which we call $U(\mathbf{a})$. It is clear that the correspondence $\mathcal{G} \leftrightarrow U$ must preserve the group relations; thus if $U(\mathbf{a})$, $U(\mathbf{a}')$, correspond to the translations $\mathcal{G}(\mathbf{a})$, $\mathcal{G}(\mathbf{a}')$, respectively, then the translation $\mathcal{G}(\mathbf{a})\mathcal{G}(\mathbf{a}')$ corresponds to $U(\mathbf{a} + \mathbf{a}')$. Hence

$$U(\mathbf{a})U(\mathbf{a}') = U(\mathbf{a} + \mathbf{a}'), \tag{1}$$

and

$$U(\mathbf{a})U(-\mathbf{a}) = U(-\mathbf{a})U(\mathbf{a}) = 1,$$

or

$$U(-\mathbf{a}) = U^{-1}(\mathbf{a}) = U^{\dagger}(\mathbf{a}). \tag{1'}$$

Just as in the time displacement case we can consider infinitesimal transformations

$$U(\delta\mathbf{a}) = 1 + \delta U,$$

and once more the group property and unitarity imply that $\delta U = \delta\mathbf{a} \cdot \mathbf{F}$, where $\mathbf{F}$ is an anti-Hermitian operator† that does not depend on $\mathbf{a}$.

* A similar argument can be constructed in the case of time displacements, but the operations involved are difficult to visualize. For this reason we have based the development of the preceding section on the invariance of the spectrum under time translations.

† To be precise, we have not proven that this operator is a vector, because we have not even discussed rotations. All we can really say is that there are three operators F_x, F_y, and F_z.

Thus we write

$$U(\delta\mathbf{a}) = 1 - i\,\delta\mathbf{a}\cdot\mathbf{P}/\hbar. \tag{2}$$

The Hermitian operator $\mathbf{P}$ is called *the total momentum* of the system.

When we translate an object along the y-direction through a distance a_y, and then along the x-direction through a_x, we arrive at the same resultant configuration as when we carry these displacements out in the opposite order. Hence $[\mathcal{G}(a_x, 0, 0), \mathcal{G}(0, a_y, 0)] = 0$, and it then follows that the infinitesimal generators must also commute:

$$[P_x, P_y] = 0, \quad \text{etc.} \tag{3}$$

A group whose generators commute is called an Abelian group. As a consequence of (3) we can integrate (2) to obtain the form of the unitary operator when the displacement is finite. The argument is the one that led to (28.9), and we merely quote the result:

$$U(\mathbf{a}) = e^{-i\mathbf{P}\cdot\mathbf{a}/\hbar}. \tag{4}$$

The kets belonging to the translated observer are therefore given by

$$|a'; \mathbf{a}\rangle = e^{-i\mathbf{P}\cdot\mathbf{a}/\hbar}|a'\rangle. \tag{5}$$

If the compatible set A includes $\mathbf{P}$, then (5) is merely a phase change and $|a'\rangle$ and $|a'; \mathbf{a}\rangle$ belong to the same ray and represent the same physical state. To be precise, if both observers prepare states that belong to some particular eigenvalue $\mathbf{P}'$ of $\mathbf{P}$, and some other set of eigenvalues, they are preparing identical states in the sense that one of the observers will not be able to distinguish his own state from that of his colleague. Thus *eigenstates of* $\mathbf{P}$ *are invariant under translations*. In particular, if $[H, \mathbf{P}] = 0$, there exist states that are simultaneously invariant under displacements in space and time. In this situation the Heisenberg-picture momentum operator does not depend on time; in the Schrödinger picture the expectation value of $\mathbf{P}$ in *any* state that satisfies the Schrödinger equation is conserved. This is the momentum conservation law. Some reflection will also reveal that the experimental definition of $\mathbf{P}$ depends on the existence of the conservation law.

Let the *first* observer measure the expectation value of an observable X in a state $|a'; \mathbf{a}\rangle$ prepared by the *displaced* observer*:

$$\langle a'; \mathbf{a}|X|a'; \mathbf{a}\rangle = \langle a'|e^{i\mathbf{P}\cdot\mathbf{a}/\hbar}Xe^{-i\mathbf{P}\cdot\mathbf{a}/\hbar}|a'\rangle. \tag{6}$$

* That is, as far as the first observer is concerned, the system is translated through $\mathbf{a}$.

If a is infinitesimal, (6) reads

$$\langle a'; \delta\mathbf{a}|X|a'; \delta\mathbf{a}\rangle = \langle a'|X + i\,\delta\mathbf{a}\cdot[\mathbf{P}, X]/\hbar|a'\rangle. \tag{7}$$

Thus if $[\mathbf{P}, X] = 0$, the result of a measurement of the property X does not depend on the choice of origin of the coordinate frame: *an observable is translation invariant if it commutes with* $\mathbf{P}$. (If X commutes only with one of the P_i, it is only invariant under translations along the corresponding axis.) It is clear that H must be translation invariant if $\mathbf{P}$ is to be a constant of the motion.

We also note from (6) that the freedom of mathematical description that existed in the time displacement case exists here too. Thus the observer measuring properties of the displaced system can use his observable X and the displaced kets $|a'; \mathbf{a}\rangle$, or he can use the undisplaced kets and the unitary transform $e^{i\mathbf{P}\cdot\mathbf{a}/\hbar}Xe^{-i\mathbf{P}\cdot\mathbf{a}/\hbar}$ of the observable. In the case of an infinitesimal translation, (7) tells us that this new observable is

$$X_{\delta\mathbf{a}} = X + \frac{i}{\hbar}\,\delta\mathbf{a}\cdot[\mathbf{P}, X]. \tag{8}$$

In this section we have not specified the operator $\mathbf{P}$ in sufficient detail to permit its actual construction. This is a question which must be attacked separately for each type of system. All we can say at the moment is that one of the observables that we must be able to construct for any realistic system is $\mathbf{P}$, because all systems can be displaced in space.

30. Systems That Correspond to Classical Point Mechanics—Galileo Invariance

We shall now investigate the quantum theory of systems that have a Correspondence Principle limit in classical point mechanics. A distinguishing feature of such systems is that they possess position observables. The existence of the momentum could be inferred from a rather general and plausible symmetry requirement, but this is not so of the position. A similar (and related) situation already prevails in classical physics: The coordinates are basic dynamical variables in point mechanics, whereas the field strengths play this role in electrodynamics. On the other hand, both the electromagnetic field and a system of particles possess a total momentum in the classical theory.

30.1 THE POSITION AND MOMENTUM AS OBSERVABLES

The notion of a position observable can only be introduced if the position is actually a measurable quantity. This question was already discussed in Sec. 2, where we concluded that the position of a particle of mass m could be determined with an uncertainty of order $\hbar/mc$, which we take to be zero in the nonrelativistic approximation. We shall therefore assume that a position observable $\mathbf{x}$ (with three components x_i) can be defined for a particle of finite rest mass m as long as we deal with momenta small compared to mc.

Our fundamental assumptions concerning the position are, (a) all components of $\mathbf{x}$ can be measured simultaneously, and (b) the expectation value of $\mathbf{x}$ in any state undergoes the translation $\mathbf{a}$ when the state in question is translated through $\mathbf{a}$. The first assumption is embodied in the commutation rule

$$[x_i, x_j] = 0. \tag{1}$$

In the notation of the preceding section, assumption (b) is

$$\langle a'; \mathbf{a}|\mathbf{x}|a'; \mathbf{a}\rangle = \langle a'|\mathbf{x}|a'\rangle + \mathbf{a}, \tag{2}$$

where the normalized ket $|a'\rangle$ represents any state of a one-particle system. Specializing (2) to an infinitesimal translation $\delta\mathbf{a}$, and using (29.7), we obtain $i\hbar\, \delta\mathbf{a} = \langle a'|[\mathbf{x}, \delta\mathbf{a}\cdot\mathbf{p}]|a'\rangle$, where we have designated the momentum observable of the particle by $\mathbf{p}$. By hypothesis this equality holds for all $\delta\mathbf{a}$ and all $|a'\rangle$, and we must therefore have

$$[x_i, p_j] = i\hbar\, \delta_{ij}. \tag{3}$$

As $\mathbf{p}$ is the total momentum of the particle, it satisfies (29.3), or

$$[p_i, p_j] = 0. \tag{4}$$

The relations (1), (3) and (4) are called the *canonical commutation rules*.

The spectra of the observables $\mathbf{p}$ and $\mathbf{x}$ follow from the commutation rules. In order to determine these spectra we shall require the commutation rules

$$[x_i, G(\mathbf{p})] = i\hbar \frac{\partial G}{\partial p_i}, \tag{5}$$

$$[p_i, F(\mathbf{x})] = \frac{\hbar}{i}\frac{\partial F}{\partial x_i}. \tag{6}$$

These identities are easily derived from (3) for all functions F and G that can be expressed as power series in their arguments.* Let $|\mathbf{x}'\rangle$ be an

* These commutation rules were already derived in Sec. 6.2 by means of the differential representations of the position and momentum operators.

eigenket of $\mathbf{x}$ with the indicated eigenvalue. We shall show that the displaced ket $e^{-i\mathbf{p}\cdot\mathbf{a}/\hbar}|\mathbf{x}'\rangle$ is an eigenket of $\mathbf{x}$ with eigenvalue $\mathbf{x}' + \mathbf{a}$. Because of (5) we may write

$$\begin{aligned}\mathbf{x}\, e^{-i\mathbf{p}\cdot\mathbf{a}/\hbar}|\mathbf{x}'\rangle &= \left(e^{-i\mathbf{p}\cdot\mathbf{a}/\hbar}\mathbf{x} + i\hbar\frac{\partial}{\partial\mathbf{p}}\, e^{-i\mathbf{p}\cdot\mathbf{a}/\hbar}\right)|\mathbf{x}'\rangle \\ &= (\mathbf{x}' + \mathbf{a})e^{-i\mathbf{p}\cdot\mathbf{a}/\hbar}|\mathbf{x}'\rangle.\end{aligned}$$

This proves the assertion. As $\mathbf{a}$ was arbitrary in this calculation, all $\mathbf{x}'$ (i.e., $-\infty < x_i' < \infty$) are eigenvalues of $\mathbf{x}$. Because of the unitary nature of the operator, $e^{-i\mathbf{p}\cdot\mathbf{a}/\hbar}|\mathbf{x}'\rangle$ has the same norm as $|\mathbf{x}'\rangle$, and we are therefore entitled to write $\langle\mathbf{x}' + \mathbf{a}|\mathbf{x}'' + \mathbf{a}\rangle = \langle\mathbf{x}'|\mathbf{x}''\rangle$, and

$$|\mathbf{x}' + \mathbf{a}\rangle = e^{-i\mathbf{p}\cdot\mathbf{a}/\hbar}|\mathbf{x}'\rangle. \tag{7}$$

The same argument can be used to determine the spectrum of $\mathbf{p}$. Let $|\mathbf{p}'\rangle$ be an eigenket of $\mathbf{p}$ with eigenvalue $\mathbf{p}'$, and $\mathbf{q}$ an arbitrary momentum. Then (6) shows that $e^{i\mathbf{x}\cdot\mathbf{q}/\hbar}|\mathbf{p}'\rangle$ is an eigenket of $\mathbf{p}$ with eigenvalue $\mathbf{p}' + \mathbf{q}$. (Needless to say, in this last sentence $\mathbf{q}$ is a set of three numbers, and $\mathbf{x}$ a set of three operators.) Hence the spectrum of $\mathbf{p}$ coincides with that of $\mathbf{x}$: $-\infty < p_i' < \infty$.

The operators $\mathbf{p}$ and $\mathbf{x}$ appear symmetrically (except for $i \to -i$) in the canonical commutation rules, and for this reason both observables have the same spectrum. This leads to a rather important observation (see Pauli, p. 140) concerning the role of the time in the mathematical formulation of a relativistic quantum theory. In a relativistic theory one would expect the canonical commutation rules to involve two 4-vectors x_μ and p_μ. The fourth component of p_μ would presumably be the Hamiltonian H, and that of x_μ a "time operator" τ, with the commutation rule $[\tau, H] = i\hbar$. By analogy with the x-p commutation rule, a continuous spectrum for τ, $-\infty < t < \infty$, would then imply a similar spectrum for H. But the energy spectrum must have a lower bound, and this scheme must therefore fail.* In fact, consistent relativistic quantum theories do not promote t to an operator; they achieve Lorentz invariance by demoting $\mathbf{x}$ to the status of a parameter along with t, and introduce other objects (e.g., field strengths) as the basic dynamical variables. We shall see an important example of this in the quantization of the electromagnetic field (Chapt. VIII).

* This observation also emphasizes that the energy-time uncertainty relation is on a rather different footing from that for position and momentum.

Because the spectra of $\mathbf{p}$ and $\mathbf{x}$ are continuous, we shall normalize their eigenkets as in Sec. 23.5, viz.

$$\langle \mathbf{x}'|\mathbf{x}''\rangle = \delta(\mathbf{x}' - \mathbf{x}''), \tag{8}$$

$$\langle \mathbf{p}'|\mathbf{p}''\rangle = \delta(\mathbf{p}' - \mathbf{p}''). \tag{9}$$

We shall assume that both of these sets are complete:

$$1 = \int d^3x'|\mathbf{x}'\rangle\langle\mathbf{x}'| = \int d^3p'|\mathbf{p}'\rangle\langle\mathbf{p}'|. \tag{10}$$

When we deal with a system of N particles, we must introduce position and momentum observables $\mathbf{x}_\alpha$ and $\mathbf{p}_\alpha$ (with $\alpha = 1, 2, \ldots, N$) for each particle. Observables belonging to different particles are assumed to commute: $[\mathbf{x}_\alpha, \mathbf{p}_\beta] = 0$ if $\alpha \neq \beta$. We may then introduce simultaneous eigenkets for all the $\mathbf{x}_\alpha$, or all the $\mathbf{p}_\alpha$, and we are also allowed to introduce simultaneous eigenkets for, say, $\mathbf{x}_5$ and $\mathbf{p}_3$. Nevertheless, the most convenient sets are defined by

$$(\mathbf{x}_\alpha - \mathbf{x}_\alpha')|\mathbf{x}_1' \cdots \mathbf{x}_\alpha' \cdots \mathbf{x}_N'\rangle = 0, \tag{11a}$$

$$(\mathbf{p}_\alpha - \mathbf{p}_\alpha')|\mathbf{p}_1' \cdots \mathbf{p}_\alpha' \cdots \mathbf{p}_N'\rangle = 0, \tag{11b}$$

$$\langle\mathbf{x}_1' \cdots \mathbf{x}_N'|\mathbf{x}_1'' \cdots \mathbf{x}_N''\rangle = \delta(\mathbf{x}_1' - \mathbf{x}_1'') \cdots \delta(\mathbf{x}_N' - \mathbf{x}_N''), \tag{11c}$$

$$\langle\mathbf{p}_1' \cdots \mathbf{p}_N'|\mathbf{p}_1'' \cdots \mathbf{p}_N''\rangle = \delta(\mathbf{p}_1' - \mathbf{p}_1'') \cdots \delta(\mathbf{p}_N' - \mathbf{p}_N''), \tag{11d}$$

$$\int d^3x_1' \cdots d^3x_N'|\mathbf{x}_1' \cdots \mathbf{x}_N'\rangle\langle\mathbf{x}_1' \cdots \mathbf{x}_N'| = 1, \tag{11e}$$

$$\int d^3p_1' \cdots d^3p_N'|\mathbf{p}_1' \cdots \mathbf{p}_N'\rangle\langle\mathbf{p}_1' \cdots \mathbf{p}_N'| = 1. \tag{11f}$$

The total momentum $\mathbf{P}$ of the N-particle system must have the property that

$$e^{i\mathbf{P}\cdot\mathbf{a}/\hbar}\mathbf{x}_\alpha e^{-i\mathbf{P}\cdot\mathbf{a}/\hbar} = \mathbf{x}_\alpha + \mathbf{a}. \tag{12}$$

This will insure that (2) holds for each of the $\mathbf{x}_\alpha$. But $e^{i\mathbf{p}_\alpha\cdot\mathbf{a}/\hbar}\mathbf{x}_\alpha e^{-i\mathbf{p}_\alpha\cdot\mathbf{a}/\hbar} = \mathbf{x}_\alpha + \mathbf{a}$, and $[\mathbf{x}_\alpha, \mathbf{p}_\beta] = 0$ if $\alpha \neq \beta$; therefore the unitary operator $\Pi_\alpha e^{-i\mathbf{p}_\alpha\cdot\mathbf{a}/\hbar}$ will perform the translation demanded by (12). Hence

$$\mathbf{P} = \sum_\alpha \mathbf{p}_\alpha, \tag{13}$$

as expected.

In the laboratory the total momentum is defined by assuming that momentum is conserved and measuring the momenta of detectable particles—usually charged particles at that. For example, the momentum of the neutrino in β-decay of stationary neutrons is defined to be $-(\mathbf{p}_p + \mathbf{p}_e)$, where $\mathbf{p}_p$ and $\mathbf{p}_e$ are the momenta of the proton and electron, respectively. Similarly, the momentum acquired by the electromagnetic field in neutron-proton capture at rest is defined to be equal and opposite to that of the deuteron. The consistency of the definitions and the validity of the conservation law are confirmed by the fact that these neutrinos (or photons) can cause reactions in which detectable particles are produced, and that these particles have precisely the total momentum ascribed to the neutrinos (or photons) in the initial reaction. By the same token, the conclusion that the momentum spectrum is completely continuous for a single particle requires the spectrum of the total momentum observable for *any* system also to be continuous $(-\infty < P_i' < \infty)$.

30.2 WAVE FUNCTIONS

We may establish the detailed connection between the abstract formalism and the wave functions of Schrödinger's theory by computing the x-p transformation function. If we call $|0_x\rangle$ the simultaneous eigenket of the $\{\mathbf{x}_\alpha\}$ with all eigenvalues $\mathbf{x}_\alpha' = 0$, and $|0_p\rangle$ the simultaneous eigenket of the $\{\mathbf{p}_\alpha\}$ with all $\mathbf{p}_\alpha' = 0$, we have

$$\begin{aligned}
\langle \mathbf{x}_1' \cdots \mathbf{x}_N' | \mathbf{p}_1' \cdots \mathbf{p}_N' \rangle &= \left\langle 0_x \middle| \prod_\alpha e^{i\mathbf{p}_\alpha \cdot \mathbf{x}_\alpha'/\hbar} \middle| \mathbf{p}_1' \cdots \mathbf{p}_N' \right\rangle \\
&= \prod_\alpha e^{i\mathbf{p}_\alpha' \cdot \mathbf{x}_\alpha'/\hbar} \langle 0_x | \mathbf{p}_1' \cdots \mathbf{p}_N' \rangle \\
&= \prod_\alpha e^{i\mathbf{p}_\alpha' \cdot \mathbf{x}_\alpha'/\hbar} \left\langle 0_x \middle| \prod_\beta e^{i\mathbf{p}_\beta' \cdot \mathbf{x}_\beta/\hbar} \middle| 0_p \right\rangle \\
&= \prod_\alpha e^{i\mathbf{p}_\alpha' \cdot \mathbf{x}_\alpha'/\hbar} \langle 0_x | 0_p \rangle .
\end{aligned}$$

The constant $\langle 0_x | 0_p \rangle$ can be determined to within a phase by combining, say, (11f) and (11d):

$$\int d^3x_1' \cdots d^3x_N' \langle \mathbf{p}_1' \cdots \mathbf{p}_N' | \mathbf{x}_1' \cdots \mathbf{x}_N' \rangle \langle \mathbf{x}_1' \cdots \mathbf{x}_N' | \mathbf{p}_1'' \cdots \mathbf{p}_N'' \rangle = \prod_\alpha \delta(\mathbf{p}_\alpha' - \mathbf{p}_\alpha'').$$

But

$$\int d^3x' \, e^{i(\mathbf{p}' - \mathbf{p}'') \cdot \mathbf{x}'/\hbar} = (2\pi\hbar)^3 \, \delta(\mathbf{p}' - \mathbf{p}'').$$

Choosing $\langle 0_x | 0_p \rangle$ to be real and positive we therefore obtain

$$\langle \mathbf{x}_1' \cdots \mathbf{x}_N' | \mathbf{p}_1' \cdots \mathbf{p}_N' \rangle = (2\pi\hbar)^{-\frac{3}{2}N} \prod_{\alpha=1}^{N} e^{i\mathbf{p}_\alpha' \cdot \mathbf{x}_\alpha'/\hbar}. \tag{14}$$

This is just the coordinate space wave function of a system of N free particles with momenta $\{\mathbf{p}_\alpha'\}$, familiar to us since Sec. 4.

Let $|a'\rangle$ be an arbitrary ket of a one-particle system. (For the moment we consider a single particle, as it is abundantly clear that the extension to the case of N particles is both trivial and unsightly.) The probability amplitude for finding the system at the point $\mathbf{x}'$ is $\langle\mathbf{x}'|a'\rangle$, and that for the momentum $\mathbf{p}'$ is $\langle\mathbf{p}'|a'\rangle$. These are related by the standard transformation formula

$$\langle\mathbf{x}'|a'\rangle = \int d^3p'\langle\mathbf{x}'|\mathbf{p}'\rangle\langle\mathbf{p}'|a'\rangle.$$

With the help of (14) this becomes

$$\langle\mathbf{x}'|a'\rangle = (2\pi\hbar)^{-\frac{3}{2}} \int d^3p'\, e^{i\mathbf{p}'\cdot\mathbf{x}'/\hbar}\langle\mathbf{p}'|a'\rangle, \tag{15}$$

which is just the Fourier theorem. To translate this equation into the language of wave mechanics we need only identify the x-space and p-space wave functions as

$$\psi_{a'}(\mathbf{x}') \equiv \langle\mathbf{x}'|a'\rangle, \qquad \phi_{a'}(\mathbf{p}') \equiv \langle\mathbf{p}'|a'\rangle. \tag{16}$$

Once this is done, (15) becomes (4.20).

The wave-mechanical formulas for the action of the momentum operator on ψ and position operator on ϕ can be obtained by considering

$$\langle\mathbf{x}'|\mathbf{p}|a'\rangle = \int d^3x''\langle\mathbf{x}'|\mathbf{p}|\mathbf{x}''\rangle\langle\mathbf{x}''|a'\rangle \tag{17}$$

and

$$\langle\mathbf{p}'|\mathbf{x}|a'\rangle = \int d^3p''\langle\mathbf{p}'|\mathbf{x}|\mathbf{p}''\rangle\langle\mathbf{p}''|a'\rangle. \tag{18}$$

Consider

$$\begin{aligned}\langle\mathbf{x}'|\mathbf{p}|\mathbf{x}''\rangle &= \int d^3p'\langle\mathbf{x}'|\mathbf{p}'\rangle\mathbf{p}'\langle\mathbf{p}'|\mathbf{x}''\rangle \\ &= (2\pi\hbar)^{-3}\int d^3p'\, \mathbf{p}'e^{i\mathbf{p}'\cdot(\mathbf{x}'-\mathbf{x}'')/\hbar};\end{aligned}$$

if we recall the definition (7.23) of the gradient of the δ-function, we obtain

$$\langle\mathbf{x}'|\mathbf{p}|\mathbf{x}''\rangle = -i\hbar\,\delta^{(1)}(\mathbf{x}' - \mathbf{x}''), \tag{19}$$

and therefore (17) becomes

$$\langle\mathbf{x}'|\mathbf{p}|a'\rangle = \frac{\hbar}{i}\frac{\partial}{\partial\mathbf{x}'}\psi_{a'}(\mathbf{x}'). \tag{20}$$

In the same way one finds

$$\langle\mathbf{p}'|\mathbf{x}|\mathbf{p}''\rangle = i\hbar\,\delta^{(1)}(\mathbf{p}' - \mathbf{p}''), \tag{21}$$

and (18) reduces to

$$\langle\mathbf{p}'|\mathbf{x}|a'\rangle = i\hbar\frac{\partial}{\partial\mathbf{p}'}\phi_{a'}(\mathbf{p}'). \tag{22}$$

These are the familiar expressions of Sec. 4. They are readily generalized to

$$\langle \mathbf{x}'|p_i^n|a'\rangle = \left(\frac{\hbar}{i}\frac{\partial}{\partial x_i'}\right)^n \psi_{a'}(\mathbf{x}'),$$
$$\langle \mathbf{p}'|x_i^n|a'\rangle = \left(i\hbar\frac{\partial}{\partial p_i'}\right)^n \phi_{a'}(\mathbf{p}'). \tag{23}$$

30.3 GALILEO TRANSFORMATIONS AND GALILEO INVARIANT HAMILTONIANS

Let $|a'\rangle$ be an arbitrary N-particle state prepared by an observer O at the instant t_0, and let $|a';\mathbf{v}\rangle$ be the state having the same properties at time t_0 insofar as an observer O' who is moving with velocity $\mathbf{v}$ relative to O is concerned. We assume that at $t = 0$ the two coordinate frames coincide. The expectation values of $\mathbf{x}_\alpha$ and $\mathbf{p}_\alpha$ in these states are then related by the Galileo transformations

$$\langle a';\mathbf{v}|\mathbf{x}_\alpha|a';\mathbf{v}\rangle = \langle \mathbf{x}_\alpha\rangle_{a'} + \mathbf{v}t_0, \tag{24}$$
$$\langle a';\mathbf{v}|\mathbf{p}_\alpha|a';\mathbf{v}\rangle = \langle \mathbf{p}_\alpha\rangle_{a'} + m_\alpha\mathbf{v}, \tag{25}$$

provided $v \ll c$. By definition, the coefficient m_α in (25) is the mass of particle α.

Consider the infinitesimal Galileo transformation wherein $\mathbf{v}$ is replaced by $\delta\mathbf{v}$, and define the infinitesimal unitary operator $\Gamma(t_0, \delta\mathbf{v}) = 1 - i\,\delta\mathbf{v}\cdot\mathbf{K}/\hbar$, $\mathbf{K} = \mathbf{K}^\dagger$, by requiring $\Gamma|a'\rangle = |a';\delta\mathbf{v}\rangle$. In virtue of (24) and (25) we must have $\Gamma^\dagger\,\mathbf{x}_\alpha\,\Gamma = \mathbf{x}_\alpha + t_0\,\delta\mathbf{v}$ and $\Gamma^\dagger\,\mathbf{p}_\alpha\,\Gamma = \mathbf{p}_\alpha + m_\alpha\,\delta\mathbf{v}$, or

$$t_0\,\delta\mathbf{v} = -(i/\hbar)[\mathbf{x}_\alpha, \delta\mathbf{v}\cdot\mathbf{K}] = \frac{\partial}{\partial\mathbf{p}_\alpha}\,\delta\mathbf{v}\cdot\mathbf{K},$$
$$m_\alpha\,\delta\mathbf{v} = -(i/\hbar)[\mathbf{p}_\alpha, \delta\mathbf{v}\cdot\mathbf{K}] = -\frac{\partial}{\partial\mathbf{x}_\alpha}\,\delta\mathbf{v}\cdot\mathbf{K}.$$

As $\delta\mathbf{v}$ is arbitrary, these equations reduce to $t_0 = \partial K_x/\partial p_{x\alpha}$, etc. Hence*

$$\Gamma(t_0, \delta\mathbf{v}) = 1 + \frac{i}{\hbar}\,\delta\mathbf{v}\cdot(M\mathbf{R} - t_0\mathbf{P}), \tag{26}$$

where $M = \Sigma m_\alpha$ is the total mass of the system, and $\mathbf{R} = M^{-1}\Sigma m_\alpha\mathbf{x}_\alpha$ is the center-of-mass observable.

Having defined the Galileo transformation, we can now discuss the concept of Galileo invariance. Consider first a system of particles

* A constant vector can also be added to the coefficient of $\delta\mathbf{v}$, but this can be removed by redefining the phase of $|a';\delta\mathbf{v}\rangle$ appropriately.

in classical theory. Let $\mathbf{x}_\alpha(t_0)$, $\mathbf{p}_\alpha(t_0)$, be the initial coordinates and momenta in some reference frame, and perform the following two sequences of operations: (a) go to a frame moving with velocity $\mathbf{v}$, i.e.,

$$\mathbf{x}_\alpha(t_0) \to \mathbf{x}'_\alpha(t_0) = \mathbf{x}_\alpha(t_0) + \mathbf{v}t_0,$$
$$\mathbf{p}_\alpha(t_0) \to \mathbf{p}'_\alpha(t_0) = \mathbf{p}_\alpha(t_0) + m_\alpha\mathbf{v},$$

then wait till $t = t_1$ and measure $\mathbf{x}'_\alpha(t_1)$ and $\mathbf{p}'_\alpha(t_1)$; (b) allow the motion to evolve until $t = t_1$, i.e.,

$$\mathbf{x}_\alpha(t_0) \to \mathbf{x}_\alpha(t_1),\ \mathbf{p}_\alpha(t_0) \to \mathbf{p}_\alpha(t_1),$$

and then go to a frame moving with velocity $\mathbf{v}$:

$$\mathbf{x}_\alpha(t_1) \to \mathbf{x}''_\alpha(t_1) = \mathbf{x}_\alpha(t_1) + \mathbf{v}t_1,$$
$$\mathbf{p}_\alpha(t_1) \to \mathbf{p}''_\alpha(t_1) = \mathbf{p}_\alpha(t_1) + m_\alpha\mathbf{v}.$$

When the system in question is isolated we find that $\mathbf{x}'_\alpha(t_1) = \mathbf{x}''_\alpha(t_1)$, $\mathbf{p}'_\alpha(t_1) = \mathbf{p}''_\alpha(t_1)$, and this fact defines, for our purposes, the notion of Galileo invariance. The analogous statement of Galileo invariance in the Schrödinger picture is

$$U(t_1, t_0)\Gamma(t_0, \mathbf{v}) = \Gamma(t_1, \mathbf{v})U(t_1, t_0). \tag{27}$$

This equation imposes a condition on the Hamiltonian that we shall now exploit. For this purpose it suffices to replace $\mathbf{v}$ by the infinitesimal $\delta\mathbf{v}$. Equation (27) then reduces to

$$M[\mathbf{R}, U(t_1, t_0)] = t_1\,\mathbf{P}\,U(t_1, t_0) - t_0U(t_1, t_0)\mathbf{P}. \tag{28}$$

Because we are only concerned with translationally invariant systems, $\mathbf{P}$ commutes with $U(t_1, t_0)$. The content of (28) can be stated most succinctly in the Heisenberg picture. Let $\mathbf{R}(t) = U^\dagger(t, t_0)\mathbf{R}U(t, t_0)$ be the H-picture center-of-mass operator. Premultiplying (28) by $U^\dagger(t_1, t_0)$ immediately yields

$$\mathbf{R}(t_1) = \mathbf{R}(t_0) + \frac{t_1 - t_0}{M}\mathbf{P}, \tag{29}$$

which has the same form as the trajectory of the center of mass of an isolated classical system. It is therefore a matter of taste whether one prefers (27) or (29) as the more intuitively obvious statement of Galileo invariance. Returning to (28), and letting $t_1 - t_0$ be infinitesimal, we obtain

$$[M\mathbf{R}, H(t)] = i\hbar\mathbf{P}. \tag{30}$$

In view of the definitions of M, $\mathbf{R}$ and $\mathbf{P}$, and the commutation rule (5), we can translate (30) into the condition

$$\sum_\alpha \left(m_\alpha \frac{\partial H}{\partial \mathbf{p}_\alpha} - \mathbf{p}_\alpha \right) = 0. \tag{31}$$

This shows that the dependence of the Hamiltonian on the momenta is severely restricted. Needless to say, (31) is satisfied by the conventional time-independent and translation invariant Hamiltonian

$$H = \sum_\alpha \frac{p_\alpha^2}{2m_\alpha} + \tfrac{1}{2} \sum_{\alpha \neq \beta} V(|\mathbf{x}_\alpha - \mathbf{x}_\beta|). \tag{32}$$

Schrödinger's wave equation is obtained by taking the scalar product of $(i\hbar\, \partial_t - H)|t\rangle = 0$ with $\langle \mathbf{x}_1' \cdots \mathbf{x}_N'|$, and using (16) and (23):

$$i\hbar \frac{\partial}{\partial t} \langle \mathbf{x}_1' \cdots \mathbf{x}_N'|t\rangle = \left[- \sum_\alpha \frac{\hbar^2}{2m_\alpha} \left(\frac{\partial}{\partial \mathbf{x}_\alpha'} \right)^2 + \tfrac{1}{2} \sum_{\alpha \neq \beta} V(|\mathbf{x}_\alpha' - \mathbf{x}_\beta'|) \right] \langle \mathbf{x}_1' \cdots \mathbf{x}_N'|t\rangle.$$

30.4 THE LIPPMANN-SCHWINGER EQUATION

In order to gain more familiarity with operator methods, we shall re-derive the integral equation for scattering (12.15). Consider a single particle interacting with the static potential $V(\mathbf{x})$. The Hamiltonian is

$$\begin{aligned} H &= H_0 + V, \\ H_0 &= \mathbf{p}^2/2m. \end{aligned} \tag{33}$$

We search for a stationary solution

$$(H - E)|\mathbf{p}_0'^{(+)}\rangle = 0, \tag{34}$$

where $|\mathbf{p}_0'^{(+)}\rangle$ is an eigenket of H describing an incident beam of momentum $\mathbf{p}_0'$, and *outgoing* scattered waves. The energy eigenvalue E is therefore $(\mathbf{p}_0')^2/2m$. Let us write (34) as

$$(H_0 - E)|\mathbf{p}_0'^{(+)}\rangle = -V|\mathbf{p}_0'^{(+)}\rangle. \tag{35}$$

We should like to bring $(H_0 - E)$ to the right side of this equation, but we cannot do so because $(H_0 - E)^{-1}$ is singular when $E > 0$. As in Sec. 12.1 we replace E by the complex variable z. Since the eigenvalues of H_0 are real and positive, the resolvent $(H_0 - z)^{-1}$ exists for $\operatorname{Re} z > 0$ and $\operatorname{Im} z \neq 0$; the inversion may therefore be carried out for z arbitrarily close to the real axis. When we carry out the inversion we may add any solution $|\ \rangle$ of $(H_0 - E)|\ \rangle = 0$ to the right-hand side. We recall from Sec. 12.1 that the choice of the particular solution is dictated by the boundary conditions that one wishes to satisfy. As in Sec. 12.1 we shall choose $|\ \rangle$ as the incident momentum eigenstate $|\mathbf{p}_0'\rangle$. Thus (35) reads

$$|\mathbf{p}_0'^{(+)}\rangle = |\mathbf{p}_0'\rangle + \frac{1}{z - H_0} V |\mathbf{p}_0'^{(+)}\rangle, \tag{36}$$

with $\mathrm{Re}\, z = E$, and $\mathrm{Im}\, z$ infinitesimal. Upon applying $(H_0 - E)$ to (36) we immediately retrieve (35).

The connection between this abstract equation and our earlier work on scattering theory can be seen by taking the scalar product of (36) with $\langle \mathbf{x}'|$. We shall show that for $\mathrm{Im}\, z = 0^+$ the coordinate representation counterpart of (36) is the integral equation for scattering that we used extensively in Chapt. III, i.e., that $\langle \mathbf{x}'|\mathbf{p}_0'^{(+)}\rangle \equiv \psi_{\mathbf{p}_0'}^{(+)}(\mathbf{x}')$ is just the familiar scattering solution of Sec. 12. In the coordinate representation (36) is

$$\psi_{\mathbf{p}_0'}^{(+)}(\mathbf{x}') = (2\pi\hbar)^{-\frac{3}{2}} e^{i\mathbf{p}_0'\cdot\mathbf{x}'/\hbar} + \int d^3x''\, d^3x''' \langle \mathbf{x}'|(z - H_0)^{-1}|\mathbf{x}''\rangle\langle \mathbf{x}''|V|\mathbf{x}'''\rangle \psi_{\mathbf{p}_0'}^{(+)}(\mathbf{x}'''). \tag{37}$$

The kernel of this integral equation must be related to the Green's function for the Helmholz equation. To demonstrate this we take advantage of the fact that

$$H_0|\mathbf{p}'\rangle = (\mathbf{p}')^2/2m|\mathbf{p}'\rangle,$$

which gives

$$\left\langle \mathbf{x}' \left| \frac{1}{z - H_0} \right| \mathbf{x}'' \right\rangle = \int d^3p' \frac{\langle \mathbf{x}'|\mathbf{p}'\rangle\langle \mathbf{p}'|\mathbf{x}''\rangle}{z - (p'^2/2m)} = (2\pi\hbar)^{-3} \int d^3p' \frac{e^{i\mathbf{p}'\cdot(\mathbf{x}'-\mathbf{x}'')/\hbar}}{z - (p'^2/2m)}.$$

However, from (12.14)

$$\int \frac{d^3q}{(2\pi)^3} \frac{e^{i\mathbf{q}\cdot(\mathbf{x}'-\mathbf{x}'')}}{k^2 - q^2 \pm i\epsilon} = -\frac{1}{4\pi} \frac{e^{\pm ik|\mathbf{x}'-\mathbf{x}''|}}{|\mathbf{x}' - \mathbf{x}''|},$$

and therefore

$$\left\langle \mathbf{x}' \left| \frac{1}{E - H_0 + i\epsilon} \right| \mathbf{x}'' \right\rangle = -\frac{2m}{4\pi\hbar^2} \frac{e^{ip|\mathbf{x}'-\mathbf{x}''|/\hbar}}{|\mathbf{x}' - \mathbf{x}''|} \qquad (E = p^2/2m). \tag{38}$$

As promised, the coordinate representation matrix elements of the resolvent operator constitute the Green's function. For this reason physicists often refer to $(z - H_0)^{-1}$ as Green's operator. If the potential operator is diagonal in the coordinate representation, i.e., $\langle \mathbf{x}'|V|\mathbf{x}''\rangle = V(\mathbf{x}')\,\delta(\mathbf{x}' - \mathbf{x}'')$, then (37) is, aside from a trivial factor of $\hbar^{-\frac{3}{2}}$, identical with the integral equation on which the scattering theory of Sec. 12 was built.

Let us write (36) once more with the correct choice of z for the outgoing boundary conditions:

$$|\mathbf{p}_0'^{(+)}\rangle = |\mathbf{p}_0'\rangle + \frac{1}{E - H_0 + i\epsilon} V|\mathbf{p}_0'^{(+)}\rangle. \tag{39}$$

This is usually called the Lippmann-Schwinger equation. As we already pointed out in Sec. 23.7, the advantage of the abstract approach is that any equation is valid in all representations. To illustrate this statement, we shall write (39) in the momentum representation. One could do this in the laborious wave-mechanical way by taking the Fourier transform of (37), but the more concise way is to take the scalar product of (39) with a momentum eigenbra. One immediately finds

$$\phi_{\mathbf{p}_0'}^{(+)}(\mathbf{p}') = \delta(\mathbf{p}' - \mathbf{p}_0') + \frac{1}{E - E' + i\epsilon} \int d^3p'' \langle \mathbf{p}'|V|\mathbf{p}''\rangle \phi_{\mathbf{p}_0'}^{(+)}(\mathbf{p}''), \tag{40}$$

where $\phi_{\mathbf{p}_0'}^{(+)}(\mathbf{p}') = \langle \mathbf{p}'|\mathbf{p}_0'^{(+)}\rangle$, and $E' = p'^2/2m$.

31. The Harmonic Oscillator

The harmonic oscillator plays a very important role in quantum mechanics, and we shall therefore study it in some detail. It provides an instructive example of quantum mechanical methods. Above and beyond this it is an essential element in the theory of radiation. The reason for this is that one may view the electromagnetic field as an infinite collection of oscillators.

31.1 EQUATIONS OF MOTION

Because the Hamiltonian of the n-dimensional oscillator is a sum of n commuting operators, we may confine ourselves to the one-dimensional case. We shall designate the single position observable by q, and the canonical momentum by p. The Hamiltonian then reads

$$H = \frac{p^2}{2m} + \tfrac{1}{2} m\omega^2 q^2. \tag{1}$$

We begin the discussion in the Heisenberg picture. The equations of motion are

$$i\hbar \frac{d}{dt} p(t) = [p(t), H] = -i\hbar m\omega^2 q(t),$$

$$i\hbar \frac{d}{dt} q(t) = [q(t), H] = \frac{i\hbar}{m} p(t).$$

These actually reduce to one equation if one introduces the non-Hermitian operator

$$a(t) = \frac{p(t) - im\omega q(t)}{\sqrt{2m\hbar\omega}}, \tag{2}$$

because it satisfies

$$\left(\frac{d}{dt} + i\omega\right) a(t) = 0.$$

This integrates to

$$a(t) = ae^{-i\omega t}, \tag{3}$$

where a is (3) evaluated at $t = 0$, at which time the S- and H-pictures are identical.

We may now solve for p and q:

$$p(t) = \left(\frac{m\hbar\omega}{2}\right)^{\frac{1}{2}} [a(t) + a^\dagger(t)] \tag{4}$$

$$= p \cos \omega t - m\omega q \sin \omega t, \tag{5}$$

$$q(t) = i\left(\frac{\hbar}{2m\omega}\right)^{\frac{1}{2}} [a(t) - a^\dagger(t)] \tag{6}$$

$$= q \cos \omega t + \frac{1}{m\omega} p \sin \omega t. \tag{7}$$

These expressions are precisely the same as in the classical theory. Note that $q(t)$ and $q(t')$ are not compatible observables: From (7) one easily verifies that

$$[q, q(t)] = \frac{i\hbar}{m\omega} \sin \omega t.$$

This expresses the fact that the eigenstates of q are not stationary.

31.2 ENERGY EIGENFUNCTIONS

In terms of the operators a and $a^\dagger$ the Hamiltonian is

$$H = \frac{\hbar\omega}{2} (aa^\dagger + a^\dagger a). \tag{8}$$

The commutation rule for a and $a^\dagger$ is easily evaluated:

$$[a, a^\dagger] = \frac{1}{2m\hbar\omega}[p - im\omega q, p + im\omega q]$$
$$= -\frac{i}{2\hbar}[q, p] + \frac{i}{2\hbar}[p, q];$$

thus

$$[a, a^\dagger] = 1. \tag{9}$$

This basic commutation rule will be used repeatedly in this section, and should be committed to memory. With its help we can rewrite (8) as

$$H = \tfrac{1}{2}\hbar\omega + \hbar\omega a^\dagger a. \tag{10}$$

The Hermitian operator $a^\dagger a$ is positive definite, and the spectrum of H therefore has $\frac{1}{2}\hbar\omega$ as a lower bound.

We define a new operator N by

$$N = a^\dagger a, \tag{11}$$

and its eigenkets by $(N - n)|n\rangle = 0$. Then

$$Na^\dagger|n\rangle = (a^\dagger N + [N, a^\dagger])|n\rangle,$$
$$Na|n\rangle = (aN + [N, a])|n\rangle.$$

However $[N, a^\dagger] = a^\dagger[a, a^\dagger]$; using (9) we have

$$[N, a^\dagger] = a^\dagger, \tag{12}$$

and similarly

$$[N, a] = -a. \tag{12$'$}$$

Hence

$$[N - (n + 1)]a^\dagger|n\rangle = 0,$$
$$[N - (n - 1)]a|n\rangle = 0. \tag{13}$$

Therefore $a^\dagger|n\rangle$ is proportional to $|n + 1\rangle$, and $a|n\rangle$ to $|n - 1\rangle$. The proportionality factors are determined by computing the norms of $a|n\rangle$ and $a^\dagger|n\rangle$. Using (9) we find that these are $\langle n|a^\dagger a|n\rangle = n$, and $\langle n|aa^\dagger|n\rangle = 1 + n$, assuming that $\langle n|n\rangle = 1$. We shall choose the arbitrary phase that remains so that

$$a^\dagger|n\rangle = \sqrt{n+1}\,|n + 1\rangle, \tag{14}$$
$$a|n\rangle = \sqrt{n}\,|n - 1\rangle. \tag{15}$$

Equation (15) shows that if we apply the operator a^k to $|n\rangle$, with $k > n$, we are guaranteed to obtain an eigenstate of N with a negative eigenvalue. On the other hand, we have already noted that N is a positive-definite operator, and therefore its eigenvalues cannot be negative. Consistency can only be achieved by requiring $a|n\rangle$ to be a null vector whenever $n < 1$. Equation (15) also shows that this can only happen when $n = 0$. Consequently *the eigenvalues n of the operator $a^\dagger a$ are the positive integers and zero.* By the same token, the energy eigenvalues are

$$E_n = (n + \tfrac{1}{2})\hbar\omega. \tag{16}$$

The entire energy spectrum of the oscillator is therefore discrete. This example illustrates assertion (a) of Sec. 7.2.

The nonvanishing matrix elements of a and $a^\dagger$ can be obtained directly from (14) and (15), and the orthogonality condition $\langle n|n'\rangle = \delta_{nn'}$:

$$\begin{aligned} \langle n|a|n'\rangle &= \sqrt{n+1}\,\delta_{n,n'-1}, \\ \langle n|a^\dagger|n'\rangle &= \sqrt{n}\,\delta_{n,n'+1}. \end{aligned} \tag{17}$$

The matrix elements of p and q can be evaluated from (4) and (6), viz.

$$\begin{aligned} \langle n|p|n'\rangle &= \left(\frac{m\hbar\omega}{2}\right)^{\frac{1}{2}} [\sqrt{n+1}\,\delta_{n,n'-1} + \sqrt{n}\,\delta_{n,n'+1}], \\ \langle n|q|n'\rangle &= i\left(\frac{\hbar}{2m\omega}\right)^{\frac{1}{2}} [\sqrt{n+1}\,\delta_{n,n'-1} - \sqrt{n}\,\delta_{n,n'+1}]. \end{aligned} \tag{18}$$

Observe that p and q have no diagonal elements. Furthermore

$$\langle n|q^2|n\rangle = \frac{\hbar}{2m\omega}\langle (a - a^\dagger)(a^\dagger - a)\rangle_n = \frac{\hbar}{2m\omega}\langle aa^\dagger + a^\dagger a\rangle_n,$$

$$\langle n|p^2|n\rangle = \frac{m\hbar\omega}{2}\langle (a + a^\dagger)^2\rangle_n = \frac{m\hbar\omega}{2}\langle aa^\dagger + a^\dagger a\rangle_n.$$

Therefore

$$\langle q^2\rangle_n = \frac{E_n}{m\omega^2}, \qquad \langle p^2\rangle_n = mE_n, \tag{18'}$$

or

$$\Delta p\,\Delta q = (n + \tfrac{1}{2})\hbar. \tag{19}$$

This shows that the ground state of the oscillator actually attains the minimum possible value of the uncertainty product.

To complete the story we still need the wave functions $\langle q'|n\rangle$. Because $a|0\rangle = 0$,

$$\begin{aligned} \langle q'|a|0\rangle &= (2m\hbar\omega)^{-\frac{1}{2}}\langle q'|p - im\omega q|0\rangle \\ &= (2m\hbar\omega)^{-\frac{1}{2}}\left(\frac{\hbar}{i}\frac{\partial}{\partial q'} - im\omega q'\right)\langle q'|0\rangle = 0, \end{aligned}$$

where we used (2) and (30.20). Defining

$$\psi_n(q') = \langle q'|n\rangle, \tag{20}$$

we see that the ground state wavefunction ψ_0 satisfies the differential equation

$$\left(\frac{\partial}{\partial q'} + \frac{m\omega}{\hbar}\, q'\right) \psi_0(q') = 0. \tag{21}$$

Here it is convenient to introduce the dimensionless variable

$$\xi = \left(\frac{m\omega}{\hbar}\right)^{\frac{1}{2}} q', \tag{22}$$

in terms of which (21) reads

$$\left(\frac{\partial}{\partial \xi} + \xi\right) \psi_0(\xi) = 0.$$

The solution, normalized to

$$\left(\frac{\hbar}{m\omega}\right)^{\frac{1}{2}} \int_{-\infty}^{\infty} d\xi\, |\psi_0(\xi)|^2 = 1,$$

is

$$\psi_0(q') = \left(\frac{m\omega}{\pi\hbar}\right)^{\frac{1}{4}} e^{-\frac{1}{2}\xi^2}. \tag{23}$$

We can then employ (14) to generate the wavefunction of the nth excited state $\psi_n(q')$:

$$|n\rangle = (n!)^{-\frac{1}{2}} (a^\dagger)^n |0\rangle, \tag{24}$$

and therefore

$$\begin{aligned} \langle q'|n\rangle &= (n!)^{-\frac{1}{2}} (2m\hbar\omega)^{-\frac{1}{2}n} \langle q'|(p + im\omega q)^n|0\rangle \\ &= [n!(2m\hbar\omega)^n]^{-\frac{1}{2}} \left(\frac{\hbar}{i}\frac{\partial}{\partial q'} + im\omega q'\right)^n \psi_0(q'), \end{aligned}$$

or

$$\psi_n(q') = i^n (2^n n!)^{-\frac{1}{2}} \left(\frac{m\omega}{\hbar\pi}\right)^{\frac{1}{4}} \left(\xi - \frac{\partial}{\partial \xi}\right)^n e^{-\frac{1}{2}\xi^2}. \tag{25}$$

These wave functions are of the form $e^{-\frac{1}{2}\xi^2} H_n(\xi)$, where H_n is a polynomial of degree n (a so-called Hermite polynomial).

31.3 THE MOTION OF WAVE PACKETS

We now turn to the question of relating these results to those of classical mechanics. Consider a classical trajectory specified by the initial coordinate and momentum q_0 and p_0. At times $t > 0$, $q(t) = q_0 \cos \omega t + (p_0/m\omega) \sin \omega t$, and $p(t) = m\dot{q}(t)$. The analogous situation in quantum mechanics is described by a wave packet localized to within Δq about q_0 and Δp about p_0 at $t = 0$. To begin with, let us confine our discussion to the case where the initial velocity $p_0/m = 0$. If $|q_0| \gg \Delta q$ and the potential varies slowly within Δq, one would expect the packet to adhere to the classical trajectory for a considerable length of time. For an oscillator potential a stronger result actually holds, because the solutions of Heisenberg's equations coincide in form with the classical solutions. Consequently the expectation values of p and q for an arbitrary packet will follow the classical trajectory, a result that we had already obtained in Sec. 8. In general the shape of the packet will change with time, and for a sufficiently pathological initial state (e.g., a delta function) the motion will bear no resemblance to a classical motion. There is, however, a very special and interesting class of nonstationary states that do not spread in time. With the appropriate choice of initial conditions these can be arranged to look very classical indeed. We shall study these here.

In order to simplify the subsequent formulas, let us introduce dimensionless variables. These can be obtained by dividing the Hamiltonian by $\hbar\omega$, i.e., by setting

$$\hat{H} \equiv H/\hbar\omega = \tfrac{1}{2}(P^2 + Q^2),$$

with

$$P = \frac{p}{\sqrt{m\hbar\omega}}, \qquad Q = \sqrt{\frac{m\omega}{\hbar}}\, q.$$

Let $|\bar{n}\rangle$ be the nth state of the oscillator displaced through the (dimensionless) distance* Q_0:

$$|\bar{n}\rangle = e^{-iPQ_0}|n\rangle. \tag{26}$$

When $Q_0 \neq 0$, this is no longer a stationary state. For $t > 0$ it evolves into

$$|\bar{n}t\rangle = e^{-i\hat{H}\omega t}e^{-iPQ_0}|n\rangle. \tag{27}$$

We shall now show that $|\bar{n}t\rangle$ has the same dispersion in both P and Q as does $|n\rangle$ itself. According to (27), $\langle P\rangle_{\bar{n}t} = \langle \exp(iPQ_0)P(t)\exp(-iPQ_0)\rangle_n$. But (5) tells us that $P(t) = P\cos \omega t - Q \sin \omega t$. We are then faced with the expectation values of P and $\exp(iPQ_0)Q\exp(-iPQ_0)$; the latter operator is simply $Q + Q_0$, and as the expectation values of P and Q

* Bear in mind that in the subsequent equations Q_0 and P_0 are numbers specifying the initial conditions, whereas P and Q are operators.

in $|n\rangle$ both vanish, we are finally left with

$$\langle P\rangle_{\bar{n}t} = -Q_0 \sin \omega t.$$

This furnishes an explicit proof that the expectation value follows the classical trajectory. A similar calculation results in

$$\langle Q\rangle_{\bar{n}t} = Q_0 \cos \omega t.$$

To evaluate the dispersion we also require

$$\begin{aligned}\langle P^2\rangle_{\bar{n}t} &= \langle P^2 \cos^2 \omega t + Q^2 \sin^2 \omega t - (PQ + QP) \cos \omega t \sin \omega t\rangle_{\bar{n}} \\ &= \langle P^2\rangle_n \cos^2 \omega t + (Q_0^2 + \langle Q^2\rangle_n) \sin^2 \omega t - \langle PQ + QP\rangle_n \sin \omega t \cos \omega t.\end{aligned}$$

But $PQ + QP$ has no diagonal elements in the $|n\rangle$-representation, as one easily shows with the help of $a^\dagger$ and a. The momentum dispersion is therefore

$$(\Delta P)_t^2 = \langle P^2\rangle_{\bar{n}t} - [\langle P\rangle_{\bar{n}t}]^2 = \langle P^2\rangle_n \cos^2 \omega t + \langle Q^2\rangle_n \sin^2 \omega t.$$

By following the same route one also finds

$$(\Delta Q)_t^2 = \langle Q^2\rangle_n \cos^2 \omega t + \langle P^2\rangle_n \sin^2 \omega t.$$

Upon substituting the values of $\langle P^2\rangle_n$ and $\langle Q^2\rangle_n$ from (18′), one obtains

$$(\Delta P)_t^2 = (\Delta Q)_t^2 = n + \tfrac{1}{2}.$$

These are the same as for the stationary states $|n\rangle$, and we have therefore shown that the packets $|\bar{n}t\rangle$ retain their shape.

One may therefore prepare a state $\exp(-iPQ_0)|n\rangle$, with $Q_0 \gg 1$, which will oscillate back and forth in accordance with the classical laws. The spread of this packet in coordinate space will always remain small compared to the oscillation amplitude Q_0, and the momentum spread will also be small compared to the mean momentum $\langle P(t)\rangle_{\bar{n}}$ except in the neighborhood of the classical turning points.

Let us investigate the wave packet formed by displacing the ground state $|0\rangle$ in more detail. First of all, we remove the restriction that the initial momentum is zero. If we want the initial momentum to be P_0, we merely append the momentum displacement operator $\exp(iP_0Q)$ to (26). The resultant time-dependent state is

$$|Q_0P_0; t\rangle = e^{-i\bar{H}\omega t} e^{iP_0Q} e^{-iPQ_0}|0\rangle. \tag{28}$$

In the sequel we shall use the identity

$$e^{A+B} = e^A e^B e^{-\frac{1}{2}[A,B]}, \tag{29}$$

which holds for any pair of operators A and B that commute with $[A, B]$. A brute-force proof of (29) can be obtained by expanding the exponentials, but a more elegant derivation can be found in R. J. Glauber, *Phys. Rev.* **84**, 399 (1951).

We wish to compute the probability of finding the state $|n\rangle$ in the wave packet (28) at time t. An expansion of (28) in terms of stationary states is therefore indicated. The most economical way of doing this calculation is to express the unitary operators in (28) in terms of a and $a^\dagger$, to bring all the a's to the right where they can act on and destroy $|0\rangle$, and then to expand the remaining function of the operator $a^\dagger$. As a first step we use (29) to combine the two displacement operators:

$$e^{iP_0Q}e^{-iPQ_0} = e^{\frac{1}{2}iP_0Q_0}e^{i(P_0Q-PQ_0)}.$$

Recalling (4) and (6) we obtain

$$i(P_0Q - PQ_0) = a^\dagger\lambda - a\lambda^*,$$

where

$$\lambda = \frac{1}{\sqrt{2}}(P_0 - iQ_0). \tag{30}$$

The square of the modulus of this complex number,

$$\hat{E} = \tfrac{1}{2}(P_0^2 + Q_0^2), \tag{31}$$

is just the energy (in units of $\hbar\omega$) of a classical particle moving with momentum P_0 through the point Q_0. The phase of λ is merely the phase of the motion in the classical sense.* Our new expression for $|Q_0P_0; t\rangle$ is then

$$|Q_0P_0; t\rangle = e^{\frac{1}{2}iP_0Q_0}e^{-i\hat{H}\omega t}e^{(a^\dagger\lambda - a\lambda^*)}|0\rangle. \tag{32}$$

Employing (29) once more, we can split the displacement operator in (32) into two factors as follows:

$$e^{(a^\dagger\lambda - a\lambda^*)} = e^{a^\dagger\lambda}e^{-a\lambda^*}e^{-\frac{1}{2}\hat{E}}.$$

When inserted into (32), $e^{-a\lambda^*}$ gives unity because $a|0\rangle = 0$. Hence

$$|Q_0P_0; t\rangle = e^{\frac{1}{2}iP_0Q_0}e^{-\frac{1}{2}\hat{E}}e^{-i\hat{H}\omega t}e^{a^\dagger\lambda}|0\rangle. \tag{33}$$

The desired expansion is then obtained with the help of (24):

$$|Q_0P_0; t\rangle = e^{\frac{1}{2}iP_0Q_0}e^{-\frac{1}{2}\hat{E}}\sum_{n=0}^{\infty}\frac{e^{-i(n+\frac{1}{2})\omega t}\lambda^n}{\sqrt{n!}}|n\rangle. \tag{34}$$

* Detailed discussions of the phase of the quantum oscillator can be found in L. Susskind and J. Glowgower, *Physics* **1**, 49 (1964), and P. Carruthers and M. M. Nieto, *Phys. Rev. Letters* **14**, 387 (1965).

The probability of finding the wave packet (34) in the nth stationary state is thus

$$p_n \equiv |\langle n|Q_0P_0;t\rangle|^2 = \frac{\hat{E}^n}{n!}e^{-\hat{E}}. \tag{35}$$

This is a Poisson distribution in the variable $\hat{E}$. The most probable value of n is determined by finding the maximum of p_n. When the "classical" energy is large, i.e., $\hat{E} \gg 1$, we can treat p_n as a continuous function of n and obtain the maximum by setting the logarithmic derivative equal to zero. By using Stirling's formula for $n!$ we then find that the most probable value of n is simply equal to $\hat{E}$. Translated into energies, this statement means that the state most likely to be found in the wave packet has an excitation energy equal to the energy of the equivalent classical motion.

The dispersion in energy of the packet is obtained by computing $\sqrt{\langle\hat{H}^2\rangle - \langle\hat{H}\rangle^2}$. We find

$$\begin{aligned}\langle\hat{H}^2\rangle &= \sum_n (n+\tfrac{1}{2})^2 p_n \\ &= e^{-\hat{E}}\left(\hat{E}^2\frac{d^2}{d\hat{E}^2} + 2\hat{E}\frac{d}{d\hat{E}} + \tfrac{1}{4}\right)\sum_n \frac{\hat{E}^n}{n!} \\ &= \hat{E}^2 + 2\hat{E} + \tfrac{1}{4}.\end{aligned}$$

On the other hand, $\langle\hat{H}\rangle^2 = (\hat{E}+\frac{1}{2})^2$, and therefore

$$\frac{\Delta\hat{H}}{\langle\hat{H}\rangle} = \frac{\sqrt{\hat{E}}}{\frac{1}{2}+\hat{E}} \simeq \frac{1}{\sqrt{\hat{E}}}, \tag{36}$$

when $\hat{E} \gg 1$. The dispersion therefore has the form familiar from statistical mechanics. It becomes negligibly small when the amplitude of the motion is large compared to the size (i.e., $\sqrt{\hbar/m\omega}$) of the wave packet.

32. Rotations and Angular Momentum

Let $R_1, R_2, \ldots$ stand for arbitrary proper rotations of a rigid coordinate frame in ordinary three-dimensional space. By "proper" we mean that the rotation does not include a reflection, i.e., that both the original and final frame are either right- or left-handed. The symbols R_i stand for any convenient set of three real and continuous parameters; the most popular are the three Euler angles, and the orientation of an axis plus an angle of rotation about that axis. If two rotations are carried out in the sequence R_1 first and then R_2, we shall designate the resulting rotation as R_2R_1. Note here that R_2 is a rotation away from the orientation previously achieved in the first rotation; it is not a rotation referred to some fixed reference frame. The Euler angles that correspond to R_2R_1 are complicated functions of those which separately parametrize R_2 and R_1, but we shall not require this relation. If Θ stands for no rotation whatever, then the inverse rotation to R, designated by R^{-1}, is defined by $RR^{-1} = R^{-1}R = \Theta$. The set of all proper rotations constitutes *the rotation group*.

32.1 THE ROTATION OPERATORS

We may think of the variously oriented coordinate systems as being attached to the measurement apparatus used in preparing states. According to Sec. 27 the equivalence of the descriptions associated with these differently oriented frames implies that to each rotation R there corresponds a unitary operator on the Hilbert space $D(R)$, with the properties

$$D(R_2)D(R_1) = D(R_2R_1), \tag{1}$$
$$D(R) = D^\dagger(R^{-1}). \tag{2}$$

The set of objects $\{D(R)\}$ constitutes a group of unitary operators. When $D(R)$ is applied to a ket $|a'\rangle$, it produces a ket

$$|a'; R\rangle = D(R)|a'\rangle \tag{3}$$

whose properties are identical to $|a'\rangle$ as far as an observer using a frame rotated by R is concerned. Put more concisely, the configuration $|a'; R\rangle$ is obtained by rotating $|a'\rangle$ through R.

Let us begin by considering a sub-group, the rotations about an axis—say the z-axis. These rotations are parametrized by one angle α, and the unitary operators that perform the corresponding mappings on the Hilbert space are defined by

$$D_z(\alpha_1)D_z(\alpha_2) = D_z(\alpha_1 + \alpha_2), \tag{4}$$
$$D_z(\alpha) = D_z^\dagger(-\alpha). \tag{5}$$

This sub-group is Abelian, and therefore has a very much simpler structure than the full group. In fact, the same argument as we used in the displacement case shows that

$$D_z(\alpha) = e^{-i\alpha J_z}, \tag{6}$$

where the Hermitian operator $\hbar J_z$ is called the z-component of the total angular momentum. The analogy with displacements is not as strong as one might think, however, because α is a confined parameter: $0 \leqslant \alpha \leqslant 2\pi$, whereas the displacements are unrestricted. Mathematicians call a group with a confined range of the parameter compact; as we shall see (or as we know) the spectrum of the generators (e.g., J_z) of a compact group is discrete, while that of an open—noncompact—group is, at least in part, continuous.

Consider two successive rotations about *different* axes. It is clear that a rotation through α' about the y axis is given by $D_y(\alpha') = e^{-i\alpha' J_y}$. Previously we knew from geometry that displacements along different axes commute, and therefore concluded that $[P_x, P_y] = 0$, etc. Let us apply a similar argument to the rotations. As the physical object that we rotate we take a vector $\mathbf{r}$, which can be thought of as specifying the position of a piece of measurement apparatus. We can describe the rotations conveniently by writing $\mathbf{r}$ as a column vector

$$\mathbf{r} = \begin{pmatrix} x \\ y \\ z \end{pmatrix}.$$

We also introduce the 3×3 real antisymmetric matrices

$$I_x = \begin{pmatrix} 0 & 0 & 0 \\ 0 & 0 & -1 \\ 0 & 1 & 0 \end{pmatrix}, \quad I_y = \begin{pmatrix} 0 & 0 & 1 \\ 0 & 0 & 0 \\ -1 & 0 & 0 \end{pmatrix}, \quad I_z = \begin{pmatrix} 0 & -1 & 0 \\ 1 & 0 & 0 \\ 0 & 0 & 0 \end{pmatrix}. \tag{7}$$

One can easily verify that these matrices obey the commutation rules

$$[I_x, I_y] = I_z, \tag{8}$$

and cyclic permutations. These matrices can be used to describe rotations about the indicated axis, and by using all three of them in the appropriate combination any proper rotation can be carried out. Thus if we rotate $\mathbf{r}$ through φ about the x-axis,

$$\mathbf{r} \to \mathbf{r}' = e^{\varphi I_x}\,\mathbf{r}. \tag{9}$$

The column vector on the right-hand side of (9) gives the components of the rotated vector with respect to the *original* coordinate system. For an infinitesimal rotation, (9) simplifies to

$$\mathbf{r} \to (1 + \delta\varphi\, I_x + \cdots)\mathbf{r} = \begin{pmatrix} x \\ y - z\,\delta\varphi \\ z + y\,\delta\varphi \end{pmatrix} + O(\delta\varphi)^2. \tag{10}$$

It should be noted that for positive φ the sense of the rotation is counter-clockwise when one looks towards the origin from the positive x-axis. We always use this counter-clockwise convention.

An arbitrary rotation can be obtained by applying the orthogonal matrix

$$e^{\varphi_x I_x} e^{\varphi_y I_y} e^{\varphi_z I_z} \tag{9'}$$

to $\mathbf{r}$, but the three angles introduced thereby do not constitute a convenient parametrization, and will rarely be used here. It is important to note that (9') is characterized by the fact that it leaves the quadratic form $\mathbf{r} \cdot \mathbf{r}$ invariant, and has determinant $+1$.

We shall now carry out two different sequences of rotations on the vector $\mathbf{r}$. The first sequence consists of the rotation $\delta\varphi_x$ about the x-axis, followed by a rotation $\delta\varphi_y$ about the *original* y-axis; it yields the vector

$$\mathbf{r}' = (1 + \delta\varphi_y\, I_y + \cdots)(1 + \delta\varphi_x\, I_x + \cdots)\mathbf{r}. \tag{11'}$$

In the second sequence these rotations are performed in the opposite order, and we obtain instead

$$\mathbf{r}'' = (1 + \delta\varphi_x\, I_x + \cdots)(1 + \delta\varphi_y\, I_y + \cdots)\mathbf{r}, \tag{11''}$$

which does not equal $\mathbf{r}'$. In fact, because of (8)

$$\begin{aligned} \mathbf{r}'' - \mathbf{r}' &= [\delta\varphi_x\, \delta\varphi_y (I_x I_y - I_y I_x) + \cdots]\mathbf{r} \\ &= \delta\varphi_x\, \delta\varphi_y\, I_z \mathbf{r} + O(\delta\varphi^3). \end{aligned} \tag{12}$$

As we have said, to each rotation of the measurement apparatus (i.e., rotation of $\mathbf{r}$) there corresponds, in a one-to-one fashion, a unitary mapping in the Hilbert space. Because we have already found that the unitary operators for rotation about a single axis are given by (6), we require the analog to (12) to be

$$[e^{-i\delta\varphi_x J_x}, e^{-i\delta\varphi_y J_y}] = (e^{-i\delta\varphi_x \delta\varphi_y J_z} - 1) + O(\delta\varphi^3).$$

This condition determines the angular momentum commutation rules:

$$[J_x, J_y] = iJ_z, \text{ etc.} \tag{13}$$

This is then the algebra of the infinitesimal generators; it is obtained formally from the I's by the substitution $I_x \to -iJ_x$, but we should remember that the J's work in the Hilbert space, whereas the I's work on objects in three dimensions. The argument leading to (13) also shows that the angular momentum commutation rules depend only on the properties of rotations, and not at all on the detailed structure of the dynamical system in question. The derivation of (13) given in Sec. 6.2 does not reveal this very important fact.

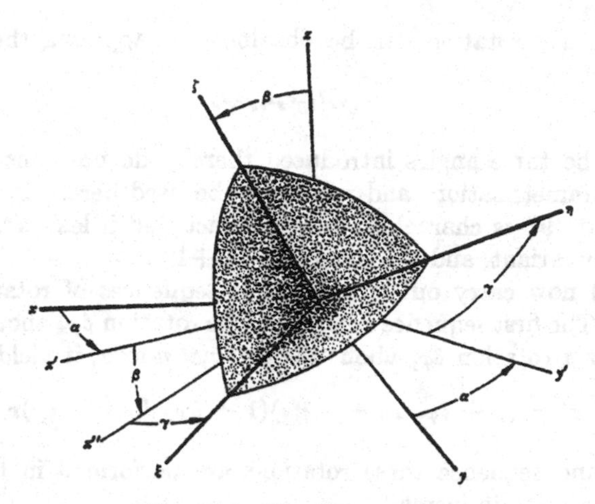

Fig. 32.1. The Euler angles.

We shall now construct the general rotation operator $D(R)$ parametrized with the Euler angles (see Fig. 32.1). These angles are defined by three successive rotations:

(a) a rotation through α about the z-axis, to which there corresponds the unitary operator $e^{-i\alpha J_z}$;
(b) a rotation through β about the y'-axis, generated by the operator $e^{-i\beta J_{y'}}$;
(c) finally, a rotation through γ about the ζ-axis, and generated by $e^{-i\gamma J_\zeta}$.

All these rotations are counterclockwise when one is looking "down" the rotation axis towards the origin. The ranges of the Euler angles are

$$\begin{aligned} 0 &\leqslant \alpha \leqslant 2\pi, \\ 0 &\leqslant \beta \leqslant \pi, \\ 0 &\leqslant \gamma \leqslant 2\pi. \end{aligned} \tag{14}$$

Upon putting the three operators just introduced together we obtain the unitary operator for an arbitrary rotation

$$D(\alpha, \beta, \gamma) = e^{-i\gamma J_\zeta} e^{-i\beta J_{y'}} e^{-i\alpha J_z}.$$

This expression is not convenient because it refers to several coordinate frames. We may easily amend this by noting that J'_y is related to J_y by a unitary transformation. If we call $\{|a'; \alpha\rangle\}$ a complete set prepared in the coordinate system rotated through α about the z-axis, and $\{|a'\rangle\}$ the corresponding set as prepared in the original frame, we have $|a';\alpha\rangle = e^{-i\alpha J_z}|a'\rangle$ and $\langle a';\alpha|J_{y'}|a'';\alpha\rangle = \langle a'|J_y|a''\rangle$. Hence

$$J_{y'} = e^{-i\alpha J_z}J_y e^{i\alpha J_z}, \tag{15}$$

and of course

$$e^{-i\beta J_{y'}} = e^{-i\alpha J_z}e^{-i\beta J_y}e^{i\alpha J_z}. \tag{16}$$

Repeated application of this result yields

$$\begin{aligned} J_\zeta &= e^{-i\beta J_{y'}}J_z e^{i\beta J_{y'}} \\ &= e^{-i\alpha J_z}e^{-i\beta J_y}J_z e^{i\beta J_y}e^{i\alpha J_z}. \end{aligned}$$

Upon putting all this together we find the important and simple result

$$D(\alpha, \beta, \gamma) = e^{-i\alpha J_z}e^{-i\beta J_y}e^{-i\gamma J_z}. \tag{17}$$

Again letting $R \equiv (\alpha, \beta, \gamma)$, we therefore have

$$|a'; R\rangle = e^{-i\alpha J_z}e^{-i\beta J_y}e^{-i\gamma J_z}|a'\rangle. \tag{18}$$

32.2 ROTATION MATRICES

The noncommutative algebra of the generators means that the determination of the matrix elements of D is not a trivial task, as it was for the Abelian groups. There we found that in the representation which diagonalizes the generators (e.g., $\mathbf{P}$), the unitary transformation reduces to a phase factor. Here no representation can diagonalize all the generators because of (13). If we choose a representation that diagonalizes J^2 and J_z, we can make progress, nevertheless. The reason for this is that J^2 is a scalar, and hence rotations of its eigenvectors do not change the eigenvalue of J^2. Let $|jm\rangle$ be a simultaneous eigenket of J^2 and J_z. The other quantum numbers are now irrelevant; they belong to observables that commute with $\mathbf{J}$, and we suppress them. Because $[D(R), J^2] = 0$, the matrix of D in this representation is diagonal in j. We can therefore confine ourselves to the matrix elements

$$\begin{aligned} D^{(j)}_{mm'}(\alpha, \beta, \gamma) &\equiv \langle jm|jm'; R\rangle = \langle jm|D(\alpha, \beta, \gamma)|jm'\rangle \\ &= e^{-i\alpha m}\, d^{(j)}_{mm'}(\beta)e^{-i\gamma m'}, \end{aligned} \tag{19}$$

where

$$d^{(j)}_{mm'}(\beta) = \langle jm|e^{-i\beta J_y}|jm'\rangle. \tag{20}$$

The matrix $D^{(j)}$ has dimension $(2j + 1)$, and is called the $(2j + 1)$-dimensional *irreducible representation of the rotation group.* It is called a representation because the product of two matrices corresponding to the rotations R_1 and R_2 is precisely* the matrix corresponding to the rotation R_1R_2:

$$D^{(j)}_{mm'}(R_1R_2) = \sum_{m''} D^{(j)}_{mm''}(R_1)D^{(j)}_{m''m'}(R_2). \tag{21}$$

This can be demonstrated immediately by taking the matrix elements of (1). Furthermore, the matrices associated with R and R^{-1} are related by

$$D^{(j)}_{mm'}(R) = D^{(j)}_{m'm}(R^{-1})^*, \tag{22}$$

as one can see from (2). As an important special case of (21) we may note

$$\sum_{m''} D^{(j)}_{mm''}(R)D^{(j)}_{m''m'}(R^{-1}) = \delta_{mm'}, \tag{23}$$

since

$$D^{(j)}_{mm'}(\Theta) = \delta_{mm'}, \tag{24}$$

where Θ stands for no rotation whatsoever (the identity element in the language of group theory). Equation (23) can be written in a somewhat more concrete fashion as

$$\sum_{m''} D^{(j)}_{mm''}(\alpha, \beta, \gamma)D^{(j)}_{m''m'}(-\gamma, -\beta, -\alpha) = \delta_{mm'}, \tag{25}$$

because one observes from Fig. 32.1 (or, if one prefers, from Eq. (17)) that

$$D(\alpha, \beta, \gamma) = D^{-1}(-\gamma, -\beta, -\alpha). \tag{26}$$

By combining (22) and (23) we also obtain the unitarity relations

$$\sum_{m''} D^{(j)}_{mm''}(R)D^{(j)}_{m'm''}(R)^* = \sum_{m''} D^{(j)}_{m''m}(R)^*D^{(j)}_{m''m'}(R) = \delta_{mm'}. \tag{27}$$

The adjective irreducible is used because it is not possible, for arbitrary R, to reduce the matrices $D^{(j)}$ to diagonal blocks of dimension smaller than $(2j + 1)$ by a change of basis. That this is so follows merely from the fact that $[J_y, J_z] \neq 0$.

Our next task is the construction of the matrices $D^{(j)}$. We shall first take up the simplest nontrivial case, the two-dimensional $D^{(\frac{1}{2})}$; later we shall actually discover that all the higher-dimensional representations can be constructed in terms of $D^{(\frac{1}{2})}$.

* This is actually an oversimplification, because there is nothing that prevents us from placing an arbitrary phase in front of (1), and therefore also in (21). Representations for which the phase is one are called *vector representations*, and those for which it is not are called *ray representations* (see Hamermesh). We shall not enter into this matter here.

33. Spin

The machinery for computing the two-dimensional representation is completely at hand now, and we could proceed directly to the construction of these matrices. Nevertheless, the fundamental importance of $j = \frac{1}{2}$, and the fact that half-integral values of j were not admissible in the ordinary Schrödinger theory, make it imperative that we first discuss the "physics" associated with the even-dimensional representations.

33.1 INTRINSIC SPIN

Consider an isolated system at rest, and call the eigenvalue of its total angular momentum in this frame $j_0(j_0 + 1)$. In "naive" wave mechanics two possibilities occur:

(a) for a one-body system, $j_0 = 0$;

(b) for a many-body system j_0 may be any integer or zero.

This does not agree with observations, because we find systems that appear to be structureless in all low energy interactions, and yet have $j_0 \neq 0$; even worse, systems where j_0 has half-integer values abound in nature. It is clear that a theory that hews so closely to the classical limit is insufficient, and must therefore be extended. To do so one introduces (Goudsmit and Uhlenbeck, 1925) the notion of an internal, purely quantum mechanical, rotational motion called the *spin*, and thereby increases the number of observables beyond the three of wave mechanics. *For a particle with spin, the total angular momentum in the rest frame is nonvanishing.* This rest-frame angular momentum will be designated by the vector operator $\mathbf{s}$. As with any other angular momentum

$$[s_x, s_y] = is_z \quad \text{(and cyclical permutations);} \tag{1}$$

the operators $(\mathbf{s}^2, s_z)$ are a complete set of compatible spin observables. The eigenvalue of $\mathbf{s}^2$, $s(s+1)$, is characteristic of a particle and does not change *by definition*. Thus for a lepton or nucleon, $s = \frac{1}{2}$, and for the deuteron $s = 1$. If, as a result of some reaction, a particle acquires a new value of s, we call this system by another name—in any case, it will inevitably have a different mass. On the other hand the orientation of $\mathbf{s}$ can change. Assuming that the Hamiltonian determining the properties of the particle is rotationally invariant, the mass of an isolated particle does not depend on the eigenvalue of s_z. This eigenvalue is called m_s, and it takes on the values $s, s - 1, \ldots, -s$. Because the length of $\mathbf{s}$ is fixed, there is no classical limit to the internal rotational motion.

For a particle in motion, a complete set of observables consists of four operators, for example, the momentum $\mathbf{p}$, and s_z ($\mathbf{s}^2$ need not be specified because its eigenvalue is already implied by the name of the particle). There are of course an infinity of other sets of compatible observables. The most obvious is the set containing the three coordinates and s_z. Less obvious, but often more useful, are the two sets (H, L^2, L_z, s_z), and (H, L^2, j^2, j_z), where $\mathbf{j}$ is the total angular momentum,

$$\mathbf{j} = \mathbf{L} + \mathbf{s}. \tag{2}$$

The Hilbert space for a particle with spin is obtained by taking the direct product of the usual infinite-dimensional space associated with the three classical degrees of freedom, and a $(2s + 1)$-dimensional complex vector space, which we call $\mathfrak{H}_s$. The operators $\mathbf{s}$ perform all possible manipulations in $\mathfrak{H}_s$, and the vectors that span this space can be taken to be the eigenvectors $|m_s\rangle$ of s_z. The basis for the full Hilbert space is then obtained by multiplying $|m_s\rangle$ into the kets we are already familiar with. For the observables $(\mathbf{p}, s_z)$ we write the basis as $|\mathbf{p}m_s\rangle \equiv |\mathbf{p}\rangle \cdot |m_s\rangle$, and for (H, L^2, L_z, s_z) the eigenkets are written as $|Elm_lm_s\rangle \equiv |Elm_l\rangle \cdot |m_s\rangle$. In order to construct the eigenstates associated with (H, j^2, L^2, j_z) one must use the vector addition techniques of Sec. 25:

$$|Eljm\rangle = \sum_{m_l m_s} |Elm_lm_s\rangle\langle lm_lsm_s|jm\rangle. \tag{2'}$$

33.2 WHAT IS A PARTICLE?

The word particle has been used frequently in the preceding discussion, and requires elucidation. The essential remark is that the particle concept is energy dependent, and the best way to appreciate this is via the following example. Assume we are carrying out experiments on a fluid such as liquid helium (He^4 nuclei). In all the usual experiments of low temperature physics, the whole He atom may be treated as an "elementary" spin zero particle, because the collisions in the fluid are not sufficiently energetic to excite the atom out of its ground state. When we irradiate the liquid with ultraviolet light, the excited states of the atom will make their appearance in the absorption spectrum, and we would no longer be able to say that the atom is an "elementary particle." We know very well that the data can then be understood in terms of a model that employs three particles: a spin zero nucleus—the α-particle, and two spin $\frac{1}{2}$ particles, the electrons. When we raise the energy of the probe even further ($\gtrsim$25 Mev) by using γ-rays, we discover that the previously given description also breaks down, because the α-particle can dissociate. We are then in the realm of nuclear physics, where the elementary particles are the nucleons. At even higher energies ($>$140 Mev) pions are produced and the description of the system has to be enriched even further. At the highest energies presently available, the nucleon itself already displays considerable structure; it has a charge distribution of radius $\sim 0.8 \times 10^{-13}$ cm, and a variety of excited states. This story should make it amply clear that the term "particle" does not necessarily have a universal meaning. At the moment there is as yet no evidence for internal structure of leptons, but further experimental work may perhaps reveal such structure. In spite of these reservations, the notion of particle usually has a well-defined meaning at some given energy.*

•In formal terms, the significance of the preceding paragraph is that the "size" of the Hilbert space necessary for an adequate description of a system depends on the energy domain of interest. For instance, let us be ambitious and describe a spectroscopic experiment by treating the He atom as a system of two electrons, two neutrons, and two protons. Let $\{|\Phi_n\rangle\}$ be a complete set of energy eigenstates for two electrons interacting with each other and an α-particle, and $\{|\chi_\mu\rangle\}$ be a complete set of stationary states for the four-nucleon system. Any energy eigenstate of the entire atom $|\Psi_n\rangle$ can be written as

$$|\Psi_n\rangle = c_0^{(n)}|\Phi_n\rangle|\chi_0\rangle + \sum_{m\mu} c_{m\mu}^{(n)}|\Phi_m\rangle|\chi_\mu\rangle,$$

where $|\chi_0\rangle$ is the ground state of the α-particle; needless to say, the sum does not include the term $m = n, \mu = 0$. For any state of interest in atomic spectroscopy, perturbation theory reveals that the coefficients $c_{m\mu}^{(n)}$ are of order 10^{-6}, and therefore the excited states of the nucleus play no role whatsoever in the problem. The ground state of the nucleus enters into essentially all calculations of atomic spectroscopy via $\langle\chi_0|\chi_0\rangle$, and therefore we can forget about the existence of nuclear structure in atomic physics, except under very special circumstances. If, however, the nucleus has a spin (e.g., the isotope He^3 with $s = \frac{1}{2}$), we have several degenerate states of the nucleus at our disposal, and then the spin degree of freedom must be incorporated into the theory (hyperfine structure). No further degrees of freedom describing the structure of the He^3 nucleus need be considered, however.

A remark concerning the invariance of the theory under Galileo transformations should be made here. The spin was originally defined in the rest frame, and just before (2′) we had asserted that the state of a spinning particle can be

* The evolution of physics would have been exceedingly slow if this hierarchial energy structure did not exist. Because of it physicists at all times in the past have been able to lump their ignorance concerning higher energy (or equivalently, smaller distance) phenomena into a small number of empirical parameters, while they systematically resolved the problems that were revealed by the energies available to them in their laboratories. Thus Boltzmann and his contemporaries were able to develop the kinetic theory without having to understand the detailed structure of molecules. (As they did not know that energies can be quantized, they actually did worry about molecular structure!)

written as $|\mathbf{p}\rangle|m_s\rangle$. This last statement is only consistent with the definition if the spin is invariant under Galileo transformations. As we have just argued, the notion of spin should not depend on whether the system in question is assumed to be elementary or composite. Let us therefore assume that the system is made up of several particles without spin having the masses m_α. In the rest frame, with the origin placed at the center of mass, we have

$$\sum_\alpha \mathbf{p}_\alpha = 0,$$

$$\sum_\alpha m_\alpha \mathbf{x}_\alpha = 0,$$

and the spin is

$$\mathbf{J}_0 = \sum_\alpha \mathbf{x}_\alpha \times \mathbf{p}_\alpha.$$

In a frame moving with velocity $\mathbf{v}$, the coordinates and momenta are given by (30.24) and (30.25),

$$\mathbf{p}_\alpha \to \mathbf{p}'_\alpha = \mathbf{p}_\alpha + m_\alpha \mathbf{v},$$
$$\mathbf{x}_\alpha \to \mathbf{x}'_\alpha = \mathbf{x}_\alpha + \mathbf{v}t,$$

and therefore

$$\mathbf{J}'_0 = \sum_\alpha \mathbf{x}'_\alpha \times \mathbf{p}'_\alpha = \mathbf{J}_0.$$

This argument shows that under a Galileo transformation $|\mathbf{p}\rangle|m_s\rangle \to |\mathbf{p}'\rangle|m_s'\rangle = |\mathbf{p} + M\mathbf{v}\rangle|m_s\rangle$, where M is the mass of the object. The formulation of spin given here is therefore sensible within the framework of nonrelativistic physics. Were we to demand invariance under Lorentz transformations the situation would be more complicated because in relativity theory the orientation of the internal angular momentum depends on the motion of the observer. In the Dirac theory of the electron one therefore finds that the dependence of the states on the spin and momentum variables is interwoven.•

33.3 SPIN $\frac{1}{2}$

After this "physical" interlude, we are ready to turn to the construction of the matrix $D^{(\frac{1}{2})}$. Let us first derive some special properties of the operators s_i $(i = x, y, z)$ for this case. Because the only eigenvalues are $\pm\frac{1}{2}$, the Cayley-Hamilton theorem implies $s_i^2 = \frac{1}{4}$. Furthermore we have

$$(s_x \pm is_y)^2 = 0 = (s_x^2 - s_y^2) \pm i(s_x s_y + s_y s_x),$$

and therefore

$$s_x s_y + s_y s_x = 0 \qquad \text{(and cyclical permutations)}, \tag{3}$$

or from (1),

$$s_x s_y = \tfrac{1}{2} i s_z \qquad \text{(and cyclical permutations).} \tag{4}$$

It is actually more convenient to work with *Pauli matrices* σ_i, defined by

$$s_i = \tfrac{1}{2}\sigma_i; \tag{5}$$

clearly σ_i has the eigenvalues ± 1.

In terms of σ_i the properties (3) and (4) are

$$\sigma_i^2 = 1, \tag{6}$$

$$\sigma_i\sigma_j + \sigma_j\sigma_i = 0 \qquad (i \neq j), \tag{7}$$

$$\sigma_i\sigma_j = i\sigma_k \qquad (i, j, k, \text{cyclic}). \tag{8}$$

From (8) and (6) we note that

$$\sigma_x\sigma_y\sigma_z = i. \tag{9}$$

Because of the spectrum of σ_i (i.e., ± 1), we also have

$$\operatorname{tr} \sigma_i = 0, \qquad \det \sigma_i = -1. \tag{10}$$

Another extremely useful identity (whose proof is left as an exercise) is

$$(\boldsymbol{\sigma} \cdot \mathbf{A})(\boldsymbol{\sigma} \cdot \mathbf{B}) = \mathbf{A} \cdot \mathbf{B} + i\boldsymbol{\sigma} \cdot (\mathbf{A} \times \mathbf{B}), \tag{11}$$

where **A** and **B** are two vectors that commute with $\boldsymbol{\sigma}$, but not necessarily with each other.

For some rare purposes, and in particular for what follows, an explicit representation of $\boldsymbol{\sigma}$ is necessary. With our phase convention (i.e., $(s_x \pm is_y)|m\rangle = \sqrt{\tfrac{3}{4} - m(m \pm 1)}\,|m \pm 1\rangle$) we have

$$\sigma_x = \begin{pmatrix} 0 & 1 \\ 1 & 0 \end{pmatrix}, \qquad \sigma_y = \begin{pmatrix} 0 & -i \\ i & 0 \end{pmatrix}, \qquad \sigma_z = \begin{pmatrix} 1 & 0 \\ 0 & -1 \end{pmatrix}. \tag{12}$$

Consider the matrix

$$A = a_0 1 + \mathbf{a} \cdot \boldsymbol{\sigma} = \begin{pmatrix} a_0 + a_z & a_x - ia_y \\ a_x + ia_y & a_0 - a_z \end{pmatrix}.$$

It is clear that an arbitrary 2×2 matrix can be written in this fashion. We therefore have the useful theorem: *The Pauli matrices, together with the two-dimensional unit matrix, form a complete set of* 2×2 *matrices.* Equation (11) is a reflection of this fact, as is (14) below.

We can now construct $D^{(\frac{1}{2})}(\alpha, \beta, \gamma)$. Because this matrix only works in the 2×2 subspace belonging to $j = \frac{1}{2}$, we can use (5) to write (32.17) as

$$D^{(\frac{1}{2})}(\alpha, \beta, \gamma) = e^{-\frac{1}{2}i\alpha\sigma_z} e^{-\frac{1}{2}i\beta\sigma_y} e^{-\frac{1}{2}i\gamma\sigma_z}. \tag{13}$$

In virtue of (6)

$$e^{-\frac{1}{2}i\beta\sigma_y} = \sum_{n=0}^{\infty} \frac{(-i\beta/2)^n}{n!}\sigma_y^n$$
$$= \sum_{n \text{ even}} \frac{1}{n!}\left(-\frac{i\beta}{2}\right)^n + \sigma_y \sum_{n \text{ odd}} \frac{1}{n!}\left(-\frac{i\beta}{2}\right)^n.$$

Hence

$$e^{-\frac{1}{2}i\beta\sigma_y} = \cos\tfrac{1}{2}\beta - i\sigma_y \sin\tfrac{1}{2}\beta, \tag{14}$$

which gives us the matrix $D^{(\frac{1}{2})}(R)$:

$$D^{(\frac{1}{2})}(\alpha, \beta, \gamma) = \begin{pmatrix} e^{-\frac{1}{2}i(\alpha+\gamma)}\cos\tfrac{1}{2}\beta & -e^{-\frac{1}{2}i(\alpha-\gamma)}\sin\tfrac{1}{2}\beta \\ e^{\frac{1}{2}i(\alpha-\gamma)}\sin\tfrac{1}{2}\beta & e^{\frac{1}{2}i(\alpha+\gamma)}\cos\tfrac{1}{2}\beta \end{pmatrix}. \tag{15}$$

Note the strange fact that if we perform a rotation of 2π about any axis, e.g., the z-axis, we do not obtain the matrix $D^{(\frac{1}{2})}(0,0,0) = 1$, but rather

$$D^{(\frac{1}{2})}(2\pi, 0, 0) = \begin{pmatrix} -1 & 0 \\ 0 & -1 \end{pmatrix} = -1. \tag{16}$$

The representation $D^{(\frac{1}{2})}$ *is double valued*; in particular, $D^{(\frac{1}{2})}(\Theta)|\tfrac{1}{2}m\rangle = \pm|\tfrac{1}{2}m\rangle$. Because these kets belong to the same ray, this does not matter. On the other hand, observables must be single valued, and the consequence of this will be discussed in Sec. 36. If we now refer back to (32.19) we see that for any half-integral j, $D^{(j)}$ is a double-valued representation. We shall have more to say about this in the next section.

If we wish to rotate a general state representing a particle with spin $\frac{1}{2}$, $|a'ljm\rangle$ say, then we must use the operator

$$D(R) = e^{-i\alpha(L_z+s_z)}e^{-i\beta(L_y+s_y)}e^{-i\gamma(L_z+s_z)}.$$

Because $\mathbf{s}$ and $\mathbf{L}$ commute, this can be written as

$$D(R)_J = D^{(\frac{1}{2})}(R)D(R)_{\text{orb}}, \tag{17}$$

where $D(R)_{\text{orb}} = e^{-i\alpha L_z}e^{-i\beta L_y}e^{-i\gamma L_z}$. We now compute the matrix elements of (17) with the help of (2′). In doing so we must remember that the matrix elements of an angular momentum operator depend only on its *own* quantum numbers in a representation where J^2 and J_z are diagonal, and not on *any* other quantum numbers. Thus quite generally

$$\langle a'jm|D(R)|a''jm'\rangle = \delta(a', a'')D^{(j)}_{mm'}(R),$$

where a' and a'' are eigenvalues of the operators beyond J^2 and J_z that serve to specify the state. Therefore (17) becomes

$$D^{(j)}_{mm'}(R) = \langle ljm|D^{(\frac{1}{2})}(R)D^{(l)}(R)|ljm'\rangle$$
$$= \sum_{\substack{m_l m_s \\ m_l' m_s'}} \langle jm|lm_l\tfrac{1}{2}m_s\rangle D^{(\frac{1}{2})}_{m_s m_s'}(R)D^{(l)}_{m_l m_l'}(R)\,\langle lm_l'\tfrac{1}{2}m_s'|jm'\rangle. \quad (18)$$

This formula shows how one may generate representations from those of lower dimensionality, and therefore offers a hint as to how one may hope to compute the general matrix $D^{(j)}$ in terms of $D^{(\frac{1}{2})}$.

34. The Irreducible Representations of the Rotation Group

34.1 THE KRONECKER PRODUCT

The argument that led to (33.18) may be generalized as follows. Let $\mathbf{J}_1$ and $\mathbf{J}_2$ be two *commuting* angular momentum operators, and $\mathbf{J}$ their resultant,

$$\mathbf{J} = \mathbf{J}_1 + \mathbf{J}_2. \quad (1)$$

Then

$$D(R) = (e^{-i\alpha J_{1z}}e^{-i\beta J_{1y}}e^{-i\gamma J_{1z}})(e^{-i\alpha J_{2z}}e^{-i\beta J_{2y}}e^{-i\gamma J_{2z}})$$
$$\equiv D(R)_1 \cdot D(R)_2. \quad (2)$$

In terms of the sets $\{|j_1m_1j_2m_2\rangle\}$ and $\{|j_1j_2jm\rangle\}$ we therefore have

$$\langle j_1j_2jm|D(R)|j_1j_2jm'\rangle = \sum_{\substack{m_1m_1' \\ m_2m_2'}} \langle jm|j_1m_1j_2m_2\rangle\langle j_1m_1|D(R)_1|j_1m_1'\rangle \cdot \langle j_2m_2|D(R)_2|j_2m_2'\rangle\langle j_1m_1'j_2m_2'|jm'\rangle,$$

or*

$$D^{(j)}_{mm'}(R) = \sum_{\substack{m_1m_1' \\ m_2m_2'}} \langle jm|j_1m_1j_2m_2\rangle D^{(j_1)}_{m_1m_1'}(R)D^{(j_2)}_{m_2m_2'}(R)\langle j_1m_1'j_2m_2'|jm'\rangle. \quad (3)$$

This formula has the interesting property that j_1 and j_2 appear on the right side, but not on the left. Note also the important "inverse" relation obtained by constructing $\langle j_1m_1j_2m_2|D(R)|j_1m_1'j_2m_2'\rangle$:

$$D^{(j_1)}_{m_1m_1'}(R)D^{(j_2)}_{m_2m_2'}(R) = \sum_{jmm'} \langle j_1m_1j_2m_2|jm\rangle D^{(j)}_{mm'}(R)\langle jm'|j_1m_1'j_2m_2'\rangle. \quad (4)$$

* Recall here the second sentence following (33.17).

Mathematicians call this the Clebsch-Gordan decomposition of the Kronecker (or direct) product, and write it as

$$D^{(j_1)} \otimes D^{(j_2)} = \sum_{=|j_1-j_2|}^{j_1+j_2} D^{(j)}.$$

Equation (3) allows us to construct all the matrices $D^{(j)}$ from $D^{(\frac{1}{2})}$. We merely insert the known elements of $D^{(\frac{1}{2})}$ into the right-hand side and generate $D^{(1)}$, and then repeat the procedure to determine $D^{(\frac{3}{2})}$ from $D^{(\frac{1}{2})}$ and $D^{(1)}$, and so on. This is certainly not a practical way of evaluating $D^{(21)}$, say, but it serves to show that the $j = \frac{1}{2}$ representation is the basic object in the theory. One might have thought that the three-dimensional matrix $D^{(1)}$, which describes the rotation of a vector, plays the basic role, but we now see that this is not the case. By starting with $D^{(1)}$, and using (3) repeatedly, one can only generate the odd-dimensional representations.

34.2 EXPLICIT FORMULAS FOR $d_{mm'}^{(j)}$

There are several methods that yield explicit formulas for $d^{(j)}_{mm'}$. We shall give a "group-theoretic" derivation in Sec. 34.4. A more elementary, though also more laborious procedure, consists of deriving a differential equation for $d^{(j)}_{mm'}$. The derivation of this equation is outlined in Prob. 4. The equation reads

$$\left\{\frac{d^2}{d\beta^2} + \cot\beta\frac{d}{d\beta} - \frac{m^2 + m'^2 - 2mm'\cos\beta}{\sin^2\beta} + j(j+1)\right\} d^{(j)}_{mm'} = 0. \tag{5}$$

This equation is of the hypergeometric type. The regular solution has the form $(\sin\frac{1}{2}\beta)^{2j}(\cot\frac{1}{2}\beta)^{m+m'}F(\cot^2\frac{1}{2}\beta)$, where F is a hypergeometric function. Because m, m' and j are all integers or half integers, it turns out that F is actually a polynomial. The normalization of the solution is determined by our phase convention

$$J_\pm|jm\rangle = \sqrt{j(j+1) - m(m\pm 1)}\,|jm\pm 1\rangle,$$

and the requirement that $d^{(j)}$ must be a unitary matrix. The solution that satisfies all of these requirements is

$$d^{(j)}_{mm'}(\beta) = \frac{(-1)^{j-m'}}{(m+m')!}\sqrt{\frac{(j+m)!(j+m')!}{(j-m)!(j-m')!}}(\sin\tfrac{1}{2}\beta)^{2j}(\cot\tfrac{1}{2}\beta)^{m+m'}$$
$$\cdot\, {}_2F_1(m-j, m'-j; m+m'+1; -\cot^2\tfrac{1}{2}\beta), \tag{6}$$

when $m \geqslant m'$; for other values of m and m' one uses

$$d^{(j)}_{mm'}(\beta) = (-1)^{m-m'}d^{(j)}_{m'm}(\beta). \tag{7}$$

We should point out that $D^{(l)}_{mm'}{}^{*}$ is the wave function of the symmetric top, with m' being the component of angular momentum along the body fixed symmetry axis, m the component along an axis fixed in space (see Landau and Lifshitz, *QM*, pp. 280, 373; van der Waerden; Kronig). The functions $D^{(j)}_{mm'}(R)$ are therefore of great importance in molecular theory. They are also employed in the theory of deformed nuclei.

34.3 RELATIONSHIP BETWEEN THE ROTATION GROUP AND *SU(2)*

Here we shall delve somewhat more deeply into the theory of the rotation group.* Among other things, we shall construct all the representations directly from $D^{(\frac{1}{2})}$, and study various integrals involving the matrices $D^{(j)}$ that are of importance in applications.

Consider the unitary matrix (33.15) once more; we observe that it has the form

$$D^{(\frac{1}{2})}(a, b) = \begin{pmatrix} a & b \\ -b^* & a^* \end{pmatrix}, \tag{8}$$

with

$$\begin{aligned} a &= e^{-\frac{1}{2}i(\alpha+\gamma)} \cos \tfrac{1}{2}\beta, \\ b &= -e^{-\frac{1}{2}i(\alpha-\gamma)} \sin \tfrac{1}{2}\beta. \end{aligned} \tag{9}$$

These are known as the Cayley-Klein parameters. The determinant of (8) equals one because

$$|a|^2 + |b|^2 = 1. \tag{10}$$

A matrix with unit determinant is said to be unimodular. We note that (8) is the most general two-dimensional unitary unimodular matrix if the parameters a and b are restricted in accordance with (10).

The set of all 2×2 unitary unimodular matrices forms a continuous group, as is clear from the identity of $D^{(\frac{1}{2})}(a, b)$ and $D^{(\frac{1}{2})}(R)$. A quasi-geometric significance can be given to this group as follows. Consider a two-dimensional complex vector space (the spin space) spanned by vectors of the type

$$\begin{pmatrix} u_1 \\ u_2 \end{pmatrix},$$

where u_1 and u_2 are complex numbers. These 2-vectors are called *spinors*. With the matrices $D^{(\frac{1}{2})}(a, b)$ we can carry out linear transfor-

* More complete and rigorous accounts of the theory can be found in Hamermesh, Secs. 9.2, 9.6; Gel'fand, Minlos, and Shapiro, Part I; Schwinger, *AM*; van der Waerden, Chapts. 3, 4; Wigner, Chapts. 10, 14, 15; Boerner, Chapts. 7, 8; Tinkham, Chapt. 5. At the other extreme, some readers might like to know that one can proceed directly to the beginning of Sec. 34.5 with but little loss in continuity.

mations on the spin space:

$$u_i \to u_i' = \sum_j u_j D_{ji}^{(\frac{1}{2})}(a, b). \tag{11}$$

These transformations leave the Hermitian form

$$Q = |u_1|^2 + |u_2|^2 \tag{12}$$

invariant.

We shall now generalize the discussion in the same fashion as for the rotation group. There, we recall, one begins by considering transformations on a real three-dimensional vector space that leave the quadratic form $(x^2 + y^2 + z^2)$ invariant. Aside from reflections, these transformations are carried out by a continuous group of 3×3 orthogonal matrices that we call $M(R)$, and which can be parametrized by three angles as in (32.9′). The theory is then vastly generalized by the introduction of a set of objects $\mathfrak{D}(R)$ that stand in one-to-one correspondence with the $M(R)$, in the sense that if $M(R_2)M(R_1) = M(R_3)$, then $\mathfrak{D}(R_2)\mathfrak{D}(R_1) = \mathfrak{D}(R_3)$. The $\{\mathfrak{D}(R)\}$ are *not* 3×3 matrices. Rather, they are abstract objects whose significance lies in the fact that they form a group with precisely the same relationships as the group of 3×3 matrices $M(R)$. The abstract group $\{\mathfrak{D}(R)\}$ is called the three-dimensional orthogonal group, and is designated by $O(3)$. The matrices $M(R)$ are said to constitute a three-dimensional representation of $O(3)$. Furthermore, any set of concrete objects (e.g., matrices, unitary operators on a Hilbert space such as $D(R)$) whose algebra (e.g., matrix multiplication) allows them to be put into one-to-one correspondence with $\{\mathfrak{D}(R)\}$ are said to be a representation of the group $O(3)$. When we speak of representations we shall, however, always mean matrices of finite rank. In particular, we recall that an irreducible representation is one whose matrices cannot be reduced to diagonal blocks of smaller dimensionality by a constant similarity transformation.

Consider now the analogous generalization for the group that leaves (12) invariant. We define the abstract group $SU(2)$ (the two-dimensional special unitary group) as the set of objects $\mathfrak{U}(a, b)$ that satisfy

$$\begin{aligned} \mathfrak{U}(a_2, b_2)\mathfrak{U}(a_1, b_1) &= \mathfrak{U}(a_2a_1 - b_2b_1^*, a_2b_1 + b_2a_1^*), \\ \mathfrak{U}^{-1}(a, b) &= \mathfrak{U}(a^*, -b), \end{aligned} \tag{13}$$

where a and b are any pair of complex numbers that satisfy (10). By construction the objects $\mathfrak{U}(a, b)$ have the same relationships among themselves as $\{D^{(\frac{1}{2})}(a, b)\}$, and hence these 2×2 matrices form an (irreducible) representation of $SU(2)$.

The groups $SU(2)$ and $O(3)$ are intimately related. By definition, in fact, the group $SU(2)$ has the same structure as the group $\{D^{(\frac{1}{2})}(a, b)\}$, whose elements supposedly constitute a two-dimensional representation of the rotation group. Consequently the generators of $SU(2)$ and $O(3)$ have the same commutation rules*; this is easily verified by computing the relevant infinitesimal transformations, or by simply noting that the Pauli matrices and the matrices (I_x, I_y, I_z) of Sec. 32.1 have the same commutation rules (to within factors of $2i$). One might therefore think that $O(3)$ and $SU(2)$ are identical. That this is not quite so is revealed by the observation that, whereas $(\alpha = \beta = \gamma = 0)$ and $(\alpha = 2\pi, \beta = \gamma = 0)$ both parametrize the identity element of $O(3)$, they correspond to $(a = 1, b = 0)$ and $(a = -1, b = 0)$, respectively. Thus $(\alpha = 2\pi, \beta = \gamma = 0)$ does not parametrize the identity element of $SU(2)$. This fact is completely equivalent to our earlier observation (33.16) that $D^{(\frac{1}{2})}(2\pi, 0, 0) = -1$. Thus $D^{(\frac{1}{2})}$ is, strictly speaking, not a representation of $O(3)$ because it stands in two-to-one correspondence with the abstract group. On the other hand, $D^{(\frac{1}{2})}$ is a single-valued representation of $SU(2)$. In classical physics only single-valued representations are admissible, but in quantum mechanics we must also entertain the possibility that physical systems are described by states belonging to multivalued representations.

•The multivalued correspondence between the representations of $SU(2)$ and the group $O(3)$ is closely connected with the topological properties of the manifold generated by the continuous parameters that specify the elements of these groups. Thus the elements of $SU(2)$ are specified by the pair of complex numbers a and b, which are subjected to the sole condition (10). Hence the parameter manifold of $SU(2)$ is the surface of a real 4-sphere. Every point on this sphere corresponds uniquely to a "rotation" in the spin space. Any closed path on this sphere can be shrunk continuously to a point, and one therefore says that $SU(2)$ is simply connected.

The topological structure of the parameter manifold belonging to the more familiar group $O(3)$ is actually more complicated. To see this we use as parameters the axis of rotation specified by the angles ϑ, φ, and an angle of rotation about the axis, ω; the ranges of the parameters are

$$\begin{aligned} 0 &\leqslant \vartheta \leqslant \pi, \\ 0 &\leqslant \varphi \leqslant 2\pi, \\ 0 &\leqslant \omega \leqslant \pi. \end{aligned} \tag{14}$$

This parametrization is not single valued, however, because the rotations $(\vartheta, \varphi, \omega = \pi)$ and $(\pi - \vartheta, \varphi + \pi, \omega = \pi)$ are identical. We can therefore represent the parameter manifold by a sphere of radius π, provided we *identify* diametrically opposite points on the surface of the sphere. Any point in the sphere then has the polar coordinates $(r = \omega, \vartheta, \varphi)$, and corresponds to a rotation, but points on

* In mathematical terms, $SU(2)$ and $O(3)$ have the same Lie algebra.

the opposite ends of a diameter actually represent the *same* element of the group. By a closed curve we now mean a path in the space of the group elements that starts out at some element, and then returns by a continuous route to the same element. There are then two (and only two) types of topologically distinct *closed* curves in the $O(3)$ parameter manifold: (a) curves C_0 that can be shrunk continuously to a point, and (b) curves C_d that can be continuously deformed to a line along some diameter. These two classes of curves are illustrated in Fig. 34.1. Observe that paths with an even number of jumps belong to C_0, while those with an odd number of jumps are in C_d.

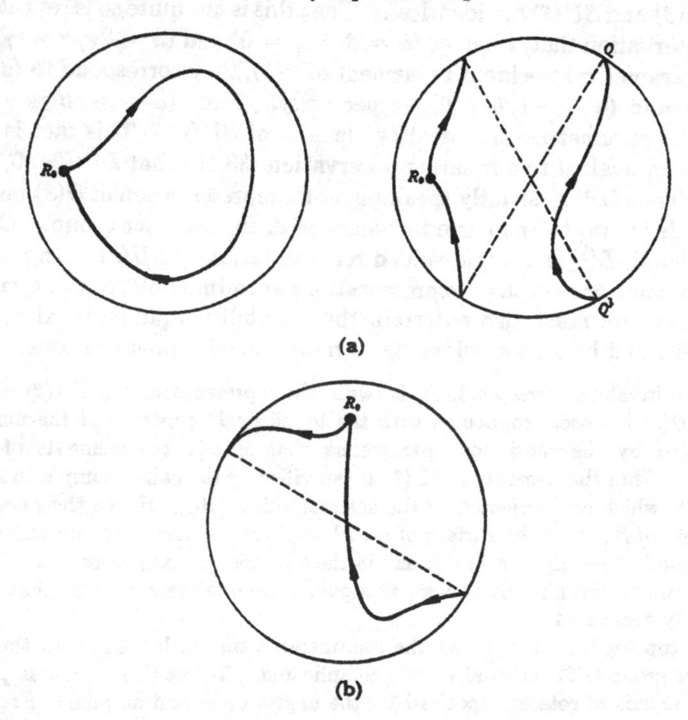

Fig. 34.1. (a) Two *closed* curves of the type C_0. Note that the dashed lines are not part of the closed curve. To shrink the second curve to a point, rotate the two dashed lines until Q and Q' coincide. (b) A closed curve C_d that cannot be shrunk to a point. Here, also, the dashed line along the diameter is not part of the curve.

We now seek to determine how many representation matrices $D^{(j)}(R)_n$ ($n = 1, 2, \ldots$) can be associated with a single group element R of $O(3)$. The basic notion used in the argument is that the continuous nature of the group requires that any representation $D^{(j)}(R)_n$ changes continuously if R is varied in a continuous fashion. This continuity requirement does not, of itself, preclude the existence of multivalued representations; this point, and many of those which follow, will be readily understood if one bears in mind the multivalued functions of a single complex variable. The continuity requirement does, however, imply that when R is in the neighborhood of R_0, each and every matrix element $D^{(j)}_{mm'}(R)_n$ remains, roughly speaking, close in *value* to $D^{(j)}_{mm'}(R_0)_n$. Consider now a matrix $D^{(j)}(R_0)_n$, and follow its value as we move along some path in the space of group elements. $D^{(j)}(R)_n$ will, as we have stated, vary continuously along such a path, but when the path returns to the starting point R_0, $D^{(j)}$ could assume any one of the values $D^{(j)}(R)_{n'}$, i.e., with n' not necessarily equal to n. The possibility $n \neq n'$ can be ruled out on the basis of continuity if the path in question can be continuously deformed until it is shrunk to a point. If the path cannot be shrunk down, however, the continuity argument does not apply because we cannot stay in the neighborhood of the starting point on such a curve; hence on paths of this type $D^{(j)}$ may (it need not) end up having a different value when we return to the beginning of the circuit. This argument indicates * that a group whose parameter manifold is simply connected and of finite extent (i.e., compact) only has single-valued representations; $SU(2)$ is an example of such a group. On the other hand, a compact group with a multiply connected parameter manifold can be expected to have multivalued representations, and the maximum degree of multiplicity is clearly given by the number of topologically distinct path classes. Because $O(3)$ is doubly connected, its representations can be at most double valued.*

The multivalued character of $O(3)$ has somewhat unpleasant mathematical consequences, even though it is the *raison d'être* of half-integral spin. It is actually preferable to work with a slightly different group, one that has the same properties for infinitesimal rotations but is simply connected. Such a group is called *the universal covering group*. By construction, $SU(2)$ has the same local properties as $O(3)$, and is simply connected. Therefore it is the universal covering group for $O(3)$. The correspondence between $SU(2)$ and $O(3)$ is that $\mathfrak{U}(\pm a, \pm b)$ correspond to one and the same rotation R, and in particular, $\mathfrak{U}(\pm 1, 0)$ correspond to the identity element $\mathfrak{e}$ of $O(3)$. The correspondence is therefore two-to-one. For the representations $D^{(j)}(a, b)$ of $SU(2)$, the correspondence is double valued if j is a half integral, and single valued if j is an integer. This last remark was already obvious from (32.19).

The parameter manifold of $O(3)$ was defined by (14), and in terms of the Euler angles by (32.14):

$$\begin{aligned} 0 &\leqslant \alpha \leqslant 2\pi, \\ 0 &\leqslant \beta \leqslant \pi, \\ 0 &\leqslant \gamma \leqslant 2\pi. \end{aligned} \tag{32.14}$$

* The argument is too loosely phrased to permit us to call this a proof.

Because $\alpha = 0$ and $\alpha = 2\pi$ are the same point in the space of group elements of $O(3)$, the double-valued representations are discontinuous at these points. Hence (32.14) cannot be the manifold for the universal covering group, because all its representations $D^{(j)}(a, b)$ are everywhere continuous and single valued. Instead of (32.14) the correct manifold for the covering group $SU(2)$ is *

$$\begin{aligned} 0 &\leqslant \alpha \leqslant 4\pi, \\ 0 &\leqslant \beta \leqslant \pi, \\ 0 &\leqslant \gamma \leqslant 4\pi. \end{aligned} \tag{15}$$

34.4 IRREDUCIBLE REPRESENTATIONS OF $SU(2)$

To construct the representations of $SU(2)$ we shall ape the approach used to study tensors in 3-space that transform according to various representations of $O(3)$. As the basic object in the latter case one takes the vector with components (x_1, x_2, x_3). This object transforms according to the three-dimensional representation

$$x_i \to x_i' = \sum_j x_j M_{ji}(R). \tag{16}$$

The orthogonal matrix $M(R)$ could, for example, be written as in (32.9′). A Cartesian tensor of rank k is defined as the set of 3^k quantities†

$$\Xi^{(k)}_{i_1 \cdots i_k} = x_{i_1} x_{i_2} \cdots x_{i_k} \tag{17}$$

that transform according to

$$\Xi^{(k)}_{i_1 \cdots i_k} \to \Xi'^{(k)}_{i_1 \cdots i_k} = \sum_{j_1 \cdots j_k} \Xi^{(k)}_{j_1 \cdots j_k} M_{j_1 i_1}(R) \cdots M_{j_k i_k}(R) \tag{18}$$

under the rotation R. The $3^k \times 3^k$ matrix that appears in (18) is not, however, one of the irreducible representations of $O(3)$. This is familiar from the case $k = 2$; the nine quantities $\Xi^{(2)}_{ij}$ transform with a 9×9 matrix that is reducible to $1 \times 1, 3 \times 3$, and 5×5 blocks. The so-called irreducible tensors can be constructed as follows: (a) a scalar $\xi = \sum_i \Xi^{(2)}_{ii}$ that transforms according to the one-dimensional representation; (b) a vector $\xi_i = \epsilon_{ijk} \Xi^{(2)}_{jk}$ that transforms with the three-dimensional represen-

* The manifold is actually covered completely if one only doubles the range of either α or γ. We use the more symmetric ranges (15) for later convenience.

† This is certainly not the most general Cartesian tensor of rank k. Nevertheless, it will suffice for our purpose because *any* Cartesian tensor of rank k satisfies the transformation law (18), by definition.

tation; and (c) a symmetric traceless tensor $\Xi_{ij}^{(2)} + \Xi_{ji}^{(2)} - \frac{2}{3}\delta_{ij}\xi$ that transforms with the five-dimensional representation. The messy way in which the Cartesian tensors (17) transform makes them rather awkward to use, and in Sec. 36 we shall introduce tensors that transform neatly under rotations. Nevertheless, one can see that in principle one could construct the representations of $O(3)$—i.e., the odd-dimensional representations of $SU(2)$—by laboriously inserting the explicit form of $M_{ij}(R)$ into (18).

Let us then apply these ideas to $SU(2)$. Now our basic quantity is the spinor (u_1, u_2), and we shall construct higher rank tensors in the spin space from this. It is obvious from the preceding paragraph that we shall encounter difficulty in evaluating the irreducible representations if we blindly study polynomials built out of u_1 and u_2. What is clearly required are polynomials that are constructed so as to belong to one irreducible representation. By using the knowledge we already possess concerning angular momentum eigenstates we shall find it rather easy to construct such polynomials. We begin with an intuitive argument, and then we shall verify our conjectures. The essential fact used in the argument is that under a rotation u_1 and u_2 transform like eigenstates with the quantum numbers $j = \frac{1}{2}$, $m = +\frac{1}{2}$, and $j = \frac{1}{2}$, $m = -\frac{1}{2}$, respectively. Consider first the polynomial $U_{i_1i_2} = u_{i_1}u_{i_2}$. The vector model implies that U_{11} and U_{22} transform like the states $j = 1$, $m = 1$, and $m = -1$, respectively. Furthermore, the vector model suggests that in general the quantities u_1^{2j} and u_2^{2j} transform under rotations like states having total angular momentum j with $m = +j$ and $m = -j$, respectively. If this is so, we can obtain a set of polynomials that belong to the $(2j + 1)$-dimensional irreducible representation by repeated application of J_- to u_1^{2j}. The irreducible representations of $SU(2)$ can then be determined by applying the transformation (11) to these polynomials.

In order to carry this program out we must have a representation of the angular momentum operators appropriate to the variables u_1 and u_2. One can obtain these operators by studying (11) for the appropriate infinitesimal rotations. In preparation for this, let us write (11) out explicitly:

$$\begin{aligned} u_1' &= au_1 - b^*u_2, \\ u_2' &= a^*u_2 + bu_1. \end{aligned} \tag{19}$$

In an infinitesimal rotation through the angle ϵ about the y-axis, the Euler angles are $\alpha = \gamma = 0$, $\beta = \epsilon$, and therefore $a = 1$, $b = -\frac{1}{2}\epsilon$; thus (19) for this case reads

$$\begin{aligned} u_1' &= u_1 + \tfrac{1}{2}\epsilon u_2, \\ u_2' &= u_2 - \tfrac{1}{2}\epsilon u_1. \end{aligned} \tag{20}$$

Under the same rotation an arbitrary function $f(u_1, u_2)$ undergoes the transformation

$$f(u_1, u_2) \to f(u_1', u_2') = \left[1 + \tfrac{1}{2}\epsilon\left(u_2 \frac{\partial}{\partial u_1} - u_1 \frac{\partial}{\partial u_2}\right)\right] f(u_1, u_2). \quad (21)$$

On the other hand, this can also be expressed as

$$f(u_1, u_2) \to f(u_1', u_2') = (1 - i\epsilon J_y) f(u_1, u_2),$$

which gives us the desired representation of J_y:

$$J_y = \frac{i}{2}\left(u_2 \frac{\partial}{\partial u_1} - u_1 \frac{\partial}{\partial u_2}\right). \quad (22)$$

A similar calculation involving an infinitesimal rotation about the x-axis leads to*

$$J_x = \frac{1}{2}\left(u_2 \frac{\partial}{\partial u_1} + u_1 \frac{\partial}{\partial u_2}\right). \quad (23)$$

The ladder operators $J_\pm$ therefore have the exceedingly simple form

$$J_+ = u_1 \frac{\partial}{\partial u_2}, \quad (24)$$

$$J_- = u_2 \frac{\partial}{\partial u_1}. \quad (25)$$

Applying $(J_-)^n$ to u_1^{2j} will produce the required polynomials. Instead of working this out we shall write down the eigenfunctions of J^2 and J_z, and verify that they are correct:

$$\Phi_{jm}(u_1, u_2) = (2j!)^{\frac{1}{2}} \frac{u_1^{j+m} u_2^{j-m}}{\sqrt{(j+m)!(j-m)!}}. \quad (26)$$

The proof that these are the eigenfunctions with the indicated eigenvalues follows from applying $J_\pm$ to them. This gives

$$u_1 \frac{\partial}{\partial u_2} \Phi_{jm} = (2j!)^{\frac{1}{2}} (j-m) \frac{u_1^{j+(m+1)} u_2^{j-(m+1)}}{\sqrt{(j+m)!(j-m)!}}$$

$$= \sqrt{(j-m)(j+m+1)} \cdot (2j!)^{\frac{1}{2}} \frac{u_1^{j+(m+1)} u_2^{j-(m+1)}}{\sqrt{(j+m+1)!(j-m-1)!}}$$

* In carrying this out one must take care to use the Euler angles $\alpha = -\frac{1}{2}\pi$, $\beta = \epsilon$, $\gamma = \frac{1}{2}\pi$, and not $\alpha = 3\pi/2$. The trouble with the latter is that it brings us to a point in the $SU(2)$ manifold that is not near the identity, and therefore it is not an infinitesimal transformation.

and similarly

$$u_2 \frac{\partial}{\partial u_1} \Phi_{jm} = \sqrt{(j+m)(j-m+1)}\, \Phi_{jm-1}.$$

Therefore

$$J_\pm \Phi_{jm}(u_1, u_2) = \sqrt{j(j+1) - m(m \pm 1)}\, \Phi_{jm\pm 1}(u_1, u_2).$$

QED. The factor $(2j!)^{\frac{1}{2}}$ in (26) is of no great significance; it was chosen so that

$$\sum_m |\Phi_{jm}|^2 = (|u_1|^2 + |u_2|^2)^{2j}.$$

The essential remaining point is to show that the set of functions Φ_{jm} for fixed j transform linearly among themselves under a rotation. To do this we substitute (19) into (26) and apply the binomial theorem:

$$\Phi_{jm}(u_1', u_2') = \sqrt{\frac{2j!}{(j+m)!(j-m)!}} \sum_{\mu\nu} \binom{j+m}{\mu} \binom{j-m}{\nu} \cdot (au_1)^{j+m-\mu}(-b^*u_2)^\mu (a^*u_2)^\nu (bu_1)^{j-m-\nu}.$$

Noting that the power of u_1 is $2j - \mu - \nu$, we make the substitution $j + m' = 2j - \mu - \nu$. This gives us

$$\Phi_{jm}' = \sqrt{\frac{2j!}{(j+m)!(j-m)!}} \sum_{m'\mu} \binom{j+m}{\mu} \binom{j-m}{j-\mu-m'} u_1^{j+m'} u_2^{j-m'} \cdot a^{j+m-\mu}(-b^*)^\mu (a^*)^{j-\mu-m'} b^{m'-m+\mu}, \tag{27}$$

which shows that the forms $u_1^{j+m}u_2^{j-m}$ indeed transform linearly among themselves under rotations. We can, in fact, obtain the representations themselves from (27). To do so we note that the basic definition (32.3) can be written as

$$|jm; R\rangle = \sum_{m'} |jm'\rangle D_{m'm}^{(j)}(R),$$

and therefore $D_{mm'}^{(j)}$ can be read off from

$$\Phi_{jm}(u_1', u_2') = \sum_{m'} \Phi_{jm'}(u_1, u_2) D_{m'm}^{(j)}(R). \tag{28}$$

(For $j = \frac{1}{2}$, this reduces to (11).) Comparing (28) with (27), and taking the definition of Φ_{jm} into account, we therefore find the following expression for the irreducible representations of $SU(2)$:

$$D^{(j)}_{mm'}(a, b) = \sqrt{\frac{(j+m)!(j-m)!}{(j+m')!(j-m')!}} \sum_{\mu} \binom{j+m'}{\mu} \binom{j-m'}{j-\mu-m} \cdot a^{j+m'-\mu}(a^*)^{j-\mu-m}(-b^*)^{\mu} b^{m-m'+\mu}. \quad (29)$$

The matrix $d^{(j)}_{mm'}(\beta)$ that appears in

$$D^{(j)}_{mm'}(\alpha, \beta, \gamma) = e^{-im\alpha} d^{(j)}_{mm'}(\beta) e^{-im'\gamma} \quad (32.19)$$

is therefore

$$d^{(j)}_{mm'}(\beta) = (-1)^{m-m'} \sqrt{\frac{(j+m)!(j-m)!}{(j+m')!(j-m')!}} (\cos \tfrac{1}{2}\beta)^{2j} \sum_{\mu} (-1)^{\mu} \binom{j+m'}{\mu} \binom{j-m'}{j-\mu-m} (\tan \tfrac{1}{2}\beta)^{m-m'+2\mu}. \quad (30)$$

This completes the determination of the irreducible representations of the rotation group.*

•The rotation matrices satisfy a number of useful symmetry relations. Equation 30 is invariant under the substitution $m \to -m'$, $m' \to -m$, or

$$d^{(j)}_{mm'}(\beta) = d^{(j)}_{-m',-m}(\beta). \quad (31)$$

Further identities follow from the group properties. Because $d^{(j)}(\beta)d^{(j)}(-\beta) = 1$, and $d^{(j)}$ is an orthogonal matrix, we have

$$d^{(j)}_{mm'}(\beta) = d^{(j)}_{m'm}(-\beta). \quad (32a)$$

Replacing β by $-\beta$ in (30) leads to the result

$$d^{(j)}_{mm'}(-\beta) = (-1)^{m-m'} d^{(j)}_{mm'}(\beta). \quad (32b)$$

Combining this with (32a) gives (7):

$$d^{(j)}_{mm'}(\beta) = (-1)^{m-m'} d^{(j)}_{m'm}(\beta) \quad (32c)$$

$$= (-1)^{m-m'} d^{(j)}_{-m,-m'}(\beta). \quad (32d)$$

Consequently

$$D^{(j)}_{mm'}(R)^* = (-1)^{m-m'} D^{(j)}_{-m,-m'}(R). \quad (32e)$$

A relationship between $d^{(j)}(\pi - \beta)$ and $d^{(j)}(\beta)$ can be derived as follows. When $\alpha = \gamma = 0$, $\beta = \pi$, $a = 0$ and $b = -1$. Inspection of (29) then yields

$$d^{(j)}_{mm'}(\pi) = (-1)^{j+m} \delta_{m,-m'}. \quad (32f)$$

But $d^{(j)}(\pi - \beta) = d^{(j)}(\pi)\, d^{(j)}(-\beta)$, and therefore

$$d^{(j)}_{mm'}(\pi - \beta) = (-1)^{j+m} d^{(j)}_{m,-m'}(\beta). \bullet \quad (32g)$$

* Compilations of useful formulas for $d^{(j)}_{mm'}$ can be found in M. Jacob and G. C. Wick, *Ann. of Phys.* **7**, 404 (1959); S. Berman and M. Jacob, *Phys. Rev.* **139**, B1023 (1965).

34.5 INTEGRALS INVOLVING THE ROTATION MATRICES

Finally we shall evaluate some integrals involving the representation matrices. The point of departure is the observation that if we average an arbitrary state over all orientations, only the spherically symmetric ($j = 0$) portion of it will survive. Hence we expect that

$$\int D(R)\, dR = P_0, \tag{33}$$

where P_0 is the projection operator onto the $j = 0$ subspace of the full Hilbert space. In the $|jm\rangle$-representation, (33) therefore becomes

$$\int D^{(j)}_{mm'}(R)\, dR = \delta_{m,0}\, \delta_{m',0}\, \delta_{j,0}. \tag{33'}$$

It is clear that (33) can only be true if the integration weights all orientations equally. In order to average over the orientation of the ζ-axis in Fig. 32.1, we must perform the integration

$$\int_0^{2\pi} d\alpha \int_0^{\pi} \sin\beta\, d\beta.$$

A further integration over γ from 0 to 2π is then required to average over the position of the ξ- and η-axes. The integration that has just been spelled out extends over the parameter manifold of the rotation group, as given by (32.14). Because of the occurrence of double-valued representations, it is necessary to integrate over the covering group, whose parameter manifold is defined by (15). In fact, from (32.19) we note that $\int_0^{4\pi} d\alpha$ insures that $m = 0$ in (33′), but that this is not true when the upper limit is only 2π. Our choice for dR is therefore

$$dR = \tfrac{1}{2} \sin\beta\, d\beta \frac{d\alpha}{4\pi} \frac{d\gamma}{4\pi}, \tag{34}$$
$$0 \leqslant \alpha, \gamma \leqslant 4\pi, \qquad 0 \leqslant \beta \leqslant \pi.$$

The arbitrary constant in dR has been chosen so that the coefficient of P_0 in (33) is 1; equivalently, $\int dR = 1$.

The integral formulas are all obtained by integrating the Kronecker-product relations over dR. By combining (4) and (33′) we obtain

$$\int dR\, D^{(j_1)}_{m_1m_1'}(R) D^{(j_2)}_{m_2m_2'}(R) = \langle j_1m_1j_2m_2|00\rangle\langle 00|j_1m_1'j_2m_2'\rangle.$$

The recursion formulas for the CG-coefficients (25.17) easily yield

$$\langle j_1m_1j_2m_2|00\rangle = (2j_1+1)^{-\frac{1}{2}}(-1)^{j_1-m_1}\, \delta_{j_1j_2}\, \delta_{m_1,-m_2}, \tag{35}$$

which, together with (32e), gives us the orthogonality relation for the representation matrices:

$$\int dR\, D^{(j_1)}_{m_1m_1'}(R)^* D^{(j_2)}_{m_2m_2'}(R) = \frac{\delta_{j_1j_2}\,\delta_{m_1m_2}\,\delta_{m_1'm_2'}}{2j_1+1}. \tag{36}$$

A very useful formula involving three rotation matrices can now be obtained by combining (36) with (4):

$$\int dR\, D^{(j_1)}_{m_1m_1'}(R)^* D^{(j_2)}_{m_2m_2'}(R) D^{(j_3)}_{m_3m_3'}(R) = \frac{\langle j_2m_2j_3m_3|j_1m_1\rangle\langle j_2m_2'j_3m_3'|j_1m_1'\rangle}{2j_1+1}. \tag{37}$$

This integral is often used in the theory of molecular and nuclear spectra. An even more frequently employed formula involving three spherical harmonics can be obtained from it as well, because for a special choice of the quantum numbers the rotation matrix reduces to a spherical harmonic. This can be seen from (5), which becomes the differential equation for the associated Legendre function when we set $m' = 0$ and j equal to an integer l. Hence one concludes that $D^{(l)}_{m0}(\alpha, \beta)^* = \text{const.}\; Y_{lm}(\beta, \alpha)$. The constant of proportionality is determined by setting $\vartheta = 0$ in (35.4). Using $Y_{lm}(0, \varphi) = \delta_{m0}[(2l+1)/4\pi]^{\frac{1}{2}}$ and Eqs. (32) we obtain

$$D^{(l)}_{m0}(\alpha, \beta)^* = \sqrt{\frac{4\pi}{2l+1}}\, Y_{lm}(\beta, \alpha). \tag{38}$$

The desired special case of (37) is therefore

$$\int d\Omega\, Y^*_{l_1m_1}(\Omega) Y_{l_2m_2}(\Omega) Y_{l_3m_3}(\Omega) = \sqrt{\frac{(2l_2+1)(2l_3+1)}{4\pi(2l_1+1)}}\, \langle l_2m_2l_3m_3|l_1m_1\rangle\langle l_20l_30|l_10\rangle. \tag{39}$$

There is a closed formula for $\langle l_20l_30|l_10\rangle$:

$$\langle l_20l_30|l_10\rangle = 0 \qquad \text{for} \quad (l_1+l_2+l_3) \text{ odd},$$

$$\langle l_20l_30|l_10\rangle = (-1)^{l_1+L} \frac{L!\sqrt{2l_1+1}}{(L-l_1)!(L-l_2)!(L-l_3)!} \times \sqrt{\frac{(l_1+l_2-l_3)!(l_2+l_3-l_1)!(l_3+l_1-l_2)!}{(l_1+l_2+l_3+1)!}}, \tag{40}$$

if $(l_1+l_2+l_3)$ is even, where $2L = l_1+l_2+l_3$. In (39) $\Omega \equiv (\alpha,\beta)$, $d\Omega = \sin\alpha\, d\alpha\, d\beta$, and the integration extends only over the unit sphere.

35. Transformation of States under Rotations

In this section we shall explore the basic relation

$$|a'; R\rangle = D(R)|a'\rangle \tag{1}$$

in greater detail. We begin with the states of a spin 0 particle.

35.1 ROTATIONS APPLIED TO SPHERICAL HARMONICS

Consider the specialization of (1) to a one-particle eigenstate of L^2 and L_z:

$$\begin{aligned} |lm; R\rangle &= \sum_{m'} |lm'\rangle\langle lm'|D(R)|lm\rangle \\ &= \sum_{m'} |lm'\rangle D^{(l)}_{m'm}(R). \end{aligned} \tag{2}$$

To make this quite concrete, we compute the amplitude for finding the particles in the direction $\hat{\mathbf{n}}$ as measured in the original coordinate system (x, y, z). In that system let $\hat{\mathbf{n}}$ be specified by the polar coordinates ϑ, φ. Because $\langle\hat{\mathbf{n}}|lm\rangle = Y_{lm}(\vartheta, \varphi)$, (2) reads

$$\langle\hat{\mathbf{n}}|lm; R\rangle = \sum_{m'} Y_{lm'}(\vartheta, \varphi)D^{(l)}_{m'm}(R). \tag{3}$$

Note that $\langle\hat{\mathbf{n}}|lm; R\rangle$ is not a spherical harmonic in (ϑ, φ) as it stands; thus when $m = 0$ it will in general depend on φ, unless the new z-axis (ζ of Sec. 32) is parallel to $\hat{\mathbf{n}}$. On the other hand, if $\hat{\mathbf{n}}$ is specified by the angles Θ, Φ in the (ξ, η, ζ) system, we surely can write $\langle\hat{\mathbf{n}}|lm; R\rangle = Y_{lm}(\Theta, \Phi)$ because $|lm; R\rangle$ is an eigenstate of L_ζ with eigenvalue m, and therefore a spherical harmonic in the coordinate representation referred to the new frame. Therefore

$$Y_{lm}(\Theta, \Phi) = \sum_{m'} Y_{lm'}(\vartheta, \varphi)D^{(l)}_{m'm}(R). \tag{4}$$

The orientation of $\hat{\mathbf{n}}$ is shown in Fig. 35.1. For our later work it will pay to complicate this argument somewhat. Note that

$$\langle\hat{\mathbf{n}}|lm; R\rangle = \langle\hat{\mathbf{n}}'|lm\rangle, \tag{5}$$

where $|\hat{\mathbf{n}}'\rangle = D^{-1}(R)|\hat{\mathbf{n}}\rangle$ is obtained from $|\hat{\mathbf{n}}\rangle$ by performing the inverse rotation. Clearly this is an eigenket of the position (or better, direction) operator with eigenvalue $\hat{\mathbf{n}}' = R^{-1}\hat{\mathbf{n}}$. Here the notation $R\mathbf{a}$ for an arbitrary 3-vector $\mathbf{a}$ designates the vector obtained by applying the rotation

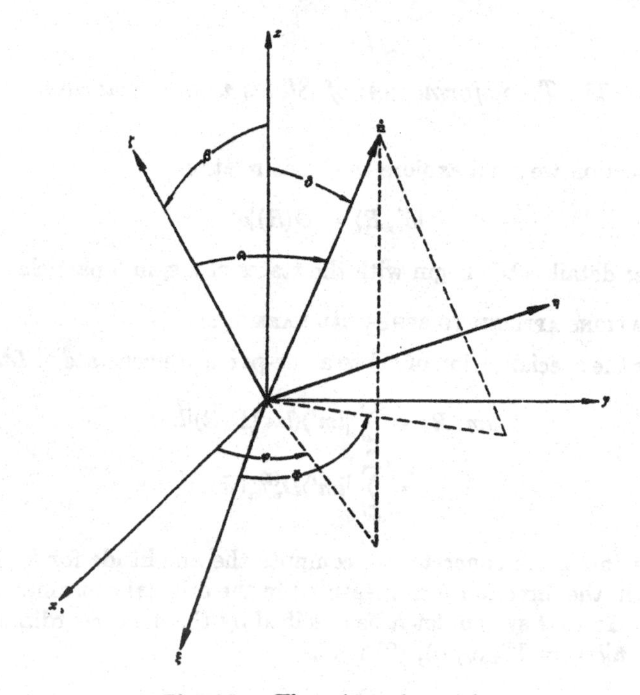

Fig. 35.1. The orientation of $\hat{\mathbf{n}}$.

R to $\mathbf{a}$; this can, for example, be achieved by use of (32.9′). The proof of

$$D(R)|\hat{\mathbf{n}}\rangle = |R\hat{\mathbf{n}}\rangle$$

is elementary and is left to the reader. The geometric significance of $\hat{\mathbf{n}}'$ can be seen by noting that R maps $\hat{\mathbf{n}}' \to \hat{\mathbf{n}}$ and $(xyz) \to (\xi\eta\zeta)$. Hence the orientation of $\hat{\mathbf{n}}'$ in (xyz) is the same as $\hat{\mathbf{n}}$ in $(\xi\eta\zeta)$, i.e., Θ, Φ. Thus the relation

$$Y_{lm}(R^{-1}\hat{\mathbf{n}}) = \sum_{m'} Y_{lm'}(\hat{\mathbf{n}})D^{(l)}_{m'm}(R), \tag{6}$$

which follows from (5) and (2), is the same as (4), as it must be. Even though (4) and (6) are mathematically identical, they admit to two different "interpretations": In (6) one can think of $\hat{\mathbf{n}}$ and $R^{-1}\hat{\mathbf{n}}$ as being measured by the xyz observer, in which case they refer to *different* directions in space, but m and m' are L_z eigenvalues, whereas in (4) m is an L_ζ-eigenvalue if one there thinks of (Θ, Φ) and (ϑ, φ) as referring to the *same* direction in space.

Consider the special case $m = 0$; then $D^l_{m'0}$ does not depend on γ, and

$$Y_{l0}(\Theta) = \sum_m Y_{lm}(\vartheta, \varphi) D^{(l)}_{m0}(\alpha, \beta, 0),$$

according to (4). Upon using (34.38) this becomes

$$Y_{l0}(\Theta) = \sqrt{\frac{4\pi}{2l+1}} \sum_m Y_{lm}(\vartheta, \varphi) Y^*_{lm}(\beta, \alpha), \tag{7}$$

which is *the addition theorem for spherical harmonics.*

35.2 SPINOR FIELDS

Another type of question we can now treat with ease is the following. An electron is prepared in a state where its spin is "up" along the z-axis: What is the probability amplitude for finding its spin pointing along the unit vector $\hat{\mathbf{n}}$? If $\hat{\mathbf{n}}$ has the polar angles (ϑ, φ), the Euler angles are $\alpha = \varphi, \beta = \vartheta, \gamma = 0$. The state we are seeking to find the particle in is $|\frac{1}{2}; R\rangle$. The required amplitude is then evaluated by consulting (33.15):

$$\langle \tfrac{1}{2} | \tfrac{1}{2}; R\rangle = D^{(\frac{1}{2})}_{\frac{1}{2}\frac{1}{2}}(R) = e^{-\frac{1}{2}i\varphi} \cos \tfrac{1}{2}\vartheta. \tag{8}$$

We can generalize this to an arbitrary state $|\psi\rangle$ of a spin $\frac{1}{2}$ particle. Such a state can always be written as a superposition of a ket in which the spin is definitely "up," and one in which it is "down":

$$|\psi\rangle = \sum_{\mu=-\frac{1}{2}}^{\frac{1}{2}} |\psi_\mu\rangle|\mu\rangle. \tag{9}$$

Here $|\psi_\mu\rangle$ is a ket in the usual Hilbert space of the classical variables, and $|\mu\rangle$ is a vector in the two-dimensional (complex) spin space. To go to the coordinate representation, we take the scalar product of (9) with $\langle \mathbf{x}|$:

$$\boldsymbol{\psi}(\mathbf{x}) = \sum_\mu \psi_\mu(\mathbf{x})|\mu\rangle. \tag{10}$$

The quantities $\boldsymbol{\psi}(\mathbf{x})$ are called *Pauli two-component wave functions.* As our notation emphasizes, they are vectors in the (two-dimensional) spin space.

When rotated through R, $|\psi\rangle$ becomes $|\psi'\rangle \equiv D(R)|\psi\rangle$, or

$$\boldsymbol{\psi}'(\mathbf{x}) = \langle \mathbf{x}|D(R)_{\text{orb}} D^{(\frac{1}{2})}(R)|\psi\rangle.$$

As in (33.17), D_{orb} only acts on the spatial degrees of freedom in this equation. Thus we have

$$\Psi'(\mathbf{x}) = \sum_{\mu} \langle \mathbf{x}|D_{\text{orb}}|\psi_\mu\rangle D^{(\frac{1}{2})}(R)|\mu\rangle$$
$$= \sum_{\mu\mu'} \psi_\mu(R^{-1}\mathbf{x})|\mu'\rangle D^{(\frac{1}{2})}_{\mu'\mu}(R).$$

The transformation law for the Pauli wavefunction is therefore

$$\Psi(\mathbf{x}) \to \Psi'(\mathbf{x}) \equiv \sum_{\mu} \psi'_\mu(\mathbf{x})|\mu\rangle, \tag{11}$$

where

$$\psi'_\mu(\mathbf{x}) = \sum_{\mu'} D^{(\frac{1}{2})}_{\mu\mu'}(R)\psi_{\mu'}(R^{-1}\mathbf{x}). \tag{12}$$

In contrast to this, under a rotation R of the system, the wave function for a spinless particle undergoes the transformation

$$\psi(\mathbf{x}) \to \psi'(\mathbf{x}) = \psi(R^{-1}\mathbf{x}). \tag{13}$$

A single function that transforms in accordance with (13) is called a *scalar field*. As we already knew, the state of a single spin zero particle is described by a scalar field function, i.e., the Schrödinger wave function. For a spin $\frac{1}{2}$ particle we must specify the amplitude for finding the spin projection "up" or "down" at each point $\mathbf{x}$ in space, and therefore two functions are required to specify its state completely. These two functions $\psi_\mu(\mathbf{x})$ are *not* scalar fields, however. Under a rotation they transform linearly among themselves as shown by (12). A pair of functions that transforms under rotations in the manner specified by (12) is said to constitute a *spinor field*—more precisely, a spin-$\frac{1}{2}$ field.

These formulas are trivially generalized to particles of higher spin. A particle with intrinsic spin s is described by three classical observables (such as the position or momentum) and one further quantum number μ that can take on the $2s + 1$ values $s, s - 1, \ldots, -s$. The Hilbert space is the direct product of the Hilbert space of ordinary wave mechanics and a $(2s + 1)$-dimensional complex vector space. A ket for such a system also has the form (9), but the sum runs over $-s \leqslant \mu \leqslant s$. If the system in question is rotated through R, the functions $\psi_\mu(\mathbf{x})$ transform linearly among themselves:

$$\psi_\mu(\mathbf{x}) \to \psi'_\mu(\mathbf{x}) = \sum_{\mu'=-s}^{s} D^{(s)}_{\mu\mu'}(R)\psi_{\mu'}(R^{-1}\mathbf{x}). \tag{14}$$

A set of functions that transform according to (14) constitutes a spin-s field.

One should try to achieve an intuitive grasp of these superficially obscure formulas. Consider first the scalar case, Eq. (13). We can visualize a scalar function in two dimensions as a mountain whose height we call $\psi(\mathbf{x})$. If the mountain is rotated rigidly through an angle ω about some fixed point P in the plane, and we continue to observe its height at some other fixed point O, then the height will change because of the rotation. The content of Eq. (13) is that we would obtain the same height if we left the mountain alone, and instead rotated our observation point through $-\omega$ about P. Next consider the case $s = 1$. There are now three functions at each point in space, and as we shall see in the following section, these three functions constitute a vector field (e.g., a magnetic field). The burden of Eq. (14) is that the separate components of a vector field (or any $s \neq 0$ field) do *not* transform like the "mountains," i.e., when we perform the inverse rotation of the observation point we do not obtain the same values for the separate components as when one stays put at O and rotates the field. This is quite familiar to us from electromagnetism, and Eq. (14) is merely a generalization of this concept.

35.3 HELICITY STATES

In Sec. 33.1 (see Eq. (2′)) we mentioned a very obvious technique for constructing one-particle eigenstates of j^2, L^2 and j_z, where $\mathbf{j} = \mathbf{L} + \mathbf{s}$ is the total angular momentum. In this method one first constructs eigenstates of (L^2, L_z, s_z), and then combines these with Clebsch-Gordan coefficients. We shall now discuss an important alternative method * for constructing one-particle total angular momentum states. These will not be eigenstates of L^2. Instead they will be eigenstates of the component of $\mathbf{s}$ along the direction of motion, $\hat{\mathbf{p}} \cdot \mathbf{s}$. The eigenvalue λ of $\hat{\mathbf{p}} \cdot \mathbf{s}$ is called *the helicity*. The possible values of the helicity are obviously

$$\lambda = s, s-1, \ldots, -s. \tag{15}$$

Although $\hat{\mathbf{p}} \cdot \mathbf{s}$ does not commute with $\mathbf{L}$ or $\mathbf{s}$, it is invariant under rotations, and therefore commutes with $\mathbf{j}$. Hence we can specify one-particle states by the eigenvalues p, $j(j+1)$, m and λ of the operators $|\mathbf{p}|$, j^2, j_z and $\hat{\mathbf{p}} \cdot \mathbf{s}$, respectively.

The helicity states enjoy several important advantages over the more familiar $|ljm\rangle$: (a) they can be used in situations where relativistic effects are important, and (b) they lead to a very elegant theory of collisions involving particles with spin. Because of (a), photon states can be classified conveniently in terms of the helicity, and we shall take advantage of

* M. Jacob, *Nuovo Cimento* 9, 826 (1958); M. Jacob and G. C. Wick, *Ann. Physics* 7, 404 (1959).

this in Sec. 52.3. A discussion of collisions involving particles with spin will be given in Chapt. XIII.

Let $\hat{\mathbf{z}}$ be a unit vector along the z-direction. In the sequel the momentum of a particle will be written as $p\hat{\mathbf{n}}$, where the orientation of $\hat{\mathbf{n}}$ is specified by the polar angle θ measured from $\hat{\mathbf{z}}$ and the azimuthal angle ϕ. Let $|p\hat{\mathbf{z}}; \lambda\rangle$ be a state of helicity λ and linear momentum $p\hat{\mathbf{z}}$. A state $|p\hat{\mathbf{n}}, \lambda\rangle$ in which the momentum is of the same magnitude but is in the direction $\hat{\mathbf{n}}$ is then given by

$$|p\hat{\mathbf{n}}, \lambda\rangle = D(\phi, \theta, 0)|p\hat{\mathbf{z}}, \lambda\rangle = e^{-i\phi j_z}e^{-i\theta j_y}|p\hat{\mathbf{z}}, \lambda\rangle. \tag{16}$$

The choice $\alpha = \phi$ and $\beta = \theta$ of the Euler angles is dictated by our conventions of Fig. 32.1, but γ is arbitrary, and any other choice of γ merely amounts to a change of phase. Because $D(\phi, \theta, 0)$ appears frequently in the subsequent formulas, we shall use the convenient abbreviations

$$D(\hat{\mathbf{n}}) \equiv D(\phi, \theta, 0), \qquad D^{(j)}_{mm'}(\hat{\mathbf{n}}) \equiv D^{(j)}_{mm'}(\phi, \theta, 0).$$

The helicity quantum number is the same on both sides of (16) because $\hat{\mathbf{p}} \cdot \mathbf{s}$ is rotationally invariant, i.e., because $[\hat{\mathbf{p}} \cdot \mathbf{s}, D(\hat{\mathbf{n}})] = 0$.

As it stands $|p\hat{\mathbf{z}}, \lambda\rangle$ is already an eigenstate of j_z with eigenvalue λ. We can therefore decompose it into simultaneous eigenstates $|p, jm\rangle$ of $|\mathbf{p}|$, j^2 and j_z as follows:

$$|p\hat{\mathbf{z}}, \lambda\rangle = \sum_{j=|\lambda|}^{\infty} |p, j\lambda\rangle\langle p, j\lambda|p\hat{\mathbf{z}}, \lambda\rangle. \tag{17}$$

If s is an integer (half-integer), the sum in (17) is taken over the integers (half-integers). The rotated state $|p\hat{\mathbf{n}}, \lambda\rangle$ is obtained by applying $D(\hat{\mathbf{n}})$ to (17):

$$|p\hat{\mathbf{n}}, \lambda\rangle = \sum_{j=|\lambda|}^{\infty} \sum_{m=-j}^{j} |p, jm\rangle D^{(j)}_{m\lambda}(\hat{\mathbf{n}})\langle p, j\lambda|p\hat{\mathbf{z}}, \lambda\rangle. \tag{18}$$

Before proceeding further, let us apply this relation to a spinless particle (i.e., $\lambda \equiv 0$). Replacing j by l in this case, and recalling (34.38), we obtain

$$|p\hat{\mathbf{n}}\rangle = \sum_{l=0}^{\infty} \sum_{m=-l}^{l} |p, lm\rangle \sqrt{\frac{4\pi}{2l+1}}\, \langle p, l|p\hat{\mathbf{z}}\rangle Y^*_{lm}(\theta, \phi). \tag{19}$$

This is just the expansion of a plane wave in terms of spherical waves. To verify this assertion in complete detail, one takes the scalar product of (19) with a coordinate eigenbra, and notes that $\langle \mathbf{x}|p, lm\rangle = c_l j_l(px/\hbar) Y_{lm}(\hat{\mathbf{x}})$. The result is then Eq. (11.25), except for possible differences in normalization conventions. It is therefore clear that (18) is merely the generalization to particles with spin of the spherical wave decomposition of a plane wave.

We may extract the sought-after helicity state from (18) by taking advantage of the orthogonality relations satisfied by elements of the rotation matrices. From (34.36) we easily obtain

$$\int d\hat{n}\, D^{(j)}_{m\lambda}(\hat{n}) D^{(j')}_{m'\lambda}(\hat{n})^* = \frac{4\pi}{2j+1}\,\delta_{jj'}\,\delta_{mm'}, \tag{20}$$

where $d\hat{n} \equiv \sin\theta\, d\theta\, d\phi$, and the integration is over the unit sphere.*
Applying this to (18) yields

$$|p, jm\lambda\rangle = N \int d\hat{n}\, D^{(j)}_{m\lambda}(\hat{n})^* |p\hat{n}, \lambda\rangle. \tag{21}$$

This state obviously has helicity λ, and we have therefore inserted this quantum number into the ket symbol. The constant in (21) is given by

$$N^{-1} = 4\pi(2j+1)^{-1}\langle p, j\lambda|p\hat{z}, \lambda\rangle,$$

and is determined once the normalization conventions are stated. Let us demand that

$$\langle p, jm\lambda|p', j'm'\lambda'\rangle = \frac{\delta(p-p')}{pp'}\,\delta_{jj'}\,\delta_{mm'}\,\delta_{\lambda\lambda'}. \tag{22}$$

The completeness relation corresponding to this convention is

$$1 = \sum_{jm\lambda} \int_0^\infty p^2\, dp |p, jm\lambda\rangle\langle p, jm\lambda|. \tag{22'}$$

The momentum eigenstates satisfy the usual orthonormality condition

$$\langle p\hat{n}, \lambda|p'\hat{n}', \lambda'\rangle = \delta(p\hat{n} - p'\hat{n}')\,\delta_{\lambda\lambda'}. \tag{23}$$

When we insert (21) into (22), and use (23), we obtain

$$\delta_{jj'}\,\delta_{mm'}\,\delta_{\lambda\lambda'}\,\delta(p-p')/pp' \\ = N^2 \int d\hat{n}\, d\hat{n}'\, D^{(j)}_{m\lambda}(\hat{n})^* D^{(j')}_{m'\lambda'}(\hat{n}')\,\delta(p\hat{n} - p'\hat{n}')\,\delta_{\lambda\lambda'}.$$

Because $\delta(p\hat{n} - p'\hat{n}') \equiv \delta(p-p')\,\delta(\hat{n} - \hat{n}')/pp'$, the remaining integral is just (20). N is thereby evaluated as $\sqrt{(2j+1)/4\pi}$. The com-

* Because both the $D^{(j)}$ in (20) are either even- or odd-dimensional representations, it is not necessary to integrate over the covering group.

plete formula for the total angular momentum eigenstate in the helicity representation is therefore

$$|p, jm\lambda\rangle = \sqrt{\frac{2j+1}{4\pi}} \int d\hat{n}\, D^{(j)}_{m\lambda}(\hat{n})^* |p\hat{n}, \lambda\rangle. \tag{24}$$

The transformation function from the linear to the angular momentum representation is thus

$$\langle p', jm\lambda' | p\hat{n}, \text{-}\lambda\rangle = \frac{\delta(p - p')}{pp'}\, \delta_{\lambda\lambda'} \sqrt{\frac{2j+1}{4\pi}}\, D^{(j)}_{m\lambda}(\hat{n}). \tag{25}$$

One should bear in mind that the state $|p, jm\lambda\rangle$ has the remarkable property that both the direction of the momentum and any fixed component of the spin have large dispersions, but the projection of the spin onto the linear momentum has the precise value λ. As we have already said, this is possible because λ is the eigenvalue of a rotationally invariant observable.

36. Tensor Operators

Let us now turn to the question of how rotations affect observables. In view of $|b'; R\rangle \equiv D(R)|b'\rangle$, we can write the matrix element of an arbitrary observable A between rotated states as

$$\langle b'; R|A|c'; R\rangle = \langle b'|D^\dagger(R) A D(R)|c'\rangle. \tag{1}$$

Instead of rotating the states, we can therefore rotate the observable:

$$A_R \equiv D^\dagger(R) A D(R). \tag{2}$$

Hence an observable is invariant under rotations if it commutes with all the $D(R)$, i.e., with the *total* angular momentum $\mathbf{J}$. If A only commutes with the component $\hat{n} \cdot \mathbf{J}$ of $\mathbf{J}$, it is cylindrically symmetric about $\hat{n}$. An observable that commutes with $\mathbf{J}$ is called a *scalar observable*. Thus H, J^2, and J_z can be used as (part of) a compatible set if the Hamiltonian is a scalar.

We shall analyze (2) in more detail in terms of infinitesimal rotations. Such a rotation is most conveniently specified by an axis of rotation $\hat{n}$, and an angle $\delta\theta$:

$$D_{\hat{n}}(\delta\theta) = 1 - i\, \delta\theta\, \hat{n} \cdot \mathbf{J}. \tag{3}$$

Thus under an infinitesimal rotation

$$A \to A' = A + i\,\delta\theta[\hat{\mathbf{n}} \cdot \mathbf{J}, A] \tag{4}$$

(compare to (29.8)).

In the translational case we merely separated observables into two classes: those that are and those that are not translation invariant. In the rotational case it is very useful to classify the nonscalar observables still further. This is already done in classical physics where one speaks of vectors and tensors of various rank; we are merely extending this classification to the quantum theory.

36.1 VECTOR OPERATORS

Consider first a set of observables (V_x, V_y, V_z) that transform under rotations like a vector. The rotation of a vector can be carried out with the matrices I_x, I_y, I_z, of (32.7). We designate the components of $\exp \phi I_x$ by $(\exp \phi I_x)_{ij}$, and those of $\mathbf{V}$ by V_i. The transformation law for $\mathbf{V}$ is then precisely the same as the one for $\mathbf{r}$, as given by (32.9):

$$D_x^{\dagger}(\phi) V_i D_x(\phi) = \sum_j (e^{\phi I_x})_{ij} V_j, \tag{5}$$

and similarly for the other axes. Allowing the rotation angle in (5) to be an infinitesimal, and comparing with (4), we obtain the conditions*

$$\begin{aligned} V_x + i\,\delta\phi[J_x, V_x] &= V_x, \\ V_y + i\,\delta\phi[J_x, V_y] &= V_y - \delta\phi\, V_z, \\ V_z + i\,\delta\phi[J_x, V_z] &= V_z + \delta\phi\, V_y. \end{aligned} \tag{6}$$

Because $\delta\phi$ is arbitrary, (6) implies the commutation rules

$$[J_x, V_x] = 0, \qquad [J_x, V_y] = iV_z, \qquad [J_x, V_z] = -iV_y, \tag{7}$$

etc. We note that the angular momentum commutation relations are a special case of (7), as they must be. Any operator that has the commutation relations (7) with $\mathbf{J}$ is called a *vector operator*. (One can show that (7) will guarantee (5) for finite rotations by using the group properties.)

36.2 IRREDUCIBLE TENSOR OPERATORS

We could extend (5) to tensor operators $T_{ijk\ldots}$ of arbitrary rank by demanding

$$D_x^{\dagger}(\phi) T_{ijk\ldots} D_x(\phi) = \sum_{i'j'k'\ldots} (e^{\phi I_x})_{ii'} (e^{\phi I_x})_{jj'} \cdots T_{i'j'k'\ldots}$$

* Note carefully that we have rotated $\mathbf{V}$ in the counter-clockwise sense. The right-hand side of (6) gives the components of the rotated vector in the *original* coordinate frame.

as in (34.18). As we already pointed out in Sec. 34.4, Cartesian tensors do not transform with one particular irreducible representation of the rotation group. It will be recalled that for the second-rank tensor the transformation matrix $A_{ij;nm}$ in $T'_{ij} = \Sigma A_{ij;nm} T_{nm}$ can be reduced to the form

$$(A_{ij;nm}) \to \begin{pmatrix} D^{(0)} & & 0 \\ & D^{(1)} & \\ 0 & & D^{(2)} \end{pmatrix}$$

by a linear transformation. Similar reductions apply to tensors of higher rank. In atomic and nuclear physics, where we usually use representations wherein J^2 and J_z are diagonal, it is more convenient to deal with observables that transform according to irreducible representations. As the Cartesian tensor classification does not achieve this, we shall introduce another one that leads to so-called irreducible tensor operators. This classification is also complete, in the sense that an arbitrary function of a vector operator can be uniquely decomposed into a sum of irreducible tensors. Furthermore, we need not concern ourselves with the problem of constructing irreducible from Cartesian tensors, because we shall always be led directly to the irreducible ones.

Let $\mathbf{V}$ be a vector operator with a continuous spectrum (e.g., the momentum or the position). A continuous and single-valued function $F(\mathbf{V})$ can then be decomposed into a unique linear combination of operators $Y_{lm}(\mathbf{V})$. The problem of determining the behavior of F under rotations is thereby reduced to the evaluation of $D^\dagger(R) Y_{lm}(\mathbf{V}) D(R) = Y_{lm}(D^\dagger \mathbf{V} D)$. As we have just seen, the rotated vector operator $D^\dagger(R)\mathbf{V}D(R)$ is precisely the linear combination of the three operators V_i that we obtain when we rotate an ordinary, numerical, vector through R. Let us therefore adopt the notation of Sec. 35.1 for the moment, viz. $D^\dagger(R)\mathbf{V}D(R) \equiv R\mathbf{V}$. With the help of (35.6) and (32.22) we therefore obtain

$$\begin{aligned} D^\dagger(R) Y_{lm}(\mathbf{V}) D(R) = Y_{lm}(R\mathbf{V}) &= \sum_{m'} Y_{lm'}(\mathbf{V}) D^{(l)}_{m'm}(R^{-1}) \\ &= \sum_{m'} D^{(l)}_{mm'}(R)^* Y_{lm'}(\mathbf{V}). \end{aligned} \tag{8}$$

Guided by these considerations we shall now *define* an irreducible tensor operator of rank k as a set of $(2k + 1)$ operators $T^{(k)}_\kappa$ (with $\kappa = k, k - 1, \ldots, -k$) which transform under rotations according to

$$D^\dagger(R) T^{(k)}_\kappa D(R) = \sum_{\kappa'} D^{(k)}_{\kappa\kappa'}(R)^* T^{(k)}_{\kappa'}. \tag{9}$$

It should be stressed that the results that we shall obtain in the remainder of this section depend *only* on this definition. The manner in which the operators are constructed from other observables will be seen to be quite irrelevant. In particular, the construction of the tensors from vector operators used in (8) shall not be assumed in our derivations.

Before we discuss the general case, let us briefly examine the operators of rank one. There are three operators, $T^{(1)}_{\pm 1,0}$, and they are simply related to the Cartesian components of a vector operator. This relationship must be the same as the one between a unit vector and the spherical harmonics with $l = 1$, viz.

$$T^{(1)}_{\pm 1} = \mp \frac{1}{\sqrt{2}} (V_x \pm iV_y), \qquad T^{(1)}_0 = V_z. \tag{10}$$

Commutation rules between the total angular momentum $\mathbf{J}$ and tensor operators can be obtained from (9), just as (7) followed from (5). By writing (9) for an infinitesimal rotation, we find

$$T^{(k)}_\kappa + i\,\delta\theta[\hat{\mathbf{n}} \cdot \mathbf{J}, T^{(k)}_\kappa] = \sum_{\kappa'} \langle k\kappa|1 - i\,\delta\theta\,\hat{\mathbf{n}} \cdot \mathbf{J}|k\kappa'\rangle^* T^{(k)}_{\kappa'},$$

which immediately leads to

$$[J_z, T^{(k)}_\kappa] = \kappa T^{(k)}_\kappa, \tag{11}$$

and

$$[J_\pm, T^{(k)}_\kappa] = \sqrt{k(k+1) - \kappa(\kappa \pm 1)}\; T^{(k)}_{\kappa \pm 1}. \tag{12}$$

As expected, (7) is the $k = 1$ case of these commutation rules. Equations (11) and (12) may be used to test a set of objects for their "tensorial character."

Tensor operators can be combined by vector addition to give further tensor operators. Let $U^{(k_1)}$ and $V^{(k_2)}$ be two tensor operators of the indicated rank; then

$$T^{(k)}_\kappa = \sum_{\kappa_1 \kappa_2} \langle k\kappa|k_1\kappa_1 k_2\kappa_2\rangle U^{(k_1)}_{\kappa_1} V^{(k_2)}_{\kappa_2} \tag{13}$$

is a tensor operator of rank k, and conversely

$$U^{(k_1)}_{\kappa_1} V^{(k_2)}_{\kappa_2} = \sum_{k\kappa} \langle k_1\kappa_1 k_2\kappa_2|k\kappa\rangle T^{(k)}_\kappa. \tag{14}$$

To verify that $T^{(k)}$ in (13) is indeed what it is purported to be, one carries out a rotation, uses the known transformation law (9) for U and V, and finally combines the rotation matrices with the help of (34.4). From this proof it is clear that it does not matter whether U and V commute with each other or not. Perhaps the most frequently occurring tensor product is *the scalar product of two tensors*, which we *define* as

$$U^{(k)} \cdot V^{(k)} \equiv (-1)^k \sqrt{2k+1} \sum_{\kappa_1\kappa_2} \langle 00|k\kappa_1 k\kappa_2\rangle U^{(k)}_{\kappa_1} V^{(k)}_{\kappa_2}$$

$$= \sum_{\kappa} (-1)^{\kappa} U^{(k)}_{\kappa} V^{(k)}_{-\kappa}. \tag{15}$$

The factors in (15) have been chosen so that it reduces to the familiar scalar product in the case when U and V are vectors defined as in (10):

$$U^{(1)} \cdot V^{(1)} = U_x V_x + U_y V_y + U_z V_z. \tag{16}$$

In connection with the transformation law under rotations, we observe that if k is a half integer, the observable $T^{(k)}$ would be double valued. This is clearly inadmissible, and therefore *only tensor operators of integer rank can be observables.* This restriction did not apply to kets, because it is the rays, and not the kets, that are of physical significance in the theory.

36.3 THE WIGNER-ECKART THEOREM

In atomic and nuclear problems one is frequently faced with a Hamiltonian of the type $H = H_0 + H_1$, where H_0 is spherically symmetric, and H_1 is a small asymmetric perturbation. The operator H_1 may be decomposed into a set of tensor operators $T^{(k)}$, and one must then evaluate the matrix elements of $T^{(k)}$ between the eigenstates of H_0. These may be designated by $|njm\rangle$, where j and m are the eigenvalues of the total angular momentum, and n designates all the other quantum numbers required to specify the states. As a consequence, one is interested in matrix elements of the type $\langle njm|T^{(k)}_{\kappa}|n'j'm'\rangle$. We note that m, m', and κ can only be given when the orientation of the coordinate frame is specified, whereas n, n', j, j', and k do not depend on this orientation. One might therefore suspect that the dependence of the matrix element on the three magnetic quantum numbers can be determined from symmetry considerations alone. This conjecture is correct, and is embodied in *the Wigner-Eckart theorem:* The matrix element of an arbitrary tensor operator between angular momentum eigenstates, $\langle njm|T^{(k)}_{\kappa}|n'j'm'\rangle$, has the form $C(nj, n'j'; k)\langle j'm'k\kappa|jm\rangle$, i.e., the dependence on the magnetic quantum numbers is given by a Clebsch-Gordan coefficient.

To prove the theorem we first rewrite (9) as

$$T^{(k)}_{\kappa} = \sum_{\kappa'} D^{(k)}_{\kappa\kappa'}(R)^* D(R) T^{(k)}_{\kappa'} D^{\dagger}(R),$$

and then take the matrix element of this relation in the $|njm\rangle$-representation:

$$\langle njm|T_\kappa^{(k)}|n'j'm'\rangle = \sum_{\kappa'\mu\mu'} D_{\kappa\kappa'}^{(k)}(R)^* D_{m\mu}^{(j)}(R) \langle nj\mu|T_{\kappa'}^{(k)}|n'j'\mu'\rangle D_{m'\mu'}^{(j')}(R)^*.$$

If we integrate this expression over R and use (34.37) we obtain

$$\langle njm|T_\kappa^{(k)}|n'j'm'\rangle = \frac{\langle j'm'k\kappa|jm\rangle}{2j+1} \sum_{\kappa'\mu\mu'} \langle nj\mu|T_{\kappa'}^{(k)}|n'j'\mu'\rangle\langle j'\mu'k\kappa'|j\mu\rangle. \quad (17)$$

This shows quite explicitly that the CG-coefficient $\langle j'm'k\kappa|jm\rangle$ carries the entire dependence on m, m', and κ. According to a time-honored tradition one writes (17) as

$$\langle njm|T_\kappa^{(k)}|n'j'm'\rangle = \frac{\langle nj\|T^{(k)}\|n'j'\rangle}{\sqrt{2j+1}} \langle j'm'k\kappa|jm\rangle, \quad (18)$$

where $\langle nj\|T^{(k)}\|n'j'\rangle$ is called the reduced matrix element. This reduced element can be evaluated by computing the left-hand side of (18) for some one choice of m, m', and κ, and then dividing by the appropriate CG-coefficient. The Wigner-Eckart theorem shows that the detailed nature of the states and the operator $T^{(k)}$ enters into the matrix element only through the factor $\langle nj\|T^{(k)}\|n'j'\rangle$.

An important example is afforded by the tensor operator $Y_{k\kappa}(\mathbf{x})$, where $\mathbf{x}$ is the position operator of a single particle. If $|lm\rangle$ is a one-particle orbital angular momentum eigenket, then

$$\langle lm|Y_{k\kappa}(\mathbf{x})|l'm'\rangle = \int d\Omega\, Y_{lm}^*(\Omega) Y_{k\kappa}(\Omega) Y_{l'm'}(\Omega).$$

This integral was evaluated in (34.39), and the reduced matrix element of the spherical harmonic operator is therefore

$$\langle l\|Y_k\|l'\rangle = \sqrt{\frac{(2l'+1)(2k+1)}{4\pi}} \langle k0l'0|l0\rangle. \quad (19)$$

The Wigner-Eckart theorem provides us with two *selection rules* of very great importance. The occurrence of the CG-coefficient in (18) insures that the matrix element *vanishes* unless the angular momentum quantum numbers satisfy

$$m - m' = \kappa, \quad (20)$$

and

$$|j - j'| \leqslant k \leqslant j + j'. \quad (21)$$

These selection rules play an essential role in atomic and nuclear spectroscopy.

In the preceding sub-section we had noted that there cannot be any observables that are tensor operators of half-integer rank. When combined with the Wigner-Eckart theorem, this tells us that there are no observables that connect the integer and half-integer angular momentum subspaces of the Hilbert space. This is an example of a *superselection rule.** A superselection rule occurs when observables which should connect different parts of the Hilbert space fail to exist. They differ from selection rules like (20) and (21) related to constants of the motion such as the total angular or linear momentum, because there certainly exist many observables which, for example, connect the subspaces belonging to different eigenvalues of the total momentum.

Aside from providing us with the selection rules, the Wigner-Eckart theorem serves as a very efficient calculational aid. In the first instance, it greatly reduces the labor in the evaluation of matrix elements, because only one unknown number appears in (18) for fixed values of j, j', and k. In radiation theory one also needs sums over the magnetic quantum numbers of squares of matrix elements. Such sums can be carried out with ease if one recalls the orthogonality relations (25.13) for the CG-coefficients.

36.4 RACAH COEFFICIENTS †

•More complicated matrix elements than (18) can also be simplified by using the Wigner-Eckart theorem. Consider a system having two independent angular momenta $\mathbf{J}_1$ and $\mathbf{J}_2$, and let $U^{(k)}$, $V^{(k)}$, be tensor operators with respect to $\mathbf{J}_1$ and $\mathbf{J}_2$, respectively, i.e., $[U^{(k)}, \mathbf{J}_2] = [V^{(k)}, \mathbf{J}_1] = 0$. We shall now evaluate the matrix element of the scalar product $U^{(k)} \cdot V^{(k)}$ in the representation $|j_1 j_2 jm\rangle$, where j is the eigenvalue of $(\mathbf{J}_1 + \mathbf{J}_2)^2$. Because $U^{(k)} \cdot V^{(k)}$ is a scalar under rotations of the entire system (though not its constituents), we know from (18) that the matrix is diagonal in j and m, and independent of m; this is also intuitively obvious. Upon decomposing the states with CG-coefficients, we obtain

$$\langle j_1 j_2 jm | U^{(k)} \cdot V^{(k)} | j_1' j_2' jm\rangle = \sum_\kappa (-1)^\kappa \sum_{\substack{m_1 m_2 \\ m_1' m_2'}} \langle jm | j_1 m_1 j_2 m_2\rangle \times \langle j_1 m_1 | U_\kappa^{(k)} | j_1' m_1'\rangle \langle j_2 m_2 | V_{-\kappa}^{(k)} | j_2' m_2'\rangle \langle j_1' m_1' j_2' m_2' | jm\rangle.$$

After applying the Wigner-Eckart theorem, this becomes

$$\langle j_1 j_2 jm | U^{(k)} \cdot V^{(k)} | j_1' j_2' jm\rangle = \frac{\langle j_1 \| U^{(k)} \| j_1'\rangle \langle j_2 \| V^{(k)} \| j_2'\rangle}{\sqrt{(2j_1 + 1)(2j_2 + 1)}} \times \sum (-1)^\kappa \langle jm | j_1 m_1 j_2 m_2\rangle \langle j_1' m_1' k\kappa | j_1 m_1\rangle \langle j_2' m_2' k\, -\kappa | j_2 m_2\rangle \langle j_1' m_1' j_2 m_2' | jm\rangle.$$

* For a detailed discussion see G. C. Wick, A. S. Wightman, and E. P. Wigner, *Phys. Rev.* **88**, 101 (1952); Sakurai, p. 187.

† The remainder of this section is of a rather specialized nature. Except for a few remarks in Chapt. XI, it will not be referred to again in this book.

Here all the magnetic quantum numbers except m are summed over, and we know that the result is independent of m. We may therefore write

$$\begin{aligned}\langle j_1 j_2 jm | U^{(k)} \cdot V^{(k)} | j_1' j_2' jm \rangle \\ = (-1)^{j_1'+j_2+j} \begin{Bmatrix} j_1 & j_2 & j \\ j_2' & j_1' & k \end{Bmatrix} \langle j_1 \| U^{(k)} \| j_1' \rangle \langle j_2 \| V^{(k)} \| j_2' \rangle,\end{aligned} \tag{22}$$

where

$$\begin{aligned}\begin{Bmatrix} j_1 & j_2 & j_3 \\ j_1' & j_2' & j_3' \end{Bmatrix} = \sum_{m_i, m_i'} (-1)^S \begin{pmatrix} j_1 & j_2 & j_3 \\ m_1 & m_2 & m_3 \end{pmatrix} \begin{pmatrix} j_1 & j_2' & j_3' \\ -m_1 & m_2' & -m_3' \end{pmatrix} \\ \times \begin{pmatrix} j_1' & j_2 & j_3' \\ -m_1' & -m_2 & m_3' \end{pmatrix} \begin{pmatrix} j_1' & j_2' & j_3 \\ m_1' & -m_2' & -m_3 \end{pmatrix},\end{aligned} \tag{23}$$

is called a 6-j symbol,* and $S = \Sigma_{i=1}^{3} (j_i + j_i' + m_i')$. (Here we have switched from CG- to 3-j symbols; cf. Sec. 25.) The 6-j symbol was originally studied in detail by Racah, and in the literature it often bears his name. Racah's original W-coefficient was defined somewhat differently from (23):

$$W(j_1 j_2 j_2' j_1'; j_3 j_3') = (-1)^{j_1+j_2+j_1'+j_2'} \begin{Bmatrix} j_1 & j_2 & j_3 \\ j_1' & j_2' & j_3' \end{Bmatrix}.$$

From the derivation, and also from the definition, it is clear that the 6-j symbol vanishes unless all of the triads $(j_1 j_2 j_3)$, $(j_1 j_2' j_3')$, $(j_1' j_2 j_3')$, $(j_1' j_2' j_3)$ simultaneously satisfy triangular inequalities.

Another matrix element that can be reduced with Racah coefficients is that of an operator $T^{(k)}$ that only acts on part of a system. Assume that $T^{(k)}$ is a tensor operator with respect to $\mathbf{J}_1$, and that the system in question possesses some other angular momentum $\mathbf{J}_2$ such that the total angular momentum is $\mathbf{J} = \mathbf{J}_1 + \mathbf{J}_2$. Then by the very same technique one finds

$$\begin{aligned}\langle j_1 j_2 j \| T^{(k)} \| j_1' j_2 j' \rangle = (-1)^{j_1+j_2+j'+k} \sqrt{(2j+1)(2j'+1)} \\ \times \begin{Bmatrix} j_1 & j_1' & k \\ j' & j & j_2 \end{Bmatrix} \langle j_1 \| T^{(k)} \| j_1' \rangle.\end{aligned} \tag{24}$$

Note that in this formula, as well as in (22), the dependence on the quantum number j and j' is completely contained in the 6-j symbol.

The 6-j symbols also play a role in the addition of three angular momenta $\mathbf{J}_1 + \mathbf{J}_2 + \mathbf{J}_3 = \mathbf{J}$. This addition can be carried out in several distinct ways. Thus we could first add $\mathbf{J}_1$ and $\mathbf{J}_2$ to give $\mathbf{J}_{12}$, and then add $\mathbf{J}_3$:

$$\begin{aligned}|j_1 j_2 (j_{12}) j_3 jm\rangle &= \sum |j_1 j_2 j_{12} m_{12} j_3 m_3\rangle \langle j_{12} m_{12} j_3 m_3 | jm \rangle \\ &= \sum |j_1 m_1 j_2 m_2 j_3 m_3\rangle \langle j_1 m_1 j_2 m_2 | j_{12} m_{12} \rangle \langle j_{12} m_{12} j_3 m_3 | jm \rangle.\end{aligned}$$

* Numerical tables of the 6-j symbols are also provided in M. Rotenberg *et al.*, *loc. cit.* p. 220.

On the other hand we could first form $\mathbf{J}_{13}$, and then add $\mathbf{J}_2$, which gives

$$|j_1j_3(j_{13})j_2jm\rangle = \sum |j_1m_1j_2m_2j_3m_3\rangle\langle j_1m_1j_3m_3|j_{13}m_{13}\rangle\langle j_{13}m_{13}j_2m_2|jm\rangle.$$

These states are not identical. Their scalar product is independent of m, and because it involves a sum over the magnetic quantum numbers of a product of four CG-coefficients, it is hardly surprising that it is related to a 6-j symbol. After some tedious algebra one in fact obtains

$$\begin{aligned}&\langle j_1j_3(j_{13})j_2j|j_1j_2(j_{12})j_3j\rangle \\ &\quad = (-1)^{j_2+j_3+j_{12}+j_{13}}\sqrt{(2j_{12}+1)(2j_{13}+1)}\begin{Bmatrix} j_1 & j_2 & j_{12} \\ j & j_3 & j_{13}\end{Bmatrix}.\end{aligned} \tag{25}$$

Transformations of this type between different angular momentum coupling schemes occur often in spectroscopy, and we shall actually encounter a somewhat more complicated example in Chapt. XI.

Because the 6-j symbols are now related to a transformation function, one can easily derive orthogonality relations. Thus one has

$$\sum_j (2j+1)\begin{Bmatrix} j_1 & j_2 & j \\ j_3 & j_4 & j'\end{Bmatrix}\begin{Bmatrix} j_1 & j_2 & j \\ j_3 & j_4 & j''\end{Bmatrix} = \frac{\delta_{j'j''}}{2j'+1}. \tag{26}$$

There are many other identities that 6-j symbols satisfy, but we shall refrain from deriving these here as they do not involve any new point of principle, and are only of interest to specialists.*

37. *Reflections in Space: Parity*

Consider two observers with coordinate systems (x, y, z) and (ξ, η, ζ), where $\xi = -x$, $\eta = -y$, $\zeta = -z$. The second frame is obtained from the first by an inversion through the origin. The matrix which performs the inversion of a vector,

$$\begin{pmatrix} -1 & 0 & 0 \\ 0 & -1 & 0 \\ 0 & 0 & -1\end{pmatrix},$$

has determinant -1, whereas the rotations discussed earlier are generated by matrices with determinant 1. This is due to the fact that the rotations can be obtained by a continuous transformation from unity, whereas the inversion cannot. (Note that inversion through a plane,

* See de Shalit and Talmi, Secs. 14 and 15, for an exhaustive treatment; also Edmonds, Sec. 6.2; Schwinger, *AM;* T. Regge, *Nuovo Cimento* 11, 116 (1959); Fano and Racah.

e.g.; $x \to -x$, $y \to y$, $z \to z$, is also a discrete transformation, whereas $x \to -x$, $y \to -y$, $z \to z$ is not, as is of course clear since it is a rotation of π about the z-axis.) Inversion through a plane is actually more visualizable than the complete inversion $\mathbf{r} \to -\mathbf{r}$; in fact, by observing a physical phenomenon as seen in a mirror in the y-z plane one is doing the same thing as using a left-handed coordinate system ($x \to -x$, $y \to y$, $z \to z$).

Let $\mathcal{R}$ be the operator in the Hilbert space that connects an arbitrary ket $|a'\rangle$ to its reflected counterpart, $|a'; \text{ref}\rangle$: $\mathcal{R}|a'\rangle = |a'; \text{ref}\rangle$. Since $\mathcal{R}^2$ must (to within a phase) equal one, we are now dealing with a group of two elements, and so we must first decide whether we are to use a unitary or anti-unitary operator $\mathcal{R}$.

This question is answered once we set up the mappings in the Hilbert space that $\mathcal{R}$ is supposed to perform. Let us consider a single spinless particle, and the position and momentum eigenkets $|\mathbf{x}'\rangle$, $|\mathbf{p}'\rangle$. In order to be consistent with our definitions of the observables $\mathbf{x}$ and $\mathbf{p}$ we must have

$$\mathcal{R}|\mathbf{x}'\rangle = |-\mathbf{x}'\rangle, \qquad \mathcal{R}|\mathbf{p}'\rangle = |-\mathbf{p}'\rangle. \tag{1}$$

In writing (1) we have adopted a convenient phase convention; as we know from Sec. 27, we are always at liberty to do this. Let us now apply (1) to $(\mathbf{x} - \mathbf{x}')|\mathbf{x}'\rangle = 0$: $(\mathcal{R}\,\mathbf{x}\,\mathcal{R}^{-1})\mathcal{R}|\mathbf{x}'\rangle = \mathbf{x}'|-\mathbf{x}'\rangle = -\mathbf{x}|-\mathbf{x}'\rangle$. Hence

$$\mathcal{R}\,\mathbf{x}\,\mathcal{R}^{-1} = -\mathbf{x}, \tag{2}$$

and the identical argument shows that

$$\mathcal{R}\,\mathbf{p}\,\mathcal{R}^{-1} = -\mathbf{p}. \tag{3}$$

Vectors that change sign under reflection are called *polar vectors*. From (2) and (3), on the other hand, the transformation law for (orbital) angular momentum is

$$\mathcal{R}(\mathbf{x} \times \mathbf{p})\mathcal{R}^{-1} = \mathbf{x} \times \mathbf{p}; \tag{4}$$

a vector that is invariant under reflection is called an *axial or pseudo-vector*.

To answer the question of whether $\mathcal{R}$ is unitary or not, we must bring in a further physical requirement. For example, we may use the geometric observation that a spatial translation through $\mathbf{a}$ followed by a reflection is the same as a reflection followed by translation through $-\mathbf{a}$:

$$\mathcal{R}e^{i\mathbf{a}\cdot\mathbf{P}} = e^{-i\mathbf{a}\cdot\mathbf{P}}\mathcal{R}, \tag{5}$$

where $\mathbf{P}$ is the total momentum operator, which of course satisfies

$$\mathcal{R}\,\mathbf{P}\,\mathcal{R}^{-1} = -\mathbf{P}. \tag{6}$$

We note that (5) and (6) are consistent if $\mathcal{R}$ is unitary, whereas if $\mathcal{R}$ were anti-unitary, we would have

$$\mathcal{R}e^{i\mathbf{a}\cdot\mathbf{P}}\mathcal{R}^{-1} = e^{-i\mathbf{a}\cdot(\mathcal{R}\mathbf{P}\mathcal{R}^{-1})},$$

which is inconsistent. Hence $\mathcal{R}$ is a unitary operator.

Let us choose the phases such that $\mathcal{R}^2 = 1$. The eigenvalues $\mathcal{R}'$ of $\mathcal{R}$ are therefore $\mathcal{R}' = \pm 1$. These eigenvalues are called the *parity*. If the Hamiltonian is reflection invariant, $\mathcal{R}H\mathcal{R}^{-1} = H$, or

$$[\mathcal{R}, H] = 0.$$

$\mathcal{R}$ is then a constant of the motion, and the parity $\mathcal{R}'$ may be chosen as a member of the complete set of quantum numbers.

The behavior of the one-particle orbital angular momentum eigenstates under reflection is of considerable importance in many applications. In view of (1) we have $\langle \mathbf{x}'|\mathcal{R}|lm\rangle = \langle -\mathbf{x}'|lm\rangle = Y_{lm}(-\hat{\mathbf{x}}')$. But $Y_{lm}(\pi - \theta, \phi + \pi) = (-1)^l Y_{lm}(\theta, \phi)$, and therefore

$$\langle \mathbf{x}'|lm; \text{ref}\rangle = (-1)^l\langle \mathbf{x}'|lm\rangle. \tag{7}$$

The parity * of a one-particle orbital angular momentum eigenstate is therefore $(-1)^l$.

We may now draw some important conclusions concerning the matrix elements of operators between states of definite parity. Let $\mathcal{E}$ be an even, $\mathcal{O}$ an odd observable under reflection, i.e., $\mathcal{R}\mathcal{E}\mathcal{R}^{-1} = \mathcal{E}$, $\mathcal{R}\mathcal{O}\mathcal{R}^{-1} = -\mathcal{O}$. Then $\langle \mathcal{R}'|\mathcal{E}|\mathcal{R}''\rangle = \langle \mathcal{R}'|\mathcal{R}^{-1}(\mathcal{R}\mathcal{E}\mathcal{R}^{-1})\mathcal{R}|\mathcal{R}''\rangle = \mathcal{R}'\mathcal{R}''\langle \mathcal{R}'|\mathcal{E}|\mathcal{R}''\rangle$, and similarly $\langle \mathcal{R}'|\mathcal{O}|\mathcal{R}''\rangle = -\mathcal{R}'\mathcal{R}''\langle \mathcal{R}'|\mathcal{O}|\mathcal{R}''\rangle$. *Hence an observable that is even under reflections only has nonzero matrix elements between states of the same parity, while an odd observable only has nonvanishing matrix elements between states*

* In this discussion we have passed over the question of *intrinsic parity*. The existence of this notion can be recognized in the two-nucleon system. As we shall see in Sec. 42, the deuteron is a ^3S-state, and therefore has the parity $+1$ in its center-of-mass system, if we *define* the parity of the neutron and proton as $+1$ in their respective center-of-mass frames. One could, however, imagine that the deuteron is in the state $^3P_1$, in which case the c.o.m. (or intrinsic) parity would be -1. If the deuteron is treated as an elementary particle and placed into a state of orbital angular momentum l, the parity of this state would be $\pm(-1)^l$, depending on the internal structure of the deuteron. These two possibilities also exist for spinless particles, as one realizes by constructing the n-p states ^1S and $^3P_0$. The formulas should therefore contain a further factor ± 1 to account for the intrinsic parity of the particle in question. The intrinsic parity of a nuclear or atomic system is known once the coupling scheme is known, because the intrinsic parity of the electron and nucleon can be set equal to $+1$ by definition (just as the charge on the electron is defined to be negative). The intrinsic parity of certain other "elementary particles," such as the pion, are no longer arbitrary once an intrinsic parity has been assigned to the nucleons, and these must then be determined by experiment. For further details see Dalitz, Chapt. 12, Källén, Sec. 3-5, and Sakurai, Chapt. 3.

of opposite parity. We shall encounter many important applications of this result in the remainder of this book.

The detailed properties of the reflection operator depend on the system to which it pertains. We shall illustrate this by constructing $\mathcal{R}$ in the spin space of a single spin $\frac{1}{2}$ particle. We assume that the spin, like an ordinary angular momentum, is an axial vector, or *

$$\mathcal{R}\,\boldsymbol{\sigma}\,\mathcal{R}^{-1} = \boldsymbol{\sigma}. \tag{8}$$

The only 2×2 matrix that commutes with all three Pauli matrices is a multiple of the unit matrix. Hence we may choose the phases so that $\mathcal{R} = 1$. A reflection in a plane, instead of the reflection through the origin, is a more interesting operation. The reflection in the x-y plane produces the following transformation of the orbital angular momentum:

$$\begin{aligned} \mathcal{R}_{xy}L_x\mathcal{R}_{xy}^{-1} &= -L_x, \\ \mathcal{R}_{xy}L_y\mathcal{R}_{xy}^{-1} &= -L_y, \\ \mathcal{R}_{xy}L_z\mathcal{R}_{xy}^{-1} &= L_z, \end{aligned} \tag{9}$$

where $\mathcal{R}_{xy}$ is the unitary operator associated with this reflection. These relations must hold for any angular momentum operator. In the spin space we therefore require an operator that anti-commutes with σ_x and σ_y, and commutes with σ_z. These requirements are fulfilled by

$$\mathcal{R}_{xy} = \xi_z\sigma_z, \tag{10}$$

where ξ_z is a number of modulus 1. Let us briefly verify that (10) is consistent with some other facts at our disposal. The reflection through the origin, $\mathcal{R}$, is related to the reflections in a plane by $\mathcal{R} = \mathcal{R}_{xy}\mathcal{R}_{yz}\mathcal{R}_{zx}$; from (10) and (33.9) we obtain $\mathcal{R} = i\xi_x\xi_y\xi_z$. The sequence of reflections $\mathcal{R}_{yz}\mathcal{R}_{zx}$ is equivalent to a rotation through π about the z-axis; this rotation is effected by the operator $\exp(-i\pi\sigma_z/2) = -i\sigma_z$, whereas $\mathcal{R}_{yz}\mathcal{R}_{zx} = i\xi_x\xi_y\sigma_z$. A consistent set of definitions for these operators can therefore be attained by setting $\xi_x = \xi_y = \xi_z = i$.

38. Static Electromagnetic Moments and Selection Rules

The interaction between the electromagnetic moments of atoms and nuclei determines the hyperfine structure of the spectrum, and because of this hyperfine spectroscopy is a powerful technique for studying the electromagnetic properties of nuclei. In this section we shall give the

* If we did not do this, we would find that the total angular momentum $\mathbf{L} + \frac{1}{2}\boldsymbol{\sigma}$ becomes $\mathbf{L} - \frac{1}{2}\boldsymbol{\sigma}$ after reflection, which would lead to nonsense.

quantum mechanical definition of the multipole moments, and study those of their properties that follow from symmetry considerations. This will illustrate both the power and limitations of symmetry arguments.

38.1 ELECTRIC MULTIPOLES

In classical physics a charge distribution $\rho_{cl}(\mathbf{r})$ gives rise to the electrostatic potential

$$\phi(\mathbf{r}) = \int \frac{\rho_{cl}(\mathbf{r}')\, d^3r'}{|\mathbf{r} - \mathbf{r}'|}. \tag{1}$$

If $\mathbf{r}$ is outside the distribution we can expand $|\mathbf{r} - \mathbf{r}'|^{-1}$ in spherical harmonics to obtain

$$\phi(\mathbf{r}) = 4\pi \sum_{lm} (2l + 1)^{-1} Y^*_{lm}(\hat{\mathbf{r}}) Q^{(cl)}_{lm} |\mathbf{r}|^{-l-1}, \tag{2}$$

where

$$Q^{(cl)}_{lm} = \int d^3r\, |\mathbf{r}|^l Y_{lm}(\hat{\mathbf{r}}) \rho_{cl}(\mathbf{r}) \tag{3}$$

is the classical 2^l-pole.

Consider now a quantum mechanical system of N particles with charges e_i ($i = 1, 2, \ldots, N$). The charge-density operator, $\rho(\mathbf{r})$, is then given by (5.22), viz.

$$\rho(\mathbf{r}) = \sum_{i=1}^{N} e_i\, \delta(\mathbf{r} - \mathbf{x}_i). \tag{4}$$

Here $\mathbf{x}_i$ is the position operator for the ith particle, but $\mathbf{r}$ is merely a numerical vector. To obtain the charge density in a state, we must take the expectation value of (4). In this section we shall only study diagonal matrix elements of $\rho(\mathbf{r})$, but in radiation problems off-diagonal elements play an essential role.

The electrostatic potential arising from the state $|\psi\rangle$ is simply

$$\begin{aligned} \phi_\psi(\mathbf{r}) &= \int d^3r' \frac{\langle\psi|\rho(\mathbf{r}')|\psi\rangle}{|\mathbf{r} - \mathbf{r}'|} \\ &= \sum_{lm} \frac{4\pi}{(2l + 1)|\mathbf{r}|^{l+1}} Y^*_{lm}(\hat{\mathbf{r}}) \langle\psi|Q_{lm}|\psi\rangle, \end{aligned} \tag{5}$$

where Q_{lm} is the electric multipole operator

$$Q_{lm} = \int d^3r\, r^l Y_{lm}(\hat{\mathbf{r}}) \rho(\mathbf{r}). \tag{6}$$

Consider the reflected operator $\mathcal{R}Q_{lm}\mathcal{R}^{-1}$; the only thing that is effected by the unitary transformation is the operator ρ. According to (37.2) and (4)

$$\begin{aligned}\mathcal{R}\rho(\mathbf{r})\mathcal{R}^{-1} &= \sum_i e_i\,\delta(\mathbf{r} - \mathcal{R}\mathbf{x}_i\mathcal{R}^{-1}) = \sum_i e_i\,\delta(\mathbf{r} + \mathbf{x}_i)\\ &= \rho(-\mathbf{r}).\end{aligned} \tag{7}$$

When we put this into $\mathcal{R}Q_{lm}\mathcal{R}^{-1}$ we can reflect the integration variable to obtain

$$\mathcal{R}Q_{lm}\mathcal{R}^{-1} = \int d^3r\, r^l Y_{lm}(-\hat{\mathbf{r}})\rho(\mathbf{r}). \tag{8}$$

But $Y_{lm}(-\hat{\mathbf{r}}) = (-1)^l Y_{lm}(\hat{\mathbf{r}})$, and therefore

$$\mathcal{R}Q_{lm}\mathcal{R}^{-1} = (-1)^l Q_{lm}. \tag{9}$$

In Sec. 37 we proved that an observable that is odd under reflections only has nonvanishing matrix elements between states of opposite parity. Hence *the expectation values of the odd-order electric moments vanish in any state of definite parity.* A particularly noteworthy special case of this theorem is that the electric dipole moment is identically zero in any state having a definite parity. If a system possesses a reflection-invariant Hamiltonian, it can only have a non-zero electric dipole moment if two states of opposite parity are accidentally degenerate.

The proof that Q_{lm} is a tensor operator of rank l is also instructive. When we perform the rotation R on (6) we obtain

$$D^\dagger(R)Q_{lm}D(R) = \int d^3r\, r^l Y_{lm}(\hat{\mathbf{r}}) \sum_i e_i\,\delta(\mathbf{r} - R\mathbf{x}_i).$$

Let us introduce the new integration variable $\mathbf{r}' = R^{-1}\mathbf{r}$. The Jacobian of this transformation is one because the matrix that relates $\mathbf{r}$ and $\mathbf{r}'$ is orthogonal. Furthermore, $\delta(R\mathbf{r}' - R\mathbf{x}_i) = \delta(\mathbf{r}' - \mathbf{x}_i)$, and therefore

$$\begin{aligned}D^\dagger(R)Q_{lm}D(R) &= \int d^3r'\, r'^l Y_{lm}(R\hat{\mathbf{r}}')\rho(\mathbf{r}')\\ &= \sum_{m'} D^{(l)}_{mm'}(R)^* \int d^3r\, r^l Y_{lm'}(\hat{\mathbf{r}})\rho(\mathbf{r}),\end{aligned}$$

where we have used (35.6). A comparison with (36.9) shows that Q_{lm} is a tensor operator of the indicated rank.*

* The vanishing of the integral (34.39) when $l_1 + l_2 + l_3$ is odd is now understandable as a parity selection rule. Note that it is not the CG-coefficient in the Wigner-Eckart theorem, but the reduced matrix element (36.19) that embodies the parity selection rule.

The Wigner-Eckart theorem may therefore be applied to the evaluation of the matrix elements of Q_{lm} between states of definite angular momentum, $|njm\rangle$. We confine our remarks once more to expectation values: $\langle Q_{l\lambda}\rangle_{njm}$. Bearing in mind the selection rules (36.21) we can make the following statements: (a) one number, the reduced matrix element $\langle nj\|Q_l\|nj\rangle$, serves to determine all the lth-order multipole moments * in a multiplet of angular momentum j; (b) the largest order electric multipole moment possible in a state of angular momentum j is $l = 2j$.

38.2 MAGNETIC MULTIPOLES

In classical electromagnetic theory (cf. e.g., Jackson, p. 156) the magnetic field $\mathcal{H}(\mathbf{r})$ arising from a current $\mathbf{j}(\mathbf{r})$ is given by

$$\mathcal{H}(\mathbf{r}) = -\nabla \int \frac{\bar{\rho}_{cl}(\mathbf{r}')\, d^3r'}{|\mathbf{r} - \mathbf{r}'|}, \tag{10}$$

where $\bar{\rho}_{cl}$ is the so-called magnetic charge density. This quantity is defined by

$$\bar{\rho}_{cl}(\mathbf{r}) = -\nabla \cdot \mathfrak{M}(\mathbf{r}),$$

where $\mathbf{j}(\mathbf{r}) = c\nabla \times \mathfrak{M}(\mathbf{r})$, and $\mathfrak{M}(\mathbf{r})$ is the magnetization density. It should be noted that (10) is only valid outside the source, i.e., in those regions of space where $\mathbf{j}(\mathbf{r}) = 0$. Because $\mathbf{j}$ is a polar vector, $\mathfrak{M}$ is an axial vector, and $\bar{\rho}_{cl}$ is therefore a pseudoscalar. The magnetization density operator, $\bar{\rho}(\mathbf{r})$, therefore has the reflection property

$$\mathcal{R}\bar{\rho}(\mathbf{r})\mathcal{R}^{-1} = -\bar{\rho}(-\mathbf{r}). \tag{11}$$

The multipole expansion proceeds from (10) as in the electrostatic case. One is led thereby to the introduction of the magnetostatic multipole operators

$$M_{lm} = \int d^3r\, r^l Y_{lm}(\hat{\mathbf{r}})\bar{\rho}(\mathbf{r}). \tag{12}$$

M_0 can be converted into a surface integral which vanishes when $\mathfrak{M}$ is spatially confined, i.e., there is no magnetic monopole moment. It is clear that M_{lm} is a tensor operator of rank l with the reflection property

$$\mathcal{R}M_{lm}\mathcal{R}^{-1} = -(-1)^l M_{lm}. \tag{13}$$

The angular momentum selection rules for the diagonal matrix elements of M_{lm} are therefore the same as for Q_{lm}. On the other hand, *a state of definite parity only possesses odd magnetic multipole moments.*

* By convention the quadrupole moment Q of a degenerate multiplet is equal to $(16\pi/5)^{\frac{1}{2}}\langle njj|Q_{20}|njj\rangle$. See Blatt and Weisskopf, p. 26 *et seq.*

On the basis of all our results we can make a table that shows which electromagnetic multipole moments have nonvanishing expectation values in an angular momentum eigenstate of definite parity. The numeral following "E" and "M" indicates the order l of the multipole.

j	0	$\frac{1}{2}$	1	$\frac{3}{2}$	2
E0	yes	yes	yes	yes	yes
M1	no	yes	yes	yes	yes
E2	no	no	yes	yes	yes
M3	no	no	no	yes	yes

This table is a striking example of how symmetry considerations can reduce the complexity of a problem. For example, it says that if one believes in rotational invariance, there is no point in looking for the electric quadrupole moment of the proton.

The magnetic dipole moment,* which we write as $\boldsymbol{\mu}$ instead of M_{1m}, plays an important role in many applications, and we therefore give a more explicit expression for it:

$$\boldsymbol{\mu} = \frac{1}{2c}\int d^3r\, \mathbf{r} \times \mathbf{j}(\mathbf{r}). \tag{14}$$

For a single spin 0 particle,

$$\mathbf{j}(\mathbf{r}) = \frac{e}{2m}[\delta(\mathbf{r} - \mathbf{x})\mathbf{p} + \text{h.c.}],$$

where h.c. stands for Hermitian conjugate. Hence

$$\boldsymbol{\mu} = \frac{e\hbar}{2mc}\mathbf{L}, \tag{15}$$

where $\mathbf{L}$ is the orbital angular momentum operator (in units of $\hbar$).

For a particle with spin, there is a further contribution to (15) due to the "intrinsic" motion, which gives a magnetic moment even when the particle is at rest. One then writes

$$\boldsymbol{\mu} = \frac{e\hbar}{2mc}(\mathbf{L} + g\mathbf{s}), \tag{16}$$

* By convention the magnetic dipole moment μ of a degenerate multiplet is equal to $\langle njj|\mu_z|njj\rangle = (4\pi/3)^{\frac{1}{2}}\langle njj|M_{10}|njj\rangle$. See Blatt and Weisskopf, p. 35.

where g is the *gyromagnetic ratio.* For the electron g is given to very high accuracy by

$$g_e = 2\left(1 + \frac{1}{2\pi}\frac{e^2}{\hbar c}\right); \tag{17}$$

the factor 2 is predicted by Dirac's theory of the electron, and the small term $[2\pi \cdot 137]^{-1}$ is the Schwinger correction as computed from quantum electrodynamics. This theoretical formula is in excellent agreement with the data, and in fact even smaller correction terms have been verified experimentally. For the proton the g-factor is measured to be $g_p = 5.59$. The deviation from 2 appears to be due to the charged pion cloud surrounding the proton, but no detailed theory of this effect exists at present. The magnetic moment of the neutron is not zero for similar reasons; its value is

$$\mu_n = -1.92\frac{e\hbar}{2Mc},$$

where M is the nucleon mass and e the charge on the proton.

39. Time Reversal

39.1 DEFINITION OF TIME REVERSAL AND THE TIME REVERSAL OPERATOR

There are no mystical notions involved in time reversal. One need not be equipped with clocks that suddenly run backwards to comprehend it. Consider a classical particle subjected to some (unknown) static forces. Let $\mathbf{r}(t_0)$, $\mathbf{v}(t_0)$ be the position and velocity of the particle at $t = t_0$, and allow it to proceed undisturbed for a time t_1, when its position and velocity have become $\mathbf{r}(t_0 + t_1)$, $\mathbf{v}(t_0 + t_1)$. Now (i.e., at $t = t_0 + t_1$) start another identical particle off at $\mathbf{r}(t_0 + t_1) \equiv \mathbf{r}'$ with velocity $-\mathbf{v}(t_0 + t_1) \equiv \mathbf{v}'$. Then at a later time $t_0 + 2t_1$, we will either find that the new position and velocity equal $\mathbf{r}(t_0)$, $-\mathbf{v}(t_0)$, or we will not. In the first case we say that the basic laws (equation of motion) are invariant under time reversal; this is clearly the case for motion in a force field derivable from a potential. If, on the other hand, we consider the motion of a charged particle in a prescribed magnetic field, there is no symmetry of this kind (see Fig. 39.1). (The symmetry is restored if we incorporate the charged particles whose motion produces the field into the system; however, this is not what one wants if one studies motion in a given "external" field.)

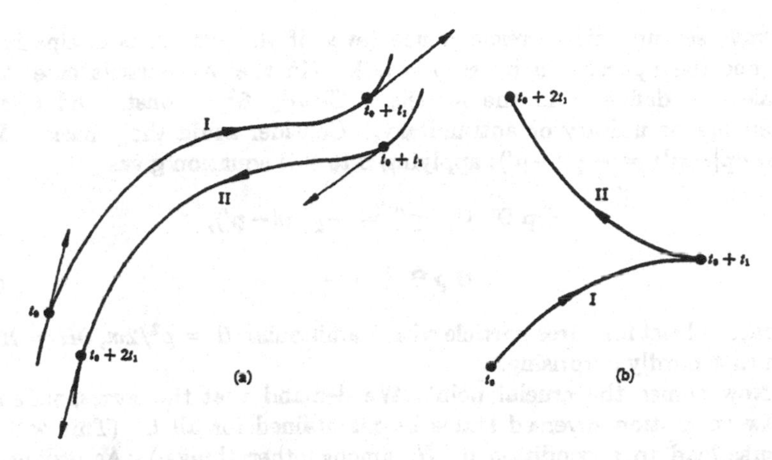

Fig. 39.1. Time reversed orbits. (a) Time reversed orbits in the case where the equations of motion are invariant under time reversal. The two trajectories really coincide; they have been separated only for clarity. The tangential arrows indicate the velocities. (b) Time reversed orbits for a charged particle moving in a magnetic field. The field direction is normal to the plane of the page. It may be noted that orbit II coincides with I if we also change the sign of the charge. The equations of nonrelativistic quantum mechanics do not automatically provide us with solutions of opposite charge, however.

To summarize, the classical equations of motion are said to be invariant under time reversal if one can separate the manifold of all motions into two subsets between which there is a one-to-one correspondence that is effected by changing the signs of all the momenta at some instant. Let $q_{\text{I}}(t)$ and $p_{\text{I}}(t)$ be the canonical variables of the motions belonging to subset I, and $q_{\text{II}}(t)$ and $p_{\text{II}}(t)$ the variables of subset II. The one-to-one correspondence just referred to is then

$$\begin{aligned} q_{\text{I}}(t_0 + t) &= q_{\text{II}}(t_0 - t), \\ p_{\text{I}}(t_0 + t) &= -p_{\text{II}}(t_0 - t). \end{aligned} \tag{1}$$

In the case of motion in a magnetic field, the correspondence (1) does not exist. Instead one has

$$\begin{aligned} q_{\text{I}}(\mathcal{H}, t_0 + t) &= q_{\text{II}}(-\mathcal{H}, t_0 - t), \\ p_{\text{I}}(\mathcal{H}, t_0 + t) &= -p_{\text{II}}(-\mathcal{H}, t_0 - t). \end{aligned} \tag{2}$$

Let $|\psi\rangle$ be a physical state, and $|\psi'\rangle$ the motion-reversed state, both at $t = 0$, say. For example if $|\psi\rangle$ is the momentum eigenstate of a single particle $|\mathbf{p}'\rangle$, then $|\psi'\rangle$ is $|-\mathbf{p}'\rangle$. Here it is important to notice that we have not yet assumed any invariance because at any instant we can always set up this correspondence (even if the system is dissipative). Define the operator Θ by $\Theta|\psi\rangle = |\psi'\rangle$. (In the one-particle case it is easiest to define it in the p-basis.) Clearly $\Theta^2 = \text{const.}$, and Θ can therefore be unitary or anti-unitary. Consider again the p-basis. We have $\mathbf{p}|-\mathbf{p}'\rangle = -\mathbf{p}'|-\mathbf{p}'\rangle$; applying Θ to this equation gives

$$\Theta\,\mathbf{p}\,\Theta^{-1}\Theta|-\mathbf{p}'\rangle = -\mathbf{p}'\,\Theta|-\mathbf{p}'\rangle,$$

or

$$\Theta\,\mathbf{p}\,\Theta^{-1} = -\mathbf{p}. \tag{3}$$

Hence at least for a free particle with Hamiltonian $H = p^2/2m$, $\Theta H = H\Theta$, which is hardly surprising.

Now comes the crucial point: We demand that the correspondence between motion reversed states be maintained for all t. (This will of course lead to a condition on H, among other things.) According to Fig. 39.1a, time reversal invariance requires that two different sequences of operations applied to an arbitrary state $\psi(t_0)$ lead to the *same* state. In the first sequence we allow $\psi(t_0)$ to evolve for a time t, whereupon we reverse all momenta, and then permit a further evolution for a time t. In the second we merely reverse all momenta in $\psi(t_0)$. Thus we demand that

$$e^{-iHt/\hbar}\Theta e^{-iHt/\hbar} = \Theta. \tag{4}$$

There are again two choices:

$$\Theta \text{ is unitary, and} \qquad H\Theta + \Theta H = 0, \tag{5a}$$

or

$$\Theta \text{ is anti-unitary, and} \qquad H\Theta - \Theta H = 0. \tag{5b}$$

But if (5a) were true, the spectrum of H would have no lower bound, and therefore Θ *must be anti-unitary*. This important fact was discovered by Wigner in 1931.

The preceding argument illustrates the contention of Sec. 27 that the choice between a unitary and anti-unitary transformation is dictated by the "physics," namely by the selection of rays that we wish to put in one-to-one correspondence. Thus we have just seen that if we wish to associate the rays $|\mathbf{p}'\rangle$ and $|-\mathbf{p}'\rangle$, and those on both sides of (4) when H is unchanged by the reversal of all momenta, the operator Θ must be anti-unitary.

Let us now consider the other observables. We certainly want Θ to leave the coordinate unchanged:

$$\Theta\,\mathbf{x}\,\Theta^{-1} = \mathbf{x}. \tag{6}$$

Furthermore, Θ must commute with rotations, and therefore

$$\Theta \mathbf{J} \Theta^{-1} = -\mathbf{J} \tag{7}$$

for any angular momentum operator, *including* $\boldsymbol{\sigma}$. Observe that (3), (6), and (7) are both internally consistent, and intuitively obvious.

If we demand invariance under t-reversal, we can discard certain kinds of Hamiltonians. For example, in the case of one particle, H cannot contain the term $\mathbf{x} \cdot \boldsymbol{\sigma}$, and this cannot be saved by putting $i\mathbf{x} \cdot \boldsymbol{\sigma}$, because H must be Hermitian. Of course $\mathbf{x} \cdot \boldsymbol{\sigma}$ could have been discarded by the requirement of space inversion invariance. Quite generally, the Hamiltonian of a spin $\frac{1}{2}$ particle in a central field cannot be a true scalar and odd under time reversal. There is such a term, however, in the interaction of two spin $\frac{1}{2}$ particles, namely

$$H' = C(\boldsymbol{\sigma}_1 \times \boldsymbol{\sigma}_2) \cdot \mathbf{L}_{\text{rel.}}, \tag{8}$$

where $\mathbf{L}_{\text{rel.}}$ is the relative angular momentum. Clearly $\mathcal{R}H'\mathcal{R}^{-1} = H'$, $\Theta H' \Theta^{-1} = -H'$. Experiments that look for such a term in nucleon-nucleon scattering have been done, but at the moment all results are consistent with invariance under t-reversal.

•Before going on, we should clear a few formal points out from underfoot. We recall from Sec. 27 that if Θ is anti-unitary, and

$$|\psi_1'\rangle = \Theta|\psi_1\rangle, \qquad |\psi_2'\rangle = \Theta|\psi_2\rangle, \tag{9}$$

then

$$\langle\psi_2'|\psi_1'\rangle = \langle\psi_2|\psi_1\rangle^* = \langle\psi_1|\psi_2\rangle, \tag{10}$$

and

$$\Theta(\lambda_1|\psi_1\rangle + \lambda_2|\psi_2\rangle) = \lambda_1^*\Theta|\psi_1\rangle + \lambda_2^*\Theta|\psi_2\rangle. \tag{11}$$

We still have to define how Θ acts on bras. Let $\langle\chi_1|$ be the dual to $|\psi_1\rangle$, and let $\langle\chi_1'|$ be dual to $|\psi_1'\rangle$. Then the dual to $\Theta(\lambda_1|\psi_1\rangle + \lambda_2|\psi_2\rangle)$ is $(\langle\chi_1'|\lambda_1 + \langle\chi_2'|\lambda_2)$. Hence we define

$$\langle\chi_1|\Theta = \langle\chi_1'|, \qquad \langle\chi_2|\Theta = \langle\chi_2'|, \tag{12}$$

$$(\langle\chi_1|\lambda_1^* + \langle\chi_2|\lambda_2^*)\Theta = \langle\chi_1'|\lambda_1 + \langle\chi_2'|\lambda_2. \tag{13}$$

We now see that

$$\langle\chi|(\Theta|\psi\rangle) = [(\langle\chi|\Theta)|\psi\rangle]^*. \tag{14}$$

For a linear operator $\mathcal{O}$, on the other hand (whether it is Hermitian or not), we never had to make a distinction between $\langle\chi|(\mathcal{O}|\psi\rangle)$ and $(\langle\chi|\mathcal{O})|\psi\rangle$.

As we have seen, Θ maps any state vector into its motion-reversed counterpart, and takes the complex conjugate of any number that may happen to multiply the ket. It is sometimes helpful to separate these two jobs of Θ, and for this purpose one must explicitly introduce a basis. Let $\{|n\rangle\}$ be a complete set of basis vectors (e.g., the $|\mathbf{p}'\rangle$'s), and let $\{|\bar{n}\rangle\}$ be the motion reversed set (i.e. the $|-\mathbf{p}'\rangle$'s in the previous example of a spin zero particle). This change of the *basis*, $\{|n\rangle\} \to \{|\bar{n}\rangle\}$, can of course be done by means of a unitary operator U:

$$U|n\rangle = |\bar{n}\rangle, \qquad U = \sum_n |\bar{n}\rangle\langle n|. \tag{15}$$

Working on an arbitrary ket $|\psi\rangle$, Θ then performs according to (11):

$$\Theta|\psi\rangle = \sum_{n'} U|n'\rangle\langle n'|\psi\rangle^*. \tag{16}$$

One therefore frequently writes

$$\Theta = UK, \tag{17}$$

by which one means that K is to take the complex conjugate of all the expansion coefficients of the arbitrary ket in terms of the particular basis with which one is working, and which must be known if U is to be specified. Note that $K^2 = 1$, and therefore

$$\Theta^{-1} = KU^\dagger. \tag{18}$$

Consider a change of basis from $\{|n\rangle\}$ to $\{|\nu\rangle\}$, and let $|\bar{\nu}\rangle$ be the time reverse of $|\nu\rangle$, [e.g., instead of the set $|\mathbf{p}'\rangle$ we can use the angular momentum eigenstates $|lm\rangle$, and their motion reversed counterparts $|l, -m\rangle$]. Let U' be the unitary operator

$$U' = \sum_\nu |\bar{\nu}\rangle\langle\nu|. \tag{19}$$

Now we again compute

$$\begin{aligned}
\Theta|\psi\rangle &= \sum_\nu U'|\nu\rangle\langle\nu|\psi\rangle^* \\
&= \sum_{\nu\nu'} |\nu'\rangle\langle\nu'|U'|\nu\rangle\langle\nu|\psi\rangle^* \\
&= \sum_{nn'\nu\nu'} |n\rangle\langle n|\nu'\rangle\langle\nu'|U'|\nu\rangle\langle\nu|n'\rangle^*\langle n'|\psi\rangle^*.
\end{aligned}$$

Comparing with (16), we have

$$\langle n|U|n'\rangle = \sum_{\nu\nu'} \langle n|\nu'\rangle\langle\nu'|U'|\nu\rangle\langle\nu|n'\rangle^*, \tag{20}$$

which is therefore *not* a unitary transformation from the operator U to U'. If W is the unitary transformation which maps $\{|n\rangle\} \to \{|\nu\rangle\}$, one might have expected $U = WU'W^\dagger$, but (20) demonstrates that this is wrong and that the correct connection is

$$U = WU'W^T. \tag{21}$$

Only if the numbers $\langle\nu|n\rangle$ are real (i.e., W is orthogonal) do we get the similarity transformation. This shows that one must be careful in using transformation theory in this problem, as could have been foreseen from the fact that the transformation $\mathbf{x} \to \mathbf{x}$, $\mathbf{p} \to -\mathbf{p}$ does not leave the canonical commutation rules invariant.*

39.2 SPINLESS PARTICLES

Because of (6), the coordinate eigenkets are unaffected by Θ. If $|\psi'\rangle$ is the time-reversed state of a single spin 0 particle, its wave function is therefore given by* $\psi'(\mathbf{x}) \equiv \langle\mathbf{x}|\psi'\rangle = \langle\psi|\mathbf{x}\rangle$, where the last equality follows from (10). Hence

$$\psi'(\mathbf{x}) = \psi(\mathbf{x})^*. \tag{22}$$

This conclusion can be drawn directly from the Schrödinger equation when the Hamiltonian is real. By taking the complex conjugate of $(i\hbar\,\partial_t - H)\psi(\mathbf{x}, t) = 0$ we obtain

$$\left(i\hbar\frac{\partial}{\partial t} + H\right)\psi^*(\mathbf{x}, t) = 0.$$

Therefore $\psi(\mathbf{x}, t)$ and $\psi^*(\mathbf{x}, -t)$ satisfy the same equation. From this argument we also conclude that in the x-basis the time reversal operator Θ is simply the complex conjugation operator K.

The significance of time reversal is best appreciated when one also considers the initial condition, and not only the evolution of the system in t. The scattering problem is ideally suited to this. Consider a particle of momentum $\mathbf{p}$ incident on the potential V. The state vector that satisfies outgoing-wave boundary conditions is $|\mathbf{p}^{(+)}\rangle$; according to (30.39) it satisfies the Lippmann-Schwinger equation

$$|\mathbf{p}^{(+)}\rangle = |\mathbf{p}\rangle + \frac{1}{E - H_0 + i\epsilon}\,V|\mathbf{p}^{(+)}\rangle. \tag{23}$$

We now assume that $\Theta V \Theta^{-1} = V$, and investigate the time reversed situation, $\Theta|\mathbf{p}^{(+)}\rangle$. Because H_0 commutes with Θ,

$$\Theta|\mathbf{p}^{(+)}\rangle = |-\mathbf{p}\rangle + \frac{1}{E - H_0 - i\epsilon}\,V\Theta|\mathbf{p}^{(+)}\rangle. \tag{24}$$

Two things have now happened: the "incident" state has momentum $-\mathbf{p}$, and the Green's operator is $(E - H_0 - i\epsilon)^{-1}$, and represents *incoming* spherical waves, because

$$\langle\mathbf{x}|\,\frac{1}{E - H_0 - i\epsilon}\,|\mathbf{x}'\rangle = -\frac{2m}{4\pi\hbar^2}\,\frac{e^{-ip|\mathbf{x}-\mathbf{x}'|/\hbar}}{|\mathbf{x} - \mathbf{x}'|} \tag{25}$$

* Henceforth we shall delete primes from the eigenvalues of $\mathbf{x}$ and $\mathbf{p}$, unless ambiguities would result therefrom.

in contrast to (30.38). All this is just as it should be. The state $|\mathbf{p}^{(+)}\rangle$ is a stationary-state idealization of the situation actually attained in the laboratory when we send in a beam with momentum $\mathbf{p}$ and eventually observe it scattered outwards from the target. The motion reversed situation is of course impossible to achieve in practice; here we start with a spherical wave converging on the target, which, by an incredible feat of coherence, then combines to form a collimated beam of momentum $-\mathbf{p}$ subsequent to the collision. Equation (24) shows that $\Theta|\mathbf{p}^{(+)}\rangle$ describes precisely this situation.

A more trivial example is a particle in an angular momentum eigenstate, wherein its wave function contains the factor $Y_{lm}(\theta, \phi)$. The time reversed state is $Y_{lm}^*(\theta, \phi) = (-1)^m Y_{l,-m}(\theta, \phi)$. In the time-reversed situation the particle travels about in the opposite direction and therefore has its component of angular momentum about *any* axis reversed.

Another simple but important consequence of time reversal is the following. If $\psi_E(\mathbf{x})$ is an energy eigenfunction, and $\Theta H \Theta^{-1} = H$, then $\psi_E^*(\mathbf{x})$ is also an eigenfunction of energy E. There are then two possibilities: Either the level is degenerate and $\psi_E^*(\mathbf{x})$ and $\psi_E(\mathbf{x})$ are *really* different (i.e., belong to different rays), or there is no degeneracy and $\psi_E \propto \psi_E^*$. Hence for a time-reversal invariant H the nondegenerate eigenfunctions can always be chosen to be real.

39.3 SPIN ½ PARTICLES

We begin with a single spin $\frac{1}{2}$ particle. The p-space and x-space kets will be written as $|\mathbf{p}m_s\rangle$ and $|\mathbf{x}m_s\rangle$, respectively, where $m_s = \pm\frac{1}{2}$. As the spin is an angular momentum, we tentatively set

$$\Theta|\mathbf{p}m_s\rangle = |-\mathbf{p}\ -m_s\rangle, \tag{26}$$

$$\Theta|\mathbf{x}m_s\rangle = |\mathbf{x}\ -m_s\rangle. \tag{27}$$

Let us work in the $|\mathbf{x}m_s\rangle$-basis. We write Θ in the form (17) and seek U. For the position kets $U = 1$, and we need only determine U in the spin-space. According to (7) we require

$$UK\boldsymbol{\sigma}KU^\dagger = -\boldsymbol{\sigma} = U\boldsymbol{\sigma}^*U^\dagger. \tag{28}$$

In the representation (33.12)

$$\sigma_x^* = \sigma_x, \qquad \sigma_y = -\sigma_y^*, \qquad \sigma_z = \sigma_z^*,$$

and therefore we need

$$U\sigma_x U^\dagger = -\sigma_x,$$
$$U\sigma_y U^\dagger = \sigma_y,$$
$$U\sigma_z U^\dagger = -\sigma_z,$$

which is satisfied by

$$U = e^{i\delta}\sigma_y. \tag{29}$$

(Im $\delta = 0$). Note that (29) represents a rotation of π about the y-axis, which certainly will make a state running clockwise about the z-axis run counter-clockwise. We also note from (33.12) that

$$\Theta|\tfrac{1}{2}\rangle = -ie^{i\delta}|-\tfrac{1}{2}\rangle, \qquad \Theta|-\tfrac{1}{2}\rangle = ie^{i\delta}|\tfrac{1}{2}\rangle,$$

and therefore no choice of δ will satisfy (27) with the phase $+1$ throughout. This is because we have held to our convention that $J_{\pm}|jm\rangle = \sqrt{j(j+1) - m(m \pm 1)}\,|jm \pm 1\rangle$. The phases in these relations are also at our disposal and could have been chosen in accordance with (27). In the same way the Y_{lm}'s could be redefined so that $Y^*_{lm} = Y_{l,-m}$. For some purposes the phases that simplify t-reversal considerations are convenient, but we shall not bother to go into this because our present convention is perfectly consistent and actually easier to remember for most purposes. With our phase conventions the correct relations are not (26) and (27), but

$$\Theta|\mathbf{p}m_s\rangle = e^{i\delta}(-i)^{2m_s}|-\mathbf{p}\ -m_s\rangle, \tag{26'}$$
$$\Theta|\mathbf{x}m_s\rangle = e^{i\delta}(-i)^{2m_s}|\mathbf{x}\ -m_s\rangle. \tag{27'}$$

Our conclusion is then that the operator for a single spin-$\tfrac{1}{2}$ particle is

$$\Theta = e^{i\delta}\sigma_y K. \tag{30}$$

The square of this operator turns out to have considerable significance. Its evaluation proceeds as follows: $\Theta^2 = e^{i\delta}\sigma_y e^{-i\delta}\sigma_y^* K^2 = -\sigma_y^2 = -1$. This result does not depend on the phase conventions, nor on the representation.* Thus

$$\begin{aligned} \Theta^2 &= +1 \quad \text{for a spin-zero particle,} \\ \Theta^2 &= -1 \quad \text{for a spin-}\tfrac{1}{2}\text{ particle.} \end{aligned} \tag{31}$$

For a system of N spin $\tfrac{1}{2}$ particles we clearly have

$$\Theta = \prod_{i=1}^{N} \sigma_y^{(i)} K \tag{32}$$

* To show this we proceed as follows. First we prove that $\Theta^2 = \pm 1$, and then show that Θ^2 is invariant under a change of basis. The first assertion is established by noting that because $[\Theta^2, \Theta] = 0$ and $\Theta^2 = c$, where $|c| = 1$, one has $c\Theta - \Theta c = (c - c^*)\Theta = 0$, or $c = c^* = \pm 1$. Now let Θ' be the t-reversal operation in some other basis. According to (17) and (21), $(\Theta')^2 = U'KU'K = U'U'^* = W^\dagger UU^*W = W^\dagger\Theta^2W = \Theta^2$; *QED.*

(aside from an arbitrary phase), where $\sigma_y^{(i)}$ is the ith particle's spin operator. Hence for N spin $\frac{1}{2}$ particles

$$\Theta^2 = (-1)^N. \tag{33}$$

Let $\{|E, N\rangle\}$ be the energy eigenstates of a system of N spin $\frac{1}{2}$ particles. If the Hamiltonian of this system is invariant under time reversal, $\Theta|E, N\rangle$ is also a stationary state of energy E. We may then ask whether this is a new state, i.e., one that is linearly independent of $|E, N\rangle$, or simply $|E, N\rangle$ apart from a possible phase. If $\Theta^2 = 1$ one cannot answer this question in general, i.e., the level E may or may not be degenerate. If $\Theta^2 = -1$, however (N odd), it is clear that $\Theta|E, N\rangle$ cannot be the same state (i.e., belong to the ray $|E, N\rangle$) because $\Theta|E\rangle = c|E\rangle$ implies $\Theta^2|E\rangle = |c|^2|E\rangle = -|E\rangle$, which is inconsistent.* Hence if N is odd, $\Theta|E, N\rangle$ and $|E, N\rangle$ are linearly independent, and therefore all levels are at least doubly degenerate. In fact, a bit of thought shows that the order of the degeneracy must be even. This type of degeneracy is called *Kramers degeneracy* and is a useful tool for treating complicated problems with few obvious symmetries such as occur in molecular and solid state physics.

40. Invariance Principles Applied to the Scattering of Spin $\frac{1}{2}$ Particles

We shall now apply what we have learned about invariance principles to a process of considerable interest, the scattering of particles with spin. We shall see how one may design scattering experiments to study the symmetry properties of the Hamiltonian. As a by-product, we shall also develop some very convenient techniques involving the density matrix.

40.1 THE DENSITY MATRIX PRECEDING AND FOLLOWING THE COLLISION

We confine ourselves to elastic scattering, and for simplicity consider a target of infinite mass. The latter restriction is of course trivially removed by transforming to the center-of-mass frame.† A beam of spin $\frac{1}{2}$ particles is then described by its momentum $\hbar\mathbf{k}$, and the spin state along some direction, $|\mu\rangle$, with $\mu = \pm\frac{1}{2}$. In practice, however, one never has beams 100% in a definite spin eigenstate, and so it is far more convenient to describe the spin degrees of freedom by a density matrix.

* This argument also shows that one cannot associate an eigenvalue with the operator Θ in general.

† Hence the techniques developed below can be applied to pion-nucleon scattering (cf., e.g., Källén, Chapt. 4).

This is then an operator ρ (2×2 matrix) in the spin space. Because 1 and $\boldsymbol{\sigma}$ are a complete set of 2×2 matrices, we certainly can write

$$\rho = a1 + \sum_{i=1}^{3} b_i \sigma_i. \tag{1}$$

Because tr $\rho = 1$, tr $1 = 2$, tr $\sigma_i = 0$, we choose $a = \frac{1}{2}$.* The meaning of b_i will emerge once we calculate the average value of σ_i, i.e., tr $\rho\sigma_i$. According to (33.10) tr $\sigma_i\sigma_j = 2\delta_{ij}$, and therefore $2b_i = \langle\sigma_i\rangle$. Instead of (1) we may therefore write

$$\rho = \tfrac{1}{2}(1 + \boldsymbol{\sigma} \cdot \langle\boldsymbol{\sigma}\rangle) \equiv \tfrac{1}{2}(1 + \boldsymbol{\sigma} \cdot \mathbf{P}). \tag{2}$$

The vector $\langle\boldsymbol{\sigma}\rangle$ is called the *polarization* of the beam, and shall often be written as $\mathbf{P}$. Clearly $\mathbf{P} = 0$ for an incoherent and equal mixture of $|\frac{1}{2}\rangle$ and $|-\frac{1}{2}\rangle$. When the beam is in a pure spin state, $\rho^2 = \rho$. The value of $\mathbf{P}$ associated with such a state is determined by setting ρ equal to

$$\rho^2 = \tfrac{1}{4}[1 + (\boldsymbol{\sigma} \cdot \mathbf{P})^2 + 2\boldsymbol{\sigma} \cdot \mathbf{P}] = \tfrac{1}{4}(1 + P^2 + 2\boldsymbol{\sigma} \cdot \mathbf{P})$$

(here we have used (33.11)). Consequently the spin state is pure if and only if $|\mathbf{P}| = 1$.

The representation that diagonalizes ρ must diagonalize the component of $\boldsymbol{\sigma}$ along $\mathbf{P}$. Let $|\chi_\uparrow\rangle$, $|\chi_\downarrow\rangle$ be the spin "up" and "down" kets with $\mathbf{P}$ as z-axis. Then

$$\boldsymbol{\sigma} \cdot \mathbf{P}|\chi_\uparrow\rangle = P|\chi_\uparrow\rangle, \qquad \boldsymbol{\sigma} \cdot \mathbf{P}|\chi_\downarrow\rangle = -P|\chi_\downarrow\rangle, \tag{3}$$

and therefore in this representation

$$\rho = \begin{pmatrix} \frac{1}{2}(1 + P) & 0 \\ 0 & \frac{1}{2}(1 - P) \end{pmatrix}, \tag{4}$$

which shows that $|\mathbf{P}|$ equals the probability of being in $|\chi_\uparrow\rangle$ minus that of being in $|\chi_\downarrow\rangle$.

Let us refresh our memory of scattering of spinless particles. We found in Sec. 12 that the asymptotic wave function has the form

$$\psi_{\mathbf{k}_i}(\mathbf{r}) \sim e^{i\mathbf{k}_i \cdot \mathbf{r}} + f(\mathbf{k}_f, \mathbf{k}_i)\frac{e^{ikr}}{r}, \tag{5}$$

where $\hbar\mathbf{k}_i$, $\hbar\mathbf{k}_f$ are the incident and scattered momenta, and the latter points in the direction of the observation point $\mathbf{r}$. Because we are dealing with elastic scattering,

$$|\mathbf{k}_f| = |\mathbf{k}_i| \equiv k. \tag{6}$$

* Note that these are traces in the two-dimensional space only!

If the potential is spherically symmetric, we found that the scattering amplitude f is only a function of k and the scattering angle ϑ. This follows from the fact that the incident wave is cylindrically symmetric about $\mathbf{k}_i$; because the potential also has this symmetry, the outgoing wave must retain this property and therefore does not depend on the azimuthal angle φ. We may put the matter somewhat differently as follows. We are asking for the probability that a particle incident in the state $|\mathbf{k}_i\rangle$ be found eventually in $|\mathbf{k}_f\rangle$ after collision with a rotationally symmetric potential. This probability is proportional to the differential cross section:

$$\sigma(\mathbf{k}_i \rightarrow \mathbf{k}_f) = |f(\mathbf{k}_f, \mathbf{k}_i)|^2. \tag{7}$$

Clearly this probability cannot depend on the orientation of the coordinate system, and hence $f(\mathbf{k}_f, \mathbf{k}_i)$, the scattering amplitude, must be a scalar function of $\mathbf{k}_i$ and $\mathbf{k}_f$. The only scalars one can form with these vectors are $k_f^2 = k_i^2 = k^2$, and $(\mathbf{k}_f - \mathbf{k}_i)^2$. The latter is the square of $\mathbf{q} = \mathbf{k}_i - \mathbf{k}_f$, which is the momentum transfer divided by $\hbar$:

$$|\mathbf{k}_f - \mathbf{k}_i| = q = 2k \sin^2 \tfrac{1}{2}\vartheta. \tag{8}$$

Therefore f is not a function of five variables as it would be if the scatterer were a pretzel, but only of two: k and q or k and ϑ.

Let us now turn to a spin $\frac{1}{2}$ particle of momentum $\hbar\mathbf{k}_i$ incident on a spin 0 target. Consider first some pure incident state $|\mathbf{k}_i\mu_i\rangle$, where $\mu_i = \pm\frac{1}{2}$. Again we ask for the probability that the scattered particles be in $|\mathbf{k}_f\mu_f\rangle$. For every $\mathbf{k}_f$ and $\mathbf{k}_i$ there will be four amplitudes, because there are four possible processes: $|\pm\frac{1}{2}\rangle \rightarrow |\pm\frac{1}{2}\rangle$ and $|\pm\frac{1}{2}\rangle \rightarrow |\mp\frac{1}{2}\rangle$. As a compact way of cataloging these, we introduce as our scattering amplitude a 2×2 matrix in the spin space, $M(\mathbf{k}_f, \mathbf{k}_i)$, defined so that

$$\sigma(\mu_i\mathbf{k}_i \rightarrow \mu_f\mathbf{k}_f) = |\langle\mu_f|M(\mathbf{k}_f, \mathbf{k}_i)|\mu_i\rangle|^2 \tag{9}$$

is the differential scattering cross section for the indicated process. Just as f makes its appearance in (5), so must M make its appearance in the asymptotic spinor. Let the latter be characterized by the incident spinor $e^{i\mathbf{k}_i\cdot\mathbf{r}}|\mu_i\rangle$; in the notation of (35.10), the asymptotic spinor is then

$$\psi(\mathbf{r}) \sim \left\{e^{i\mathbf{k}_i\cdot\mathbf{r}} + \frac{e^{ikr}}{r} M(\mathbf{k}_f, \mathbf{k}_i)\right\} |\mu_i\rangle, \tag{10}$$

where $(e^{ikr}/r)M|\mu_i\rangle$ represents the scattered particles. The amplitude for finding $|\mu_f\rangle$ in this beam is proportional to $\langle\mu_f|M(\mathbf{k}_f, \mathbf{k}_i)|\mu_i\rangle$, and hence (9). We note that since the Schrödinger equation is linear, an arbitrary incident spinor $e^{i\mathbf{k}_i\cdot\mathbf{r}}|\chi\rangle$ will, in the direction $\mathbf{k}_f$, have the asymptotic scattered form $(e^{ikr}/r)M(\mathbf{k}_f, \mathbf{k}_i)|\chi\rangle$. Let the incident beam have the density matrix

$$\rho_i = \sum_{n=1}^{2} |\chi_n\rangle p_{i,n}\langle\chi_n|, \tag{11}$$

where the $|\chi_n\rangle$ are two orthogonal spinors having their quantization direction along the incident polarization $\mathbf{P}_i$ (i.e., $P_i = |p_{i,1} - p_{i,2}|$). Then the scattered beam will have the density matrix that is obtained from (11) by replacing each $|\chi_n\rangle$ by $M(\mathbf{k}_f, \mathbf{k}_i)|\chi_n\rangle$. Thus the beam scattered into the direction $\mathbf{k}_f$ will be described by the density matrix

$$\rho_f = \frac{M\rho_i M^\dagger}{\text{tr } \rho_i M^\dagger M}, \tag{12}$$

which is a shorthand for

$$\rho_{\mathbf{k}_f} = \frac{M(\mathbf{k}_f, \mathbf{k}_i)\rho_{\mathbf{k}_i}M(\mathbf{k}_f, \mathbf{k}_i)^\dagger}{\text{tr } \rho_i M^\dagger M}. \tag{13}$$

The denominator in these formulas merely guarantees the normalization $\text{tr } \rho_f = 1$. It is clear that these results do not depend on the particular diagonal representation (11) used to describe the incident beam. In fact, (12) applies equally well to the scattering of particles of arbitrary spin, provided M and ρ are suitably generalized.

Very compact and convenient expressions for quantities of experimental interest can now be obtained. The cross section for scattering from $|\mathbf{k}_i\chi_n\rangle$ to $|\mathbf{k}_f\mu\rangle$ is $\sigma(\mathbf{k}_i\chi_n \rightarrow \mathbf{k}_f\mu) = |\langle\mu|M|\chi_n\rangle|^2$. The total cross section for going to both states $|\pm\frac{1}{2}\rangle$ is then $\Sigma_\mu|\langle\mu|M|\chi_n\rangle|^2$; on the other hand the probability of having the incident system in $|\chi_n\rangle$ is $p_{i,n}$, and so the cross section when we measure no actual spin orientations is therefore

$$\frac{d\sigma}{d\Omega} = \sum_{\mu n} |\langle\mu|M|\chi_n\rangle|^2 p_{i,n}.$$

But this is just

$$\frac{d\sigma}{d\Omega} = \text{tr}(M\rho_i M^\dagger), \tag{14}$$

and this is the desired expression for the cross section. Another useful quantity is the polarization of the scattered beam, which according to (13) is *

$$\mathbf{P}_f \equiv \text{tr}(\boldsymbol{\sigma}\rho_f) = \frac{\text{tr}(\boldsymbol{\sigma} M\rho_i M^\dagger)}{\text{tr}(\rho_i M^\dagger M)}. \tag{15}$$

* From the derivation it is clear that (14) and (15) apply to scattering of particles of arbitrary spin. On the other hand, the structure of M and ρ depends on the length of the spin.

40.2 THE SCATTERING AMPLITUDE

We shall now examine how the symmetries of the interaction are reflected in the scattering amplitude M.

As M is a matrix in the spin space, it can always be cast into the form (1). When the interaction is rotationally invariant, M is a scalar, and then the coefficients b_i in (1) cannot be unrelated. In fact, the b_i must form a vector so that M can be written as a scalar product of $\boldsymbol{\sigma}$ with one or more vectors. The only independent vectors that can be constructed from $\mathbf{k}_f$ and $\mathbf{k}_i$ are $\mathbf{k}_i \times \mathbf{k}_f$ and $\mathbf{k}_i \pm \mathbf{k}_f$. Therefore the most general form for M is

$$M = g_1 + \boldsymbol{\sigma} \cdot (\mathbf{k}_i \times \mathbf{k}_f)g_2 + \boldsymbol{\sigma} \cdot (\mathbf{k}_i + \mathbf{k}_f)g_3 + \boldsymbol{\sigma} \cdot (\mathbf{k}_i - \mathbf{k}_f)g_4, \quad (16)$$

where the g_i are arbitrary functions of k and ϑ (or k and q). If we make no further assumptions, this is all we can say without explicit calculation.

Let us now look at what happens to the various vectors under space and time reflection. We easily find that

$$\begin{aligned} \boldsymbol{\sigma} &\to \boldsymbol{\sigma}, \\ \mathbf{k}_i \times \mathbf{k}_f &\to \mathbf{k}_i \times \mathbf{k}_f, \\ \mathbf{k}_i \pm \mathbf{k}_f &\to -(\mathbf{k}_i \pm \mathbf{k}_f), \end{aligned} \quad (17)$$

under space reflection. Under time reversal * $\mathbf{k}_i \to -\mathbf{k}_f$, $\mathbf{k}_f \to -\mathbf{k}_i$, and therefore

$$\begin{aligned} \boldsymbol{\sigma} &\to -\boldsymbol{\sigma}, \\ \mathbf{k}_i \times \mathbf{k}_f &\to -(\mathbf{k}_i \times \mathbf{k}_f), \\ \mathbf{k}_i + \mathbf{k}_f &\to -(\mathbf{k}_i + \mathbf{k}_f), \\ \mathbf{k}_i - \mathbf{k}_f &\to \mathbf{k}_i - \mathbf{k}_f. \end{aligned} \quad (18)$$

Hence time reversal invariance requires that $g_4 = 0$ in (16). On the other hand, invariance under space reflection *alone* rules out *both* g_3 *and* g_4 (which agrees with the remark preceding (39.8)). Let us then assume that reflection symmetry applies. We then rewrite (16) as

$$M = g(k, \vartheta) + (\boldsymbol{\sigma} \cdot \hat{\mathbf{n}})\, h(k, \vartheta), \quad (19)$$

where $\hat{\mathbf{n}} = (\mathbf{k}_i \times \mathbf{k}_f)/|\mathbf{k}_i \times \mathbf{k}_f|$ is a unit vector normal to the scattering plane. This is the form for M that we shall use in most of the calculations that follow. The scalar functions g and h can only be computed if the interaction Hamiltonian is known. Further remarks concerning this point will be made in Sec. 40.4, and in Prob. 8.

* It must be admitted that (18) has been written on the basis of plausibility. A formal proof will be given in Sec. 40.5.

40.3 POLARIZATION PRODUCED IN SCATTERING. TESTS OF REFLECTION INVARIANCE

We first apply our formulas to the scattering of an unpolarized beam, $\rho_i = \frac{1}{2}$. Using the form (19) in (14) we find

$$\frac{d\sigma}{d\Omega} = \tfrac{1}{2}\,\mathrm{tr}\,MM^\dagger = \mathrm{tr}\,\tfrac{1}{2}(g + \boldsymbol{\sigma}\cdot\hat{\mathbf{n}}h)(g^* + \boldsymbol{\sigma}\cdot\hat{\mathbf{n}}h^*),$$

because $\boldsymbol{\sigma} = \boldsymbol{\sigma}^\dagger$. Using (33.11),

$$(\boldsymbol{\sigma}\cdot\mathbf{A})(\boldsymbol{\sigma}\cdot\mathbf{B}) = \mathbf{A}\cdot\mathbf{B} + i\boldsymbol{\sigma}\cdot(\mathbf{A}\times\mathbf{B}), \tag{20}$$

we obtain

$$\frac{d\sigma}{d\Omega} = [|g|^2 + |h|^2], \tag{21}$$

because $\mathrm{tr}\,\boldsymbol{\sigma} = 0$. Thus *for an unpolarized incident beam the differential cross section does not depend on the azimuthal angle.* This result is obvious from cylindrical symmetry, and in fact one easily sees from (14) that it holds true even if space inversion and time reversal are not symmetries. Furthermore, it clearly applies to the scattering between particles of arbitrary spin provided neither target nor projectile is polarized initially.

Next we compute the polarization of the scattered beam. Because $\rho_i = \frac{1}{2}$, the numerator of (15) is

$$\begin{aligned}\tfrac{1}{2}\,\mathrm{tr}\,\boldsymbol{\sigma}(g + \boldsymbol{\sigma}\cdot\hat{\mathbf{n}}h)(g^* + \boldsymbol{\sigma}\cdot\hat{\mathbf{n}}h^*) &= \tfrac{1}{2}\,\mathrm{tr}\,\boldsymbol{\sigma}(\boldsymbol{\sigma}\cdot\hat{\mathbf{n}})(gh^* + g^*h)\\ &= \hat{\mathbf{n}}(gh^* + g^*h),\end{aligned}$$

and therefore

$$\mathbf{P}_f = \frac{2\hat{\mathbf{n}}\,\mathrm{Re}\,gh^*}{|g|^2 + |h|^2}. \tag{22}$$

This result shows that *the polarization after scattering,* $\mathbf{P}_f$, *is perpendicular to the scattering plane.* It takes but a minute to demonstrate that *parity nonconserving terms in the scattering amplitude will produce components of* $\mathbf{P}_f$ *in the scattering plane.*

These results can be understood in terms of a simple geometrical argument, and one really does not need the complicated machinery just developed to conclude that $\mathbf{P}_f$ must lie along $\hat{\mathbf{n}}$ when parity is conserved. Consider the scattering experiment depicted in Fig. 40.1(a). The z-axis is towards the reader, unmarked arrows are components of $\mathbf{P}_f$ in the scattering plane, and the circular arrows give the sense of $\mathbf{P}_f$ along the z-axis (i.e., the sign of $\hat{\mathbf{n}}\cdot\mathbf{P}_f$). We then perform the following sequence of operations: (i) a reflection in the y-z plane; (ii) a rotation through π about the z-axis; (iii) a rotation through π about the x-axis. If reflection and rotation invariance holds, the probabilities for the collision processes shown in these four figures *must all be equal.* Observe that the components of $\mathbf{P}_f$ in the scattering plane have changed sign in going from (a) to (d), but that the component $\hat{\mathbf{n}} \cdot \mathbf{P}_f$ has not. Hence $\mathbf{P}_f$ cannot have any components in the scattering plane. Comparing (c) and (d) we also conclude that the probability of scattering to the right through ϑ and finding $\mathbf{P}_f$ up must equal that for scattering through the same angle to the left and finding $\mathbf{P}_f$ down. Equation 22 incorporates both of these statements, as it must.

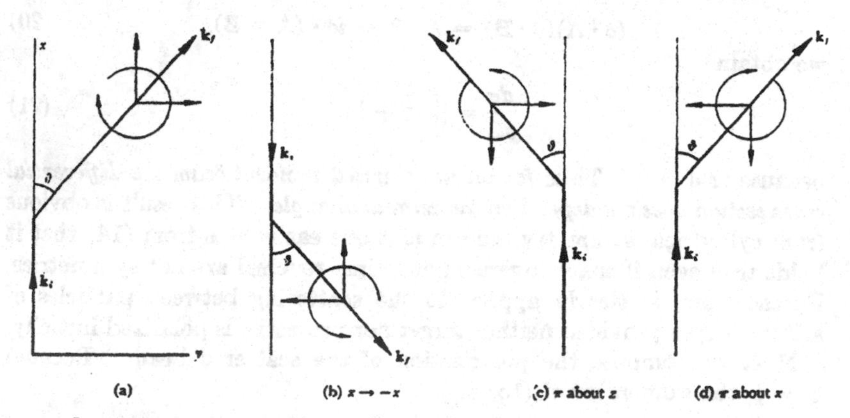

Fig. 40.1. Symmetry operations applied to a scattering experiment.

From these diagrams we can derive a more general result: *If the target is an assembly of particles of arbitrary spin that is completely unpolarized (i.e., all $|\mu\rangle$-states equally populated), and we scatter an unpolarized beam of particles off it, the final polarization is along* $\hat{\mathbf{n}}$, *provided the interaction is reflection invariant.*

Finally there comes the question of detecting the polarization. The obvious way is to measure it directly by some magnetic effect. This is frequently not practical, however, and instead one uses the fact that if a polarized beam is scattered, there is an *azimuthal* dependence in the cross section even when no spins are measured directly. To see this in detail, we simply compute (14) when ρ_i has a polarization $\mathbf{P}_i \neq 0$. Then

$$\frac{d\sigma}{d\Omega} = \tfrac{1}{2}\,\mathrm{tr}(1 + \mathbf{P}_i \cdot \boldsymbol{\sigma})(g^* + \boldsymbol{\sigma} \cdot \hat{\mathbf{n}} h^*)(g + \boldsymbol{\sigma} \cdot \hat{\mathbf{n}} h)$$
$$= \tfrac{1}{2}\,\mathrm{tr}(1 + \mathbf{P}_i \cdot \boldsymbol{\sigma})(|g|^2 + |h|^2 + 2\boldsymbol{\sigma} \cdot \hat{\mathbf{n}}\ \mathrm{Re}\ gh^*),$$

or

$$\frac{d\sigma}{d\Omega} = |g|^2 + |h|^2 + 2\mathbf{P}_i \cdot \hat{\mathbf{n}}\ \mathrm{Re}\ gh^*. \tag{23}$$

If the scattering plane is perpendicular to $\mathbf{P}_i$, there is a difference in $d\sigma/d\Omega$ depending on whether the scattering is through $\varphi = \pm\frac{1}{2}\pi$, i.e., right or left, where the azimuthal angle φ is measured from $\mathbf{P}_i$. In fact

$$\frac{\left.\dfrac{d\sigma}{d\Omega}\right|_{\text{right}} - \left.\dfrac{d\sigma}{d\Omega}\right|_{\text{left}}}{\left.\dfrac{d\sigma}{d\Omega}\right|_{\text{right}} + \left.\dfrac{d\sigma}{d\Omega}\right|_{\text{left}}} = \frac{2|\text{Re}\, gh^*|}{|g|^2 + |h|^2} P_i. \tag{24}$$

If $|\mathbf{P}_i| = 1$, this is just the magnitude of the polarization produced by scattering off an unpolarized beam, (22). The equality between left-right asymmetry and polarization is a general result that depends only on space reflection and time reversal, not on the spin of the target, provided it is unpolarized.*

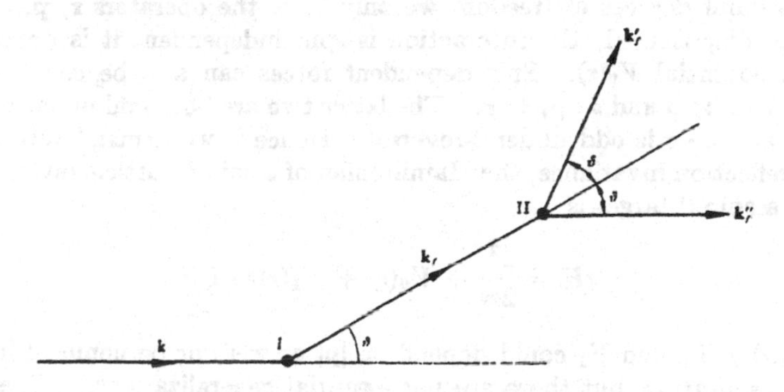

Fig. 40.2. A double-scattering experiment.

A measurement of Re$|gh^*|$ can be achieved by a double scattering experiment (see Fig. 40.2). In the first experiment one scatters through ϑ in the plane of the paper, producing a polarized beam (this target acts as a polarizer). The scattered beam is scattered a second time from an identical target through ϑ in the same plane, both to the right and left. (Now the second target acts as an analyzer.) According to (21), the differential cross section in the first scattering process gives us $(|g|^2 + |h|^2)$, while (24) and (22) tell us that the left-right asymmetry in the second process equals

$$4\left(\frac{\text{Re}\, gh^*}{|g|^2 + |h|^2}\right)^2.$$

* See L. Wolfenstein, *Ann. Rev. Nucl. Sci.* **6**, 43 (1956).

Hence $|\text{Re } gh^*|$ is determined. The sign of Re gh^* cannot be measured in this way, and one must resort to interference with some other process (e.g., Coulomb scattering) to determine it.

As we mentioned in connection with Eq. (22), the presence of parity nonconserving terms in M would produce a polarization in the plane of the paper in Fig. 40.2, and therefore leads to an up-down asymmetry in process II. An asymmetry of the latter type cannot occur if M is invariant under reflection, as one can verify from (23). By the same token, any up-down asymmetry constitutes an experimental proof that the interaction responsible for the collision is not reflection invariant.*

40.4 IMPOSITION OF SYMMETRIES ON THE HAMILTONIAN

By using symmetry considerations we can also restrict the form of the Hamiltonian. Because the scatterer is spherically symmetric and has no internal degrees of freedom we only have the operators **r**, **p**, and $\boldsymbol{\sigma}$ at our disposal. If the interaction is spin independent it is described by a potential $V_0(r)$. Spin-dependent forces can also be constructed from $\boldsymbol{\sigma} \cdot \mathbf{r} \times \mathbf{p}$ and $\boldsymbol{\sigma} \cdot \mathbf{p}$, $\boldsymbol{\sigma} \cdot \mathbf{r}$. The latter two are both odd under reflection, and $\boldsymbol{\sigma} \cdot \mathbf{r}$ is odd under *t*-reversal. Hence if we demand rotational and reflection invariance, the Hamiltonian of a spin $\frac{1}{2}$ particle interacting with a spin 0 target is

$$H = \frac{p^2}{2m} + V_0(r) + V_1(r)\boldsymbol{\sigma} \cdot \mathbf{L}. \tag{25}$$

(Actually V_0 and V_1 could depend on $|\mathbf{p}|$ as well, or be nonlocal in the **r**-representation, but these are not essential generalizations.) The second term in (25) is called a ***spin-orbit coupling***. For a charged particle with a magnetic moment (e.g., an electron) in the field of a fixed charge, relativistic corrections lead to such a term (see the discussion of hydrogen fine structure, Sec. 46); it is a small correction in atomic spectroscopy. In nuclear physics such spin-orbit forces are of much greater importance.

If we only require invariance under time reversal we could add a term

$$\tfrac{1}{2}[V_2(r)\boldsymbol{\sigma} \cdot \mathbf{p} + \boldsymbol{\sigma} \cdot \mathbf{p}\, V_2(r)]$$

to (25), while if we surrender all the discrete symmetries, we may also add

$$V_3(r)\boldsymbol{\sigma} \cdot \mathbf{r}.$$

It is obvious that when $V_2 = V_3 = 0$, h in (19) is only nonzero if the spin-orbit term is present. Furthermore, g_3 in (16) depends on the exist-

* For a survey of experiments of this type see Wilson, Chap. 9.

ence of V_2, and g_4 on the presence of V_3 in the Hamiltonian. The actual construction of the scattering amplitude from the potential is left for Prob. 8.

40.5 TIME REVERSAL INVARIANCE OF THE SCATTERING AMPLITUDE

We must still prove that under time reversal

$$M(\mathbf{k}_f, \mathbf{k}_i, \boldsymbol{\sigma}) \rightarrow M(-\mathbf{k}_i, -\mathbf{k}_f, -\boldsymbol{\sigma}). \tag{26}$$

A precise definition of time reversal invariance in scattering is a prerequisite to such a proof. Let $|\Phi_i\rangle$ and $|\Phi_f\rangle$ be a pair of kets that describe the system before and after the collision, and define an operator T such that $\langle\Phi_f|T|\Phi_i\rangle$ is the amplitude for the scattering process $\Phi_i \rightarrow \Phi_f$. Let $|\Phi_i'\rangle = \Theta|\Phi_i\rangle$ and $|\Phi_f'\rangle = \Theta|\Phi_f\rangle$ be the motion-reversed states. An interaction is said to be time reversal invariant if for *all* i and f

$$\langle\Phi_f|T|\Phi_i\rangle = \langle\Phi_i'|T|\Phi_f'\rangle. \tag{27}$$

In our case the states $|\Phi\rangle$ are of the form $|\mathbf{k}\chi\rangle$, where $|\chi\rangle$ is a vector in the two-dimensional spin space. Furthermore, T is such that

$$\langle\mathbf{k}_f\chi_f|T|\mathbf{k}_i\chi_i\rangle = \langle\chi_f|M(\mathbf{k}_f, \mathbf{k}_i, \boldsymbol{\sigma})|\chi_i\rangle,$$

where $M(\mathbf{k}_f, \mathbf{k}_i, \boldsymbol{\sigma})$ is the 2×2 matrix defined in Sec. 40.1. Under time reversal, $|\mathbf{k}\chi\rangle \rightarrow \Theta|\mathbf{k}\chi\rangle = |-\mathbf{k}\chi'\rangle$. Condition (27) therefore reads

$$\langle\chi_f|M(\mathbf{k}_f, \mathbf{k}_i, \boldsymbol{\sigma})|\chi_i\rangle = \langle\chi_i'|M(-\mathbf{k}_i, -\mathbf{k}_f, \boldsymbol{\sigma})|\chi_f'\rangle.$$

Upon using (39.10), the right-hand side of this equation becomes

$$(\langle\chi_i'|M)|\chi_f'\rangle = \langle\chi_f|\Theta_s(M^\dagger|\chi_i'\rangle) = \langle\chi_f|\Theta_s M^\dagger \Theta_s|\chi_i\rangle$$

where Θ_s is the time reversal operator in the spin space, Eq. (39.30). Therefore

$$\langle\chi_f|M(\mathbf{k}_f, \mathbf{k}_i, \boldsymbol{\sigma})|\chi_i\rangle = \langle\chi_f|M(-\mathbf{k}_i, -\mathbf{k}_f, -\boldsymbol{\sigma}^\dagger)|\chi_i\rangle.$$

But $\boldsymbol{\sigma}$ is Hermitian, and we have therefore proven that time reversal invariance requires

$$M(\mathbf{k}_f, \mathbf{k}_i, \boldsymbol{\sigma}) = M(-\mathbf{k}_i, -\mathbf{k}_f, -\boldsymbol{\sigma}). \tag{28}$$

An important remark concerning inelastic scattering should be made here. Assume that our spin-zero target has several excited states which also have $J = 0$. (As an example of such a situation we may think of neutron scattering by O^{16}, which has a first excited $J = 0$ state.) If α, β, . . . designate the various states of the target, time reversal invariance implies that the amplitude for the transition $\alpha \rightarrow \beta$ satisfies

$$M_{\alpha\beta}(\mathbf{k}_f, \mathbf{k}_i, \boldsymbol{\sigma}) = M_{\beta\alpha}(-\mathbf{k}_i, -\mathbf{k}_f, -\boldsymbol{\sigma}).$$

Hence a term such as $g_{\alpha\beta}(k, \theta)\, \boldsymbol{\sigma} \cdot (\mathbf{k}_i - \mathbf{k}_f)$, with $g_{\alpha\beta} = -g_{\beta\alpha}$, can appear in the scattering amplitude even when the Hamiltonian is invariant under time reversal. A time-reversal invariant interaction that leads to such a term in M is $i(\Lambda_{\alpha\beta} - \Lambda_{\beta\alpha})\boldsymbol{\sigma} \cdot \mathbf{r}$, where $\Lambda_{\alpha\beta} = |\alpha\rangle\langle\beta|$. This example illustrates the rather obvious fact that it is much more difficult to verify time-reversal invariance in inelastic scattering.

41. Indistinguishable Particles—Spin and Statistics

As Dirac and Heisenberg discovered in 1926, the fact that two electrons (or protons, etc.) cannot be distinguished leads to consequences of the utmost importance and generality. These considerations constitute one of the most impressive triumphs of quantum mechanics.

41.1 PERMUTATIONS AS SYMMETRY OPERATORS

Consider a system of N identical particles, e.g., electrons. Each electron has 4 degrees of freedom, $\mathbf{r}_i$ and σ_{zi} say, and it is not possible, conceptually, to give any more information than that which pertains to these observables: one cannot put a blue spot on one, and a yellow one on another. Hence the Hamiltonian of this system is a *symmetric* function of all the variables $\mathbf{r}_i$, $\boldsymbol{\sigma}_i$, and *all* other observables must also be symmetric functions of the basic dynamical variables. As an example, recall that the charge density operator for a set of equal charges is $e\Sigma_i\delta(\mathbf{x} - \mathbf{r}_i)$. Although one can write down operators referring to only one particle, e.g., $\mathbf{r}_{17}$, it is not possible to distinguish any observation of $\mathbf{r}_{17}$ from $\mathbf{r}_{82}$. It is of course true that the classical Hamiltonian of this system would also be symmetric, but one cannot draw any surprising or profound conclusions from this symmetry in the classical case. We shall see that the deductions that follow depend crucially on the linearity of the Schrödinger equation, a property that has no counterpart in classical mechanics.

Let the Hamiltonian be $H = H(1, 2, 3, \ldots, N)$, and let P_{12} be the operator that permutes the variables of particles 1 and 2:

$$P_{12}H(1, 2, 3, \ldots N)P_{12}^{-1} = H(2, 1, 3, \ldots N) = H(1, 2, 3, \ldots N),$$

or more generally

$$P_{ij}HP_{ij}^{-1} = H. \tag{1}$$

Since P_{ij} also commutes with the total momentum and angular momentum, P_{ij} must be unitary if it is to commute with space displacement or rotations. Thus we can choose $P_{ij}^2 = 1$, and hence the eigenvalues of each permutation operator P_{ij} are $P'_{ij} = \pm 1$.

Consider the action of $\{P_{ij}\}$ on a complete set of N-particle states. Such sets of states can be constructed as follows. Let $\{|\xi'\rangle\}$ be a complete set for one particle. Then the set of all product states of the type

$$|\xi_1'\xi_2' \cdots \xi_N'\rangle \equiv |\xi_1'\rangle_1|\xi_2'\rangle_2 \cdots |\xi_N'\rangle_N, \tag{2}$$

where $|\xi_i'\rangle_i$ means that the ith particle is in the state $|\xi_i'\rangle$, is a complete set of N-particle states. The action of P_{12} on (2) is

$$\begin{aligned} P_{12}|\xi_1'\xi_2'\xi_3' \cdots \xi_N'\rangle &= |\xi_2'\xi_1'\xi_3' \cdots \xi_N'\rangle \\ &\equiv |\xi_2'\rangle_1|\xi_1'\rangle_2|\xi_3'\rangle_3 \cdots |\xi_N'\rangle_N. \end{aligned} \tag{3}$$

It is easy to see that

$$[P_{ij}, P_{kl}] \neq 0 \tag{4}$$

if one of the particles labeled (k, l) is also one of those in (i, j), and therefore only a limited set of permutations can be diagonalized simultaneously with H and the other symmetric observables.

Consider the cases $N = 2$ and $N = 3$ in detail. For $N = 2$ we form the two linearly independent and normalized sets

$$|\xi_1'\xi_2'\rangle_\pm = \frac{1}{\sqrt{2}}[|\xi_1'\xi_2'\rangle \pm |\xi_2', \xi_1'\rangle], \tag{5}$$

where "+" designates the symmetric and "−" the antisymmetric states. In this case there is only one permutation operator, P_{12}, and the situation summarized in (4) cannot arise. An arbitrary symmetric or antisymmetric state is then given by

$$|\psi_\pm\rangle = \sum_{\xi_1'\xi_2'} C(\xi_1'\xi_2')|\xi_1'\xi_2'\rangle_\pm. \tag{6}$$

$N = 3$ is not so trivial, because there are three noncommuting P_{ij}. Again we have the totally symmetric and antisymmetric cases. These special kets have the property $P_{ij}|\ \rangle_\pm = \pm|\ \rangle_\pm$ for *all* the permutation operators. Their explicit form is

$$\begin{aligned} |\xi_1'\xi_2'\xi_3'\rangle_\pm = \frac{1}{\sqrt{6}}\{&|\xi_1'\xi_2'\xi_3'\rangle \pm |\xi_2'\xi_1'\xi_3'\rangle \pm |\xi_1'\xi_3'\xi_2'\rangle \pm |\xi_3'\xi_2'\xi_1'\rangle \\ &+ |\xi_3'\xi_1'\xi_2'\rangle + |\xi_2'\xi_3'\xi_1'\rangle\}. \end{aligned} \tag{7}$$

As there must be 3! independent states, there are four further states with more complicated permutation properties than (7), and we may call them $|\xi_1'\xi_2'\xi_3'\rangle_K$, with $K = 1, \ldots, 4$. In the states (7) all three P_{ij} are diagonal, but this is not true of the states $|\ \rangle_K$. That states like $|\ \rangle_K$ should occur is not surprising, of course, because (4) told us that $\{P_{ij}\}$

cannot be diagonalized in general. What is perhaps surprising is that the especially simple sets $|\ \rangle_\pm$ exist. This situation has its counterpart in the rotation group, where in general J_x, J_y, and J_z cannot be simultaneously diagonalized. But in the one-dimensional representation ($j = 0$) all the J_i are diagonal (they have the eigenvalue zero). In a similar fashion (7) belongs to the two distinct one-dimensional representations of the permutation group on three objects, whereas the states $|\ \rangle_K$ belong to higher dimensional representations (two distinct two-dimensional ones).

For general N the situation is similar. There always exist the symmetric and antisymmetric states $|\xi_1'\xi_2'\xi_3' \cdots \xi_N'\rangle_\pm$, and $(N! - 2)$ states $|\xi_1'\xi_2'\xi_3' \cdots \xi_N'\rangle_K$ that belong to various higher dimensional irreducible representations. The $|\ \rangle_\pm$ states may be constructed as follows

$$|\xi_1' \cdots \xi_N'\rangle_\pm = \frac{1}{\sqrt{N!}} \begin{Bmatrix} \text{perm.} \\ \text{det.} \end{Bmatrix} \begin{vmatrix} |\xi_1'\rangle_1 & |\xi_1'\rangle_2 & \cdots & |\xi_1'\rangle_N \\ |\xi_2'\rangle_1 & |\xi_2'\rangle_2 & \cdots & |\xi_2'\rangle_N \\ \cdot & \cdot & & \cdot \\ \cdot & \cdot & & \cdot \\ \cdot & \cdot & & \cdot \\ |\xi_N'\rangle_1 & |\xi_N'\rangle_2 & \cdots & |\xi_N'\rangle_N \end{vmatrix}, \tag{8}$$

where det. stands for determinant, and perm. for the same linear combination except that all the signs are positive. In the antisymmetric case, (8) is called a *Slater determinant.*

Let $\mathfrak{H}_\pm$ and $\mathfrak{H}_K$ stand for the subspaces spanned by the states $|\ \rangle_\pm$ and $|\ \rangle_K$ respectively, and let A be any symmetric observable for N particles (e.g., the Hamiltonian). Then A has no matrix elements that connect the subspaces $\mathfrak{H}_+$, $\mathfrak{H}_-$ and $\mathfrak{H}_K$. (The subspace $\mathfrak{H}_K$ can always be reduced further to smaller diagonal blocks, as in the case $N = 3$ where the four-dimensional representation reduces to two two-dimensional ones; we shall not enter into this here.) The eigenvalues A_+', A_-' are in general distinct, and since all the P_{ij} are already diagonal in $\mathfrak{H}_\pm$, these eigenvalues need not correspond to degenerate sets of states. In $\mathfrak{H}_K$ the situation is necessarily more complicated, however. Here the fact that all the P_{ij} are not diagonal but commute with A means that the spectrum of A in $\mathfrak{H}_K$ is always degenerate; the degree of the degeneracy is determined by the particular irreducible representation in question. That this must be so can be seen by recalling the rotational case: if a Hamiltonian is spherically symmetric, i.e., commutes with all three J_i, then the energy levels are degenerate since the application of J_i to any state must produce another one with the same energy eigenvalue, unless all the J_i have eigenvalue zero (an S-state).

If we prepare a state of the N-particle system belonging to $\mathfrak{H}_+$ (or $\mathfrak{H}_-$) at $t = 0$, and allow it to evolve according to the Schrödinger equation with an arbitrary (time-dependent) symmetric Hamiltonian, it will always remain in $\mathfrak{H}_+$ ($\mathfrak{H}_-$). The same remark applies to any state that belongs to one irreducible representation in $\mathfrak{H}_K$: for later times it will always belong to the same irreducible representation.

41.2 FERMI-DIRAC AND BOSE-EINSTEIN STATISTICS

Fortunately nature is kind, because it appears that we need not concern ourselves with the states in $\mathfrak{H}_K$. Experimentally we know that *for each system of N identical particles the states are either even or odd under all permutations of two particles, and that this property depends only on the species of particles and not the number N.* Note therefore that if $|\psi\rangle$ is a ket for N indistinguishable particles, then $P_{ij}|\psi\rangle$ belongs to the same ray. This is, of course, not true in the spaces $\mathfrak{H}_K$. The advantages of not having to worry about $\mathfrak{H}_K$ are very appreciable because the types of higher dimensional representations depend on N. This leads to a very complicated situation in collision problems, which is discussed briefly in Pauli, Sec. 14.

We now observe a remarkable property of the state $|\xi_1'\xi_2' \cdots \xi_N'\rangle_-$: it vanishes when any two (or more) of the quantum numbers are equal. This is just the *Pauli principle*, and shows that *particles that obey the exclusion principle are described by antisymmetric kets.*

The statistical mechanics of systems obeying the exclusion principle was worked out by Fermi and Dirac and is called Fermi-Dirac statistics; particles that obey this type of statistics are called *fermions*. On the other hand particles with symmetric kets obey Bose-Einstein statistics, and are called *bosons*.* It appears that there is a unique connection between the spin of a particle and the type of statistics that one must use to describe ensembles of such particles: *half-integral spin particles are fermions, integral spin particles are bosons.* Relativistic quantum mechanics offers an explanation of the connection between spin and statistics, and this is actually a very important achievement of that theory.†

The remarks in Sec. 33.2 concerning the meaning of the term "particle" apply equally well to the present discussion. In low-temperature physics,

* Bose and Fermi statistics are discussed in some detail in Chapt. XI. Recently there has been some interest in the statistics of (hypothetical) particles that have wave functions belonging to the nontrivial Hilbert space $\mathfrak{H}_K$ introduced above. See A.M.L. Messiah and O. W. Greenberg, *Phys. Rev.* **135**, B1447 (1964).

† This explanation was first proposed by Pauli in 1940. For a modern treatment, see Streater and Wightman.

liquid helium is treated as a many-boson system if the nuclei are He^4, and as a many-fermion system if the nuclei are He^3 ($s = \frac{1}{2}$). The dramatically different properties of these two fluids are almost entirely due to this difference in statistics, because the interaction potentials between He atoms do not depend on the spins of the nuclei. In nuclear physics, on the other hand, the He^4 atom must be thought of as a system of six fermions with a spin zero ground state. Incidentally, a description that works entirely in terms of "elementary" particles can always be given. To do so one merely calls every state of a system by a different name, and describes it in terms of a particle of the appropriate mass and spin. Any process that causes an excitation of the system is then viewed as a reaction in which some particles are transformed into others. Such a zoological formulation hardly serves to elucidate the structure of matter, however.

At the moment the most fundamental * building blocks of nature appear to be † the following:

Fermions (all $s = \frac{1}{2}$)	*Bosons* ($s = 0$)
neutrinos (ν_e, ν_μ)	π-meson
electron	K-meson
μ-meson	
	($s = 1$)
nucleon	
Λ, Σ, Ξ	photon

42. Symmetries of the Two-Nucleon System

In this section we shall study the symmetry properties of two-nucleon systems. The power of symmetry considerations will be brought out by the fact that we shall accumulate a considerable amount of information without detailed calculations.

Let us first consider the p-p system. These are two identical fermions, and therefore the states must be antisymmetric. In the c.o.m. frame (where we always work) the degrees of freedom apart from the energy are the relative orbital angular momentum **L**, and the spins $\boldsymbol{\sigma}_1$, $\boldsymbol{\sigma}_2$. The eigenfunctions of L^2, L_z are $Y_{lm}(\hat{\mathbf{r}})$, where **r** is the separation of the two

* This list does not include "particles" that decay via strong or electromagnetic interactions, such as the ρ-meson or the η-meson. By this we do not seek to imply that the pion is more "fundamental" than η, etc.

† There is actually no experimental proof that the strange particles satisfy the spin-statistics relationship stated previously. For a discussion of this question and references to the literature, see Greenberg and Messiah, *loc. cit.*

protons. The spin eigenkets are $|\mu_1\mu_2\rangle$, with $\mu_1 = \pm\frac{1}{2}$, $\mu_2 = \pm\frac{1}{2}$; frequently we shall write these as $|+,+\rangle$, $|+,-\rangle$, etc. Under a permutation of 1 and 2, $\mathbf{r} \to -\mathbf{r}$, $\mu_1 \leftrightarrow \mu_2$. Hence the orbital state acquires the factor $(-1)^l$. This must be compensated by a symmetric spin function if l is odd, and an antisymmetric spin function if l is even, because the overall function must be odd. Out of $\{|\mu_1\mu_2\rangle\}$ we can construct the four linearly independent kets

$$|\chi_0\rangle = \frac{1}{\sqrt{2}}(|+,-\rangle - |-,+\rangle), \tag{1}$$

$$|\chi_1^1\rangle = |+,+\rangle, \tag{2}$$

$$|\chi_0^1\rangle = \frac{1}{\sqrt{2}}(|+,-\rangle + |-,+\rangle), \tag{3}$$

$$|\chi_{-1}^1\rangle = |-,-\rangle; \tag{4}$$

$|\chi_0\rangle$ is antisymmetric in μ_1 and μ_2, whereas $|\chi_\nu^1\rangle$ is symmetric. Let

$$\mathbf{S} = \tfrac{1}{2}(\boldsymbol{\sigma}_1 + \boldsymbol{\sigma}_2) \tag{5}$$

be the total spin. Then

$$S_z|\chi_0\rangle = 0, \qquad S_z|\chi_\nu^1\rangle = \nu|\chi_\nu^1\rangle. \tag{6}$$

By applying the operators $S_\pm = (S_x \pm iS_y)$ to the kets $|\chi_\nu^1\rangle$ one easily shows that these states have all the properties pertaining to the three members of a multiplet having angular momentum unity. Furthermore, $S_\pm|\chi_0\rangle = 0$. Let us designate the eigenvalue of $\mathbf{S}^2$ by $S(S+1)$. We have therefore shown that the three kets $|\chi_\nu^1\rangle$ are states with $S = 1$, whereas $|\chi_0\rangle$ is an $S = 0$ state. We shall refer to these as spin triplets and singlets henceforth.

Because of the exclusion principle the possible states of the p-p system are therefore

$$\begin{array}{ll} \text{odd } l, & \text{spin triplet,} \\ \text{even } l, & \text{spin singlet.} \end{array} \tag{7}$$

A simple example of a complete set of antisymmetric two-body states is provided by the spherical waves:

$$j_l(kr)Y_{lm}(\theta\phi) \times \begin{cases} |\chi_0\rangle & (l \text{ even}) \\ |\chi_\nu^1\rangle & (l \text{ odd}). \end{cases} \tag{8}$$

The Pauli principle does not apply to the n-p system. All spin configurations can therefore occur for every value of l.

Henceforth we shall use the conventional spectroscopic notation for angular momentum eigenstates. A state having the total spin S, total orbital angular momentum L, and total angular momentum J (i.e., $\mathbf{J} = \mathbf{L} + \mathbf{S}$) shall be designated by * ${}^{2S+1}L_J$, where the letters S, P, D, F, G, etc., are used for the values $L = 0, 1, 2, \ldots$. Needless to say $|L - S| \leqslant J \leqslant L + S$. The magnetic quantum numbers are not stated in this shorthand; there are therefore $(2J + 1)$ states for each symbol. In this notation the possible p-p states are ${}^1S_0$, ${}^3P_{0,1,2}$, ${}^1D_2$, etc. The n-p system can also occur in the states ${}^3S_1$, ${}^1P_1$, etc. Frequently the quantum number J shall be deleted from the symbol; in this case the discussion applies to all states having the quoted values of S and L.

The qualitative properties of nuclear systems indicate that nuclear forces must be dominantly attractive and of short range. The simplest possible assumption that we could make is that they are central and spin independent, i.e., that the spin operators do not appear in the interaction Hamiltonian. As an example of such an interaction potential, one could take $V(r) = ge^{-r/a}/r$, the Yukawa potential, where g is a number that characterizes the strength of the force, and the range parameter a is of order 10^{-13} cm. As the spins do not appear in the Hamiltonian, the energy must be independent of S. Consequently a bound state of the n-p system having orbital angular momentum L must be $4(2L + 1)$-fold degenerate, because there can be no energy difference between ${}^3L$ and ${}^1L$.

The only bound two-nucleon configuration that occurs in nature is the deuteron. It is made up of a neutron and proton, has a binding energy of -2.22 Mev, and a total angular momentum $J = 1$. The last fact proves that the n-p force must be spin dependent, because we have just shown that a spin-independent Hamiltonian must lead to bound states that have at least a fourfold degeneracy. The simplest generalization of the interaction that can hope to account for the data would therefore have an explicit dependence on $\mathbf{S} = \frac{1}{2}(\boldsymbol{\sigma}_1 + \boldsymbol{\sigma}_2)$. Rotational invariance requires that only scalars appear in the potential, and we therefore suppose the interaction V_{np} to be $V_a(r) + \mathbf{S}^2 V_b(r)$. The observables $\mathbf{S}^2$, S_z, $\mathbf{L}^2$, and L_z are all conserved when V_{np} has this form, and the states ${}^{2S+1}L_J$ are therefore energy eigenfunctions. The triplet state now experience the interaction $V_a + 2V_b$, whereas the singlet interaction is V_a. By making V_b sufficiently attractive we can therefore bind one ${}^3S_1$-state, while leaving all ${}^1S$-states (not to speak of states with $L \neq 0$) in the continuum. The assignment of ${}^3S$ to the deuteron receives strong support from the observation that the magnetic moment of the deuteron is almost exactly the sum of the neutron and proton moments. In a

* The state ${}^{2S+1}L_J$ with eigenvalue M of J_z can be constructed by means of the CG-coefficients: $|{}^{2S+1}L_J; M\rangle = \Sigma|LM_LSM_S\rangle\langle LM_LSM_S|JM\rangle$, where M_L and M_S are the eigenvalues of L_z and S_z.

^3S-state the spins are parallel and there is no convection current from the motion of the proton; the magnetic moment is then precisely the sum of the moments belonging to the constituents. (The mathematical proof of this statement is left as an exercise for the reader.)

It is customary to write V_{np} in a slightly different form. We note that $\mathbf{S}^2 = S(S+1) = \frac{3}{2} + \frac{1}{2}\boldsymbol{\sigma}_1 \cdot \boldsymbol{\sigma}_2$, because $\boldsymbol{\sigma}_i^2 = 3$. Hence we can express V_{np} in the equivalent form

$$V_{np} = V_1(r) + \boldsymbol{\sigma}_1 \cdot \boldsymbol{\sigma}_2 V_2(r). \tag{9}$$

The argument preceding Eq. (9) shows that $V_{np} = V_1 - 3V_2$ in singlet states and $V_{np} = V_1 + V_2$ in triplet states.

This was roughly the state of the theory before Rabi *et al.* determined the hyperfine structure of heavy hydrogen and discovered that the deuteron must have a nonvanishing electric quadrupole moment Q. This immediately shows that the interaction must be more complicated than (9). If the ground state were really ^3S, Q would vanish because the charge distribution in such a state is spherically symmetric. The ^1P-state is also ruled out because (a) a potential such as (9) cannot have a bound P-state without a bound S-state, and (b) the magnetic moment of such a state would certainly disagree with the data already quoted. The measured value of Q is actually quite small in the sense that it is much smaller than eR^2, where R is some dimension typical of the deuteron (i.e., $R \sim 2 \times 10^{-13}$ cm). The most plausible explanation of the data was put forward by Rarita and Schwinger, who observed that rotational invariance only requires $\mathbf{J} = \mathbf{L} + \mathbf{S}$ to be conserved, but not $\mathbf{L}$ and $\mathbf{S}$ separately. For example, the state $a|^3S_1\rangle + b|^3D_1\rangle + c|^3P_1\rangle + d|^1P_1\rangle$ would have a nonzero Q, and by making $|b|$, $|c|$, and $|d|$ small compared to $|a|$ the magnetic moment would still be close to the sum of the neutron and proton moments. On the other hand, this state is not an eigenstate of the reflection operator, because the P-states have the opposite parity to the S- and D-states. We therefore assume that the deuteron ground state is $a|^3S_1\rangle + b|^3D_1\rangle$, with $|a| \gg |b|$.

Let us construct a force that will produce such a ground state. As it does not mix singlets and triplets, we assume that it is symmetric in $\boldsymbol{\sigma}_n$ and $\boldsymbol{\sigma}_p$. If it is to produce a nonzero matrix element between S- and D-states it must be a second rank tensor in the orbital variables. Furthermore, it must be invariant under rotations and reflections. If one also insists on time-reversal invariance, the only possible combinations are

$$(\boldsymbol{\sigma}_1 \cdot \mathbf{r})(\boldsymbol{\sigma}_2 \cdot \mathbf{r}), \tag{10}$$

$$(\boldsymbol{\sigma}_1 \cdot \mathbf{p})(\boldsymbol{\sigma}_2 \cdot \mathbf{p}), \tag{11}$$

$$(\boldsymbol{\sigma}_1 \cdot \mathbf{L})(\boldsymbol{\sigma}_2 \cdot \mathbf{L}), \tag{12}$$

multiplied by central potentials. Here $\mathbf{p}$ is the relative momentum operator. The simplest of these is $(\boldsymbol{\sigma}_1 \cdot \mathbf{r})(\boldsymbol{\sigma}_2 \cdot \mathbf{r})$, the so-called tensor force.

On the basis of the deuteron data alone one could not distinguish very easily between (10) and (11), but (12) would be ruled out because it has no matrix element between the ^3S- and ^3D-states. The variation of the scattering phase shifts at low energies is not compatible with (11), and so (10) is usually employed. If time reversal invariance is abandoned, $[(\boldsymbol{\sigma}_1 \cdot \mathbf{r})(\boldsymbol{\sigma}_2 \cdot \mathbf{p}) + (\boldsymbol{\sigma}_2 \cdot \mathbf{r})(\boldsymbol{\sigma}_1 \cdot \mathbf{p})]$ would be a possible form.

With any one of these tensor type forces, the eigenstates of H are

$$\text{n-p:}\quad \begin{aligned} J &= 0: && {}^1S_0,\ {}^3P_0, \\ J &= 1: && {}^3S_1 + {}^3D_1,\ {}^3P_1,\ {}^1P_1, \\ J &= 2: && {}^1D_2,\ {}^3P_2 + {}^3F_2,\ {}^3D_2, \end{aligned} \tag{13}$$

where by ${}^3S_1 + {}^3D_1$ we mean some unknown linear combination of $|{}^3S_1\rangle$ and $|{}^3D_1\rangle$, etc. In scattering there will be an energy-dependent phase shift for each of these eigenstates. Thus $|{}^3D_1\rangle$ will no longer be a "eigen partial wave," instead two (energy-dependent) combinations of $|{}^3D_1\rangle$ and $|{}^3S_1\rangle$ will be, with two separate phase shifts. As the tensor force is "turned off" one of these will tend to $\delta({}^3S_1)$, the other to $\delta({}^3D_1)$.

We may note that there is no symmetry principle stated so far which prevents the existence of a force that mixes ${}^3P_1$ and ${}^1P_1$. Such a force must be antisymmetric in $\boldsymbol{\sigma}_1$ and $\boldsymbol{\sigma}_2$, e.g.,

$$(\boldsymbol{\sigma}_1 - \boldsymbol{\sigma}_2) \cdot \mathbf{L}, \tag{14}$$

$$(\boldsymbol{\sigma}_1 \times \boldsymbol{\sigma}_2) \cdot \mathbf{L}. \tag{15}$$

Of these (15) is ruled out by time reversal. Because (14) is odd under interchange of the two particles, it cannot appear in the p-p case.

Let us turn to the p-p system now. If we assume that V_{pp} is of the form (9) plus a tensor force, the possible states are not (13), but

$$\text{p-p:}\quad \begin{aligned} J &= 0: && {}^1S_0,\ {}^3P_0, \\ J &= 1: && {}^3P_1, \\ J &= 2: && {}^1D_2,\ {}^3P_2 + {}^3F_2. \end{aligned} \tag{16}$$

Note that triplet-singlet mixing is not possible here if one wants the parity to be conserved.

Empirically we now believe that there exists another important symmetry, namely that the phase shifts of (16) equal those of (13) when the quantum numbers coincide; i.e., $\delta_{np}({}^3P_1) = \delta_{pp}({}^3P_1)$, etc. This is known as the charge independence of nuclear forces, and has very far reaching consequences. The principle of charge independence states that the forces between nucleons (neutron *or* proton) are independent of their charges, except for the small effects of Coulomb interaction. This principle also appears to hold for pion phenomena, i.e., the Hamiltonian describing the pion-nucleon system seems to be invariant under transformations like $\pi^+ \leftrightarrow \pi^-$, p $\leftrightarrow$ n, etc. This leads to further groups of symmetries that are in fact isomorphic to the rotation group (isotopic spin), but these questions are outside the scope of this book.* If charge independence is assumed, then the absence of (14) in the n-p interaction is guaranteed.

43. Scattering of Identical Particles

If one wishes to treat collisions between indistinguishable particles, one must modify somewhat the scattering theory of Chapt. III. We gave this matter a cursory glance when we discussed p-p scattering in the preceding section, but we must still give explicit formulas for the scattering cross section and related quantities.

43.1 BOSON-BOSON SCATTERING

Consider a collision between identical spin 0 bosons as viewed in the c.o.m. frame. We shall use the relative and c.o.m. coordinates and momenta defined on p. 44 (with $m_1 = m_2 = m$). Such a collision can be described by the symmetrized packet

$$\Xi(\mathbf{r}, \mathbf{R}, t) = 2^{-\frac{1}{2}} u(\mathbf{R}, t)[\Psi(\mathbf{r}, t) + \Psi(-\mathbf{r}, t)], \tag{1}$$

where Ψ is described at length in (12.24) *et seq.*, and u is a normalized packet that keeps $\mathbf{R}$ localized near $\mathbf{R} = 0$ throughout.

In a collision experiment we place a detector at $\mathbf{r}_0$ and count particles. As we are dealing with a two-body wave function, some care must be taken in defining the current that is to be used in computing this counting rate. By hypothesis we cannot distinguish between particles that were originally in the accelerator, and those that were in the target. To make the calculation as clear as possible, let us assume that the bosons have a charge Ze. The detector could then measure the rate at which charge impinges on it, $\dot{Q}$. From this information we would then conclude that the counting rate $\dot{N}_{sc}$ is $\dot{Q}/Ze$. The current associated with $\dot{Q}$

* See G. C. Wick, *Ann. Rev. Nucl. Sci.* **9**, 1 (1959).

is not the probability current in configuration space, as was to be expected in view of the remarks following (5.21). What we require is the electric current (5.23):

$$\mathbf{i}(\mathbf{r}_0, t) = (2Ze/m)\,\mathrm{Re}\int \Xi^* \frac{\hbar}{i}\left(\frac{1}{2}\frac{\partial}{\partial\mathbf{R}} + \frac{\partial}{\partial\mathbf{r}}\right)\Xi\;\delta(\tfrac{1}{2}\mathbf{r} + \mathbf{R} - \mathbf{r}_2)\,d^3R\,d^3r. \quad (2)$$

In writing (2) we have already taken advantage of the symmetry of Ξ.

In our case, (2) can be greatly simplified. Because the center of mass does not move, $\mathbf{P}u \simeq 0$; furthermore, $u \simeq 0$ unless $\mathbf{R} \simeq 0$. In the asymptotic region r and r_0 are both large, and therefore

$$\mathbf{i}(\mathbf{r}_0, t) \sim \tfrac{1}{2}Ze\,\mathrm{Re}\int \Psi_S^*(\mathbf{r}, t)\;\mathbf{v}\;\Psi_S(\mathbf{r}, t)\delta(\tfrac{1}{2}\mathbf{r} - \mathbf{r}_0)\,d^3r, \quad (3)$$

where $\mathbf{v}$ is the relative velocity, and $\Psi_S = \Psi(\mathbf{r}, t) + \Psi(-\mathbf{r}, t)$.

The scattered current, $\mathbf{i}_{sc}$, is computed from (3) by replacing Ψ by the scattered packet (12.32):

$$\Psi_S(\mathbf{r}, t) = [f(k\hat{\mathbf{r}}, \mathbf{k}) + f(-k\hat{\mathbf{r}}, \mathbf{k})]e^{i\omega_k t}\Phi(\mathbf{k}r - \mathbf{v}_k t, 0)/r. \quad (4)$$

Hence

$$\mathbf{i}_{sc}(\mathbf{r}_0, t) \sim \tfrac{1}{2}Zev_k\hat{\mathbf{r}}_0|f(\mathbf{k}', \mathbf{k}) + f(-\mathbf{k}', \mathbf{k})|^2 I(t), \quad (5)$$

$$I(t) = \int |\Phi(\hat{\mathbf{k}}r - \mathbf{v}_k t, 0)|^2\delta(\tfrac{1}{2}\mathbf{r} - \mathbf{r}_0)r^{-2}\,d^3r, \quad (6)$$

where $\hbar\mathbf{k}' = \hbar k\hat{\mathbf{r}}_0$ is the relative momentum after scattering. Clearly

$$I(t) = (2/r_0^2)|\Phi(2\hat{\mathbf{k}}r_0 - \mathbf{v}_k t, 0)|^2. \quad (7)$$

To find $dN_{sc}/d\Omega$, we must divide $\mathbf{i}_{sc}$ by Ze, and follow the steps used in connection with (12.36):

$$\begin{aligned} dN_{sc}/d\Omega &= (r_0^2/Ze)\int_{-\infty}^{\infty} dt\,|\mathbf{i}_{sc}(\mathbf{r}_0, t)| \\ &= |f(\mathbf{k}', \mathbf{k}) + f(-\mathbf{k}', \mathbf{k})|^2\int_{-\infty}^{\infty} dz\,|\Phi(0, 0, z, 0)|^2. \end{aligned} \quad (8)$$

Initially there is no overlap between the target and the beam, and no ambiguity in counting the beam particles can arise. Hence dN_{inc}/dA is given by (12.34). But (12.34) is precisely the integral in (8), and therefore

$$\frac{d\sigma}{d\Omega} = |f(\mathbf{k}', \mathbf{k}) + f(-\mathbf{k}', \mathbf{k})|^2 \quad (9)$$

is the theoretical expression for the cross section. It is important to note that with this definition of the differential cross section, the total cross section

$$\sigma = \int d\Omega \, |f(\mathbf{k}', \mathbf{k}) + f(-\mathbf{k}', \mathbf{k})|^2 \tag{10}$$

equals *twice*(no. of particles taken out of the incident beam/incident flux). We may also express (9) in terms of the wave number k and the scattering angle θ:

$$\frac{d\sigma}{d\Omega} = |f(k, \theta) + f(k, \pi - \theta)|^2. \tag{11}$$

If one wishes to transform these results to the laboratory frame, one should use the formulas already given on p. 105.

Let us examine (11) more closely:

$$\frac{d\sigma}{d\Omega} = |f(k, \theta)|^2 + |f(k, \pi - \theta)|^2 + 2 \operatorname{Re} f(k, \theta) f^*(k, \pi - \theta). \tag{12}$$

The first two terms were to be expected, because in the classical theory of collisions between identical particles one has

$$\left(\frac{d\sigma^{\text{id}}}{d\Omega}\right)_\theta = \left(\frac{d\sigma^{\text{dist}}}{d\Omega}\right)_\theta + \left(\frac{d\sigma^{\text{dist}}}{d\Omega}\right)_{\pi-\theta}, \tag{12'}$$

where σ^{id} and σ^{dist} are the classical cross sections when the particles are assumed to be identical and distinguishable, respectively. Equation (12) differs from the classical formula by the interference term $2 \operatorname{Re} f(\theta) f^*(\pi - \theta)$. This is a typically quantum mechanical effect because it is a reflection of the fact that the wave function, as the fundamental object in the theory, has symmetry requirements imposed on it. The effects of this interference are illustrated by the example of Coulomb scattering. According to (17.16)

$$\begin{aligned}
&|f_C(k, \theta) + f_C(k, \pi - \theta)|^2 \\
&\quad = \left(\frac{\gamma}{2k}\right)^2 \left| \frac{e^{2i\gamma \ln \sin \frac{1}{2}\theta}}{\sin^2 \frac{1}{2}\theta} + \frac{e^{2i\gamma \ln \cos \frac{1}{2}\theta}}{\cos^2 \frac{1}{2}\theta} \right|^2 \\
&\quad = \left(\frac{\gamma}{2k}\right)^2 \left\{ \frac{1}{\sin^4 \frac{1}{2}\theta} + \frac{1}{\cos^4 \frac{1}{2}\theta} + 8 \operatorname{cosec}^2 \theta \cos(2\gamma \ln \tan \tfrac{1}{2}\theta) \right\}
\end{aligned} \tag{13}$$

Unlike the first two terms, the interference term depends explicitly on $\hbar$ because $\gamma = Z_1 Z_2 e^2/\hbar v$. The results of an experiment that demonstrates the interference term are shown in Fig. 43.1.

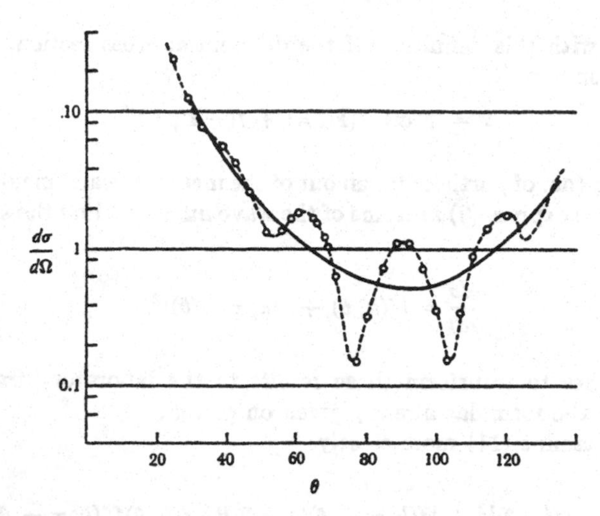

Fig. 43.1. The center-of-mass differential cross section for the elastic scattering of C^{12}-nuclei by C^{12}. The energy in the center-of-mass frame is 5 Mev. The classical cross section (i.e., the symmetrized Rutherford formula) is shown as a heavy line, the experimental points are the open circles, and the dashed line is the Mott formula, Eq. (13). The data is taken from D. A. Bromley, J. A. Kuehner, and E. Almqvist, *Phys. Rev. Lett.* **4**, 365 (1960). Observe that the vertical scale is logarithmic and that the quantum mechanical interference effect is rather immense in the vicinity of 90°. The actual cross section is larger than the classical formula at $\theta = \frac{1}{2}\pi$ because the wave function of two indistinguishable bosons is symmetric. For fermion-fermion scattering, on the other hand, there is a minimum in the cross section at $\theta = \frac{1}{2}\pi$ (see Eq. (27)).

The partial wave expansion of (11) also reveals some interesting features. The expansion of $f(k, \theta)$ is given by (14.13):

$$f(k, \theta) = \frac{1}{k}\sum_l \sqrt{4\pi(2l+1)}\, e^{i\delta_l} \sin\delta_l\, Y_{l0}(\theta).$$

But

$$Y_{l0}(\pi - \theta) = (-1)^l Y_{l0}(\theta),$$

and therefore

$$\frac{d\sigma}{d\Omega} = \frac{4}{k^2}\left| \sum_{l=0,2,4,\ldots} \sqrt{4\pi(2l+1)}\, e^{i\delta_l} \sin\delta_l\, Y_{l0}(\theta)\right|^2. \tag{14}$$

The symmetry requirement has eliminated all partial waves of odd angular momentum.

•So far we have only discussed elastic scattering, but we should like to point out that the statement in italics can be extended to reactions involving identical bosons in the incident state. Assume that we are concerned with α-α scattering in the region of several hundreds of Mev, so that reactions like

$$\begin{aligned} &\alpha + \alpha \rightarrow \alpha + d + d, \\ &\alpha + \alpha \rightarrow He^3 + H^3 + n + p, \\ &\alpha + \alpha \rightarrow 4n + 4p, \end{aligned} \tag{15}$$

are all energetically possible. Clearly we cannot treat the α-particle as a structureless boson if we wish to understand these reactions, and therefore the question arises as to whether any of our previous conclusions still apply. Obviously the formulas for the cross section cannot, but we shall show that it is still rigorously true that all the final states on the right-hand side of (15) will have even total angular momentum if the interaction causing the reactions is rotationally invariant. To prove this we use a time-dependent description, where at time t_0, long before the collision, α-particle I is in a wave packet localized about $\mathbf{r}$ and has mean momentum $\mathbf{P}$, and particle II is localized about $-\mathbf{r}$ and moves with average momentum $-\mathbf{P}$. Let ξ_1, ξ_2, be the space and spin variables of the neutrons in α-particle I, ξ_3, ξ_4 those of the neutrons in II, and let η_i ($i = 1, \ldots, 4$) be the variables pertaining to the protons defined in a similar fashion. If we were to ignore the indistinguishability of the particles, our initial state would be represented by

$$\phi_{\rm I}(\xi_1\xi_1\eta_2\eta_2, t)\phi_{\rm II}(\xi_3\xi_4\eta_3\eta_4, t),$$

where $\phi_{\rm I}$ and $\phi_{\rm II}$ are the two wave packets described above. Note that in $\phi_{\rm I}$ the c.o.m. variable is localized near $\mathbf{r}$ when $t \approx t_0$, and that the lengths of the relative coordinates of the nucleons are restricted to be smaller than 10^{-12} cm in the sense that the wave function decreases exponentially as soon as they become bigger than this. Because we are actually dealing with a system of several indistinguishable fermions, the correct wave function is

$$\Psi = \text{const} \sum \epsilon(P^{(p)})\epsilon(P^{(n)})P^{(p)}P^{n}\phi_{\rm I}(\xi_1\xi_2\eta_1\eta_2, t)\phi_{\rm II}(\xi_3\xi_4\eta_3\eta_4, t), \tag{16}$$

where $P^{(n)}$ and $P^{(p)}$ are permutation operators acting on the neutrons and protons, respectively, $\epsilon(P)$ is the signature of the indicated permutation, and the sum goes over all $(4!)^2$ permutations of the identical nucleons. Because of the exponential fall-off just referred to, most permutations in (16) actually yield incredibly small terms (i.e., of order $\exp(-r/10^{-12}\text{ cm})$, with r macroscopic). The only permutation that does not is the one where *all* the nucleons in I are permuted with *all* of those present in II. Therefore (16) is actually equal to

$$\Psi = \text{const.}\{\phi_{\rm I}(\xi_1\xi_2\eta_1\eta_2, t)\phi_{\rm II}(\xi_3\xi_4\eta_3\eta_4, t) + \phi_{\rm II}(\xi_1\xi_2\eta_1\eta_2, t)\phi_{\rm I}(\xi_3\xi_4\eta_3\eta_4, t)\}.$$

At times before the collision the correctly antisymmetrized wave function is therefore *symmetric* in the c.o.m. variables, and the incident state only contains even partial waves. By hypothesis angular momentum is conserved, and therefore all the final states, no matter how complex, must also have even total angular momentum. *QED.*•

43.2 FERMION-FERMION SCATTERING

The scattering of indistinguishable fermions is necessarily more complicated than that of bosons because of the spin. Here we shall confine ourselves to the most important, and fortunately the simplest case, spin $\frac{1}{2}$. In Sec. 42—cf. (42.16)—we had shown that reflection invariance implies that the length of the total spin S ($S = 0$ or 1) is conserved. Let us retain this assumption here. First we treat the scattering as if the particles were distinguishable. The scattering amplitude is then a matrix in the four-dimensional spin space of the two particles, and can, by hypothesis, be written as

$$M(\mathbf{k}', \mathbf{k}) = M_s(\mathbf{k}', \mathbf{k}) + M_t(\mathbf{k}', \mathbf{k}), \tag{17}$$

where s and t refer to singlet and triplet spin states, respectively. Because the singlet (triplet) state is antisymmetric (symmetric) in the spin variables, the correctly antisymmetrized amplitude is

$$\mathfrak{M}(\mathbf{k}', \mathbf{k}) = [M_s(\mathbf{k}', \mathbf{k}) + M_s(-\mathbf{k}', \mathbf{k})] + [M_t(\mathbf{k}', \mathbf{k}) - M_t(-\mathbf{k}', \mathbf{k})]. \tag{18}$$

The differential cross section when the initial state has the spin-space density matrix ρ_i is then given by (40.14):

$$\frac{d\sigma}{d\Omega} = \text{tr}\, \rho_i \mathfrak{M}^\dagger \mathfrak{M}. \tag{19}$$

Note that ρ_i not only describes the polarization of the incident beam, but also the spin state of the target before the collision. If neither the beam nor the target is polarized, $\rho_i = \frac{1}{4}$. The most general form that ρ can have is

$$\rho = \tfrac{1}{4}\Big[1 + \boldsymbol{\sigma}_1 \cdot \mathbf{P}_1 + \boldsymbol{\sigma}_2 \cdot \mathbf{P}_2 + \sum_{\alpha\beta} \sigma_{1\alpha}\sigma_{2\beta} Q_{\alpha\beta}\Big],$$

where $\sigma_{1\alpha}$ is the αth component of the spin operator belonging to the first particle, etc., $\mathbf{P}_1$ and $\mathbf{P}_2$ are polarization vectors, and $Q_{\alpha\beta}$ is a tensor that describes correlations between the spins of the two particles. Such correlations cannot exist in any initial state that we can hope to prepare, but they will usually exist after the collision.* This spin correlation between the scattered particles can be computed in the same way as we evaluated the polarization after collision in (40.15), i.e., when $\rho_i = \frac{1}{4}$

$$Q_{\alpha\beta} = \frac{\text{tr}\, \sigma_{1\alpha}\sigma_{2\beta}\mathfrak{M}^\dagger\mathfrak{M}}{\text{tr}\, \mathfrak{M}^\dagger\mathfrak{M}}.$$

* Such correlations have been measured in nucleon-nucleon scattering. See Wilson.

One may go somewhat further in the analysis without making detailed assumptions by following the procedure of (40.16) *et seq.*, i.e., by cataloging all the invariant operators in the spin-space that can be constructed from $\boldsymbol{\sigma}_1$, $\boldsymbol{\sigma}_2$, $\mathbf{k}$ and $\mathbf{k}'$, and multiplying each such object by an arbitrary function of k and θ; the sum of these quantities is then the most general form that (17) can have. The two pieces M_s and M_t can then be obtained by use of the operators P_s and P_t that project onto the singlet and triplet states, respectively. These are constructed by noting that $(\mathbf{s}_1 + \mathbf{s}_2)^2 = \frac{3}{2} + 2\mathbf{s}_1 \cdot \mathbf{s}_2$, and therefore

$$\boldsymbol{\sigma}_1 \cdot \boldsymbol{\sigma}_2 = \begin{cases} -3 & (S = 0) \\ 1 & (S = 1) \end{cases},$$

or

$$\begin{aligned} P_s &= \tfrac{1}{4}(1 - \boldsymbol{\sigma}_1 \cdot \boldsymbol{\sigma}_2), \\ P_t &= \tfrac{1}{4}(3 + \boldsymbol{\sigma}_1 \cdot \boldsymbol{\sigma}_2). \end{aligned} \tag{20}$$

These operators have the properties

$$P_s P_t = 0, \qquad P_s + P_t = 1, \tag{21}$$

$$\operatorname{tr} P_s = 1, \qquad \operatorname{tr} P_t = 3. \tag{22}$$

Then

$$\begin{aligned} M_t &= P_t M P_t, \\ M_s &= P_s M P_s. \end{aligned}$$

We shall not spell out this complete catalogue here *; instead we discuss the simple situation which occurs if there are no tensor or spin-orbit forces, i.e., when the potential has the form (42.9):

$$V = V_1(r) + \boldsymbol{\sigma}_1 \cdot \boldsymbol{\sigma}_2 V_2(r). \tag{23}$$

If this is the case M can be expressed as

$$M = f_s(k, \theta) P_s + f_t(k, \theta) P_t, \tag{24}$$

where $f_{s,t}$ are scalars in the spin space. More complicated objects like $(\boldsymbol{\sigma}_1 \cdot \mathbf{k} \times \mathbf{k}')(\boldsymbol{\sigma}_2 \cdot \mathbf{k} \times \mathbf{k}')$ cannot be produced by the interaction potential (23). With this form for the amplitude, (19) becomes

$$\frac{d\sigma}{d\Omega} = |f_s(k, \theta) + f_s(k, \pi - \theta)|^2 \operatorname{tr} \rho_i P_s + |f_t(k, \theta) - f_t(k, \pi - \theta)|^2 \operatorname{tr} \rho_i P_t \tag{25}$$

in virtue of (21). This reduces to

$$\frac{d\sigma}{d\Omega} = \tfrac{1}{4}|f_s(k, \theta) + f_s(k, \pi - \theta)|^2 + \tfrac{3}{4}|f_t(k, \theta) - f_t(k, \pi - \theta)|^2 \tag{26}$$

* This is actually necessary if one wishes to treat nucleon-nucleon scattering above approximately 20 Mev. See Hamilton, Chapt. 8.

in the case of a completely unpolarized initial state ($\rho_i = \frac{1}{4}$). If the forces are completely spin independent as in nonrelativistic electron-electron scattering, $V_2 = 0$, and $f_s = f_t = f$. Equation (26) then reduces to

$$\frac{d\sigma}{d\Omega} = |f(k, \theta)|^2 + |f(k, \pi - \theta)|^2 - \text{Re}\, f^*(k, \theta)f(k, \pi - \theta), \quad (27)$$

which, it should be noted, differs from the cross section (12) for boson-boson scattering. In contrast to (12), the antisymmetrization forces (27) to be *smaller* than the classical value (12′) at $\theta = \frac{1}{2}\pi$.

Finally we investigate the correlations between spins resulting from collisions between completely unpolarized beams; we continue to assume that M has the form (24). Because $\text{tr}\, \sigma_{1\alpha}\sigma_{2\beta}(\boldsymbol{\sigma}_1 \cdot \boldsymbol{\sigma}_2) = 4\delta_{\alpha\beta}$, we have $\text{tr}\, P_s\sigma_{1\alpha}\sigma_{2\beta} = -\delta_{\alpha\beta}$, and $\text{tr}\, P_t\sigma_{1\alpha}\sigma_{2\beta} = \delta_{\alpha\beta}$. Therefore the correlation tensor $Q_{\alpha\beta}$ is diagonal,

$$Q_{\alpha\beta} = \delta_{\alpha\beta}\, Q,$$

with

$$Q(k, \theta) = \frac{|f_t(k, \theta) - f_t(k, \pi - \theta)|^2 - |f_s(k, \theta) + f_s(k, \pi - \theta)|^2}{3|f_t(k, \theta) - f_t(k, \pi - \theta)|^2 + |f_s(k, \theta) + f_s(k, \pi - \theta)|^2}. \quad (28)$$

If the scattering only takes place in one of the two spin states, (28) becomes

$$\begin{aligned} Q(k, \theta) &= \tfrac{1}{3} \qquad \text{(pure triplet)}, \\ &= -1 \qquad \text{(pure singlet)}, \end{aligned} \quad (29)$$

and in general

$$\tfrac{1}{3} \geqslant Q \geqslant -1. \quad (30)$$

The limiting values (29) can be understood as follows. The general formula for $Q_{\alpha\beta}$ reduces to $3Q = \text{tr}\, \rho_f(\boldsymbol{\sigma}_1 \cdot \boldsymbol{\sigma}_2) \equiv \langle \boldsymbol{\sigma}_1 \cdot \boldsymbol{\sigma}_2 \rangle_f$ in the simple example of interest to us now, where ρ_f is the density matrix of the final state. But the eigenvalues of $\boldsymbol{\sigma}_1 \cdot \boldsymbol{\sigma}_2$ are -3 and 1 in the singlet and triplet states, respectively, and this explains the limits (29). (Crudely speaking, the signs of these eigenvalues are "due" to the parallel configuration of the spins in $S = 1$ states, and the antiparallel configuration in the $S = 0$ state.)

Another remarkable result of the exclusion principle is that the spin correlation is nonvanishing even in the absence of spin-dependent forces. Thus with $f_s = f_t = f$, (28) reduces to

$$Q(k, \theta) = -\frac{\text{Re}\, f(k, \theta)f^*(k, \pi - \theta)}{d\sigma/d\Omega}. \quad (31)$$

That such correlations must exist even with spin-independent forces is already clear from the low energy limit where the scattered wave is purely ${}^1S$, and therefore contains only antiparallel spin configurations, i.e., $Q = -1$.

PROBLEMS

1. The interaction picture is distinguished from that of Heisenberg and Schrödinger in that *both* states and dynamical observables move in time. Thus if a system's Hamiltonian is

$$H = H_0 + H'(t),$$

the interaction picture kets, $|a't\rangle_I$, are defined by

$$|a't\rangle_I = e^{iH_0t/\hbar}|a't\rangle_S, \tag{1}$$

where $|a't\rangle_S$ is the ket that represents the same physical state in the Schrödinger picture.

(a) Show that if A is an observable in the S-picture, then in the I-picture it is

$$A_I(t) = e^{iH_0t/\hbar}Ae^{-iH_0t/\hbar}.$$

(b) Show that

$$\left[i\hbar\frac{d}{dt} - H_I'(t)\right]|a't\rangle_I = 0.$$

(c) Let

$$|a't_2\rangle_I = U_I(t_2, t_1)|a't_1\rangle_I.$$

Show that

$$U_I(t_2, t_1) = 1 - \frac{i}{\hbar}\int_{t_1}^{t_2} H_I'(t')U_I(t', t_1)\,dt'. \tag{2}$$

(d) If H' is not an explicit function of t, show directly from (1) that

$$U_I(t_2, t_1) = e^{iH_0t_2/\hbar}e^{-iH(t_2-t_1)/\hbar}e^{-iH_0t_1/\hbar}.$$

(e) A system is prepared in the state $|a'\rangle$ at $t = 0$; show that the probability of finding it in the state $|b'\rangle$ at time t is

$$|\langle b'|e^{-iH_0t/\hbar}U_I(t, 0)|a'\rangle|^2.$$

2. The interaction picture is usually used in computing transition probabilities in systems that are weakly perturbed. As an illustration, consider a one-dimensional oscillator that is perturbed by a weak, time-dependent, but spatially uniform force $K(t)$, where $K(t) = 0$ for $t < 0$. Initially (i.e., $t < 0$) the system is in its ground state $|0\rangle$, and for $t > 0$ we assume that the response can be treated by the expansion of (2) in a power series in $K(t)$.

(a) Show that in the approximation where we truncate the expansion at K^n, the system cannot be excited beyond the nth excited state $|n\rangle$.

(b) To first order in K, show that the probability of finding the system in $|1\rangle$ at time t is

$$p_1(t) = \frac{1}{2m\hbar\omega} |\mathcal{K}(t)|^2,$$

where

$$\mathcal{K}(t) = \int_0^t K(t')e^{i\omega t'}\, dt'.$$

(c) For $t \geqslant 0$, let

$$K(t) = K_0(1 - e^{-t/\tau}).$$

Show that for large times

$$p_1(t) \simeq \frac{K_0^2}{2m\hbar\omega^3}, \tag{3}$$

if $\omega\tau \gg 1$. If $\tau = 0$, show that $p_1(t)$ oscillates indefinitely. Compare the time average of this oscillating function with (3).

(d) Consider an oscillator which is in the uniform, time-independent force field K, and call its ground state $|\bar{0}\rangle$. If $|1\rangle$ is the first excited state of the same oscillator when $K = 0$, show that $|\langle 1|\bar{0}\rangle|^2$ is given by (3) in the lowest nonvanishing approximation, and comment.

3. Let s_i (with $i = x, y, z$) be the angular momentum matrices for $j = 1$. Show that $s_i^3 = s_i$, and from this deduce

$$D^{(1)}(\alpha\beta\gamma) = e^{-i\alpha s_z}[1 - i \sin\beta\, s_y - (1 - \cos\beta)s_y^2]e^{-i\gamma s_z}.$$

Compare this result with the matrix that induces the rotations of a vector.

4. From the definitions given in Sec. 32, derive in succession the following relationships:

(a)
$$\frac{\partial}{\partial\alpha} D(R) = -iJ_z D(R), \qquad \frac{\partial}{\partial\beta} D(R) = -iJ_{y'} D(R),$$
$$\frac{\partial}{\partial\gamma} D(R) = -iJ_\zeta D(R);$$

(b)
$$J_\zeta = J_z \cos\beta + \tfrac{1}{2}(e^{-i\alpha}J_+ + e^{i\alpha}J_-)\sin\beta,$$
$$J_{y'} = \tfrac{1}{2}i(e^{i\alpha}J_- - e^{-i\alpha}J_+),$$
$$[J_\zeta, J_z] = -iJ_{y'} \sin\beta;$$

(c)
$$\mathbf{J}^2 = \operatorname{cosec}^2\beta(J_z^2 + J_\zeta^2 - 2\cos\beta\, J_\zeta J_z) + J_{y'}^2 - iJ_{y'} \cot\beta.$$

Derive the differential equation (34.5), and verify that (34.6) is a solution.

5. Show that

(a)

$$\langle njm|T^{(k)} \cdot U^{(k)}|n'j'm'\rangle = \frac{\delta_{jj'}\,\delta_{mm'}}{2j+1} \times \sum_{n''} \sum_{j''=|j-k|}^{j+k} (-1)^{j-j''}\langle nj\|T^{(k)}\|n''j''\rangle\langle n''j''\|U^{(k)}\|n'j\rangle.$$

(b) $$\langle nj \| \mathbf{J} \| n'j' \rangle = \delta_{nn'} \delta_{jj'} \sqrt{j(j+1)(2j+1)},$$

(c) $$\langle njm | V_\kappa | n'jm \rangle = [j(j+1)]^{-1} \langle njm | \mathbf{V} \cdot \mathbf{J} | n'jm \rangle \langle jm | J_\kappa | jm' \rangle,$$

where V_κ are the spherical components (36.10) of an arbitrary vector operator.

6. Show that the operators

$$\Lambda_l^+ = \frac{l + 1 + \boldsymbol{\sigma} \cdot \mathbf{L}}{2l + 1}, \qquad \Lambda_l^- = \frac{l - \boldsymbol{\sigma} \cdot \mathbf{L}}{2l + 1}$$

are projection operators onto the states $j = l \pm \frac{1}{2}$, respectively, in the subspace of orbital angular momentum l. (Here we are considering a single spin $\frac{1}{2}$ particle.)

7. A particle of spin $\frac{1}{2}$ is scattered by the potential

$$H' = V_0(r) + V_1(r) \boldsymbol{\sigma} \cdot \mathbf{L}.$$

The incident state has momentum $\hbar \mathbf{k}_i$, and is in the spin state $|\chi_i\rangle$.

(a) Show that a solution of the Schrödinger equation can be written as the spinor

$$\psi(\mathbf{r}) = \sum_{l=0}^{\infty} \sqrt{4\pi(2l+1)}\, i^l \{ C_l^+ R_l^{(+)}(r) \Lambda_l^+ + C_l^- R_l^-(r) \Lambda_l^- \} Y_{l0}(\theta) |\chi_i\rangle,$$

where $\cos\theta = \hat{\mathbf{r}} \cdot \hat{\mathbf{k}}_i$, $C_l^\pm$ are constants that are to be fixed by the boundary conditions, and the radial functions are solutions of

$$\left\{ \frac{1}{r^2} \frac{d}{dr} r^2 \frac{d}{dr} + k^2 - \frac{l(l+1)}{r^2} - \frac{2m}{\hbar^2} V_l^\pm(r) \right\} R_l^{(\pm)}(r) = 0,$$

with

$$V_l^+(r) = V_0(r) + l V_1(r) \qquad (l = 0, 1, \ldots),$$
$$V_l^-(r) = V_0(r) - (l+1) V_1(r) \qquad (l = 1, 2, \ldots).$$

(b) Let $\delta_l^\pm$ be the phase shifts for the lth partial wave due to the potentials $V_l^\pm$, i.e.,

$$R_l^{(\pm)}(r) \underset{r \to \infty}{\sim} \frac{\text{const.}}{kr} \sin(kr - \tfrac{1}{2} l\pi + \delta_l^\pm).$$

Show that the scattering matrix $M(k, \theta)$ that appears in

$$\psi(\mathbf{r}) \sim \left\{ e^{i\mathbf{k}_i \cdot \mathbf{r}} + \frac{e^{ikr}}{r} M(k, \theta) \right\} |\chi_i\rangle$$

can be written as

$$M = g(k, \theta) + \boldsymbol{\sigma} \cdot \hat{\mathbf{n}} h(k, \theta),$$

where $\hat{\mathbf{n}}$ is a unit vector along $\mathbf{k}_i \times \mathbf{k}_f$, and

$$g(k, \theta) = \frac{1}{k} \sum_{l=0}^{\infty} \sqrt{\frac{4\pi}{2l+1}} \{ (l+1) e^{i\delta_l^+} \sin \delta_l^+ + l e^{i\delta_l^-} \sin \delta_l^- \} Y_{l0}(\theta),$$

$$h(k, \theta) = \frac{1}{k} \sum_{l=1}^{\infty} \sqrt{\frac{4\pi}{2l+1}} (e^{i\delta_l^+} \sin \delta_l^+ - e^{i\delta_l^-} \sin \delta_l^-) i \sin\theta \frac{d}{d(\cos\theta)} Y_{l0}(\theta).$$

(c) Show that the total cross section is

$$\sigma = \frac{4\pi}{k^2} \sum_l \{(l+1)\sin^2\delta_l^+ + l\sin^2\delta_l^-\}.$$

(d) Apply the eikonal approximation to the spin-dependent interaction H'. Show that in this approximation the functions g and h are given by

$$g(k,\theta) = -ik\int_0^\infty b\,db\,J_0(kb\theta)\{e^{2i\Delta_0(b)}\cos[2kb\,\Delta_1(b)] - 1\},$$

$$h(k,\theta) = ik\int_0^\infty b\,db\,J_1(kb\theta)e^{2i\Delta_0(b)}\sin[2kb\,\Delta_1(b)],$$

where

$$\Delta_i(b) = -\frac{1}{4k}\int_{-\infty}^{\infty} U_i(\sqrt{b^2+z^2})\,dz,$$

and $U_i = \hbar^2 V_i/2m$. Compare these expressions with the exact formulas for h and g derived in (b).

8. Show that the Wigner-Eckart theorem follows from the commutation rules (36.11) and (36.12).

9. We wish to measure the density matrix ρ of a monoenergetic beam of spin $\frac{1}{2}$ particles. Assume that you know the functions $g(k,\theta)$ and $h(k,\theta)$ that appear in the amplitude $M(\mathbf{k}_f, \mathbf{k}_i)$ that describes the scattering of these particles by an infinitely heavy spinless object. Devise an apparatus which measures ρ. (Hint: A double scattering experiment will not suffice.)

10. A beam of neutrons moves through a homogeneous and static magnetic field $\mathbf{\mathcal{H}} = \hat{\mathbf{u}}\mathcal{H}$. At time $t = 0$ the polarization of the beam is known to be $\mathbf{P}$. Show that at time $t > 0$ the polarization is

$$\mathbf{P}(t) = \mathbf{P}\cos^2\omega t + (\mathbf{P}\times\hat{\mathbf{u}})\sin 2\omega t + [\hat{\mathbf{u}}(\hat{\mathbf{u}}\cdot\mathbf{P}) - (\hat{\mathbf{u}}\times\mathbf{P})\times\hat{\mathbf{u}}]\sin^2\omega t,$$

where $\hbar\omega = \mu_0\mathcal{H}$, and μ_0 is the neutron's magnetic moment. Compare this result with the motion of a classical magnetic moment in the same field.

VII

Stationary State Perturbation Theory

The vast majority of problems in quantum mechanics cannot be solved exactly. As in all theories of physics approximation methods are therefore of great importance. Here we shall deal with time-independent problems, which means that we seek the eigenvalues and eigenstates of H. In perturbation theory we split H into two pieces, $H = H_0 + H_1$, where we know the spectrum and eigenstates of H_0, and H_1 is, in some sense, small. Needless to say, a split of this type is often not possible, or at least unknown. In that case one frequently replaces the system by a simpler model for which perturbation theory works, and which, it is hoped, captures some of the essential features of the actual problem at hand. This art of finding models is something that can only be taken up within the context of a particular subject (e.g., solid state or nuclear physics). Here we shall treat systems where good zero-order solutions are rather obvious.

44. Symmetries and Perturbation Theory

In splitting the Hamiltonian into

$$H = H_0 + H_1 \tag{1}$$

it is tacitly assumed that H_0 is completely understood. That is to say, the eigenvectors $|a_0'E_n^0\rangle$ and eigenvalues E_n^0 are known:

$$(H - E_n^0)|a_0'E_n^0\rangle = 0. \tag{2}$$

Here a_0' stands for the eigenvalues of the remaining observables A_0 in the complete compatible set. The set $\{|a_0'E_n^0\rangle\}$ is assumed to span the space in which the complete Hamiltonian H operates.

Let $|a'E_n\rangle$ be the eigenkets of the complete compatible set of observables (A, H). As we shall see in a moment, the sets A and A_0 are not identical in general. It is the task of perturbation theory to provide practical techniques for the approximate computation of $\langle a_0'E_n^0|a'E_m\rangle$ and E_n. In principle this problem is solved by the solution to the secular equation

$$\det\|\langle a_0'E_n^0|H_1|a_0''E_m\rangle + (E_n^0 - E)\,\delta_{nm}\,\delta_{a_0'a_0''}\| = 0, \tag{3}$$

but this, as it stands, is rarely a practical solution.

The complexity of the problem is frequently reduced by symmetry considerations. To see this consider the example where H_0 is the Hamiltonian of the isotropic oscillator in three dimensions,

$$H_0 = \frac{1}{2m}(p_x^2 + p_y^2 + p_z^2) + \tfrac{1}{2}m\omega^2(x^2 + y^2 + z^2), \tag{4}$$

with eigenvalues

$$E^0_{n_xn_yn_z} = (n_x + n_y + n_z + \tfrac{3}{2})\hbar\omega, \tag{5}$$

where $n_x = 0, 1, \ldots$, etc. The energy spectrum is highly degenerate, because (5) depends only on the single quantum number $N = n_x + n_y + n_z$. We shall therefore write E_N^0 instead of (5). The degree of degeneracy g_N, i.e., the number of linearly independent states with energy E_N^0, is

$$g_N = \tfrac{1}{2}(N + 1)(N + 2). \tag{6}$$

The Schrödinger equation belonging to (4) can be solved in a variety of coordinate systems: Cartesian, polar and spherical. In Cartesian coordinates the good * quantum numbers are n_x, n_y and n_z, and the

* The colloquialism "good quantum number" refers to the eigenvalues of the conserved observables.

wave functions are products of functions of the type (31.25) for each coordinate. In spherical coordinates, l and m, the eigenvalues of $\mathbf{L}^2$ and L_z, are the good quantum numbers beside N, and the wave functions are a spherical harmonic times the solution of the radial Schrödinger equation for angular momentum l with potential $\frac{1}{2}m\omega^2r^2$. The wave functions in the $|Nlm\rangle$-representation can be expressed as linear combinations of wave functions of the type $\psi_{n_x}(x)\psi_{n_y}(y)\psi_{n_z}(z)$ provided $n_x + n_y + n_z$ also equals N.

N	g_N	l	parity
0	1	0	+
1	3	1	−
2	6	0, 2	+
3	10	1, 3	−
4	15	0, 2, 4	+

The assignment of l to the various degenerate levels can be made on the basis of a simple argument. According to (31.25), $\psi_n(-x) = (-1)^n\psi_n(x)$, and the parity of all states with energy E_N^0 is therefore $(-1)^N$. Hence l is even when N is, and vice versa. We also know that there are $(2l+1)$ orthogonal states for each value of l. Consequently $l = 1$ for $N = 1$. For $N = 2$ there are two possibilities: there are either six linearly independent states with $l = 0$, or one such state and an $l = 2$ multiplet. Higher values of l are impossible because the total multiplicity of the $N = 2$ level is only 6. The rather absurd possibility of six $l = 0$ states can be discarded because the radial Schrödinger equation for one value of l cannot produce orthogonal wave functions having the same energy. Our conclusions concerning the quantum numbers are shown in the table.

In polar coordinates the good quantum numbers are N, n_z and m, where m is again the angular momentum about the z-axis; the values these numbers can assume are obvious.

Consider now the following perturbations:

$$H_1^{(a)} = V(r), \tag{7}$$

$$H_1^{(b)} = V_x(x) + V_y(y) + V_z(z), \tag{8}$$

$$H_1^{(c)} = V_\rho(x^2 + y^2). \tag{9}$$

In the Cartesian representation the perturbation $H_1^{(a)}$ has nonvanishing matrix elements between all states for which N is either even or odd, because (7) is reflection invariant. After solving the secular problem, we would find certain peculiar degeneracies of the energy spectrum, the degree of degeneracy always being odd. In this simple example it is clear that this is an exceptionally stupid procedure, because $V(r)$ commutes with $\mathbf{L}^2$ and L_z, but not with the operators $(p_x^2/2m) + \frac{1}{2}m\omega x^2$, etc. When one uses the Cartesian basis to find the eigenvectors of $H_0 + H_1^{(a)}$ one must, by laborious diagonalization of the secular determinant, construct the spherical harmonics from Hermite polynomials in the Cartesian components of $\mathbf{r}$. Obviously the spherical basis, in which $\mathbf{L}^2$ and L_z are already diagonal, is best suited to the spherically symmetric perturbation (7). In this representation the secular determinant breaks up into diagonal blocks characterized by the symmetry quantum numbers l and m. Furthermore, it is not necessary to solve for the energy eigenvalues for different values of m because the Hamiltonian is spherically symmetric, i.e., because the eigenvalues cannot depend on m. Similar considerations immediately lead to the conclusion that the Cartesian and polar representations are the most convenient basis for the perturbations (8) and (9), respectively.

A number of general conclusions can be drawn from the simple example just discussed: (a) the secular problem is simplified if one uses a basis in which the constants of the motion are already diagonal, and (b) when a perturbation is less symmetrical than the unperturbed Hamiltonian, the degree of degeneracy of the energy spectrum is reduced. In the preceding example, H_0 is exceptionally symmetric and consequently several distinct sets of compatible observables exist. The perturbations (7)–(9) single out one of these sets.

We can also illustrate these conclusions with the important example of the two-nucleon system already discussed in Sec. 42. Let the unperturbed Hamiltonian be

$$H_0 = \frac{p_1^2}{2m_1} + \frac{p_2^2}{2m_2} + V_0(r), \tag{10}$$

and consider the perturbations

$$H_1 = \boldsymbol{\sigma}_1 \cdot \boldsymbol{\sigma}_2\, V_1(r), \tag{11}$$

$$H_2 = (\boldsymbol{\sigma}_1 + \boldsymbol{\sigma}_2) \cdot \mathbf{L}\, V_2(r), \tag{12}$$

$$H_3 = (\boldsymbol{\sigma}_1 \cdot \mathbf{r})(\boldsymbol{\sigma}_2 \cdot \mathbf{r}) V_3(r). \tag{13}$$

In the center-of-mass system the eigenfunctions of H_0 may be labeled by the energy, the eigenvalues of $\mathbf{L}^2$ and L_z, and, say, s_{1z} and s_{2z}. When the spin-spin interaction H_1 is incorporated, the latter two operators no longer commute with H, and one must use $\mathbf{S}^2$ and S_z (with $\mathbf{S} = \mathbf{s}_1 + \mathbf{s}_2$) instead. When the spin-orbit term H_2 is added to the Hamiltonian, S_z and L_z are also not constants of the motion, and one must construct eigenfunctions of the total angular momentum $\mathbf{J} = \mathbf{L} + \mathbf{S}$. The latter states are still eigenfunctions of $\mathbf{L}^2$. The tensor force H_3 does not commute with $\mathbf{L}^2$, however, and the eigenfunctions of the total Hamiltonian $(H_0 + H_1 + H_2 + H_3)$ are consequently not eigenfunctions of $\mathbf{L}^2$ (cf. (42.13)). The degeneracy of the spectrum is reduced correspondingly as the complexity of the Hamiltonian increases. For example, when the Hamiltonian is $(H_0 + H_1)$, the phase shifts * for the $J = 0, 1, 2$ states ^3P_J are all equal, but the spin-orbit interaction (12) will destroy these equalities between the phase shifts.

The foregoing discussion of the two-nucleon system is summarized in the table below. "Yes" or "no" indicates whether the observables are constants of the motion or not.

Hamiltonian	Compatible Sets of Observables			
	$(\mathbf{L}^2, L_z, s_{1z}, s_{2z})$	$(\mathbf{L}^2, L_z, \mathbf{S}^2, S_z)$	$(\mathbf{J}^2, \mathbf{L}^2, \mathbf{S}^2, J_z)$	$(\mathbf{J}^2, \mathbf{S}^2, J_z,$ parity$)$
H_0	yes	yes	yes	yes
$H_0 + H_1$	no	yes	yes	yes
$H_0 + H_1 + H_2$	no	no	yes	yes
$H_0 + H_1 + H_2 + H_3$	no	no	no	yes

The simple examples discussed here should not lead one to think that it is always an easy matter to find the representation appropriate to the perturbed Hamiltonian. It is only in the two-body problem that the generators of the symmetry transformations serve to specify the states. In more complex problems the remaining constants of the motion can only be determined by detailed dynamical calculations.

45. The Rayleigh-Schrödinger Perturbation Expansion

Let us assume now that the greatest possible advantage of symmetry has already been taken, and that we are working in a subspace belonging to definite eigenvalues of all the obvious constants of motion. For the moment we shall also assume that in this subspace the discrete part of

* The relationship between the phase shifts and energy eigenvalues will be given in Sec. 49.1. But one does not need to know this relationship to understand this sentence. When the Hamiltonian is $H_0 + H_1 + H_2$, the phase shift for the partial wave with quantum numbers (S, L, J) is computed from the radial Schrödinger equation with the potential

$$V_0(r) + [2S(S+1) - 3]V_1(r) + [J(J+1) - S(S+1) - L(L+1)]V_2(r).$$

H_0's spectrum is nondegenerate. In general, H_0 will have part of its spectrum in the continuum, but we are only concerned with computing the perturbations of the discrete spectrum here.* The assumption that there is no degeneracy shall be removed shortly.

45.1 THE NONDEGENERATE CASE

Let $\{|E_n^0\rangle\}$ be a complete set that satisfies

$$(H_0 - E_n^0)|E_n^0\rangle = 0, \tag{1}$$

and let $|E_n\rangle$ be the eigenket of H that tends to $|E_n^0\rangle$ as $H_1 \to 0$:

$$(H - E_n)|E_n\rangle = 0, \tag{2}$$

$$\lim_{H_1\to 0} |E_n\rangle = |E_n^0\rangle, \qquad \lim_{H_1\to 0} E_n = E_n^0. \tag{3}$$

Since $\langle E_n^0|H - E_n|E_n\rangle = 0 = (E_n^0 - E_n)\langle E_n^0|E_n\rangle + \langle E_n^0|H_1|E_n\rangle$, the energy shift, defined as

$$\Delta_n = E_n - E_n^0,$$

is given by

$$\Delta_n = \frac{\langle E_n^0|H_1|E_n\rangle}{\langle E_n^0|E_n\rangle}. \tag{4}$$

We now try to solve for $|E_n\rangle$ in a fashion reminiscent of the treatment of collisions in Sec. 30.4. Instead of (2) we write

$$(H_0 - E_n^0)|E_n\rangle = (\Delta_n - H_1)|E_n\rangle \tag{5}$$

and treat this as an inhomogeneous equation. Clearly the "homogeneous" part has the solution $|E_n^0\rangle$. The Green's operator $(H_0 - E_n^0)^{-1}$ is not defined, however, because of the usual singularity. In the collision problem we avoided the singularity by deforming the integration path. This cannot be done here because we only have a sum over a discrete set of energies in the vicinity of E_n^0. Nevertheless, we can do the thing closest to the principal value of integration, namely forbid the appearance of $|E_n^0\rangle$ when we insert the complete set between $(H_0 - E_n^0)^{-1}$ and $(\Delta_n - H_1)|E_n\rangle$. To this end we introduce the projection operators

$$P_n = |E_n^0\rangle\langle E_n^0|, \qquad Q_n = 1 - P_n, \tag{6}$$

and consider

$$|E_n\rangle = |E_n^0\rangle + \frac{Q_n}{H_0 - E_n^0}(\Delta_n - H_1)|E_n\rangle. \tag{7}$$

* Perturbations in the continuum are observed as scattering events. The extension of perturbation theory to the continuous part of the spectrum is taken up in Sec. 49.

We shall now show that (7) is well defined and satisfies (5); clearly it guarantees (3), and is therefore a valid replacement of the Schrödinger equation. Consider then the "scattering" term*

$$|E_n\rangle - |E_n^0\rangle = \sum_m \frac{Q_n}{H_0 - E_n} |E_m^0\rangle\langle E_m^0|\Delta_n - H_1|E_n\rangle$$

$$= \sum_m |E_m^0\rangle \frac{1 - \delta_{m,n}}{E_m^0 - E_n^0} \langle E_m^0|\Delta_n - H_1|E_n^0\rangle.$$

The $m = n$ term appears to be 0/0. However, (4) tells us that

$$\langle E_n^0|\Delta_n - H_1|E_n\rangle = 0, \tag{8}$$

and therefore the explicitly finite expression

$$|E_n\rangle = |E_n^0\rangle + \sum_{m \neq n} |E_m^0\rangle \frac{\langle E_m^0|\Delta_n - H_1|E_n\rangle}{E_m^0 - E_n^0} \tag{9}$$

is equivalent to (7). That (9) is actually a solution of (5) can be demonstrated by applying the operator $(H_0 - E_n^0)$ to (9) and using (8). We have therefore proven that (7) (or equivalently, (9)) is the desired formal solution of the Schrödinger equation.

As it stands, (7) does not provide us with a normalized solution $|E_n\rangle$. Upon taking the scalar product of (7) with $\langle E_n^0|$, we obtain

$$\langle E_n^0|E_n\rangle = 1. \tag{10}$$

Consequently $\langle E_n|E_n\rangle = \Sigma_m|\langle E_n|E_m^0\rangle|^2$ becomes

$$\langle E_n|E_n\rangle = 1 + \sum_{m \neq n} |\langle E_m^0|E_n\rangle|^2, \tag{11}$$

which necessarily exceeds unity.

In view of (4) and (10) the energy shift is

$$\Delta_n = \langle E_n^0|H_1|E_n\rangle. \tag{12}$$

A useful equation for Δ_n can be obtained by substituting (7) into this last identity:

$$\Delta_n = \langle E_n^0|H_1|E_n^0\rangle + \langle E_n^0|H_1 \frac{Q_n}{H_0 - E_n^0} (\Delta_n - H_1)|E_n\rangle. \tag{13}$$

These relations for the energy shift Δ_n and state vector $|E_n\rangle$ are exact. They only become useful if they are approximated, however. Our basic

* In using a summation symbol here we do not mean to imply that the continuous portion of the spectrum of H_0 is to be deleted.

assumptions are that H_1 is "small" in some sense, that $\langle E_n^0|H_1|E_m^0\rangle$ exists,* and that (7) and (13) can be solved by iteration.† The successive approximations can be labeled by the number of powers of H_1 that are retained. To zeroth order we have the trivial results $\Delta_n^{(0)} = 0$ and $|E_n\rangle^{(0)} = |E_n^0\rangle$. To first order in H_1 we can ignore the second term in (13):

$$\Delta_n^{(1)} = \langle E_n^0|H_1|E_n^0\rangle. \tag{14}$$

This shows that the lowest correction to the energy is given by the expectation value of H_1, a fact that should be committed to memory. To the same order one can replace $|E_n\rangle$ by $|E_n^0\rangle$ in the right-hand side of (9). The term involving Δ_n then drops out by orthogonality, and the lowest order approximation to the eigenvector is therefore

$$|E_n\rangle^{(1)} = \left(1 + \frac{Q_n}{E_n^0 - H_0} H_1\right)|E_n^0\rangle. \tag{15}$$

To first order in H_1 this ket is normalized to unity because $|E_n\rangle^{(1)} - |E_n^0\rangle$ is orthogonal to $|E_n^0\rangle$. The second order approximation to the energy shift can be obtained by inspection from (13):

$$\begin{aligned}\Delta_n^{(2)} &= \langle E_n^0|H_1\left(1 + \frac{Q_n}{E_n^0 - H_0} H_1\right)|E_n^0\rangle \\ &\equiv \Delta_n^{(1)} + \sum_{m\neq n} \frac{|\langle E_n^0|H_1|E_m^0\rangle|^2}{E_n^0 - E_m^0}.\end{aligned} \tag{16}$$

The fact that $\Delta_n^{(k)}$ is less complicated than $|E_n\rangle^{(k)}$ is explained by (12), which implies that

$$\Delta_n^{(k)} = \langle E_n^0|H_1|E_n\rangle^{(k-1)}. \tag{17}$$

The formula for $\Delta_n^{(2)}$ is of considerable importance because the perturbation H_1 frequently has no diagonal matrix elements in the unperturbed representation.‡ When this is the case (16) demonstrates that the ground state energy is depressed to leading order by any perturbation because the energy denominator $(E_0^0 - E_m^0)$ is negative definite. This fact is a special consequence of the variational principle (see Sec. 48),

* In certain problems (e.g., nuclear structure and the theory of liquids) these elements do not exist and more complicated formalisms must be resorted to. See bibliography at end of Chapt. XI.

† A plethora of noniterative solutions of Eqs. (7) and (13) have been proposed. Some of these are discussed by Morse and Feshbach, Sec. 9.1.

‡ An ingenious technique which allows one to perform the sum in (16) under certain circumstances is described by A. Dalgarno and J. T. Lewis, *Proc. Roy. Soc.* **A233**, 70, **A238**, 269 (1956), and C. Schwartz, *Ann. of Phys.* **2**, 156 (1959).

which states that the exact ground state energy always lies below the energy associated with an approximate eigenket.

Although little is known in general about the convergence of the iteration solution of Eqs. (13) or (7), it is clear that under some circumstances one can be optimistic, whereas in other cases there is no hope of success. From (15) and (16) a crude criterion for accuracy is

$$|\langle E_n^0|H_1|E_m^0\rangle| \ll |E_n^0 - E_m^0|. \tag{18}$$

This is only a necessary condition, of course. Useful sufficient conditions are hard to come by. Frequently one can use physical intuition to tell

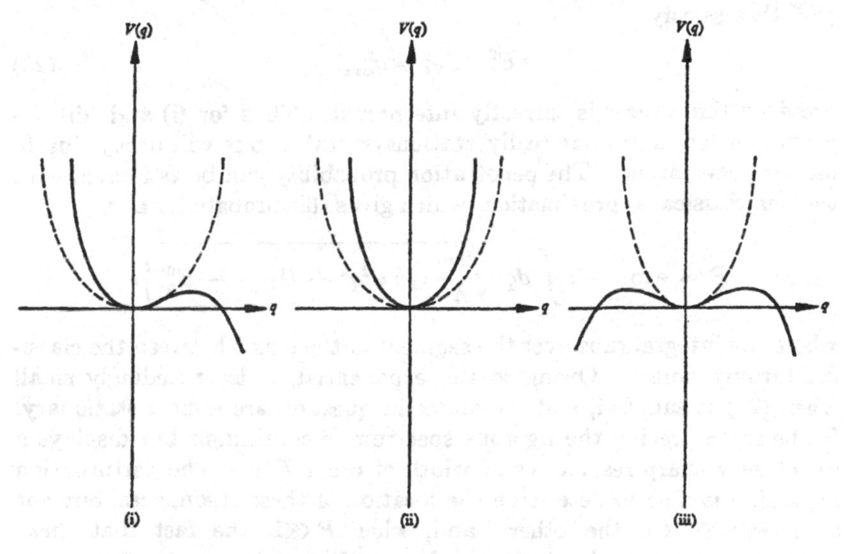

Fig. 45.1. Various perturbations of the one-dimensional harmonic oscillator. The solid lines are (i) $\frac{1}{2}m\omega^2q^2 + H_1^3$; (ii) $\frac{1}{2}m\omega^2q^2 + H_1^4$, $\gamma > 0$; and (iii) same as (ii) but $\gamma < 0$. The dashed line is $\frac{1}{2}m\omega^2q^2$ throughout.

whether (18) is enough. As an illustration, consider two different perturbations of the one-dimensional harmonic oscillator:

$$H_1^{(k)} = \gamma\hbar\omega(q\sqrt{m\omega/\hbar})^k, \tag{19}$$

with $|\gamma| \ll 1$ and $k = 3$ or 4. We note that $\langle E_0^0|H_1^{(k)}|E_n^0\rangle$ vanishes for $n > k$. By making γ sufficiently small one can make $|\langle E_n^0|H_1^{(k)}|E_m^0\rangle| \ll \hbar\omega$, and therefore (18) is always satisfied.

Reflect for a moment, however, on what (19) does to the potential energy curve (see Fig. 45.1). We observe that no matter how small γ

may be, only case (ii) is actually a slight perturbation. In particular, in (i) and (iii) the exact energy spectrum is continuous from $-\infty$ to $+\infty$. Hence it is clear that neither for $H_1^{(3)}$ nor $H_1^{(4)}$ can the expansion in powers of γ have a finite radius of convergence. The situation is not totally bleak, of course. Let us define q_{max} by

$$\tfrac{1}{2}m\omega^2 q_{max}^2 = \gamma\hbar\omega\left(q_{max}\sqrt{\frac{m\omega}{\hbar}}\right)^k.$$

Then for $|q| \ll |q_{max}|$ the potential is only slightly changed by the perturbation. We can then apply perturbation theory to those levels $|E_n^0\rangle$ that satisfy

$$E_n^0 \ll \tfrac{1}{2}m\omega^2 q_{max}^2, \tag{20}$$

provided the answer is correctly interpreted. Thus for (i) and (iii) the perturbed levels are not really stationary states, but will decay due to barrier penetration. The penetration probability can be estimated with the semiclassical approximation, which gives this probability as *

$$P = \exp\left\{-2\int dq\sqrt{\frac{2m}{\hbar^2}\left(\tfrac{1}{2}m\omega^2 q^2 + H_1(q) - E_n^0\right)}\right\},$$

where the integral runs over the segment of the q-axis between the classical turning points. Owing to the exponential, P is exceedingly small when (20) is satisfied, and the states in question are almost stationary. To be more precise, the rigorous spectrum is continuous but displays a set of very sharp resonances of width of order $P\hbar\omega$. The perturbation expansion is able to determine the location of these resonances, but not their width. On the other hand, when $P \lll 1$, the fact that these states are not precisely stationary is not of importance on a time scale small compared to $1/P\omega$. In case (ii) the spectrum is rigorously discrete, and perturbation theory will at least give qualitatively correct answers for all $\gamma > 0$.

•As we have already observed, the fact that one obtains completely different potential curves when γ goes from positive to negative values in H implies that E_n cannot be an analytic function of γ when k is even. Presumably $E_n(\gamma) = E_n^0 + f_n(\gamma) + g_n(\gamma)$, where f_n is analytic at $\gamma = 0$, and g_n is not. Perturbation theory determines f_n as a power series in γ, and completely misses the existence of g_n. Our intuitive considerations lead us to conjecture that g_n is smaller than any power of γ as $\gamma \to 0$, for example $C\exp(-1/|\gamma|)$, where C is real when $\gamma > 0$ and complex when $\gamma < 0$.•

* Cf. e.g., Merzbacher, p. 123.

45.2 THE DEGENERATE CASE

When the criterion (18) is violated, it is clear that the iterative solutions of the equations for the energy shifts and the eigenkets of H are at best unreliable. We shall not even attempt to give a general discussion of what one is to do when the matrix elements of H_1 are comparable in magnitude to the level spacing in the spectrum of H_0. Hardly anything useful is known about this problem; when one is faced with such a situation one must either find a better split of H into $H_0 + H_1$, or devise special techniques adapted to the peculiarities of the particular problem. These are usually very difficult to find. There is, however, a special type of violation of the inequality (18) that is of the utmost importance in physics, and for which a very simple modification of perturbation theory is effective. The most familiar illustration of the situation that we have in mind here is provided by the spectra of atoms, and in particular, the hydrogen atom. Let us identify H_0 with the Coulomb interaction, already treated in Sec. 17, and H_1 with the interaction between the spin of the electron and the magnetic field due to nuclear motion as seen in the electron's rest frame (spin-orbit coupling, cf. Sec. 46), relativistic corrections, and, say, the interaction with the earth's magnetic field. As we already know, the spectrum of H_0 is highly degenerate: there is the degeneracy due to the fact that H_0 does not contain any preferred direction in space, because H_0 does not differentiate between different spin states of the electron, and also the degeneracy peculiar to the Coulomb field. To summarize, the degree of degeneracy of the level with principal quantum number n is $2n^2$, and the spacing between degenerate groups of levels is of order $mc^2/(137)^2$. The perturbations that we have lumped into H_1 are all *very* small compared to this energy spacing, but the degeneracy just delineated still renders the perturbation expansion of the preceding subsection useless. The way out of this difficulty is not difficult to find: One must first diagonalize H exactly in each of the degenerate subspaces, and use perturbation theory only to evaluate the effect of distant levels on the level in question. We shall spell out the details of this procedure in a moment. Before we do so we should emphasize that the example of hydrogen is not atypical. On the contrary, rather similar situations abound in atomic, molecular, solid state and nuclear physics, and even the mass spectrum of the fundamental particles is characterized by nearly degenerate groups of levels (i.e., particles of nearly equal mass).

Let us now formulate perturbation theory for the case where H_0 has degenerate or nearly degenerate sets of levels. Let $E^0_{K,n}$ be the eigenvalues of H_0 belonging to the Kth group of nearly degenerate levels, and $|a_n E^0_{K,n}\rangle$ the associated eigenkets, where a_n are the quantum numbers of the observables that are compatible with H_0. Also define P_K, the projection operator onto the subspace spanned by these states:

$$P_K = \sum_{n=1}^{g_K} |a_n E^0_{K,n}\rangle\langle a_n E^0_{K,n}|, \tag{21}$$

where g_K is the dimensionality of the subspace. We can use the set of operators P_K to separate the perturbation H_1 into a piece H_D that only operates within the subspaces and has no matrix elements between different subspaces, and a remainder, H':

$$H_D = \sum_K P_K H_1 P_K, \tag{22}$$

$$H' = \sum_{K,K'(K\neq K')} P_K H_1 P_{K'}. \tag{23}$$

It is clear that $H = H_0 + H_D + H'$. Because of the degeneracy H_D cannot be treated by perturbation theory, but H' can because it has been constructed to have matrix elements only between different subspaces, and these are assumed to be separated by energy gaps large compared to H_1 (or equivalently, H').

In all subspaces in which g_K is finite, the diagonalization of $H_0 + H_D$ is merely an algebraic problem that is solved by finding the roots of the finite-dimensional secular equations

$$\det\{\langle a_n E^0_{K,n}|H_D|a_m E^0_{K,m}\rangle + \delta_{n,m}(E^0_{K,n} - E^{(1)}_K)\} = 0. \tag{24}$$

The solution of this set of equations provides us with new eigenvalues $E^{(1)}_{K,n}$ and eigenvectors $|b_n E^{(1)}_{K,n}\rangle$, where the quantum numbers b_n specify the eigenvalues of the observables that are compatible with $H_0 + H_D$. In solving these finite-dimensional secular problems the symmetry considerations discussed in the preceding section frequently play a decisive role. We shall discuss a number of examples of this procedure in the remainder of this chapter and in Chapt. XI.

If necessary, H' can now be handled by perturbation theory. For this purpose one transforms to the basis defined by the eigenkets of $H_0 + H_D$. The unitary transformation that does this leaves P_K, and therefore H', invariant. The leading correction to the energy eigenvalue $E^{(1)}_{K,n}$ due to the effects of distant levels is then given by (16):

$$E_{K,n} \simeq E^{(1)}_{K,n} + \sum_{K'n'(K\neq K')} \frac{|\langle b_n E^{(1)}_{K,n}|H'|b_{n'} E^{(1)}_{K',n'}\rangle|^2}{E^{(1)}_{K,n} - E^{(1)}_{K',n'}}. \tag{25}$$

To this accuracy it is legitimate to replace the energy denominator in (25) by $E^0_K - E^0_{K'}$, where these energies are the averages of the eigenvalues of $H_0 + H_D$ in the indicated subspaces. The sum over n' in (25) may then be converted to a sum over the original basis defined by H_0:

$$E_{K,n} \simeq E^{(1)}_{K,n} + \sum_{K' \neq K} \frac{1}{E^0_K - E^0_{K'}} \sum_{n'} |\langle b_n E^{(1)}_{K,n} | H_1 | a_{n'} E^0_{K',n'} \rangle|^2. \quad (26)$$

In some circumstances this formula is somewhat easier to use than (25).

The formalism just developed does not apply when H_0 possesses a degenerate spectrum, and H_1 only has nonzero matrix elements between states belonging to *different* degenerate subspaces. This situation occurs in K°-decay; the appropriate formalism is developed in Prob. 1.

46. *The Fine Structure of Hydrogen*

The zeroth order Hamiltonian for the electron in the field of the proton is

$$H_0 = \frac{p^2}{2\mu} - \frac{e^2}{r}. \quad (1)$$

The corresponding energy eigenvalues are

$$E^0_n = -\frac{\alpha^2 \mu c^2}{2n^2} \qquad (n = 1, 2, \ldots), \quad (2)$$

where

$$\alpha = \frac{e^2}{\hbar c} = \frac{1}{137.04} \quad (3)$$

is the Sommerfeld fine structure constant, and the reduced mass μ equals $0.9995 m_e$. As we already mentioned in the preceding section, the spectrum of H_0 is highly degenerate: there are $2n^2$ levels for each distinct energy eigenvalue. The ground state energy E^0_0 is -13.60 ev, and the size of the atom is of order the Bohr radius $a_0 = \hbar^2/\mu e^2 = 0.5292 \times 10^{-8}$ cm.

There are many corrections that must be added to (1). A thorough treatment of the hydrogen spectrum can only be given within the framework of quantum electrodynamics and the Dirac theory of the electron, but as Jordan and Heisenberg were the first to show, the leading corrections can be computed with the nonrelativistic theory. The most important modifications of the naive theory arise from relativistic kinematics and the electron spin. The relativistic kinetic energy is $\sqrt{(\mu c^2)^2 + p^2c^2} - \mu c^2$. In the ground state the r.m.s. velocity is of order αc, and in excited states it is even smaller. Consequently it is sufficient to expand the relativistic expression in powers of $p/\mu c$, and to approximate the kinetic energy by

$$\frac{p^2}{2\mu} - \frac{1}{2}\left(\frac{p^2}{2\mu}\right)^2 \frac{1}{\mu c^2}. \tag{4}$$

The other important correction comes about because there is a magnetic field $\mathfrak{H}$ in the electron's rest frame due to the motion of the proton as seen in that frame. The electron's magnetic moment interacts with this field with an energy given by $-\boldsymbol{\mu} \cdot \mathfrak{H}$, where

$$\boldsymbol{\mu} = \frac{e\hbar}{2m_e c} \boldsymbol{\sigma},$$

and $\boldsymbol{\sigma}$ is the Pauli spin vector. If $V = -e^2/r$ is the Coulomb energy, then

$$\mathfrak{H} = -\frac{1}{c}\mathbf{v} \times \mathcal{E} = \frac{1}{ec}\mathbf{v} \times \hat{\mathbf{r}}\frac{dV}{dr} = -\frac{\hbar}{emc}\mathbf{L}\frac{1}{r}\frac{dV}{dr}.$$

One would therefore suppose that the additional term in the Hamiltonian due to the spin interaction is

$$\frac{1}{2}\left(\frac{\hbar}{mc}\right)^2 \mathbf{L} \cdot \boldsymbol{\sigma} \frac{1}{r}\frac{dV}{dr}. \tag{5}$$

As we see, this interaction couples the spin of the electron with its own orbital angular momentum. One refers to such an interaction as a *spin-orbit coupling.* The "derivation" of the spin-orbit interaction just given is actually incorrect, because we have been too cavalier with the Lorentz transformation to the electron's rest frame. There should be a further factor of $\frac{1}{2}$ in (5) arising from the so-called Thomas precession. A detailed treatment of this point can be found in Møller, Sec. 21, or Jackson, pp. 364–365.

Our final result for the perturbation is

$$\begin{aligned} H_1 &= H_{\text{kin}} + H_{LS}, \\ H_{\text{kin}} &= -\frac{1}{8}\frac{p^4}{\mu^3 c^2}, \\ H_{LS} &= \frac{1}{2}\left(\frac{\hbar}{mc}\right)^2 \mathbf{L} \cdot \mathbf{s} \frac{1}{r}\frac{dV}{dr}. \end{aligned} \tag{6}$$

The orders of magnitude are

$$\frac{H_{\text{kin}}}{H_0} \simeq \frac{p^2}{(mc)^2} \simeq \frac{H_0}{mc^2} \simeq \alpha^2,$$

and

$$\frac{H_{LS}}{H_0} \simeq \left(\frac{\hbar}{mc}\right)^2 \frac{e^2}{r^3} \bigg/ H_0 \simeq \left(\frac{\hbar}{mca_0}\right)^2 \equiv \alpha^2,$$

which is why α is called the fine structure constant. Quantum electrodynamics shows that all other corrections are smaller than this by further powers of α.

We must, first of all, diagonalize H_1 in the degenerate subspaces belonging to a definite value of n. This will give fractional changes in energy of order α^2, i.e., change the energy by $O(\alpha^4 mc^2)$. According to (45.16) the contribution from nondegenerate levels is proportional to $(\alpha^4 mc^2)^2/\alpha^2 mc^2 \sim \alpha^6 mc^2$, and can therefore be ignored.

H_{kin} commutes with all the angular momentum operators. However, H_{LS} is only compatible with the $\{L^2, J^2, J_z\}$ set, and so we choose these instead of $\{L^2, L_z, s_z\}$. As our eigenstates* we therefore use $1s_{\frac{1}{2}}$, $2s_{\frac{1}{2}}$, $2p_{\frac{1}{2}}$, $2p_{\frac{3}{2}}$, etc., which are constructed from the orbital-spin function by vector addition as in (33.2′). In this representation we can use the fact that

$$\mathbf{j}^2 = (\mathbf{L} + \mathbf{s})^2 = L^2 + s^2 + 2\mathbf{L} \cdot \mathbf{s}$$

to write

$$\begin{aligned}\langle \mathbf{L} \cdot \mathbf{s} \rangle_{lj} &= \tfrac{1}{2}\{j(j+1) - l(l+1) - \tfrac{3}{4}\} \\ &= \begin{cases} \tfrac{1}{2}l & j = l + \tfrac{1}{2} \\ -\tfrac{1}{2}(l+1) & j = l - \tfrac{1}{2} \end{cases}\end{aligned} \tag{7}$$

for the diagonal elements (the others vanish because $\mathbf{L} \cdot \mathbf{s}$ is a scalar). Naturally $\langle \mathbf{L} \cdot \mathbf{s} \rangle = 0$ if $l = 0$. The zero order kets are then $|nljm\rangle = |nl\rangle \cdot |ljm\rangle$, where $|nl\rangle$ refers purely to the radial motion, and

$$\langle r|nl\rangle = R_{nl}(r) \tag{8}$$

are the radial wave functions normalized to

$$\int_0^\infty R_{nl}(r) R_{n'l'}(r) r^2\, dr = \delta_{nn'}\, \delta_{ll'}. \tag{9}$$

The explicit form of $R_{nl}(r)$ was given in Sec. 17.

Thus the only nonvanishing matrix elements of H_{LS} are

$$\langle nljm|H_{LS}|nljm\rangle = \frac{1}{2}\left(\frac{\hbar}{mc}\right)^2 e^2 \left\langle nl \left| \frac{1}{r^3} \right| nl \right\rangle \times \begin{cases} \tfrac{1}{2}l & j = l + \tfrac{1}{2}. \\ -\tfrac{1}{2}(l+1) & j = l - \tfrac{1}{2} \end{cases} \tag{10}$$

* Here we again use standard spectroscopic notation. A one-electron state is specified by the symbol nl_j, where n is the principal quantum number, and for $l = 0, 1, 2, 3,$ etc., we use the letters $s, p, d, f,$ etc.

Within a subspace spanned by the states belonging to one principal quantum number n, H_{kin} is also diagonal. In virtue of (4) its diagonal matrix elements can be written as

$$\langle H_{\text{kin}}\rangle_{nljm} = -\frac{1}{2\mu c^2}\left\langle nl\left|\left(E_n^0 - \frac{e^2}{r}\right)^2\right|nl\right\rangle. \tag{11}$$

If we had not used the states $|nljm\rangle$ as our basis, and had instead used eigenfunctions of L_z and s_z, we would have been forced to diagonalize the Hamiltonian in the subspace of states belonging to one principal quantum number, as discussed at the end of Sec. 45. The resultant eigenvectors are then given in terms of Clebsch-Gordan coefficients, and so we would merely be duplicating work already done. This shows that if one uses a poorly chosen basis one will not make a mistake, but one will be forced to do a lot of unnecessary labor.

Tables of radial integrals involving hydrogen wave functions can be found in Condon and Shortley, p. 117; we require

$$\left\langle nl\left|\frac{1}{r^3}\right|nl\right\rangle = \frac{1}{a_0^3}\frac{1}{n^3(l+1)(l+\frac{1}{2})l}, \tag{12}$$

$$\left\langle nl\left|\frac{1}{r}\right|nl\right\rangle = \frac{1}{a_0 n^2}, \tag{13}$$

and

$$\left\langle nl\left|\frac{1}{r^2}\right|nl\right\rangle = \frac{1}{a_0^2 n^3(l+\frac{1}{2})}. \tag{14}$$

Therefore

$$\langle H_{LS}\rangle_{nlj} = E_n^0\alpha^2\frac{1}{n(2l+1)}\times\begin{cases}\dfrac{1}{l+1}, & j = l+\frac{1}{2}\\[2ex] -\dfrac{1}{l}, & j = l-\frac{1}{2}\end{cases}, \tag{15}$$

if $l > 0$, and

$$\langle H_{\text{kin}}\rangle_{nlj} = -E_n^0\alpha^2\frac{1}{n^2}\left(\frac{n}{l+\frac{1}{2}} - \frac{3}{4}\right). \tag{16}$$

These two corrections can now be combined to give the fine structure correct to $O(\alpha^2)$:

$$E_{nlj} = -E_n^0\left\{1 + \frac{\alpha^2}{n^2}\left(\frac{n}{j+\frac{1}{2}} - \frac{3}{4}\right)\right\}. \tag{17}$$

Note that for fixed j the correction decreases with n since the excited states have lower mean velocities.

According to this formula the levels $2p_{\frac{1}{2}}$ and $2s_{\frac{1}{2}}$ are still degenerate. The same result comes out of the Dirac theory, and actually holds there to all orders of α^2! This is a celebrated failure of the theory, because Lamb and Retherford showed that the levels $2p_{\frac{1}{2}}$ and $2s_{\frac{1}{2}}$ are not degenerate (see Fig. 46.1). This discrepancy, which is known as the Lamb shift, is now completely explained by quantum electrodynamics.

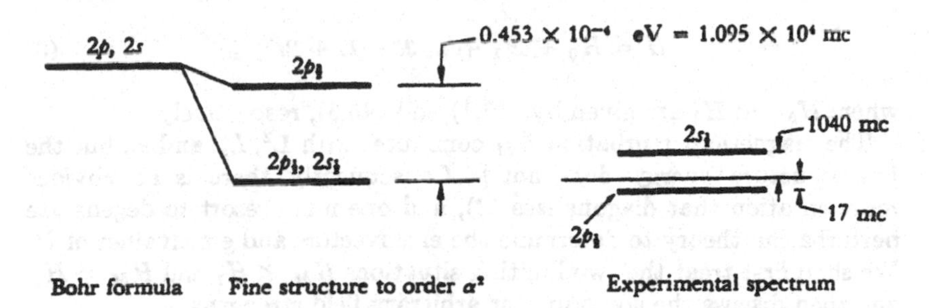

Fig. 46.1. The spectrum for the $n = 2$ states in hydrogen. The symbol "mc" stands for megacycles.

There is a bit of a swindle in the preceding calculations. One would think that $\langle H_{LS}\rangle = 0$ in s-states since $\langle \mathbf{L}\cdot\mathbf{s}\rangle_{l=0} = 0$. However, the radial integral $\langle n0|r^{-3}|n0\rangle$ diverges (see (12)). In order to arrive at (17) we have multiplied (12) by $\langle \mathbf{L}\cdot\mathbf{s}\rangle_{j=l+\frac{1}{2}} = \frac{1}{2}l$. The factor of l in the denominator of (12) therefore disappears, and as a result $\langle H_{LS}\rangle$ has a finite value in s-states. In the Dirac theory this result emerges unambiguously.

47. *The Hydrogen Atom in an External Magnetic Field*

The magnetic moment due to the orbital motion of an electron was found in Sec. 18 to be

$$\boldsymbol{\mu}_{\text{orb}} = -\mu_0 \mathbf{L}, \tag{1}$$

where $\mu_0 = e\hbar/2mc$ is the Bohr magneton. The sign is negative in (1) because we adopt the convention that the electron's charge is $-e$. A useful numerical expression for μ_0 is provided by

$$\mu_0 = 0.579 \times 10^{-8} \text{ ev/gauss.}$$

The spin gives rise to a further contribution to the electron's magnetic moment. According to (38.17) this additional moment is

$$\boldsymbol{\mu}_{\text{spin}} = -2\mu_0 \mathbf{s}. \tag{2}$$

The Hamiltonian of the hydrogen atom in a uniform applied field $\mathcal{H}$ is therefore given by *

$$H = H_0 + H_1 + \mu_0 \mathcal{H} \cdot (\mathbf{L} + 2\mathbf{s}), \tag{3}$$

where H_0 and H_1 are given by (46.1) and (46.6), respectively.

The magnetic perturbation H_M commutes with $\mathbf{L}^2$, L_z, and s_z, but the fine structure energy does not.† Consequently there is no obvious representation that diagonalizes (3), and one must resort to degenerate perturbation theory to determine the eigenvectors and eigenvalues of H. We shall first treat the two limiting situations $H_M \ll H_1$ and $H_M \gg H_1$, and then discuss the spectrum for arbitrary field strengths.

If the external field $\mathcal{H}$ is very weak compared to 10^4 gauss, the magnetic perturbation is small compared to the fine structure splitting. In this situation one first diagonalizes $H_0 + H_1$ as in the preceding section, and then takes H_M into account by means of nondegenerate perturbation theory. This last step is permissible because the degenerate states belonging to $H_0 + H_1$ (e.g., $2s_{\frac{1}{2}}$ and $2p_{\frac{1}{2}}$) have opposite parity and cannot be mixed by H_M. The shift of the fine structure levels to lowest order in $\mathcal{H}$ is therefore given by the expectation value of H_M in the $|nljm\rangle$-representation:

$$\begin{aligned}(\Delta E_M)_{nljm} &= \mu_0\mathcal{H}\langle J_z + s_z\rangle_{nljm} \\ &= \mu_0\mathcal{H}[m + \langle s_z\rangle_{nljm}].\end{aligned}$$

The radial integral in $\langle s_z\rangle$ is merely one, and therefore

$$\begin{aligned}\langle s_z\rangle_{nljm} &= \sum \langle jm|lm_l\tfrac{1}{2}m_s\rangle\langle lm_l\tfrac{1}{2}m_s|s_z|lm_l'\tfrac{1}{2}m_s'\rangle\langle lm_l'\tfrac{1}{2}m_s'|jm\rangle \\ &= \sum m_s|\langle lm_l\tfrac{1}{2}m_s|jm\rangle|^2 \\ &= \pm\frac{m}{2l+1} \qquad (j = l \pm \tfrac{1}{2}),\end{aligned}$$

* Here we ignore the proton's spin and the associated hyperfine structure. Detailed treatments of this problem can be found in C. Schwartz, *Phys. Rev.* **105**, 173 (1957), Tinkham, pp. 193–206, and Ramsay, Chapt. 3.

† Here the z-direction coincides with that of $\mathcal{H}$.

where we have consulted the table of CG-coefficients in Sec. 25. The energy shifts are therefore

$$(\Delta E_M)_{ljm} = \mu_0 \mathcal{H} m \frac{2l + 1 \pm 1}{2l + 1} \qquad (j = l \pm \tfrac{1}{2}). \tag{4}$$

This splitting of the levels is called the *Zeeman effect.*

We also consider case $H_1 \ll H_M$, the *Paschen-Back limit.* Here the zero order states are the $|nlm_l\tfrac{1}{2}m_s\rangle$, and the spectrum when H_1 is neglected is given by

$$(\Delta E_M)_{m_l m_s} = \mu_0 \mathcal{H}(m_l + 2m_s). \tag{5}$$

The fine structure can then be added by perturbation theory. Instead of doing this we go straight to the diagonalization of H in the $n = 2$ subspace, which will then give the energy levels for arbitrary fields as long as $\mathcal{H} \ll 10^7$ gauss, i.e., $\mu_0\mathcal{H} \ll \alpha^4 mc^2$.

The important property of (3) is that H_M is cylindrically symmetric and only has matrix elements between states of the same parity. If we use the set $2p_{\frac{1}{2}}$, $2p_{\frac{3}{2}}$, $2s_{\frac{1}{2}}$, there are only off-diagonal elements between $2p_{\frac{1}{2}}$ and $2p_{\frac{3}{2}}$ when both have $m = \pm\tfrac{1}{2}$. Hence (4) is already exact for both $2s_{\frac{1}{2}}$ states, and also for the $2p_{\frac{3}{2}}$ states with $m = \pm\tfrac{3}{2}$.

We must now compute the matrix elements of H_M between the appropriate states $|nljm\rangle$. We already know the diagonal elements, namely (4), and require only

$$\begin{aligned}\langle p_{\frac{1}{2}}, m|s_z|p_{\frac{3}{2}}, m\rangle &= \sum \langle \tfrac{1}{2}m|1m_l\tfrac{1}{2}m_s\rangle m_s \langle 1m_l\tfrac{1}{2}m_s|\tfrac{3}{2}m\rangle \\ &= -\tfrac{1}{3}\sqrt{(\tfrac{3}{2} + m)(\tfrac{3}{2} - m)},\end{aligned}$$

where we have again used the table of CG-coefficients. If we call Δ the fine structure splitting when $\mathcal{H} = 0$, the matrix we must diagonalize is

	$p_{\frac{3}{2}}$	$p_{\frac{1}{2}}$
$p_{\frac{3}{2}}$	$\Delta + \tfrac{4}{3}\mu_0\mathcal{H}m - \epsilon$	$-\tfrac{1}{3}\mu_0\mathcal{H}\sqrt{(\tfrac{3}{2} + m)(\tfrac{3}{2} - m)}$
$p_{\frac{1}{2}}$	$-\tfrac{1}{3}\mu_0\mathcal{H}\sqrt{(\tfrac{3}{2} + m)(\tfrac{3}{2} - m)}$	$\tfrac{2}{3}\mu_0\mathcal{H}m - \epsilon$

The eigenvalues are

$$\epsilon_m^{\pm} = \tfrac{1}{2}\Delta + \mu_0\mathcal{H}m \pm \tfrac{1}{2}\sqrt{\Delta^2 + \tfrac{4}{3}m\,\Delta\mu_0\mathcal{H} + (\mu_0\mathcal{H})^2}. \tag{6}$$

For $\Delta \gg \mu_0\mathcal{H}$ we find

$$\epsilon_m^+ = \Delta\left\{1 + \tfrac{4}{3}m\left(\frac{\mu_0\mathcal{H}}{\Delta}\right) + \left(\frac{\mu_0\mathcal{H}}{\Delta}\right)^2\left(\frac{1}{4} - \frac{m^2}{9}\right) + \cdots\right\},$$
$$\epsilon_m^- = \tfrac{2}{3}\mu_0\mathcal{H}m - \mu_0\mathcal{H}\left(\frac{\mu_0\mathcal{H}}{\Delta}\right)\left(\frac{1}{4} - \frac{m^2}{9}\right) + \cdots. \tag{7}$$

The terms linear in $\mathcal{H}$ are just those of (4); we recall that they were found by taking the expectation value of H_M in the eigenstates of $H_0 + H_1$. The quadratic terms in $\mathcal{H}$ could also be obtained from (45.16) by keeping only the fine structure partner in the sum over states, and the coefficients of the higher powers of $\mathcal{H}$ can all be computed in a similar fashion by using the more complicated formulas for the higher order energy shifts.*

The other limiting case is $\mu_0\mathcal{H} \gg \Delta$:

$$\epsilon_m^\pm = \mu_0\mathcal{H}\left\{(m \pm \tfrac{1}{2}) + \frac{\Delta}{\mu_0\mathcal{H}}(\tfrac{1}{3} \pm \tfrac{1}{3}m) \pm \left(\frac{\Delta}{\mu_0\mathcal{H}}\right)^2\left(\frac{1}{4} - \frac{m^2}{9}\right)\cdots\right\}, \tag{8}$$

or

$$\begin{aligned} \epsilon_{\frac{1}{2}}^+ &= \mu_0\mathcal{H} + \tfrac{2}{3}\Delta, \\ \epsilon_{-\frac{1}{2}}^+ &= \tfrac{1}{3}\Delta, \\ \epsilon_{\frac{1}{2}}^- &= \tfrac{1}{3}\Delta, \\ \epsilon_{-\frac{1}{2}}^- &= -\mu_0\mathcal{H} + \tfrac{2}{3}\Delta, \end{aligned} \tag{9}$$

in the Paschen-Back limit. Here the terms of order Δ are just the expectation values of H_1 in the "Paschen-Back states" $|nlm_lm_s\rangle$, and the quadratic and higher terms in Δ could be obtained by the perturbation expansion where, of course, the sum over states is restricted to the $n = 2$ subspace. A graph of the levels as given by (6) is shown in Fig. 47.1.

The eigenvectors can also be computed once the eigenvalues are known. They are certain combinations of $|p_{\frac{3}{2}}, m\rangle$ and $|p_{\frac{1}{2}}, m\rangle$. In the limit $\mathcal{H} \gg \Delta$,

* It should be noted that even in this very simple example of a two-dimensional Hilbert space, the Rayleigh-Schrödinger expansion has a finite radius of convergence. If we introduce the complex variable $z = \Delta/\mu_0\mathcal{H}$, we conclude from (6) that $\epsilon_m(z)$ has branch points at $z = -\frac{2}{3}(m \mp i\sqrt{2})$. Hence the perturbation expansion only converges if $\mu_0\mathcal{H} < 2\,\Delta/\sqrt{3}$. The eigenvalues of the Hamiltonian in an n-dimensional Hilbert space are the roots of a polynomial of degree n, and in general algebraic branch points will therefore occur. Essential singularities (whose existence we conjectured in connection with the perturbed harmonic oscillator at the very end of Sec. 45.1) can only come about if the Hilbert space is infinite dimensional.

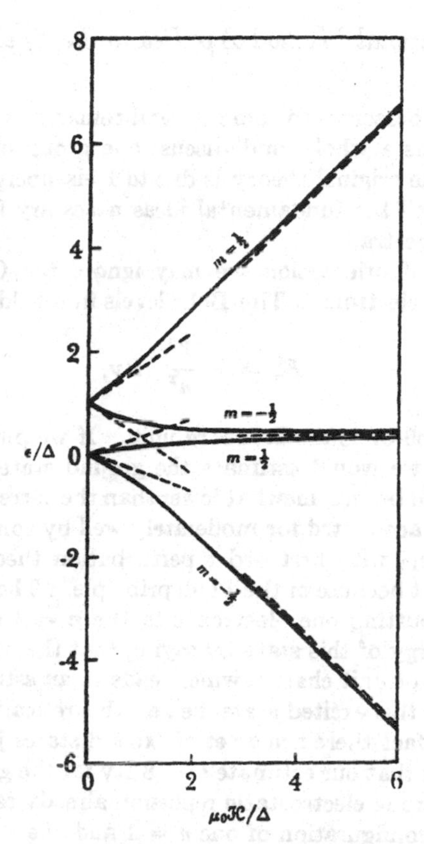

Fig. 47.1. Hydrogenic $2p$ levels in a magnetic field. Only the $m = \pm\frac{1}{2}$ levels are shown, because the $m = \pm\frac{3}{2}$ levels are given by Eq. (4) for all values of $\mathcal{H}$. The dashed lines on the left are the low-field curves as given by (4), and the dashed curves on the right are the Paschen-Back limits as given by (9). It should be noted that once the limiting curves are known, the remainder of the curves can almost be drawn in by eye. This fact is very useful in qualitative analysis of complex perturbations. A comparison between experimental data and the curves shown here can be found in Condon and Shortley, p. 155.

this linear combination is just given by the Clebsch-Gordan coefficients from $p_{\frac{3}{2}}$ and $p_{\frac{1}{2}}$ to $l = 1$, with $m_l + m_s = m$, because we know that these are the correct eigenstates in the Paschen-Back limit.

Note that when $\mathcal{H} \neq 0$ there is no degeneracy left even though we are dealing with a single spin $\frac{1}{2}$ particle. This is an example where Kramer's theorem does not apply because H is not invariant under time reversal.

48. The Variational Method Applied to the Spectrum of Helium

We shall confine ourselves to some general remarks concerning the spectrum of helium as a whole, and discuss one group of excited states in more detail. The original theory is due to Heisenberg (1926); his paper contains almost all the fundamental ideas necessary for an understanding of complex spectra.

For purposes of orientation we may ignore the Coulomb repulsion between the two electrons. The Bohr levels in a field of charge Z are

$$E_n^0 = -\frac{Z^2}{n^2} \text{ Ry}, \tag{1}$$

where Ry = 13.60 ev is the Rydberg unit. If we put both electrons in the $n = 1$ state we would estimate the ground state energy as -8 Ry, which, as it should be, is somewhat lower than the correct result, -5.8 Ry. The difference is accounted for moderately well by computing the energy shift due to e^2/r_{12} with first order perturbation theory. This state is then a spin singlet because of the Pauli principle. The lowest spin triplet is obtained by putting one electron into the $n = 2$ states. We would estimate the energy of this state by saying that the outer electron sees a screened nucleus of unit charge, which leads to an estimate of -4.25 Ry for the energy of this excited state, i.e., a "theoretical" excitation energy of 3.75 Ry. In fact there are a set of excited states just above -2 Ry. The point here is that our estimate of -8 Ry for the ground state is poor because of the strong electrostatic repulsion already referred to. On the other hand, the configuration of one $n = 1$ and one $n = 2$ electrons does not suffer so much from this error, because the electrons are further apart on the average. In fact, if one uses the experimental ground state energy (-5.8 Ry) and the "theoretical" value of -4.25 Ry for the excited state, rough agreement with the observed excitation energy is achieved.

Further excited states are obtained by leaving one electron in the $1s$ state, and putting the other into states of various n. The energy required to ionize one electron while leaving the other in $1s$ is $(5.8 - 4.0)$ Ry $\simeq 1.8$ Ry. In other words, if we measure up from the ground state the continuum should set in at 1.8 Ry (the empirical value is 1.79 Ry!). The energy required to lift *both* electrons to the $n = 2$ states is roughly $2 \times 4\,(\frac{1}{1} - \frac{1}{4}) = 6$ Ry. Thus *all* states where both electrons are excited lie in the continuum and we shall not treat them. The discrete spectrum can therefore be labeled by the quantum numbers of the excited electron, e.g., 2^3P means $(1s2p)$, spin triplet. The experimental spectrum is shown in Bethe and Salpeter (their Fig. 15).

48.1 THE VARIATIONAL PRINCIPLE

Although perturbation theory converges tolerably well for He, much more precise answers can be obtained by using the variational method. This method is also very useful in other problems (e.g., the computation of the binding energies of light nuclei), and we shall therefore discuss its application to the helium spectrum. The technique is based on the *variational principle.** The content of this principle is

(i) the functional

$$E[\psi] = \frac{\langle\psi|H|\psi\rangle}{\langle\psi|\psi\rangle} \tag{2}$$

is stationary with respect to arbitrary variations of $|\psi\rangle$ about any solution $|\psi_n\rangle$ of $(H - E_n)|\psi_n\rangle = 0$;

(ii) let $|\psi_0\rangle$ be the lowest energy eigenstate with a definite eigenvalue a' of the other constants of the motion. (Thus $|\psi_0\rangle$ could be the lowest $J = 4$ state of helium.) Then if $|\Phi\rangle$ is an arbitrary state with the same eigenvalue a',

$$E[\Phi] \geqslant E[\psi_0],$$

where the equality only applies if $|\Phi\rangle = |\psi_0\rangle$.

To prove (i) we note that $|\psi\rangle$ and $\langle\psi|$ are to be treated as independent variables in carrying out variations, because the Hilbert space is over the complex field. We therefore vary $\langle\psi|$ alone, and find

$$\begin{aligned}\delta E[\psi] &= \frac{\langle\psi|H|\psi\rangle + \langle\delta\psi|H|\psi\rangle}{\langle\psi|\psi\rangle + \langle\delta\psi|\psi\rangle} - E[\psi] \\ &= \frac{1}{\langle\psi|\psi\rangle}\{\langle\delta\psi|H|\psi\rangle - E[\psi]\langle\delta\psi|\psi\rangle\} + O(\delta\psi^2).\end{aligned}$$

Hence if $|\psi\rangle = |\psi_n\rangle$, $\delta E[\psi] = O(\delta\psi^2)$. *QED.* To prove (ii) we must introduce the eigenstates $|\psi_n\rangle$ which, together with $|\psi_0\rangle$, constitute a complete set bearing the eigenvalue a'. Then

$$\begin{aligned}E[\Phi] - E[\psi_0] &= \frac{\langle\Phi|H - E_0|\Phi\rangle}{\langle\Phi|\Phi\rangle} \\ &= \sum_n (E_n - E_0)\frac{|\langle\Phi|\psi_n\rangle|^2}{\langle\Phi|\Phi\rangle},\end{aligned}$$

which proves the theorem because $E_n > E_0$ if $n \neq 0$.

* It is also called the Rayleigh-Ritz principle.

In applying the variational method one makes as well informed a guess as possible (or practical) concerning the state vector, and expresses this as a so-called trial function. This trial function contains one or more parameters, which are then adjusted so as to minimize the functional (2).* According to (ii) this value of $E[\psi]$ is then an upper bound to the true eigenvalue. On the other hand we can say nothing about the discrepancy between the trial function and the exact solution. In general the variational evaluation of the eigenvalue is considerably more accurate than the corresponding trial function.† In this section we shall describe a variational calculation of certain of the states of He. In Chapt. XI we shall see how the variational method may be used in much more complex situations.

48.2 THE 1S AND 2P LEVELS

Let us now apply the method to the *lowest* S and P levels, the $1S$, and the $2^{1,3}P$ states. Note that though 1^1S_0 and 2^3P_0 have the same total angular momentum, they are automatically orthogonal because they have opposite parity. Hence the variational method applies to them straightforwardly.

Ignoring relativistic effects and nuclear motion, the Hamiltonian is

$$H = \frac{1}{2m}(p_1^2 + p_2^2) - 2e^2\left(\frac{1}{r_1} + \frac{1}{r_2}\right) + \frac{e^2}{r_{12}}. \tag{3}$$

As a trial function for the ground state we can choose the product of two hydrogenic $1s$ states with the charge Z treated as a variational parameter:

$$\Psi_0(\mathbf{r}_1\mathbf{r}_2) = C\exp[-Z_{\text{eff}}(r_1 + r_2)/a_0]. \tag{4}$$

(Here we suppress the singlet spin function as it does not enter into the calculation of $\langle H\rangle$.) The minimization of $\langle H\rangle$ with (4) yields $Z_{\text{eff}} = 2 - \frac{5}{16}$, which shows the effect of screening, and an energy of -77.46 ev, which is only 1% above the experimental ground state. Trial functions of slightly greater complexity (i.e., several parameters) give results of amazing accuracy.‡

* Procedures exist for modifying the trial function so as to yield systematically improved eigenvalues. These are called variation-iteration methods. They are discussed by Morse and Feshbach, pp. 1137–1152, 1698–1701.

† C. Schwartz [*Ann. of Phys.* **2**, 170 (1959)] has invented a method for improving the trial function so that its accuracy is comparable to that of the eigenvalue. The article cited also contains applications to He.

‡ For a detailed investigation of the ground state with the variational method, see T. Kinoshita, *Phys. Rev.* **115**, 366 (1959); C. L. Pekeris, *ibid.*, **115**, 1216 (1959).

The excited states are much more interesting. We follow a calculation of Eckart (*Phys. Rev.* 36, 878 (1930)). Once more we build the two-electron wave function out of hydrogenic one-electron functions with the charge as a variational parameter. The explicit forms of the normalized $1s$ and $2p$ functions are

$$1s: \qquad u(r) = (\alpha^3/a_0^3\pi)^{\frac{1}{2}} e^{-\alpha r/a_0}, \tag{5}$$

$$2p: \qquad v_{m_l}(\mathbf{r}) = \frac{\sqrt{6}}{12}\left(\frac{\beta}{a_0}\right)^{\frac{3}{2}} \cdot \frac{\beta r}{a_0} \cdot e^{-\beta r/2a_0} Y_{1m_l}(\theta\phi), \tag{6}$$

where α and β are the effective nuclear charges; in writing (5) and (6) we have consulted the table in Sec. 17.2. Our naive picture would lead us to expect $\alpha = 2, \beta = 1$. From these orbital wave functions we can form totally antisymmetric two-electron states:

$$\Psi(2^1P, m_l) = 2^{-\frac{1}{2}}[u(r_1)v_{m_l}(\mathbf{r}_2) + (\mathbf{r}_1 \leftrightarrow \mathbf{r}_2)]|\chi^0\rangle, \tag{7}$$

$$\Psi(2^3P, m_l M_S) = 2^{-\frac{1}{2}}[u(r_1)v_{m_l}(\mathbf{r}_2) - (\mathbf{r}_1 \leftrightarrow \mathbf{r}_2)]|\chi^1_{M_s}\rangle, \tag{8}$$

where $|\chi^0\rangle$ and $|\chi^1_{M_s}\rangle$ are the singlet and triplet spin states constructed in Sec. 42. Because the energies do not depend on m_l and M_S we put them equal to zero. Furthermore we write $v_0(\mathbf{r}) = R(r)Y_{10}(\theta)$. In an obvious notation we now have

$$\begin{aligned}
\langle H\rangle_\pm &= \tfrac{1}{2}\int d^3r_1\, d^3r_2 [u^*(1)v_0^*(2) \pm u^*(2)v_0^*(1)]\left\{H_0^{(1)} + H_0^{(2)} + \frac{e^2}{r_{12}}\right\} \\
&\qquad\qquad \times [u(1)v_0(2) \pm u(2)v_0(1)] \\
&= \int d^3r_1 [u^*(1)H_0^{(1)}u(1) + v_0^*(1)H^{(1)}v_0(1)] \\
&\quad + \int d^3r_1\, d^3r_2 \frac{e^2}{r_{12}} |u(1)v_0(2)|^2 \\
&\qquad\qquad \pm \int d^3r_1\, d^3r_2 \frac{e^2}{r_{12}} u^*(1)v_0^*(2)u(2)v_0(1),
\end{aligned}$$

or

$$\langle H\rangle_\pm = \langle H_0^{(1)} + H_0^{(2)}\rangle + I(\alpha, \beta) \pm J(\alpha, \beta). \tag{9}$$

The meaning of I is clear: it is the electrostatic interaction energy between the two charge distributions $e|u(1)|^2$ and $e|v_0(2)|^2$, and is just what one would expect classically. I is called a *direct* integral. J, on the other hand, is Heisenberg's startling discovery, and it has no analogue whatsoever in classical physics; it is called the *exchange integral*. Looking back we see that J comes about because of the symmetry requirement put on the wave function. In other words, the functions $\Psi(2^{3,1}P)$ have *correlations* built into them. The probability of finding one electron at point $\mathbf{r}_1$, the other at $\mathbf{r}_2$ is

$$\begin{aligned}|\Psi(2^{3,1}P)|^2 &= \tfrac{1}{2}|u(1)v_0(2) \mp u(2)v_0(1)|^2 \\ &= \tfrac{1}{2}\{|u(1)|^2|v_0(2)|^2 + |u(2)|^2|v_0(1)|^2\} \\ &\quad \mp \tfrac{1}{2}\{u^*(1)v_0^*(2)u(2)v_0(1) + \text{c.c.}\}. \end{aligned} \tag{10}$$

The first term is just what one would expect classically for the pair distribution function of a system of two noninteracting but indistinguishable particles. The last term, which arises from the interference between the two parts of Ψ, has no classical counterpart,* and in fact shows that there are correlations in the system's distribution function in spite of the fact that the wave function is the eigenfunction of a Hamiltonian $(H_0^{(1)} + H_0^{(2)})$ that neglects the interactions between the particles.† These correlations insure, for example, that in the triplet state the particles can never be found at the same point, because $|\Psi(2^3P)|^2$ vanishes when $\mathbf{r}_1 = \mathbf{r}_2$. This is just the Pauli principle in coordinate space. In the ^1P state, on the other hand, the spatial part of the wave function is symmetric, and it therefore produces a probability for finding the electrons at the same point which is *larger* than that of the product wave function. In addition to these spatial correlations, there are also spin correlations. In the ^3P state the spins are "parallel," and in the ^1P state they are "antiparallel." We already encountered such spin correlations in the scattering of indistinguishable fermions (Sec. 43.2). The correlations which arise from the permutation symmetry result in energy differences between states of different total spin (e.g., 2^3P and 2^1P) which are of order α^2mc^2, and therefore *larger* by a factor of α^{-2} than any relativistic, "truly" spin-dependent, force. Put another way, we can say that the antisymmetry of the wave function simulates spin-dependent forces that are electrostatic in magnitude, or α^{-2} times larger than one might otherwise expect. As Heisenberg later realized (1928), this offers an explanation of the amazingly large spin-spin interaction in ferromagnets.

The integrals I and J can be simplified with the help of

$$\frac{1}{|\mathbf{r}_1 - \mathbf{r}_2|} = \sum_{lm} \frac{4\pi}{2l+1} \frac{r_<^l}{r_>^{l+1}} Y_{lm}^*(\hat{\mathbf{r}}_1) Y_{lm}(\hat{\mathbf{r}}_2).$$

* A classical "interpretation" of $J(\alpha, \beta)$ can be found in Bethe and Salpeter, p. 132.

† A more elaborate treatment of such correlations can be found in Chapts. XI and XII.

For I one finds

$$I = e^2 \int d^3r_1\, d^3r_2\, r_>^{-1} |u(1)v_0(2)|^2$$
$$= 4\pi e^2 \int_0^\infty r_1^2\, dr_1 \int_0^\infty r_2^2\, dr_2\, r_>^{-1} |u(1)R(2)|^2$$

because $u(r)$ is spherically symmetric. A similar calculation results in

$$J = e^2 \sum_m \frac{4\pi}{3} \int d^3r_1\, d^3r_2 \frac{r_<}{r_>^2} Y_{1m}^* (\hat{\mathbf{r}}_1)\, Y_{1m}(\hat{\mathbf{r}}_2) u^*(1)u(2)v_0(1)v_0^*(2)$$
$$= \frac{4\pi e^2}{3} \int_0^\infty r_1^2\, dr_1 \int_0^\infty r_2^2\, dr_2 \frac{r_<}{r_>^2} u^*(1)u(2)R(1)R^*(2).$$

In the present case $J > 0$ since all the radial functions are positive. Hence the triplet state will have the lower energy. This is quite general because in the triplet state the space function is antisymmetric and therefore is not punished as much by the Coulomb repulsion. There is in fact a widely applicable recipe known as *Hund's Rule:* states of highest spin multiplicity lie lowest. The ferromagnet is the extreme illustration of Hund's rule.

The evaluation of I and J is now elementary and the minimization with respect to α and β tedious. Eckart's results are tabulated here. The

			Ionization Energy	
State	α	β	Theory	Experiment
2^3P	1.99	1.09	0.262 Ry	0.266 Ry
2^1P	2.003	0.965	0.245 Ry	0.248 Ry

agreement is very good considering the simplicity of the trial function. It is gratifying that the values of α and β are as expected. Note that the $(2^3P - 2^1P)$ separation of 0.017 Ry = 0.23 ev is entirely due to the exchange effect. This is rather small as such things go, but of course huge compared to the fine structure α^4mc^2. The small magnitude of J in this particular case is due to the very poor overlap between $u(1)$ and $R(2)$, as shown in Fig. 48.1.

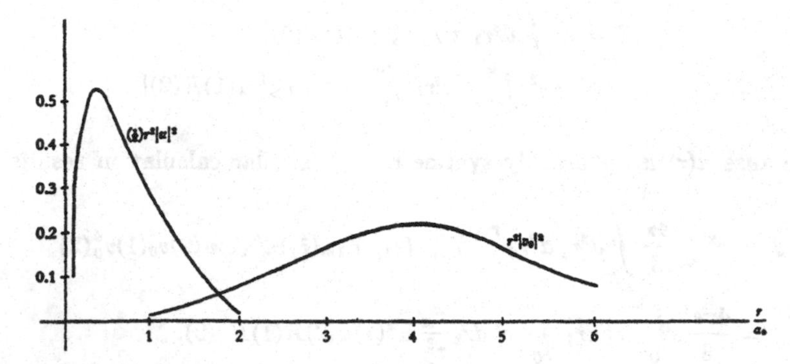

Fig. 48.1. Radial wave functions for the 2*P* state in He.

49. *Scattering Viewed as a Perturbation of the Continuum*

Before leaving the subject of stationary state perturbation theory, we shall indicate its relation to scattering theory.* That such a relationship must actually exist can be seen from the following simple argument.

49.1 RELATIONSHIP BETWEEN ENERGY AND PHASE SHIFTS

Consider a spinless particle in the presence of a fixed central potential $V(r)$, with $\lim_{r\to\infty} rV(r) = 0$. In order to render the spectrum completely discrete, we imagine the particle confined to the interior of the sphere $r = R$, with R much larger than both the range of the potential $V(r)$ and the de Broglie wavelengths of interest to us. Because of the spherical symmetry, the eigenfunctions have the form $Y_{lm}(\vartheta, \varphi)\psi_l(r)$. Although we do not know $\psi_l(r)$ everywhere, we know that for values of r in excess of the range ψ_l becomes a linear of combination of Neumann and Bessel functions which, in the asymptotic region, may be written as (recall (14.11))

$$\psi_l(r) \sim \frac{1}{r}\sin(kr - \tfrac{1}{2}l\pi + \delta_l), \tag{1}$$

* The variational method can also be applied to collision problems. See Morse and Feshbach, pp. 1123–1128, 1701–1709; Messiah, pp. 856–863; W. Kohn, *Phys. Rev.* **74**, 1763 (1948); J. M. Blatt and J. D. Jackson, *Phys. Rev.* **76**, 18 (1949); B. A. Lippmann and J. Schwinger, *Phys. Rev.* **79**, 469 (1950); B. A. Lippmann, *Phys. Rev.* **79**, 480 (1950); M. Moe and D. Saxon, *Phys. Rev.* **111**, 950 (1958).

where $\delta_l(E)$ is the phase shift at energy E. In contrast to this, the radial wave function of the free particle has the asymptotic form

$$\psi_l^0(r) \sim \frac{1}{r}\sin(kr - \tfrac{1}{2}l\pi). \tag{2}$$

Because these wave functions must vanish at $r = R$, k must be one of the values

$$\begin{aligned} \psi_l^0: &\qquad k_{l,n}^0 R - \tfrac{1}{2}l\pi = n\pi, \\ \psi_l: &\qquad k_{l,n} R - \tfrac{1}{2}l\pi + \delta_l = n\pi, \end{aligned} \tag{3}$$

with n a positive integer. The energy levels of the free ($E_{l,n}^0$) and perturbed ($E_{l,n}$) system then take on the discrete values

$$E_{l,n}^0 = \frac{1}{2m}(\hbar k_{l,n}^0)^2, \qquad E_{l,n} = \frac{1}{2m}(\hbar k_{l,n})^2. \tag{4}$$

The phase shift is only determined modulo π by (1), and so we must also specify that $\delta_l \to 0$ as $V \to 0$, and therefore $E_{l,n} \to E_{l,n}^0$ as $V \to 0$.

If we wish E_n^0 and E_n to tend to a nonzero limit as $R \to \infty$, we must keep k_n^0 and k_n finite, i.e., allow the integer n to become very large.* Therefore the spacing between neighboring values of k_n^0, which is π/R, becomes vanishingly small compared to k_n^0 itself, and the same remark applies to k_n. This is quite obvious, of course, since the spectrum becomes continuous when $R \to \infty$. Though we shall not take this limit for the time being, we must bear in mind that we are dealing with values of n for which the spacing between allowed k values is exceedingly small in comparison to k. We are therefore permitted to approximate the spacing between adjacent levels by

$$\begin{aligned} \Delta E_n \equiv E_{n+1}^0 - E_n^0 &\simeq \frac{\hbar^2 k_n^0}{m}(k_{n+1}^0 - k_n^0) \\ &= \frac{\hbar^2 k_n^0}{m}\cdot\frac{\pi}{R}. \end{aligned}$$

The difference in energy between the true and unperturbed levels is

$$\delta E_n \equiv E_n - E_n^0 = \frac{\hbar^2}{2m}(k_n + k_n^0)\left(-\frac{\delta_l}{R}\right).$$

But $k_n = k_n^0 + O(1/R)$, and therefore to $O(1/R)$

$$\frac{\delta E_n}{\Delta E_n} = -\frac{1}{\pi}\,\delta_l(E), \tag{5}$$

* Because we shall stay with one partial wave throughout, we shall usually drop the index l from now on.

where $\delta_l(E)$ is the phase shift evaluated at the energy E_n or E_n^0, the difference again being of no consequence to terms of order $1/R$. This equation therefore expresses the sought-after relationship between the scattering problem and the perturbation δE_n of the spectrum. Because $\delta E_n \leqslant 0$ if $V(r)$ is everywhere attractive, it is clear that $\delta_l(E) \geqslant 0$ for an attractive potential, and $\delta_l(E) \leqslant 0$ for a repulsive potential.

49.2 FREDHOLM THEORY

Having established the relation between the spectrum and the phase shift, it is natural to seek a method that will permit us to compute δ. It is clear that the Rayleigh-Schrödinger expansion cannot be used for this purpose, because the energy denominators appearing there are not defined as $R \to \infty$. The technique we shall employ instead is known as the Fredholm theory of integral equations. Because the original derivation * of the theory is not based on the relation (5) between the phase shift and the energy shift, we shall adopt a formulation due to Schwinger.† Were we to attack the eigenvalue problem by conventional methods, we would try to solve the secular equation $\det(E - H) = 0$. This could perhaps be a well-defined procedure when there is an upper bound to the spectrum; in our present problem there is no such bound, however. Let us instead consider

$$\frac{\det(E - H)}{\det(E - H_0)} = \prod_{l=0}^{\infty} \left[\prod_n \left(\frac{E - E_{l,n}}{E - E_{l,n}^0} \right) \right]^{2l+1}$$
$$= \prod_{l=0}^{\infty} \left[\prod_n \left(1 - \frac{\delta E_{l,n}}{E - E_{l,n}^0} \right) \right]^{2l+1} \tag{6}$$

It is clear that we can confine our attention once more to one partial wave, and we therefore introduce the function ‡

$$D_R^{(l)}(z) = \prod_n \left(\frac{z - E_{l,n}}{z - E_{l,n}^0} \right) = \prod_n \left(1 - \frac{\delta E_{l,n}}{z - E_{l,n}^0} \right), \tag{7}$$

where z is a complex variable. Because $\delta E \sim O(1/R)$, we may hope that the infinite product exists for finite values of $\operatorname{Im} z$ in the limit $R \to \infty$. Before going to this limit, and setting convergence questions aside for

* R. Jost and A. Pais, *Phys. Rev.* **82**, 840 (1951). R. Jost and W. Kohn, *Phys. Rev.* **87**, 977 (1952).

† M. Baker, *Ann. of Phys.* **4**, 27 (1958).

‡ $D_\infty^{(l)}(z)$ is called the Fredholm determinant belonging to the integral equation for the lth partial wave. The function of the complex variable k defined by $f_l(-k) \equiv D_\infty^{(l)}(\hbar^2k^2/2m)$ is known as the Jost function.

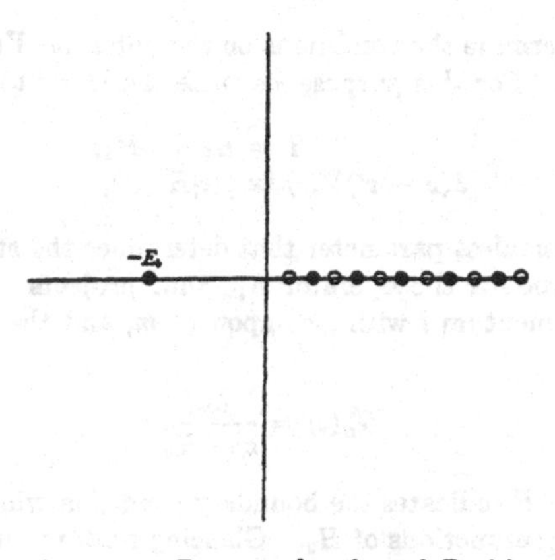

Fig. 49.1. Zeros and poles of $D_R(z)$. The poles are shown as circles, and the zeros as dots. When $R \to \infty$ the zeros and poles on Z_+ coalesce and form a branch cut.

the moment, it would appear that $D_R^{(l)}(z)$ has the following properties (see Fig. 49.1):

(a) it is an analytic function in the finite z-plane except for simple poles on the real axis at the eigenvalues $E_{l,n}^0$ of the free Hamiltonian H_0;
(b) it has simple zeros on the real axis at the eigenvalues $E_{n,l}$ of the complete Hamiltonian H.

The poles of * $D_R(z)$ therefore lie on † Z_+, their spacing being $O(1/R)$. The zeros (eigenvalues of H) usually fall into two classes:

(i) those that lie within $O(1/R)$ of a pole;
(ii) there may also be zeros which lie on Z_- at an R-independent distance from the origin.

This second class of zeros clearly corresponds to the bound states, should they exist. The first kind of zero is associated with a scattering state. In the limit $R \to \infty$, the poles and zeros on Z_+ coalesce; as we shall see, $D_\infty(z)$ has a branch cut on Z_+ arising from the scattering continuum, and one simple pole on Z_- for each bound state.

* Here we again drop the label l.
† $Z_\pm$ designates the (positive/negative) real axis of z.

•We must now determine the conditions on the potential $V(r)$ that must be met if $D_R(z)$ is to exist. For this purpose we write the Hamiltonian as

$$H = H_0 + gH_1, \qquad (8)$$
$$\delta(\mathbf{r} - \mathbf{r}')V(r) \equiv g\langle \mathbf{r}|H_1|\mathbf{r}'\rangle,$$

where g is a dimensionless parameter that determines the strength of the potential. We also introduce the operator Λ_{lm} that projects onto the subspace of orbital angular momentum l with z-component m, and the resolvent or Green's operator

$$G_R^0(z) = \frac{\Lambda_{lm}}{z - H_0}, \qquad (9)$$

where the subscript R indicates the boundary condition which has been imposed at $r = R$ on the eigenfunctions of H_0. Glancing back to our original definitions (6) and (7), we have

$$D_R^{(l)}(z) = \det[\Lambda_{lm}(1 - gG_R^0(z)H_1)],$$

or

$$D_R^{(l)}(z) = \exp\{\operatorname{tr} \ln \Lambda_{lm}(1 - gG_R^0(z)H_1)\}, \qquad (10)$$

because*

$$\det A = \exp\{\operatorname{tr} \ln A\}. \qquad (11)$$

Baker (*loc. cit.*) has established bounds on $D_R^{(l)}$ by using the form (10); from these he finds that when †

$$|g \operatorname{tr} G_R^0(z)H_1| < \infty, \qquad (12)$$

$D_R^{(l)}(z)$ is an entire function of g, the interaction strength. This statement holds for all values of z that satisfy (12). We can easily obtain conditions on $V(r)$ from (12) by evaluating the trace in the coordinate representation:

$$g \operatorname{tr} G_R^0(z)H_1 = \int_0^R r^2\, dr\, V(r)\langle r|G_R^0(z)|r\rangle.$$

But $\langle r|G_\infty^0(E + i\epsilon)|r'\rangle$, with $E > 0$, is just the radial Green's function (14.8), and therefore

$$|g \operatorname{tr} G_R^0(z)H_1|_{z=E+i\epsilon} = k\left| \int_0^\infty r^2\, dr\, V(r)j_l(kr)h_l(kr) \right|. \qquad (13)$$

Let us assume that $V(r)$ is only singular at $r = 0$. Then only too strong a singularity at $r = 0$, or too slow a fall-off of V as $r \to \infty$ can cause (13) to diverge. Using the expansions of j_l and h_l at the origin (see Sec. 11), we see that at the lower limit of integration $\int_0 r\, V(r)\, dr$ must be finite, or $V(r)$ must be less singular than $1/r^2$ as $r \to 0$. At infinity the functions j_l and h_l fall off as $1/r$, and therefore $V(r)$ must go to zero faster than a Coulomb field if $D_\infty(E + i\epsilon)$ is to exist.

* The proof of this identity is left for Prob. 6.

† Henceforth we suppress Λ_{lm}, and understand that all traces go only over states belonging to one value of l and m.

We can easily extend (13) to the whole z plane by interpreting k as $(2mz/\hbar^2)^{\frac{1}{2}}$; because $\zeta j_l(\zeta)h_l(\zeta)$ is an entire function of ζ, the convergence is still determined by the behavior of $V(r)$ at the origin and infinity. We therefore conclude that for all values of z not on Z_+, $D_\infty(z)$ is an entire function of the potential strength g provided

$$\lim_{r\to 0} r^2V(r) = 0,$$
$$\lim_{r\to\infty} rV(r) = 0. \tag{14}$$

These are seen to be rather weak conditions * on $V(r)$. The Coulomb potential fails to comply, but we already know how to solve this problem rigorously.•

Our next task is to relate the function $D_\infty(z)$ to the phase shift, which we shall do with the help of (7) and (5). We first separate the bound state and scattering terms in (7). Let us follow the behavior of the lowest eigenvalue $E_1(g)$ of H as $|g|$ is increased so as to make $V(r)$ progressively more attractive. Clearly $E_1(0) = E_1^0$, and for $|g|$ sufficiently small $E_1(g) = E_1^0 - O(1/R)$. If $|g|$ is increased beyond some critical value, $E_1(g) - E_1^0$ becomes independent of R as $R \to \infty$, and we may say that the potential has "peeled" a bound state off the bottom of the continuum. If $|g|$ becomes even larger, a second state $E_2(g)$, with $E_2^0 = E_2(0)$, may be peeled off as well. We note that as $R \to \infty$, the free eigenvalues to which the bound states tend as $|g| \to 0$ all lie at $z = 0$. Calling the bound state energies $-E_b$, $b = 1, \ldots, N$, we can therefore write (7) as

$$\begin{aligned} D_R(z) &= \prod_{b=1}^{N}\left(\frac{z+E_b}{z}\right)\exp\left\{\sum_{n>N}^{\infty}\ln\left(1-\frac{\delta E_n}{z-E_n^0}\right)\right\} + O(1/R) \\ &= \prod_{b=1}^{N}\left(\frac{z+E_b}{z}\right)\exp\left\{-\sum_{n>N}^{\infty}\frac{\delta E_n/\Delta E_n}{z-E_n}\,\Delta E_n\right\} + O(1/R). \end{aligned}$$

In the limit $R \to \infty$ the sum becomes an integral over the positive values of E,

$$D_\infty(z) = \prod_{b=1}^{N}\left(\frac{z+E_b}{z}\right)\exp\left\{\frac{1}{\pi}\int_0^\infty \frac{\delta(E')\,dE'}{z-E'}\right\}, \tag{15}$$

where we have of course used (5). An explicit relationship between $D_\infty(z)$ and $\delta(E)$ can be obtained by use of the identity

$$\lim_{z\to x\pm i\epsilon}\frac{1}{\pi}\int_a^b \frac{f(x')\,dx'}{z-x'} = \mp if(x) + \frac{\mathrm{P}}{\pi}\int_a^b \frac{f(x')\,dx'}{x-x'}, \tag{16}$$

* The condition at $r = 0$ eliminates the singular potentials that occur in the interactions between atoms.

where $a < x < b$, $f(x')$ is continuous at $x' = x$, and P denotes the Cauchy principal value of the integral. We shall verify this identity in a moment, but let us first use it to evaluate (15) as $\text{Im}\, z \to 0$. We have

$$\frac{1}{\pi}\int_0^\infty \frac{\delta(E')\,dE'}{E \pm i\epsilon - E'} = \mp i\delta(E) + \frac{\text{P}}{\pi}\int_0^\infty \frac{\delta(E')\,dE'}{E - E'}$$

which shows that $D_\infty(z)$ is discontinuous across Z_+:

$$D_\infty(E \pm i\epsilon) = e^{\mp i\delta(E)} \prod_b \left(\frac{E + E_b}{E_b}\right) \exp\left\{\frac{\text{P}}{\pi}\int_0^\infty \frac{\delta(E')\,dE'}{E - E'}\right\}, \quad (17)$$

or

$$\frac{D_\infty(E + i\epsilon)}{D_\infty(E - i\epsilon)} = e^{-2i\delta(E)}. \quad (18)$$

These formulas show that $D_\infty(z)$ is analytic in the finite z-plane except for a cut on Z_+.* As we shall show below, $\delta(E)$ tends to zero as $E \to \infty$, which means that the discontinuity vanishes as $E \to \infty$; moreover, we shall also conclude that $D_\infty(z) \to 1$ as $z \to \infty$, and therefore the word "finite" should be dropped from the first sentence of this paragraph. It is important to understand that the limiting procedure we have just outlined only makes sense if z is kept off Z_+ as $R \to \infty$. If z is placed on Z_+ first, the subsequent limit $R \to \infty$ does not exist.

•Let us now derive (16). Call

$$\Phi(z) = \frac{1}{\pi}\int_a^b \frac{f(x')\,dx'}{z - x'}; \quad (19)$$

then

$$\Phi(x + i\epsilon) - \Phi(x - i\epsilon) = -\frac{i}{\pi}\int_a^b \left\{\frac{2\epsilon}{(x - x')^2 + \epsilon^2}\right\} f(x')\,dx'.$$

The function inside the braces { } is $O(\epsilon)$ unless $|x - x'| \lesssim \epsilon$, and hence if $f(x')$ is continuous near $x' = x$ we may replace $f(x')$ by $f(x)$ as $\epsilon \to 0$. The integration limits may also be extended to $\pm\infty$ as $\epsilon \to 0$, and therefore †

$$\lim_{\epsilon \to 0^+} [\Phi(x + i\epsilon) - \Phi(x - i\epsilon)] = -2if(x). \quad (20)$$

* In the example of potential scattering treated here, this cut is of the square-root type, and consequently the Riemann surface is only two-sheeted. As we already discussed in Sec. 15.4, the introduction of multivalued functions can therefore be avoided by working in the complex k-plane. There is considerable merit in working in the E-plane, however, because the structure of the results is then similar to those that apply to inelastic collisions and relativistic theories.

† Note that this argument amounts to

$$\frac{1}{\pi}\lim_{\epsilon \to 0} \frac{\epsilon}{x^2 + \epsilon^2} = \delta(x).$$

Next we compute

$$\Phi(x + i\epsilon) + \Phi(x - i\epsilon) = \frac{2}{\pi}\int_a^b \frac{(x - x')f(x')}{(x - x')^2 + \epsilon^2}\, dx'. \tag{21}$$

As $\epsilon \to 0$, the integrand is $f(x')/(x - x')$ everywhere except in a tiny interval around $x' = x$. We can therefore approximate (21) by

$$\Phi(x + i\epsilon) + \Phi(x - i\epsilon) = \frac{2}{\pi}\left(\int_a^{x-\epsilon_1} + \int_{x+\epsilon_1}^b\right)\frac{f(x')\, dx'}{x - x'} + \frac{2}{\pi}\int_{x-\epsilon_1}^{x+\epsilon_1} \frac{f(x')(x - x')}{(x - x')^2 + \epsilon^2}\, dx' + O(\epsilon_1), \tag{21'}$$

where $\epsilon_1 \gg \epsilon$, but $\epsilon_1 \to 0$ as $\epsilon \to 0$. If $f(x')$ is continuous its variation in the interval $(x - \epsilon_1) < x' < x + \epsilon_1$ is negligible, and the last integral in (21′) vanishes by symmetry. The first two integrals in (21′) just constitute the definition of the Cauchy principal value, and we therefore have

$$\lim_{\epsilon\to 0}\, [\Phi(x + i\epsilon) + \Phi(x - i\epsilon)] = \frac{2}{\pi}\,\mathrm{P}\int_a^b \frac{f(x')\, dx'}{x - x'}. \tag{22}$$

Upon combining (20) and (22) we obtain (16).•

Although (15) and (18) establish the connection between $D_\infty(z)$ and the phase shift $\delta(E)$, we still have no practical way of computing these functions. Several methods suggest themselves, but we shall only examine one technique which is particularly well adapted to discussing the behavior of $\delta(E)$ as $E \to 0$. The technique in question is based on yet another representation for the function $D_\infty(z)$, which follows from our earlier observation that $D_\infty(z)$ is analytic in the z-plane cut from $z = 0$ to $z = +\infty$. If we know the behavior of $D_\infty(z)$ as $z \to \infty$, we can then express $D(z)$ as a Cauchy integral. From (15) we have

$$D_\infty(z) \underset{z\to\infty}{\longrightarrow} \exp\left\{\frac{1}{\pi z}\int_0^\infty \delta(E)\, dE\right\}.$$

Thus $D_\infty(z)$ tends to one exponentially if the integral exists, i.e., if $\lim_{E\to\infty} E\,\delta(E) = 0$. If this rather stringent condition is not met by $V(r)$, the determination of the asymptotic behavior by $D_\infty(z)$ is somewhat more involved. Because one can use the Born approximation to evaluate $\delta(E)$ as $E \to \infty$, this offers no real difficulty however, and so we shall not enter into a detailed discussion of this point here. Suffice it to say that for all potentials of interest to us (i.e., those that satisfy (14)) $|D_\infty(z) - 1|$ tends to zero faster than $1/z$ as $z \to \infty$. The representation of $D_\infty(z)$ that we seek is then obtained by applying the Cauchy theorem to $D_\infty(z) - 1$, namely

$$\frac{1}{2\pi i}\oint \frac{D_\infty(z') - 1}{z' - z}\, dz' = D_\infty(z) - 1,$$

where the contour encloses the pole at $z' = z$. Because $[D_\infty(z) - 1]$ is analytic everywhere except on Z_+, we can deform the contour into a large circle and an integral about the cut. The circle does not contribute because $\lim_{z\to\infty} z(D_\infty - 1) = 0$, and therefore

$$D_\infty(z) = 1 + \frac{1}{2\pi i}\int_\infty^0 \frac{D_\infty(E - i\epsilon) - 1}{E - z}\,dE + \frac{1}{2\pi i}\int_0^\infty \frac{D_\infty(E + i\epsilon) - 1}{E - z}\,dE$$

or *

$$D_\infty(z) = 1 + \int_0^\infty \frac{A(E)\,dE}{z - E}. \tag{23}$$

The so-called spectral weight $A(E)$, which plays an important role in the remainder of this section, is related to the previously defined functions by

$$\begin{aligned} A(E) &\equiv \frac{1}{2\pi i}[D_\infty(E - i\epsilon) - D_\infty(E + i\epsilon)] \\ &= \frac{1}{\pi} D_\infty(E + i\epsilon)e^{i\delta(E)} \sin \delta(E). \end{aligned} \tag{24}$$

Because of (17), $A(E)$ is real, and $D_\infty^*(z) = D_\infty(z^*)$. Note the appearance of the scattering amplitude $e^{i\delta} \sin \delta$ here. Upon using the identity (16) on (23), we obtain †

$$D_\infty(E + i\epsilon) = 1 - i\pi A(E) + \mathrm{P}\int_0^\infty \frac{A(E')\,dE'}{E - E'},$$

* That such a representation of $D_\infty(z)$ exists could have been inferred directly from (7), which reveals that $D_R(z) \underset{z\to\infty}{\longrightarrow} 1$, and that it only has simple poles:

$$D_R(z) = 1 + \sum_{n=0}^{\infty} \frac{A_n}{z - E_n^0}.$$

By taking the limit $R \to \infty$ of this expression, one can also deduce (23) *et seq.* (see Baker, *loc. cit.*).

† Because $A(E)$ is real, (24) shows that

$$A(E) = -\frac{1}{\pi}\,\mathrm{Im}\, D_\infty(E + i\epsilon).$$

The real and imaginary parts of D_∞ are therefore related by

$$\mathrm{Re}\, D_\infty(E) = 1 - \frac{\mathrm{P}}{\pi}\int_0^\infty \frac{\mathrm{Im}\, D_\infty(E + i\epsilon)}{E - E'}\,dE'.$$

This is known as a *dispersion relation*. Relations of this type play a very important role in the modern theory of collisions between "elementary" particles. Dispersion relations were first derived by Kramers and Kronig in 1926, and applied by them to which can be combined with (24) to give

$$\pi A(E) \cot \delta(E) = 1 + \mathrm{P} \int_0^\infty \frac{A(E')\, dE'}{E - E'}. \tag{25}$$

From the practical point of view, (25) is one of the most important relations in the approach under discussion. This is due to two facts:

(a) if $V(r)$ satisfied (14), $A(E)$ is an entire function of the interaction strength parameter g;
(b) for a broad class of potentials, the right-hand side of (25) can be approximated by a polynomial in E as $E \to 0$.

The significance of these mathematical statements is the following. According to (a), we may hope to approximate $A(E)$ by the first few terms of its Taylor expansion in g without committing an horrendous error. With this approximate form of $A(E)$ we could then compute $\delta(E)$ from (25) under conditions where the Born approximation surely fails, e.g., near a resonance. Furthermore, with this *same* approximate $A(E)$ we could search for the occurrence of bound states by solving $D_\infty(-E_b) = 0$, i.e., we could look for roots E_b of the implicit equation

$$1 = \int_0^\infty \frac{A(E')\, dE'}{E' + E_b}. \tag{26}$$

These remarks reveal the superiority of the Fredholm method over the Born approximation. We recall from Sec. 13 that the latter does not enable one to say anything concerning the bound states or resonances. We shall discuss the significance of the second remark, (b), in more detail in Sec. 49.3, where we shall show that it allows us to express low-energy scattering data in terms of a small number of parameters.

With these motivations in mind, let us then proceed to the evaluation of $A(E)$ as a power series in g. We shall do this by expanding (10); the first few terms are

$$D_R(z) = 1 - g\, \mathrm{tr}[G_R^0(z)H_1] + \tfrac{1}{2}g^2\{[\mathrm{tr}\, G_R^0(z)H_1]^2 - \mathrm{tr}[G_R^0(z)H_1]^2\} + \cdots. \tag{27}$$

The coefficients of this series are computable because we can evaluate the traces in the known representation $\{|n\rangle\}$ defined by

$$(H_0 - E_n^0)|n\rangle = 0.$$

---

the propagation of light through matter. A derivation of the Kramers-Kronig relation, and several applications, will be found in Chapt. XII. For applications to scattering theory, see Goldberger and Watson, Chapt. 10, Källèn, Chapt. 5.

An elementary calculation reveals that the radial wave functions $\langle r|n\rangle$ normalized to unity in the sphere of radius R are

$$\langle r|n\rangle = \left(\frac{2k_n^2}{R}\right)^{\frac{1}{2}} j_l(k_n r). \tag{28}$$

We wish to pass to the limit $R \to \infty$ in (27), however. We can do this by changing the completeness relationship in the following manner:

$$\sum_n |n\rangle\langle n| = \sum_n \Delta E_n \frac{|n\rangle}{\sqrt{\Delta E_n}} \frac{\langle n|}{\sqrt{\Delta E_n}}$$
$$\xrightarrow[R\to\infty]{} \int_0^\infty dE|E\rangle\langle E|,$$

where

$$|E\rangle = \lim_{R\to\infty} \frac{|n\rangle}{\sqrt{\Delta E_n}}. \tag{29}$$

Combining (29) with $\Delta E_n = \hbar^2 k_n \pi/mR$ gives

$$\langle r|E\rangle = \left(\frac{2km}{\pi\hbar^2}\right)^{\frac{1}{2}} j_l(kr) \tag{30}$$

as the continuum radial wave function normalized in the desired fashion.* With these functions we immediately find

$$\operatorname{tr} G_\infty^0(z) H_1 = \int_0^\infty \frac{\langle E|H_1|E\rangle}{z-E}\, dE,$$

$$[\operatorname{tr} G_\infty^0(z)H_1]^2 - \operatorname{tr}[G_\infty^0(z)H_1]^2$$
$$= \int_0^\infty \frac{dE\, dE'}{(z-E)(z-E')} \times \{\langle E|H_1|E\rangle\langle E'|H_1|E'\rangle$$
$$- \langle E|H_1|E'\rangle\langle E'|H_1|E\rangle\}$$
$$= 2\int_0^\infty \frac{dE}{z-E} \int_0^\infty \frac{dE'}{E-E'} \begin{vmatrix} \langle E|H_1|E\rangle & \langle E|H_1|E'\rangle \\ \langle E'|H_1|E\rangle & \langle E'|H_1|E'\rangle \end{vmatrix},$$

and so forth. (The last equation follows if one expresses the energy denominators in partial fractions, and notes that the expression in the braces { } is symmetric in E and E'.) Thus each term of (27) can be cast into the form (23), and we may therefore read off the expansion of the

* Observe that $\langle r|E\rangle$ is precisely the radial portion of the free particle wave function (11.15).

spectral function:

$$A(E) = -g\langle E|H_1|E\rangle + g^2 \int_0^\infty \frac{dE_1}{E - E_1} \begin{vmatrix} \langle E|H_1|E\rangle & \langle E|H_1|E_1\rangle \\ \langle E_1|H_1|E\rangle & \langle E_1|H_1|E_1\rangle \end{vmatrix}$$

$$+ \cdots + \frac{(-g)^{n+1}}{n!} \int_0^\infty \frac{dE_1 \cdots dE_n}{(E - E_1) \cdots (E - E_n)}$$

$$\times \begin{vmatrix} \langle E|H_1|E\rangle & \cdots & \langle E|H_1|E_n\rangle \\ \vdots & & \vdots \\ \langle E_n|H_1|E\rangle & \cdots & \langle E_n|H_1|E_n\rangle \end{vmatrix} + \cdots \tag{31}$$

The matrix elements that appear here are

$$g\langle E|H_1|E'\rangle = \left(\frac{2m}{\pi\hbar^2}\right)\sqrt{kk'} \int_0^\infty r^2\, dr\, j_l(kr) j_l(k'r) V(r). \tag{32}$$

The proof that (31) converges in the finite g-plane will not be given here (see Baker, *loc. cit.*).

The relation between the Fredholm method and the Born series is now apparent. According to (24) the partial wave scattering amplitude is

$$e^{i\delta} \sin \delta = \frac{\pi A(E)}{D_\infty(E + i\epsilon)}. \tag{33}$$

In the Fredholm approach, the numerator and denominator are both given by entire functions of g, and therefore (33) enables us to systematically compute the scattering amplitude for arbitrary values of g, even though the power series may not converge rapidly enough for practical purposes. The Born series is obtained from (33) by expanding $1/D_\infty(E + i\epsilon)$ as a power series in g, and it is clear that this is not permissible if $D_\infty(E + i\epsilon)$ has zeros. This last point is already well known to us,* because the delta-shell potential of Sec. 15 had a scattering amplitude of precisely the form (33), namely

$$e^{i\delta_l} \sin \delta_l = \frac{gka[j_l(ka)]^2}{1 - igkaj_l(ka)h_l(ka)}. \tag{15.4}$$

The radius of convergence of the Born approximation was then shown to be

$$g_l^{(B)}(ka) = |kaj_l(ka)h_l(ka)|^{-1}. \tag{15.13}$$

* The same remark applies to Prob. 4, Chapt. III, and Prob. 5, Chapt. VII.

In the general case (33), the radius of convergence $g_l^{(B)}(E)$ is determined by the root of $D_\infty^{(l)}(g; E + i\epsilon) = 0$ nearest the origin of the g-plane.

As a final remark concerning the Born approximation, we note that the usual Born formula for the phase shift,

$$\delta_l(k) = -\frac{2mk}{\hbar^2}\int_0^\infty [j_l(kr)]^2 V(r) r^2\,dr, \tag{14.14'}$$

can be obtained from (25) by completely dropping the integral over the spectral function, replacing $A(E)$ by $-g\langle E|H_1|E\rangle$, and making the small phase shift approximation $\cot\delta \simeq 1/\delta$.

49.3 SCATTERING LENGTH AND EFFECTIVE RANGE *

We must still substantiate our claim that (25) provides a convenient parametrization of the phase shifts at low energy. For this purpose we must show that the right-hand side of (25) is an analytic function of the energy near the origin. Consider the behavior of $A^{(l)}(E)$ as $E \to 0$. Upon using the small kr-expansion (11.6) in (32), we obtain

$$g\langle E|H_1|E'\rangle \xrightarrow[E,E'\to 0]{} \frac{2m}{\pi\hbar^2}[(2l+1)!!]^{-2}(kk')^{l+\frac{1}{2}}\int_0^\infty V(r)r^{2l+2}\,dr. \tag{34}$$

The results that we shall obtain are mainly of interest in nuclear physics where potentials fall off exponentially as $r \to \infty$. The integrals in (34) then converge for all l. If we insert (34) into (31), we observe that every term of this convergent power series varies as $E^{l+\frac{1}{2}}$ as $E \to 0$, or to be precise,

$$A^{(l)}(E) \xrightarrow[E\to 0]{} c_0 E^{l+\frac{1}{2}}(1 + c_1 E + c_2 E^2 + \cdots). \tag{35}$$

Consequently $A^{(l)}(z)$ has a branch point at $z = 0$; we shall place the cut along Z_+ in the sequel, and take (35) as the boundary value above Z_+. The analytic continuation of (25) to complex z is given by

$$\pi A^{(l)}(z)\cot\delta(z) = 1 + \int_0^\infty \frac{A^{(l)}(E')\,dE'}{z - E'} + i\pi A^{(l)}(z). \tag{36}$$

When we set $z = E + i\epsilon$ in (36) we retrieve (25). Neither the integral in (36), nor the remaining term on the right-hand side, is analytic near

* The results given in this subsection can also be derived by more elementary methods based directly on the Schrödinger equation; consult Blatt and Weisskopf, pp. 61–64, G. F. Chew and M. L. Goldberger, *Phys. Rev.* **75**, 1367 (1949), and H. A. Bethe, *Phys. Rev.* **76**, 38 (1949). The advantage of the technique employed here is that it can be extended beyond the domain of elastic potential scattering (cf. e.g., Baker, *loc. cit.*).

$z = 0$. According to (16), the integral changes by $-2\pi i A^{(l)}(E)$ as we cross the cut from above to below. On the other hand, (35) shows that when z goes from E to $Ee^{2\pi i}$, $A^{(l)}(E) \to -A^{(l)}(E)$. The two discontinuities therefore compensate exactly, and (36) is an analytic function of z when z lies inside the radius of convergence of (35). The right-hand side of (36) can therefore be expanded in a power series about $z = 0$. If we also use (35) in the left-hand side of (36) we obtain

$$\pi c_0 E^{l+\frac{1}{2}} \cot \delta(E) = 1 - \int_0^\infty \frac{A^{(l)}(E')\, dE'}{E'} + \frac{E}{W_l} + O(E^2), \tag{37}$$

where W_l is a constant having the dimension of energy. It is clearly some energy characteristic of the potential, and we therefore conclude that the convergence of (37) improves when the potential *increases* in strength.*

The expansion (37) shows that in low energy scattering the phase shift in each partial wave depends only on two parameters. This is so since (37) can be written more compactly as †

$$k^{2l+1} \cot \delta_l \underset{k \to 0}{\longrightarrow} -\frac{1}{a_l} + \tfrac{1}{2} r_l k^2. \tag{38}$$

Even though this formula looks innocent enough, it is by no means trivial. Its essential content is that $k^{2l+1} \cot \delta_l$ is a slowly varying function at low energy. This is not *a priori* obvious; for example, the scattering amplitude computed from (38),

$$\frac{1}{k} e^{i\delta_l} \sin \delta_l = \frac{k^{2l}}{(-(1/a_l) + \frac{1}{2} r_l k^2) - ik^{2l+1}}, \tag{39}$$

is a rapidly varying function if $\frac{1}{2} a_l r_l k^2 = 1$ somewhere in the low-energy region.

Let us consider the s-wave case in somewhat greater detail. Both a_0 and r_0 have the dimension of length; they are called the *scattering length* and *effective range*, respectively. [The scattering length was already introduced in our analysis of the delta-shell potential, Eq. (15.32).] In the $l = 0$ case the effective range formula (38) becomes

$$k \cot \delta_0 \underset{k \to 0}{\longrightarrow} -\frac{1}{a_0} + \tfrac{1}{2} r_0 k^2, \tag{40}$$

* It must be admitted that this important fact emerges much more clearly from the derivations referred to above.

† Note that this formula confirms the estimate (14.17), which was based on the Born approximation.

and the corresponding s-wave scattering amplitude is

$$\frac{1}{k}e^{i\delta_0}\sin\delta_0 = \frac{k^2}{-a_0^{-1} + \frac{1}{2}r_0k^2 - ik}. \tag{41}$$

There is an interesting relationship between the binding energy E_b of a weakly bound s-state (assuming that one exists), and the scattering parameters r_0 and a_0. This relationship is merely one aspect of the fact that a bound state appears as a pole in the scattering amplitude, when the latter is treated as a function of the complex variable z. That the scattering amplitude has a pole at $z = -E_b$ can be seen by using (33) to define the analytic continuation of $e^{i\delta_0(z)}\sin\delta_0(z)$. In the complex k-plane this pole appears on the positive imaginary axis, a fact which is already familiar to us from Chapt. III. Let $E_b = (\hbar\alpha)^2/2m$, and assume that the binding is sufficiently weak so that $\frac{1}{2}r_0\alpha^2 \ll |a_0^{-1}|$. In this case (41) provides us with an accurate description of the scattering amplitude in the region of the complex k-plane of interest. We therefore set $k = i\alpha$ in (41), and solve for the location of the pole:

$$\alpha = \frac{1}{a_0} + \frac{1}{2}r_0\alpha^2. \tag{42}$$

This equation must have positive solution if there is to be a bound state. As $E_b \to 0$, $\alpha \to 0^+$, and $a_0 \to \infty$ through positive values. When g is just a bit too small to bind a state, $\alpha < 0$, and a_0 is large and negative. On the other hand, if the binding energy is not small (in the sense defined above), one cannot make a general statement concerning the sign and magnitude of a_0. In any case, it is clear that a_0 is usually not a length characterizing the range of the potential; it must also depend on g in a crucial fashion (recall (15.32) in this connection). The effective range r_0, on the other hand, can be shown * to be of order the range for attractive † potentials that are monotonic functions of r.

49.4 LEVINSON'S THEOREM

We shall now establish a connection between the phase shift at zero energy and the number of bound states. The necessity of such a relationship arises from the following "paradox":

* J. M. Blatt and J. D. Jackson, *Phys. Rev.* **76**, 18 (1949).

† For repulsive interactions the effective range expansion as given here does not converge as well as for attractive potentials. A modification of the method that is more appropriate for the repulsive case is discussed by D. G. Ravenhall, *Phys. Rev. Let.* **9**, 504 (1962). A modified effective range expansion applicable to scattering between charged particles (e.g., p-p scattering) is developed in J. D. Jackson and J. M. Blatt, *Rev. Mod. Phys.* **22**, 77 (1950).

(a) if we let $z \to 0$ in (15), it would seem that $D_\infty(z)$ has the singularity z^{-N} at the origin, where N is the number of bound states;
(b) on the other hand, the low-energy expansion implies that $D_\infty(0)$ is finite, and in general nonzero.

This last observation follows upon inserting (35) into (23), which shows that

$$D_\infty(0) = 1 - \int_0^\infty \frac{A(E')\,dE'}{E'}$$

exists, and can only vanish under the very exceptional circumstance of a bound state at precisely zero energy. If this "paradox" is to be resolved, it must be that the factor

$$f(E) = \exp\left\{\frac{\mathrm{P}}{\pi}\int_0^\infty \frac{\delta(E')\,dE'}{E - E'}\right\} \tag{43}$$

in (17) removes the singularity E^{-N}. Let us therefore examine $f(E)$ more closely. Because $\delta(\infty) = 0$ with our definition of the phase shift (i.e., $\delta \to 0$ if $V \to 0$), we can perform an integration by parts in (43) to obtain

$$f(E) = \exp\left\{\frac{1}{\pi}\,\delta(0)\ln E + \frac{\mathrm{P}}{\pi}\int_0^\infty \frac{d\delta(E')}{dE'}\ln|E - E'|\,dE'\right\}. \tag{43'}$$

In virtue of the effective range formula (38), the phase shift varies as $E^{l+\frac{1}{2}}$ as $E \to 0$. Consequently the integral in (43′) tends to a finite limit at $E = 0$, whence

$$f(E) \underset{E\to 0}{\longrightarrow} f_0 E^{\delta(0)/\pi},$$

where f_0 is a constant. The condition that $E^{-N}f(E)$ should tend to a finite value as $E \to 0$ therefore requires

$$\delta(0) = N\pi. \tag{44}$$

This relationship is called *Levinson's theorem.*

Equation (44) has one immediate experimental consequence. If there are N bound states, and because $\delta(\infty) = 0$, it is clear that $\sin^2\delta(E)$ will have to pass through one at least N times in the interval $0 \leqslant E < \infty$. These resonances need not be sharp, however; a sharp resonance will only occur if $\delta(E)$ passes rapidly through an odd multiple of $\pi/2$. The best known example of a very broad resonance accompanying a bound state occurs in the ${}^3S$ n-p system, and is associated with the existence of the deuteron. The distinction between a sharp resonance and a point where $\delta(E)$ happens to pass slowly through an odd multiple of $\pi/2$ can be seen

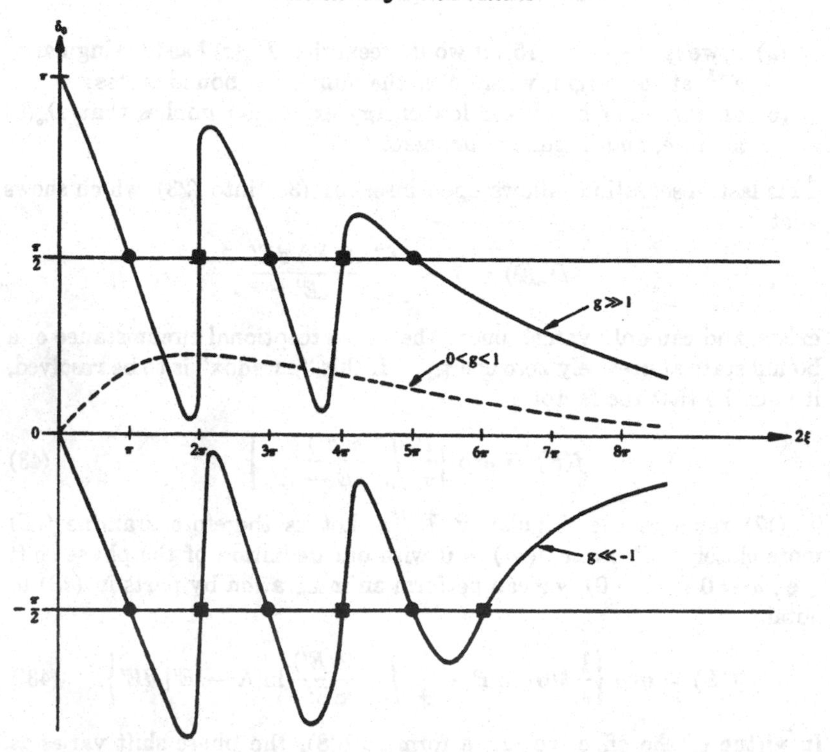

Fig. 49.2. The *s*-wave phase shift as a function of energy for the delta-shell potential of Sec. 15. Here $\xi \equiv ka$, a being the radius of the shell, and g is a dimensionless coupling constant. When $g<0$, the potential is repulsive, and if $g>0$, it is attractive. There is one bound state if $g>1$. The sharp resonances are shown as squares, and the broad resonances as dots. These curves are not the result of detailed calculations. They have been sketched by using Fig. 15.3, and $\tan \delta_0 = 2g \sin^2\xi/(2\xi - g \sin 2\xi)$.

very clearly in the example of the delta-shell potential of Sec. 15. In Fig. 49.2 we have sketched $\delta_0(k)$ for two cases: (i) when the potential is attractive and binds a state, and (ii), when it is repulsive and therefore cannot bind a state. In both cases $|g|$ is sufficiently large to lead to several sharp resonances. We note that at the broad resonances which come from "hard sphere" scattering $d\delta/dk < 0$, whereas at the sharp resonances $d\delta/dk$ is large and positive; in fact, the larger $d\delta/dk$, the narrower the resonance. The points where δ falls through $\pm\pi/2$ can never be narrow resonances since Wigner has shown * that causality

* E. P. Wigner, *Phys. Rev.* **98**, 145 (1955).

implies a lower bound on $d\delta/dk$ which is of order the negative of the range of the potential.

Finally we make some rather obvious remarks concerning the scope of the method developed in this section. It is clear that it can be generalized immediately to any problem where the states are completely specified by symmetry quantum numbers and the energy. Thus for nucleon-nucleon scattering in the absence of tensor forces, the Hilbert space can be separated into subspaces labeled by L, J, M_J, and within such a subspace the energy uniquely labels the state. The formulas given above then apply if the label l is replaced by L and J, and if H_1 designates the interaction Hamiltonian in the subspace (L, J). If tensor forces are included, however, mixing of the type $({}^3S_1 + {}^3D_1)$ occurs, as explained in Sec. 42, and there are two states to every value of E, J, M_J, and parity. Hence a (2×2) matrix generalization of the formalism would be required.* If inelastic scattering to three-body channels is also possible (e.g., $p + p \to p + p + \pi^0$), the states are labeled by more than one continuous parameter and the technique breaks down.

PROBLEMS

1. Let $\{|E^0_{K,n}\rangle\}$ be the eigenkets of H_0 (in Sec. 45.2 these were designated by the more complete symbol $|a_n E^0_{K,n}\rangle$), and let $\{|E_{K,n}\rangle\}$ be the eigenkets of $(H_0 + H_1)$. It is assumed that as $H_1 \to 0$, the subspace spanned by the $|E^0_{K,n}\rangle$ belonging to one value of K coincides with the subspace selected by the projection operator P_K.

(a) Show that

$$P_K H_1 \left(1 + \frac{Q_K}{E^0_{K,n} - H_0}(\Delta_{K,n} - H_1)\right)^{-1} P_K|E_{K,n}\rangle = \Delta_{K,n} P_K|E_{K,n}\rangle,$$

where $Q_K = 1 - P_K$.

(b) Consider the case of a degenerate two-dimensional subspace (i.e., $E^0_{K,i} = E^0_K$ with $i = 1$ or 2), and assume all matrix elements of H_1 between the unperturbed states belonging to this subspace vanish. Show that to second order in H_1 the energy shifts are the roots of the quadratic equation

$$(\Delta - M_{11})(\Delta - M_{22}) = |M_{12}|^2,$$

where

$$M_{ij} = \sum_{K',n'} \frac{\langle E^0_{K,i}|H_1|E^0_{K',n'}\rangle\langle E^0_{K',n'}|H_1|E^0_{K,j}\rangle}{E^0_K - E^0_{K',n'}}.$$

* In this connection see R. Blankenbecler in *Strong Interactions and High Energy Physics*, edited by R. G. Moorhouse, Oliver & Boyd, London, 1964.

2. The proton is actually not a point charge, but has a charge distribution whose r.m.s. radius is 0.8×10^{-13} cm. Show that electronic s-states are vastly more sensitive to the finite nuclear size than states of higher angular momentum. Estimate the energy shift for the $1s$ and $2s$ level in hydrogen by assuming the proton's charge to be spread uniformly over a sphere, and compare your result to the fine structure splitting. How big are these effects for a μ-meson bound to a proton?

3. Compute the Stark effect for the $2s_{\frac{1}{2}}$ and $2p_{\frac{1}{2}}$ levels of hydrogen for a field $\mathcal{E}$ sufficiently weak so that $e\mathcal{E}a_0$ is small compared to the fine structure, but taking the Lamb shift δ into account (i.e., ignore $2p_{\frac{3}{2}}$ in this calculation). Show that for $e\mathcal{E}a_0 \ll \delta$ the energy shifts are quadratic in $\mathcal{E}$, whereas for $e\mathcal{E}a_0 \gg \delta$ they are linear in $\mathcal{E}$. (The radial integral you need is $\langle 2s|r|2p\rangle = 3\sqrt{3}\, a_0$.) Briefly discuss the consequences (if any) of time reversal for this problem.

4. A hydrogen atom whose proton is fixed at the origin is perturbed by two fixed charges Q_1 and Q_2 placed on the z-axis at $z = \pm b$. Assume that b is much larger than the Bohr radius a_0, and that $Q_{1,2}$ are much smaller in magnitude than the electronic charge e.

(a) What can be said about the states of the system purely on the basis of symmetry considerations when (i) $Q_1 = Q_2$, (ii) $Q_1 \neq Q_2$?

(b) In the case $Q_1 = Q_2 = Q$, show that the ground state of the perturbed atom has a quadrupole moment, estimate its order of magnitude, and indicate how you would compute it rigorously to lowest order in Q/e and a_0/b. Show that the quadrupole moment can be computed from the ground state energy.

(c) Assume that the perturbation due to the two fixed charges is of order the fine structure splitting. Compute the spectrum of the $n = 3$ states as a function of $(Q_1 - Q_2)$ for a fixed value of $Q_1 + Q_2$.

(d) What are the consequences if Q_1 and/or Q_2 attain a magnitude comparable to e.

5. A spin zero particle interacts with a potential V whose coordinate space matrix elements are *

$$\langle \mathbf{r}|V|\mathbf{r}'\rangle = \delta(\mathbf{r} - \mathbf{r}')V_0(r) - \lambda v(r)v(r'),$$

where

$$\begin{aligned} V_0(r) &= 0 \qquad \text{if} \quad r > r_c, \\ &= \infty \qquad \text{if} \quad r \leqslant r_c, \end{aligned}$$

and

$$v(r) = \frac{e^{-r/b}}{r}.$$

(a) Use the techniques of Sec. 49 to show that the s-wave scattering phase shift δ is given by

$$k \cot(\delta + kr_c) = \frac{(k^2b^2 + 1)^2 + \xi(k^2b^2 - 1)}{2\xi b},$$

* The results quoted in this problem are due to C. deDominicis and P. C. Martin (unpublished).

where

$$\hbar^2\xi = 4\pi\lambda m b^3 \exp(-2r_c/b).$$

(b) Evaluate the scattering length and effective range.

(c) Determine the condition on the parameters that must be met if there is to be a bound state. Can there be more than one bound state with this potential?

(d) Sketch the phase shift as a function of k for various values of the dimensionless parameter ξ. Show that the phase shift changes sign at high energy,* and comment on this in the light of the connection between the phase and energy shifts. Study carefully the behavior of the phase shift at low energy when ξ is increased through the critical value required for binding a state.

6. Verify (49.11) when A is an operator in a finite dimensional space.

* The proton-proton S-wave phase shift is positive at low energy, but changes sign at approximately 300 Mev lab. energy. This fact leads to the conclusion that the nucleon-nucleon interaction in the ^1S state must be repulsive at short distances.

VIII

The Electromagnetic Field

Dirac invented the quantum theory of the radiation field in 1927. Since then more sophisticated formulations of the theory have been developed.* Nevertheless, we shall follow the Dirac approach here because it is both easiest to understand and adequate for our purposes. The modern formulations are indispensable if one is concerned with calculations to high order of accuracy in the fine structure constant, or with phenomena in which the sources of the field move with velocities comparable to *c*.

We first cast the classical theory into a Hamiltonian format.† Once this is done the passage to the quantum theory is straightforward.

50. Classical Electrodynamics in Hamiltonian Form

The total energy of a system of *N* charged particles in an electromagnetic field is

$$H = U + \sum_{i=1}^{N} H_i, \tag{1}$$

* The history of the subject can be traced in the reprint collection edited by Schwinger.

† A more detailed discussion of the classical theory can be found in Heitler, pp. 38–53.

where

$$H_i = \frac{1}{2m_i}\left(\mathbf{p}_i - \frac{e_i}{c}\mathbf{A}_i\right)^2 + e_i\phi_i, \tag{2}$$

$$U = \frac{1}{8\pi}\int d^3r\,(\mathcal{E}^2 + \mathcal{H}^2). \tag{3}$$

The electromagnetic field strengths, $\mathcal{E}$ and $\mathcal{H}$, are related to the potentials by

$$\mathcal{E} = -\nabla\phi - \frac{1}{c}\dot{\mathbf{A}}, \tag{4}$$

$$\mathcal{H} = \nabla \times \mathbf{A}. \tag{5}$$

The notation $(\mathbf{A}_i, \phi_i)$ indicates that the potentials are evaluated at the instantaneous position of the ith particle.

50.1 THE FREE FIELD

Consider first the free field. In this case one may choose the gauge

$$\nabla \cdot \mathbf{A} = \phi = 0. \tag{6}$$

Maxwell's equations are then satisfied if

$$\Box^2\mathbf{A} = 0, \tag{7}$$

where $\Box^2 = \nabla^2 - c^{-2}\,\partial_t^2$.

Because of (6), $\mathbf{A}$ only has two components at each point in space. To see this in detail we expand $\mathbf{A}$ in terms of plane waves satisfying periodic boundary conditions * on a large cube of volume Ω:

$$\mathbf{A}(\mathbf{r}, t) = \Omega^{-\frac{1}{2}}\sum_{\mathbf{k}} \mathbf{A}_{\mathbf{k}}(t)e^{i\mathbf{k}\cdot\mathbf{r}}. \tag{8}$$

The components of $\mathbf{k}$ are $(n_x, n_y, n_z)2\pi\Omega^{-\frac{1}{3}}$, where $n_i = 0, \pm 1, \pm 2, \ldots$. Since $\nabla \cdot \mathbf{A} = 0$,

$$\mathbf{k} \cdot \mathbf{A}_{\mathbf{k}} = 0, \tag{9}$$

which shows that the Fourier coefficients are perpendicular to the propagation vector $\mathbf{k}$. (A vector field having this property is said to be *transverse*.) Consequently $\mathbf{A}_{\mathbf{k}}$ has only *two* components $A_{\mathbf{k}\alpha}$ (with $\alpha = 1, 2$) for each $\mathbf{k}$. Equivalently we may write

$$\mathbf{A}_{\mathbf{k}} = \sum_{\alpha} \mathbf{e}_{\mathbf{k}\alpha}A_{\mathbf{k}\alpha}, \tag{10}$$

* In this connection recall the discussion at the end of Sec. 7.3.

where $\mathbf{e}_{\mathbf{k}\alpha}$ is called a polarization unit vector. There are only two independent polarization vectors because light is a purely transverse vibration of the electromagnetic field.

When we substitute the Fourier series (8) into the wave equation (7), and take the linear independence of the separate terms in (8) into account, we find that

$$\ddot{\mathbf{A}}_{\mathbf{k}}(t) + c^2k^2\mathbf{A}_{\mathbf{k}}(t) = 0. \tag{11}$$

Hence the Fourier coefficients oscillate harmonically with frequency

$$\omega_k = ck. \tag{12}$$

This is the familiar vacuum dispersion law for light.

We are not concerned with the most general solution of (7), but only with those that are real. We can find the real solutions by writing $\mathbf{A}(\mathbf{r}, t)$ as (8) plus its complex conjugate *:

$$\mathbf{A}(\mathbf{r},t) = \Omega^{-\frac{1}{2}} \sum_{\mathbf{k}} e^{i\mathbf{k}\cdot\mathbf{r}}[\mathbf{A}_{\mathbf{k}}(t) + \mathbf{A}^*_{-\mathbf{k}}(t)]. \tag{13}$$

This is the expansion we shall use henceforth. In the sequel we usually do not designate the subscript on $\mathbf{A}_{\mathbf{k}}$ and related quantities as a vector, but it should be remembered that it is one.

It will prove convenient to choose the solution

$$\mathbf{A}_k(t) = e^{-i\omega t}\mathbf{A}_k \tag{14}$$

to (11). The Fourier series for the field strengths are then computed from (4) and (5). One easily finds

$$\mathcal{E}(\mathbf{r}, t) = \frac{i}{c\sqrt{\Omega}} \sum_{\mathbf{k}} e^{i\mathbf{k}\cdot\mathbf{r}} \omega_k [\mathbf{A}_k(t) - \mathbf{A}^*_{-k}(t)], \tag{15}$$

and

$$\mathcal{H}(\mathbf{r}, t) = \frac{i}{\sqrt{\Omega}} \sum_{\mathbf{k}} e^{i\mathbf{k}\cdot\mathbf{r}} \mathbf{k} \times [\mathbf{A}_k(t) + \mathbf{A}^*_{-k}(t)]. \tag{16}$$

Therefore

$$\int d^3r\, \mathcal{E} \cdot \mathcal{E}^* = c^{-2} \sum_{\mathbf{k}} \omega_k^2 |\mathbf{A}_k(t) - \mathbf{A}^*_{-k}(t)|^2,$$

$$\int d^3r\, \mathcal{H} \cdot \mathcal{H}^* = \sum_{\mathbf{k}} k^2 |\mathbf{A}_k(t) + \mathbf{A}^*_{-k}(t)|^2,$$

* Because the free field equations are homogeneous we need not concern ourselves with multiplicative constants.

where we have used (7.41). In terms of the Fourier coefficients the energy of the free field is therefore given by

$$U = \frac{1}{2\pi} \sum_{\mathbf{k}\alpha} k^2 |A_{\mathbf{k}\alpha}|^2. \tag{17}$$

In order to quantize the electromagnetic field we shall have to express the field equations in Hamiltonian form and impose the canonical commutation rules on the canonically conjugate variables. We have already seen that we can think of the field as a collection of uncoupled oscillators, two for each value of $\mathbf{k}$, and described by the complex displacements $A_{\mathbf{k}\alpha}$. It now merely remains to replace these variables by real canonical variables. To this end we introduce

$$\mathbf{Q}_k(t) = \frac{i}{c\sqrt{4\pi}} [\mathbf{A}_k(t) - \mathbf{A}_k^*(t)], \tag{18}$$

$$\mathbf{P}_k(t) = \frac{k}{\sqrt{4\pi}} [\mathbf{A}_k(t) + \mathbf{A}_k^*(t)], \tag{19}$$

or

$$\mathbf{A}_k(t) = -ic\sqrt{\pi}\,[\mathbf{Q}_k(t) + \frac{i}{\omega_k}\mathbf{P}_k(t)]. \tag{20}$$

In terms of these variables the field energy reads

$$U = \tfrac{1}{2} \sum_{\mathbf{k}\alpha} (P_{\mathbf{k}\alpha}^2 + \omega_k^2 Q_{\mathbf{k}\alpha}^2). \tag{21}$$

As U is conserved, it is not necessary to specify the time argument of P and Q in (21). *The energy of the free field therefore has the same form as the Hamiltonian of a set of uncoupled oscillators of unit mass and frequency ω_k.*

The proof that $\mathbf{P}_k$ and $\mathbf{Q}_k$ are canonically conjugate momenta and coordinates is now elementary.* From (14) we observe that $\dot{\mathbf{A}}_k = -i\omega_k \mathbf{A}_k$. The time derivative of (18) is therefore equal to (19), or $\mathbf{P}_k = \dot{\mathbf{Q}}_k$. Comparing with (21) we therefore obtain the first of Hamilton's equations:

$$\dot{Q}_{\mathbf{k}\alpha} = \partial U/\partial P_{\mathbf{k}\alpha}. \tag{22}$$

In a similar fashion one verifies the other of Hamilton's equations:

$$\dot{P}_{\mathbf{k}\alpha} = -\partial U/\partial Q_{\mathbf{k}\alpha}. \tag{23}$$

* Although the vectors $\mathbf{P}_k$ are canonical momenta, they are not momenta in the colloquial sense. In particular, the momentum of the electromagnetic field is not $\Sigma_k \mathbf{P}_k$. The momentum of the field is discussed in Sec. 52.2.

As a consequence

$$\ddot{\mathbf{Q}}_k + \omega_k^2 \mathbf{Q}_k = \ddot{\mathbf{P}}_k + \omega_k^2 \mathbf{P}_k = 0. \tag{24}$$

This completes the proof that $\mathbf{Q}_k$ and $\mathbf{P}_k$ are canonically conjugate coordinates and momenta, and that $U(\mathbf{Q}_k, \mathbf{P}_k)$ plays the role of the Hamiltonian. Because of this we shall designate the energy of the radiation field by H_γ instead of by U henceforth.

50.2 THE FIELD IN THE PRESENCE OF SOURCES

Finally we must incorporate the sources. When this is done, the final form of the Hamiltonian has a very simple and plausible form. Because of this, and because the derivation is purely classical, we shall forego most of the proofs here; a detailed development can be found in Heitler, *loc. cit.*

We remain in the so-called *radiation gauge*, $\nabla \cdot \mathbf{A} = 0$. The field is expanded as in (13), and the expansion coefficients $\mathbf{A}_k(t)$ are again related to the $\mathbf{P}_k(t)$ and $\mathbf{Q}_k(t)$ by (20). The equations of motion for these quantities are now much more complicated. The sources are described by the usual canonical variables $\mathbf{p}_i$ and $\mathbf{r}_i$. In terms of these variables the Hamiltonian for the entire system is then

$$H = H_\gamma + H_C + \sum_{i=1}^{N} H_i. \tag{25}$$

Here H_γ is just (21), but it is no longer a conserved quantity. H_C is the instantaneous (*not* retarded) Coulomb interaction between the particles:

$$H_C = \frac{1}{2} \sum_{i \neq j} \frac{e_i e_j}{|\mathbf{r}_i - \mathbf{r}_j|} \tag{26}$$

Finally,

$$H_i = \frac{1}{2m_i} \left[\mathbf{p}_i - \frac{e_i}{c} \mathbf{A}(\mathbf{r}_i, t) \right]^2. \tag{27}$$

The independent dynamical variables are the $\mathbf{p}_i$, $\mathbf{r}_i$ of the sources, and the *transverse* vector potential.* The longitudinal part of $\mathbf{A}$ (the curl-free part) and the scalar potential do not appear here at all. In this gauge the magnetic forces on the charged particles are mediated by the radiation field, i.e., by the emission and absorption of light.

This formulation does not look Lorentz invariant. Actually it is; when one makes a Lorentz transformation one can again put H into the

* Note that the solenoidal or transverse part of $\mathbf{A}$ is not affected by gauge transformations, and therefore the formulation given here is actually gauge invariant, even though it may not appear to be so.

form (25) by carrying out a gauge transformation as well. There is actually one swindle: when one carries out the development, one finds that the term $i = j$ in H_C also contributes. One usually waves one's hand here and says that these self-energy terms are already contained in the rest mass of the particles. Unfortunately this leads to certain problems with Lorentz invariance that we shall not enter into. Clearly the present formulation is not suited, as it stands, to a discussion of self-energy problems, nor to problems where the sources move rapidly. It is very well suited for our purposes, however, because we shall confine ourselves to nonrelativistic motions of the sources.

51. Canonical Quantization

The passage to quantum mechanics is achieved by asserting that the quantities $\mathbf{Q}_k$ and $\mathbf{P}_k$ are operators satisfying the canonical commutation rules *:

$$[P_{k\alpha}, P_{k'\alpha'}] = [Q_{k\alpha}, Q_{k'\alpha'}] = 0, \tag{1}$$

$$[Q_{k\alpha}, P_{k'\alpha'}] = i\hbar\, \delta_{\mathbf{kk}'}\, \delta_{\alpha\alpha'}. \tag{2}$$

The particle variables $\mathbf{r}_i$ and $\mathbf{p}_i$ also become operators satisfying the now familiar canonical commutation rules. We shall continue to comply with the Correspondence Principle, i.e., assume that the Hamiltonian operator is still (50.25). In carrying out this transition to quantum mechanics we retain the right to add or delete terms that vanish as $\hbar \to 0$.

The states of the system are specified by the eigenvalues of a complete set of compatible observables. The set $\{\mathbf{r}_i, Q_{k\alpha}\}$ is the most obvious, but we shall see that the displacements $Q_{k\alpha}$ are not a convenient set of electromagnetic variables for many purposes. A more convenient set of field observables is related to the particle aspects of the electromagnetic field, the photons. To see how these arise we shall study the free radiation field in some detail before attacking the problem of its interaction with matter.

Although Dirac's quantization prescription is both simple and plausible, it does represent a very fundamental theoretical step. Consequently it is not possible to make an *a priori* judgment concerning the validity of the theory. In the remainder of this chapter we shall show that the theory based on recipes (1) and (2) is internally consistent and

* These equations apply to the operators in the Schrödinger picture, or to the Heisenberg operators evaluated at equal time arguments.

describes both the corpuscular and wave-like aspects of electromagnetic phenomena. An introduction to the theory of the interaction of the field with charged particles shall be given in Sec. 58 and Chapt. X. Because we shall not do justice to the theory we should stress at the outset that modern quantum electrodynamics is one of the most successful theories in the history of physics.* Its predictions have been checked experimentally to incredibly high accuracy, and at the moment the scope of its validity remains to be delineated.

52. Photons

In Sec. 50 we proved that the free electromagnetic field is equivalent to an infinite set of uncoupled one-dimensional harmonic oscillators, one oscillator being associated with each value of the propagation vector $\mathbf{k}$ and polarization mode α. As we shall see, this very simple system displays a rather wide variety of phenomena, and its theory is correspondingly rich in content.

52.1 FIELD OPERATORS

In our discussion of the oscillator in Sec. 31 we found that the theory took on a very simple form in terms of a pair of non-Hermitian operators a and $a^\dagger$. We shall follow the same route here, and define a set of non-Hermitian operators $a_{k\alpha}$ and $a_{k\alpha}^\dagger$ by

$$\begin{aligned} Q_{k\alpha} &= i\sqrt{\hbar/2\omega}\,(a_{k\alpha} - a_{k\alpha}^\dagger), \\ P_{k\alpha} &= \sqrt{\hbar\omega/2}\,(a_{k\alpha} + a_{k\alpha}^\dagger). \end{aligned} \tag{1}$$

At times we shall also employ the abbreviation

$$\mathbf{a}_k = \sum_\alpha \mathbf{e}_{k\alpha} a_{k\alpha}.$$

In this notation the transversality condition (50.9) reads

$$\mathbf{k} \cdot \mathbf{a}_k = 0. \tag{2}$$

The commutation rules for the operators $a_{k\alpha}$ and $a_{k\alpha}^\dagger$ are an immediate consequence of the canonical commutation rules:

$$\begin{aligned} [a_{k\alpha}, a_{k'\alpha'}^\dagger] &= \delta_{\alpha\alpha'}\,\delta_{\mathbf{k}\mathbf{k}'}, \\ [\mathbf{a}_k, \mathbf{a}_{k'}] &= [\mathbf{a}_k^\dagger, \mathbf{a}_{k'}^\dagger] = 0. \end{aligned} \tag{3}$$

The very close parallel to the equations of Sec. 31.1 should be noted.

* Detailed expositions can be found in the monographs by Bjorken and Drell, and by Thirring.

When we substitute the expressions (1) into the Hamiltonian H_γ, and use the commutation rules (3), we obtain the expression $\Sigma_{\mathbf{k}\alpha}\hbar\omega_k(a^\dagger_{k\alpha}a_{k\alpha} + \frac{1}{2})$. The lower bound to the spectrum of H_γ is therefore $\Sigma\frac{1}{2}\hbar\omega_k$, and this sum is clearly infinite. This is an example of a spurious difficulty arising from too literal an interpretation of the Correspondence Principle. In each mode $(\mathbf{k}, \alpha)$ the troublesome term $\frac{1}{2}\hbar\omega_k$ is negligible in the classical limit because we shall soon see that the expectation value of the operators $a^\dagger_{k\alpha}a_{k\alpha}$ is huge compared to one in a state that can be adequately described by classical electrodynamics. Put another way, we could have asserted that the quantum theory is obtained by replacing the Fourier coefficients $A_{k\alpha}$ by $(2\pi\hbar c^2/\omega_k)^{\frac{1}{2}}a_{k\alpha}$ in all the classical expressions, where the operators satisfy the commutation rules (3). We shall therefore adopt the definition

$$H_\gamma = \sum_{\mathbf{k}\alpha} \hbar\omega_k a^\dagger_{k\alpha}a_{k\alpha} \tag{4}$$

for the Hamiltonian of the free radiation field.

Because the a-operators pertaining to different modes commute, we can confine our attention to the separate modes for the moment. Within one mode we can take over the results for the one-dimensional oscillator *in toto*, because a_k has precisely the same algebraic properties as the operator a of Sec. 31. We therefore introduce the Hermitian operator

$$N_{k\alpha} = a^\dagger_{k\alpha}a_{k\alpha}. \tag{5}$$

The eigenvalues of this operator shall be designated by $n_{k\alpha}$, and as we know from Sec. 31, $n_{k\alpha} = 0, 1, 2, \ldots$. We also know from Sec. 31 that the single operator $N_{k\alpha}$ already constitutes a complete set of observables for one mode.

As a complete set of observables for the field we may take all the operators $N_{k\alpha}$.* A complete set of eigenkets can then be defined by

$$(N_{q\beta} - n_{q\beta})|\{n_{k\alpha}\}\rangle = 0, \tag{6}$$

and

$$\langle\{n_{k\alpha}\}|\{n'_{k\alpha}\}\rangle = \prod_{k\alpha} \delta_{n_{k\alpha},n'_{k\alpha}}. \tag{7}$$

The kets $|\{n_{k\alpha}\}\rangle$ play a central role in the theory because they represent the stationary states of the free field. The energy eigenvalue pertaining to $|\{n_{k\alpha}\}\rangle$ is given by

$$E(\{n_{k\alpha}\}) = \sum_{\mathbf{k}\alpha} \hbar\omega_k n_{k\alpha}. \tag{8}$$

* It is hardly necessary to point out that this is only one of an infinite variety of possible complete sets. Other obvious sets are $\{\mathbf{P}_k\}$, $\{\mathbf{Q}_k\}$, $\{N_{k1}, Q_{k2}\}$.

The lowest stationary state is specified by the eigenvalues $n_{k\alpha} = 0$ for every **k** and α. This state is called the *vacuum* state, and we shall designate it by $|0\rangle$. All the other stationary states can be constructed from $|0\rangle$ by recalling (31.24):

$$|\{n_{k\alpha}\}\rangle = \prod_{k\alpha} (n_{k\alpha}!)^{-\frac{1}{2}}(a_{k\alpha}^{\dagger})^{n_{k\alpha}}|0\rangle. \tag{9}$$

As in ordinary quantum mechanics, the dynamics can be formulated in either the Heisenberg or the Schrödinger picture. We shall usually find the H-picture to be more convenient. The Heisenberg operators satisfy

$$i\hbar \frac{d}{dt} \mathbf{a}_k(t) = [\mathbf{a}_k(t), H_\gamma] = \hbar\omega_k \mathbf{a}_k(t). \tag{10}$$

The solution of this equation is obviously

$$\mathbf{a}_k(t) = \mathbf{a}_k e^{-i\omega_k t}. \tag{11}$$

By redefining $\mathbf{Q}_k$ and $\mathbf{P}_k$ as operators, we have also turned the vector potential $\mathbf{A}(\mathbf{r}, t)$ into an operator. The form of this operator can be obtained by substituting (1) into (50.20) and (50.13):

$$\mathbf{A}(\mathbf{r}, t) = \sum_{\mathbf{k}l} \left(\frac{hc^2}{\omega\Omega}\right)^{\frac{1}{2}} e^{i\mathbf{k}\cdot\mathbf{r}}[\mathbf{a}_k(t) + \mathbf{a}_{-k}^{\dagger}(t)]. \tag{12}$$

We then define *operators for the electromagnetic field strengths* by (50.4) and (50.5):

$$\mathcal{E}(\mathbf{r}, t) = i \sum_{\mathbf{k}} \left(\frac{h\omega}{\Omega}\right)^{\frac{1}{2}} e^{i\mathbf{k}\cdot\mathbf{r}}[\mathbf{a}_k(t) - \mathbf{a}_{-k}^{\dagger}(t)], \tag{13}$$

$$\mathcal{H}(\mathbf{r}, t) = i \sum_{\mathbf{k}} \left(\frac{hc^2}{\omega\Omega}\right)^{\frac{1}{2}} e^{i\mathbf{k}\cdot\mathbf{r}}\mathbf{k} \times [\mathbf{a}_k(t) + \mathbf{a}_{-k}^{\dagger}(t)]. \tag{14}$$

Here $h = 2\pi\hbar$ is Planck's original constant. The basic mathematical object in the theory is the operator (12), because we can compute all the observables (such as $\mathcal{E}$ or $\mathcal{H}$) from it. Frequently we shall refer to $\mathbf{A}(\mathbf{r}, t)$ simply as the field operator.

It will be noted that in the quantum theory of the radiation field we associate vector operators with each point in the space-time continuum. The space-time variables $\mathbf{r}$ and t play the role of parameters, not of dynamical variables—recall the discussion on p. 248.

With the help of (11) it is now a simple matter to show that the H-picture observables $\mathbf{A}(\mathbf{r}, t)$, $\mathcal{E}(\mathbf{r}, t)$ and $\mathcal{H}(\mathbf{r}, t)$ satisfy

$$\begin{aligned} &\Box^2 \mathbf{A} = 0, \\ &\nabla \cdot \mathcal{E} = \nabla \cdot \mathcal{H} = 0, \\ &\frac{1}{c}\frac{\partial \mathcal{H}}{\partial t} + \nabla \times \mathcal{E} = 0, \\ &\frac{1}{c}\frac{\partial \mathcal{E}}{\partial t} - \nabla \times \mathcal{H} = 0, \end{aligned} \tag{15}$$

where

$$\frac{\partial \mathcal{H}}{\partial t} = \frac{1}{i\hbar}[\mathcal{H}, H_\gamma], \tag{16}$$

etc. We therefore conclude that *the field operators in the Heisenberg picture satisfy Maxwell's equations.* The analogy with classical electrodynamics is far from complete, however, because these operators no longer commute with each other. We shall derive their commutation rules in the following section.

52.2 CREATION AND DESTRUCTION OPERATORS—PHOTONS

We shall now investigate the physical significance of the states $|\{n_{k\alpha}\}\rangle$ in some detail. We already know from the theory of the oscillator that when $a_{k\alpha}^\dagger$ acts on $|\{n_{q\beta}\}\rangle$ it produces a state in which all the eigenvalues belonging to modes other than $(\mathbf{k}, \alpha)$ are unaffected, but the eigenvalue of $N_{k\alpha}$ is increased by unity. The operator $a_{k\alpha}$ has the opposite function: it produces a state in which the eigenvalue of $N_{k\alpha}$ is decreased by one. By the same token, the energy eigenvalue of $a_{k\alpha}^\dagger|\{n_{q\beta}\}\rangle$ differs from that of $|\{n_{q\beta}\}\rangle$ by $\hbar\omega_k$, and the energy difference between the states $a_{k\alpha}|\{n_{q\beta}\}\rangle$ and $|\{n_{q\beta}\}\rangle$ is $-\hbar\omega_k$.

Another significant feature emerges when one considers the momentum of the electromagnetic field. In classical theory the field momentum is given by $(1/4\pi c)\int d^3r\, \mathcal{E} \times \mathcal{H}$. Because $\mathcal{E}$ and $\mathcal{H}$ do not commute with each other, we first try the Hermitian form

$$\mathbf{P} = \frac{1}{8\pi c}\int d^3r(\mathcal{E} \times \mathcal{H} - \mathcal{H} \times \mathcal{E})$$

as the momentum operator for the electromagnetic field. Using (11) and (12), and the fact that $\mathcal{E}$ and $\mathcal{H}$ are both Hermitian, we find that

$$\begin{aligned} \mathbf{P} &= \frac{1}{8\pi c}\int d^3r(\mathcal{E}^\dagger \times \mathcal{H} - \mathcal{H}^\dagger \times \mathcal{E}) \\ &= \tfrac{1}{4}\hbar \sum_{\mathbf{k}} \{(\mathbf{a}_k^\dagger - \mathbf{a}_{-k}) \times [\mathbf{k} \times (\mathbf{a}_k + \mathbf{a}_{-k}^\dagger)] \\ &\qquad - [\mathbf{k} \times (\mathbf{a}_k^\dagger + \mathbf{a}_{-k})] \times (\mathbf{a}_k - \mathbf{a}_{-k}^\dagger)\}. \end{aligned}$$

Because $\mathbf{k} \cdot \mathbf{a}_k = 0$, the expression in curly braces reduces to

$$\mathbf{k}\{(\mathbf{a}_k^\dagger - \mathbf{a}_{-k}) \cdot (\mathbf{a}_k + \mathbf{a}_{-k}^\dagger) + (\mathbf{a}_k^\dagger + \mathbf{a}_{-k}) \cdot (\mathbf{a}_k - \mathbf{a}_{-k}^\dagger)\},$$

and $\mathbf{P}$ therefore becomes $\hbar\Sigma_{\mathbf{k}\alpha}\mathbf{k}(a_{k\alpha}^\dagger a_{k\alpha} + \frac{1}{2})$. The infinite term $\Sigma_{\mathbf{k}\alpha}\mathbf{k}$ can be dropped once more, because it does not contribute to a "classical" state, i.e., one wherein all the $n_{k\alpha}$ are huge compared to unity. (If the sum is carried out symmetrically, the expression $\Sigma_{\mathbf{k}\alpha}\mathbf{k}$ actually has the "value" zero!) We therefore redefine the momentum operator as

$$\mathbf{P} = \sum_{\mathbf{k}\alpha} \hbar\mathbf{k}\, a_{k\alpha}^\dagger a_{k\alpha}. \tag{17}$$

•An important check on the consistency of our procedure can be obtained at this point by verifying that (17) generates displacements. To this end we compute

$$\mathcal{E}_{\mathbf{x}}(\mathbf{r}) \equiv e^{i\mathbf{P}\cdot\mathbf{x}/\hbar}\mathcal{E}(\mathbf{r})e^{-i\mathbf{P}\cdot\mathbf{x}/\hbar}.$$

Because of (3) and (17)

$$e^{i\mathbf{P}\cdot\mathbf{x}/\hbar} = \prod_{\mathbf{k}\alpha} \exp(iN_{k\alpha}\mathbf{k}\cdot\mathbf{x}) \equiv \prod_{\mathbf{k}\alpha} U_{k\alpha},$$

which allows us to write

$$\mathcal{E}_{\mathbf{x}}(\mathbf{r}) = i\sum_{\mathbf{k}\alpha}\left(\frac{h\omega}{\Omega}\right)^{\frac{1}{2}} e^{i\mathbf{k}\cdot\mathbf{r}}\boldsymbol{\varepsilon}_{k\alpha}[a_{k\alpha}(\mathbf{x}) - a_{-k\alpha}^\dagger(\mathbf{x})], \tag{18}$$

where $a_{k\alpha}(\mathbf{x}) = U_{k\alpha}a_{k\alpha}U_{k\alpha}^\dagger$. But

$$\nabla a_{k\alpha}(\mathbf{x}) = i\mathbf{k}U_{k\alpha}[N_{k\alpha}, a_{k\alpha}]U_{k\alpha}^\dagger = -i\mathbf{k}a_{k\alpha}(\mathbf{x}),$$

or

$$a_{k\alpha}(\mathbf{x}) = e^{-i\mathbf{k}\cdot\mathbf{x}}a_{k\alpha}.$$

After inserting this into (18) we obtain

$$e^{i\mathbf{P}\cdot\mathbf{x}/\hbar}\mathcal{E}(\mathbf{r})e^{-i\mathbf{P}\cdot\mathbf{x}/\hbar} = \mathcal{E}(\mathbf{r} - \mathbf{x}). \tag{19}$$

This proves that $\mathbf{P}$ is indeed the generator of spatial translations.•

The stationary state $|\{n_{k\alpha}\}\rangle$ is obviously an eigenstate of $\mathbf{P}$ with eigenvalue $\mathbf{P}' = \Sigma_{\mathbf{k}\alpha}\hbar\mathbf{k}n_{k\alpha}$. The action of the a-operators on the eigenvectors of $\mathbf{P}$ is also of importance. Let $|\mathbf{P}'\rangle$ be an arbitrary eigenstate of $\mathbf{P}$, $(\mathbf{P} - \mathbf{P}')|\mathbf{P}'\rangle = 0$; we do not assume that $|\mathbf{P}'\rangle$ is a stationary state. We now ask whether $a_{k\alpha}^\dagger|\mathbf{P}'\rangle$ is also an eigenket of $\mathbf{P}$. To answer the question we apply $\mathbf{P}$ to the state:

$$\mathbf{P}a_{k\alpha}^\dagger|\mathbf{P}'\rangle = a_{k\alpha}^\dagger\mathbf{P}'|\mathbf{P}'\rangle + [\mathbf{P}, a_{k\alpha}^\dagger]|\mathbf{P}'\rangle.$$

The commutation rules between the a-operators and the momentum follow immediately from (17) and (3):

$$[a_{k\alpha}^\dagger, \mathbf{P}] = -\hbar\mathbf{k}a_{k\alpha}^\dagger,$$
$$[a_{k\alpha}, \mathbf{P}] = \hbar\mathbf{k}a_{k\alpha}. \tag{20}$$

Consequently

$$\mathbf{P}a_{k\alpha}^\dagger|\mathbf{P}'\rangle = (\mathbf{P}' + \hbar\mathbf{k})a_{k\alpha}^\dagger|\mathbf{P}'\rangle, \tag{21}$$
$$\mathbf{P}a_{k\alpha}|\mathbf{P}'\rangle = (\mathbf{P}' - \hbar\mathbf{k})a_{k\alpha}|\mathbf{P}'\rangle. \tag{22}$$

The kets $a_{k\alpha}^\dagger|\mathbf{P}'\rangle$ and $a_{k\alpha}|\mathbf{P}'\rangle$ are therefore eigenstates of momentum having the eigenvalues $\mathbf{P}' + \hbar\mathbf{k}$ and $\mathbf{P}' - \hbar\mathbf{k}$, respectively. We also recall that if $|E\rangle$ is an eigenket of H_γ, the states $a_{k\alpha}^\dagger|E\rangle$ and $a_{k\alpha}|E\rangle$ are eigenkets of H_γ having the energy eigenvalues $E + \hbar\omega_k$ and $E - \hbar\omega_k$, respectively.

It should now be apparent that the operators $a_{k\alpha}^\dagger$, $a_{k\alpha}$, and $N_{k\alpha}$ provide us with a mathematical description of the corpuscular aspects of electromagnetic phenomena, i.e., a theory of *photons*. That is to say, our results may be cast into the following language:

(a) *A stationary state of the free electromagnetic field is characterized uniquely by the photon occupation numbers $\{n_{k\alpha}\}$; such a state has the total energy $\sum_{\mathbf{k}\alpha}\hbar\omega_k n_{k\alpha}$, and the total momentum $\sum_{\mathbf{k}\alpha}\hbar\mathbf{k}n_{k\alpha}$.*

(b) *The state $a_{k\alpha}^\dagger|0\rangle$ has the energy and momentum $\hbar ck$ and $\hbar\mathbf{k}$, respectively. Hence we may say that $|\{n_{k\alpha}\}\rangle$ is a state containing $n_{k\alpha}$ photons in each of the modes $(\mathbf{k}, \alpha)$. The total number of photons is therefore $\sum_{\mathbf{k}\alpha}n_{k\alpha}$.*

(c) *We may say that the operator $a_{k\alpha}^\dagger$ creates ($a_{k\alpha}$ destroys) a single photon of momentum $\hbar\mathbf{k}$ and energy $\hbar ck$.*

(d) *Photons are bosons.*

The last statement follows from (9), which, together with the commutation rules (3), shows that $|\{n_{k\alpha}\}\rangle$ is a symmetric function of its arguments. One therefore calls the electromagnetic field a Bose field.

Because of (c) we shall usually refer to the operators $a_{k\alpha}$ and $a_{k\alpha}^\dagger$ as *destruction and creation operators*, respectively. Although the details remain to be spelled out, it is already clear that these operators must play an essential role in the theory of emission and absorption of radiation.

It should be stressed that the $|\{n_{k\alpha}\}\rangle$-representation, and therefore the photon picture, is only one of an infinite variety of possible representations in the Hilbert space pertaining to the electromagnetic field. The photon representation is particularly convenient because it utilizes a complete set of constants of the motion of the free field, the operators $N_{k\alpha}$. Because the photon states are a complete set, they can also be used to describe the field when sources are present. But other representations also play an important role; some of these shall be discussed in the following section.

52.3 THE SPIN OF THE PHOTON

We must still find the quantum mechanical significance of the polarization index α. As we shall see, it is related to the intrinsic angular momentum of the electromagnetic field. In the classical theory the total angular momentum of the field is

$$\mathbf{M} = \frac{1}{4\pi c} \int d^3r\, \mathbf{r} \times (\mathcal{E} \times \mathcal{H}). \tag{23}$$

By taking advantage of $\nabla \cdot \mathcal{E} = 0$, one can rewrite this as *

$$\mathbf{M} = \mathbf{M}_o + \mathbf{M}_s, \tag{24}$$

where

$$\mathbf{M}_o = \frac{1}{4\pi c} \sum_{i=1}^{3} \int d^3r\, \mathcal{E}_i\, \mathbf{r} \times \nabla A_i, \tag{25}$$

$$\mathbf{M}_s = \frac{1}{4\pi c} \int d^3r\, \mathcal{E} \times \mathbf{A}. \tag{26}$$

In passing to the quantum theory, we must take care to construct Hermitian forms for $\mathbf{M}$. The quantum observables are therefore assumed to be

$$\mathbf{M}_s = \frac{1}{8\pi c} \int d^3r (\mathcal{E} \times \mathbf{A} - \mathbf{A} \times \mathcal{E}), \tag{27}$$

$$\mathbf{M}_o = \frac{1}{8\pi c} \sum_i \int d^3r [\mathcal{E}_i\, \mathbf{r} \times \nabla A_i + (\mathbf{r} \times \nabla A_i)\mathcal{E}_i]. \tag{28}$$

We shall now show that $\mathbf{M}_s$ and $\mathbf{M}_o$ are to be interpreted as the intrinsic (or spin) and orbital angular momenta of the field. First note that $\mathbf{r}$ appears explicitly in $\mathbf{M}_o$, but not in $\mathbf{M}_s$. The expectation value of $\mathbf{M}_o$ in any state will therefore depend on the choice of origin, whereas the

* The proof of these relations runs as follows. In terms of the Levi-Cevita density we can write $(\mathcal{E} \times \mathcal{H})_k = \epsilon_{klm}\mathcal{E}_l\mathcal{H}_m = \epsilon_{klm}\epsilon_{mij}\mathcal{E}_l\nabla_i A_j$. But $\epsilon_{klm}\epsilon_{mij} = \delta_{ki}\,\delta_{lj} - \delta_{kj}\,\delta_{li}$, and $\nabla \cdot \mathcal{E} = 0$. Therefore

$$(\mathcal{E} \times \mathcal{H})_k = \mathcal{E}_l\nabla_k A_l - \nabla_l(\mathcal{E}_l A_k),$$

and finally

$$(\mathbf{r} \times (\mathcal{E} \times \mathcal{H}))_i = \mathcal{E}_l(\mathbf{r} \times \nabla)_i A_l - \nabla_l(\mathcal{E}_l\epsilon_{ijk}x_j A_k) + \epsilon_{ilk}\mathcal{E}_l A_k.$$

When this expression is integrated over space, the second term drops out if the fields vanish sufficiently rapidly at infinity (or if periodic boundary conditions are imposed). *QED.*

expectation value of $\mathbf{M}_s$ does not depend on this choice.* Let us construct $\mathbf{M}_s$ explicitly in terms of the destruction and creation operators. If we substitute (12) and (13) into (27), we readily find

$$\mathbf{M}_s = -i\hbar \sum_{\mathbf{k}} \mathbf{a}_k^\dagger \times \mathbf{a}_k. \tag{29}$$

The linear polarization vectors and $\mathbf{k}$ form a right-handed triad:

$$\boldsymbol{\varepsilon}_{k1} \times \boldsymbol{\varepsilon}_{k2} = \hat{\mathbf{k}}. \tag{30}$$

The operator $\mathbf{M}_s$ is therefore

$$\mathbf{M}_s = -i\hbar \sum_{\mathbf{k}} \hat{\mathbf{k}}(a_{k1}^\dagger a_{k2} - a_{k2}^\dagger a_{k1}). \tag{29'}$$

We see that the operators $N_{k\alpha}$, whose eigenvalues are the photon occupation numbers, do not commute with $\mathbf{M}_s$. The situation is easily rectified by introducing new destruction and creation operators that diagonalize the Hermitian form (29′). Stated in more physical terms, by switching from linearly to circularly polarized waves we can obtain a representation characterized by occupation number operators that commute not only with the energy and linear momentum, but also the portion $\mathbf{M}_s$ of the angular momentum. The circular polarization unit vectors are related to $\boldsymbol{\varepsilon}_{k\alpha}$ by

$$\mathbf{e}_{\mathbf{k},+1} = -\frac{1}{\sqrt{2}}(\boldsymbol{\varepsilon}_{\mathbf{k}1} + i\boldsymbol{\varepsilon}_{\mathbf{k}2}), \qquad \mathbf{e}_{\mathbf{k},-1} = \frac{1}{\sqrt{2}}(\boldsymbol{\varepsilon}_{\mathbf{k}1} - i\boldsymbol{\varepsilon}_{\mathbf{k}2}). \tag{31}$$

Note that our phase convention agrees with that adopted earlier for the "spherical" components of a vector operator, Eq. (36.10). The vectors $\mathbf{e}_{\mathbf{k}\lambda}$ (where $\lambda = \pm 1$) also satisfy the identities

$$\mathbf{e}_{\mathbf{k}\lambda}^* \cdot \mathbf{e}_{\mathbf{k}\lambda'} = \delta_{\lambda\lambda'}, \tag{32}$$
$$\mathbf{e}_{\mathbf{k}\lambda}^* \times \mathbf{e}_{\mathbf{k}\lambda'} = i\lambda\hat{\mathbf{k}}\,\delta_{\lambda\lambda'}. \tag{33}$$

New destruction and creation operators are defined by the requirement

$$\mathbf{a}_k = \sum_\alpha \boldsymbol{\varepsilon}_{\mathbf{k}\alpha} a_{k\alpha} = \sum_\lambda \mathbf{e}_{\mathbf{k}\lambda} a_{k\lambda}. \tag{34}$$

* One might object here that the split into (27) and (28) depends explicitly on **A**, and is therefore gauge dependent. This objection is unfounded, however, because by **A** we *always* mean the solenoidal (or transverse) part of the vector potential. This part of the vector potential is invariant under gauge transformations, in contrast to the irrotational part.

Consequently

$$a_{k,+1} = -\frac{1}{\sqrt{2}}(a_{k1} - ia_{k2}), \qquad a_{k,-1} = \frac{1}{\sqrt{2}}(a_{k1} + ia_{k2}). \tag{35}$$

This transformation leaves the commutation rules invariant, i.e.,

$$[a_{k\lambda}, a^\dagger_{k'\lambda'}] = \delta_{\mathbf{kk'}}\,\delta_{\lambda\lambda'},$$

etc. If we use (33) and (34) in (29), we find that the spin operator attains the desired diagonal form,

$$\mathbf{M}_s = \hbar \sum_{\mathbf{k}\lambda} \lambda\hat{\mathbf{k}}\; a^\dagger_{k\lambda} a_{k\lambda}. \tag{36}$$

The Hamiltonian and linear momentum both involve the scalar operators $\mathbf{a}^\dagger_k \cdot \mathbf{a}_k$, and are therefore unaffected by the transformation (35):

$$H_\gamma = \sum_{\mathbf{k}\lambda} \hbar ck\; a^\dagger_{k\lambda} a_{k\lambda}, \tag{37}$$

$$\mathbf{P} = \sum_{\mathbf{k}\lambda} \hbar\mathbf{k}\; a^\dagger_{k\lambda} a_{k\lambda}. \tag{38}$$

The operators $N_{k\lambda} = a^\dagger_{k\lambda} a_{k\lambda}$ therefore allow us to specify simultaneously the energy, linear momentum, and intrinsic angular momentum of the field.

One-photon states play a central role in many phenomena, and we therefore discuss them in some detail. In the sequel we shall have to integrate over the orientation of the propagation vector $\mathbf{k}$, and for this reason we must introduce one-photon states normalized in the continuous k-space, instead of the lattice associated with periodic boundary conditions. Following the recipes of Sec. 7.3, we define such one-photon kets by

$$|\mathbf{k}\lambda\rangle = \sqrt{\Omega/8\pi^3}\; a^\dagger_{k\lambda}|0\rangle, \tag{39}$$

where the limit $\Omega \to \infty$ is to be taken at the end of any calculation. The normalization of these states is

$$\langle \mathbf{k}\lambda|\mathbf{k}'\lambda'\rangle = \delta(\mathbf{k} - \mathbf{k}')\,\delta_{\lambda\lambda'}. \tag{40}$$

The energy and linear momentum of the state $|\mathbf{k}\lambda\rangle$ are $\hbar ck$ and $\hbar\mathbf{k}$, respectively. A somewhat tedious calculation (see Prob. 1) shows that $\mathbf{M}_o \cdot \mathbf{k}|\mathbf{k}\lambda\rangle = 0$, as it must be if one has properly identified the orbital portion of the angular momentum. Consequently $|\mathbf{k}\lambda\rangle$ is an eigenstate of the component $\mathbf{M}\cdot\hat{\mathbf{k}}$ of the total angular momentum:

$$\mathbf{M}\cdot\hat{\mathbf{k}}|\mathbf{k}\lambda\rangle = \hbar\lambda|\mathbf{k}\lambda\rangle. \tag{41}$$

Let us recall that the helicity is defined as the projection of the total angular momentum (in units of $\hbar$) along the direction of propagation. Equation (41) informs us that *the helicity of a photon is either* $+1$ *or* -1, *but never* 0. Because of this we say that the photon is a spin 1 particle. The absence of the helicity-0 state is a peculiarity of systems having a vanishing rest mass, and could not have been foreseen from our earlier experience with nonrelativistic quantum mechanics.* In this connection it should be noted that the whole vector operator $\mathbf{M}_s$ in a one-photon state of linear momentum $\hbar\mathbf{k}$ is along the direction $\hat{\mathbf{k}}$, which is in striking contrast to the nonrelativistic Pauli theory of intrinsic spin.

One-photon states having a definite helicity λ, as well as the total angular momentum quantum numbers (j, m), can be constructed from (39) by the techniques of Sec. 35.3. If we designate such a state by $|kjm\lambda\rangle$, Eq. (35.24) tells us that

$$|kjm\lambda\rangle = \sqrt{\frac{2j+1}{4\pi}} \int d\hat{\mathbf{k}}\, D^{(j)}_{m\lambda}(\hat{\mathbf{k}})^* |\mathbf{k}\lambda\rangle. \tag{42}$$

It will be recalled that the Euler angles in this rotation matrix are given by $\alpha = \phi$, $\beta = \theta$, $\gamma = 0$, where the orientation of $\hat{\mathbf{k}}$ with respect to the axis along which J_z is measured is specified by the polar angles (θ, ϕ). The inverse relation to (42) is given by (35.18),

$$|\mathbf{k}\lambda\rangle = \sum_{j=1}^{\infty} \sum_{m=-j}^{j} \sqrt{\frac{2j+1}{4\pi}}\, |kjm\lambda\rangle D^{(j)}_{m\lambda}(\hat{\mathbf{k}}), \tag{43}$$

and the transformation function from the angular momentum to the linear momentum representation is

$$\langle \mathbf{k}\lambda | k'jm\lambda'\rangle = \frac{\delta(k-k')}{kk'}\, \delta_{\lambda\lambda'} \sqrt{\frac{2j+1}{4\pi}}\, D^{(j)}_{m\lambda}(\hat{\mathbf{k}})^*. \tag{44}$$

An essential feature of Eq. (43) is that the sum over angular momentum starts with $j = 1$: *There are no one-photon states having zero total angular momentum.* This is another consequence of the transverse nature of the electromagnetic field. A vector field with massive quanta † (vector mesons) possesses one-particle states with helicity zero, and therefore one-particle states with vanishing total angular momentum exist in this case. In such a state the intrinsic spin of the meson is coupled with one

* A massless particle of spin s only has two helicity states, $\lambda = \pm s$. See E. P. Wigner, *Rev. Mod. Phys.* **29**, 255 (1957).

† Such a field satisfies the Klein-Gordon equation, $(\Box^2 + \kappa^2)\mathbf{A}(\mathbf{r}, t) = 0$, where the mass of the quanta is given by $\kappa\hbar/c$.

unit of orbital angular momentum to form a state with $j = 0$. Equation (43) shows that this cannot be done when the $\lambda = 0$ state is missing.

The angular momentum expansion of the one-photon states is easily extended to the entire operator $\mathbf{A}(\mathbf{r}, t)$. We first define a set of operators that create the states $|\mathbf{k}\lambda\rangle$ out of the vacuum:

$$a_\lambda^\dagger(\mathbf{k}) = \sqrt{\Omega/8\pi^3}\, a_{\mathbf{k}\lambda}^\dagger. \tag{45}$$

The commutation rules obeyed by these operators are

$$[a_\lambda(\mathbf{k}), a_{\lambda'}^\dagger(\mathbf{k}')] = \delta(\mathbf{k} - \mathbf{k}')\, \delta_{\lambda\lambda'}, \tag{46}$$

etc. Upon converting the sum in (12) into an integral we obtain

$$\mathbf{A}(\mathbf{r}, t) = \left(\frac{\hbar c^2}{2\pi^2}\right)^{\frac{1}{2}} \sum_\lambda \int \frac{d^3k}{\sqrt{2\omega}} [e^{-i(\mathbf{k}\cdot\mathbf{r}-\omega t)} \mathbf{e}_{\mathbf{k}\lambda}^* a_\lambda^\dagger(\mathbf{k}) + \text{h.c.}]. \tag{47}$$

Operators $a_{jm\lambda}^\dagger(k)$ that create photons into states of definite angular momentum, helicity, and energy, are then defined by the transformation (43):

$$a_\lambda^\dagger(\mathbf{k}) = \sum_{jm} \sqrt{\frac{2j+1}{4\pi}}\, D_{m\lambda}^{(j)}(\hat{\mathbf{k}}) a_{jm\lambda}^\dagger(k). \tag{48}$$

Clearly

$$[a_{jm\lambda}(k), a_{j'm'\lambda'}^\dagger(k')] = \frac{\delta(k - k')}{kk'}\, \delta_{jj'}\, \delta_{mm'}\, \delta_{\lambda\lambda'}. \tag{49}$$

When we insert (48) into (47) we obtain the desired expansion of the vector potential:

$$\mathbf{A}(\mathbf{r}, t) = \left(\frac{\hbar c^2}{2\pi^2}\right)^{\frac{1}{2}} \sum_{jm,\lambda=\pm 1} \left(\frac{2j+1}{4\pi}\right)^{\frac{1}{2}} \int_0^\infty \frac{k^2\, dk}{\sqrt{2\omega}} [e^{i\omega t} \mathbf{f}_{jm}^\lambda(k, \mathbf{r}) a_{jm\lambda}^\dagger(k) + \text{h.c.}], \tag{50}$$

where the expansion coefficients are given by

$$\mathbf{f}_{jm}^\lambda(k, \mathbf{r}) = \int d\hat{\mathbf{k}}\, \mathbf{e}_{\mathbf{k}\lambda}^* e^{-i\mathbf{k}\cdot\mathbf{r}} D_{m\lambda}^{(j)}(\hat{\mathbf{k}}). \tag{51}$$

In nuclear and atomic physics one is concerned with radiative transitions between states of the source (i.e., nucleus or atom) having definite angular momentum quantum numbers. As we shall see in Sec. 61, the representation (50) is ideally suited to such problems.

52.4 SPACE REFLECTION AND TIME REVERSAL

•The electric field strength $\mathcal{E}$ has the same transformation laws as a force, i.e., it is a vector field that is even under time reversal. Maxwell's equations then imply that the magnetic field is an axial vector that is odd under time reversal.

The vector potential is therefore a polar vector, odd under time reversal. In terms of the space reflection operator $\mathcal{R}$, and the time reversal operator Θ, these transformation laws read

$$\mathcal{R}\mathbf{A}(\mathbf{r}, t)\mathcal{R}^{-1} = -\mathbf{A}(-\mathbf{r}, t), \tag{52}$$

and

$$\Theta\mathbf{A}(\mathbf{r}, t)\Theta^{-1} = -\mathbf{A}(\mathbf{r}, -t). \tag{53}$$

If we carry out the spatial reflection on the Fourier representation (12), and demand that (52) be satisfied, we obtain

$$\sum_{\mathbf{k}} \sqrt{hc^2/\omega\Omega}\, e^{i\mathbf{k}\cdot\mathbf{r}}\mathcal{R}[\mathbf{a}_k + \mathbf{a}_{-k}^{\dagger}]\mathcal{R}^{-1} = -\sum_{\mathbf{k}} \sqrt{hc^2/\omega\Omega}\, e^{-i\mathbf{k}\cdot\mathbf{r}}[\mathbf{a}_k + \mathbf{a}_{-k}^{\dagger}],$$

or more succinctly,*

$$\mathcal{R}\mathbf{a}_{\mathbf{k}}^{\dagger}\mathcal{R}^{-1} = -\mathbf{a}_{-\mathbf{k}}^{\dagger}. \tag{54}$$

In performing the time reversal operation, one must bear in mind that $\Theta e^{i(\mathbf{k}\cdot\mathbf{r}-\omega t)}\Theta^{-1} = e^{-i(\mathbf{k}\cdot\mathbf{r}-\omega t)}$. One then finds

$$\Theta\mathbf{a}_{\mathbf{k}}^{\dagger}\Theta^{-1} = -\mathbf{a}_{-\mathbf{k}}^{\dagger}. \tag{55}$$

We wish to obtain the transformation laws for the helicity components of the vector creation operator $\mathbf{a}_{\mathbf{k}}^{\dagger}$. (These components are defined by (34).) It is clear from (54) and (55) that we can only do so if we define the relationship between the circular polarization vectors belonging to the momenta $\hbar\mathbf{k}$ and $-\hbar\mathbf{k}$. We therefore *define* $\mathbf{e}_{\mathbf{k}\lambda}$ for *all* $\mathbf{k}$ by †

$$\mathbf{e}_{\mathbf{k}\lambda} = \sum_{\nu} \mathbf{e}_{\nu} D_{\nu\lambda}^{(1)}(\hat{\mathbf{k}}), \tag{56}$$

where

$$\mathbf{e}_0 = \hat{\mathbf{z}}, \qquad \mathbf{e}_{\pm 1} = \mp \frac{1}{\sqrt{2}}(\hat{\mathbf{x}} \pm i\hat{\mathbf{y}}). \tag{57}$$

The orientation of $-\mathbf{k}$ is given by $\alpha = \pi + \phi, \beta = \pi - \theta, \gamma = 0$. From (34.32g) we then find that $D_{mm'}^{(1)}(-\hat{\mathbf{k}}) = -D_{m-m'}^{(1)}(\hat{\mathbf{k}})$. Consequently

$$\mathbf{e}_{-\mathbf{k}\lambda} = -\mathbf{e}_{\mathbf{k},-\lambda}. \tag{58}$$

The definition (31) also implies that $\mathbf{e}_{\mathbf{k}\lambda}^{*} = -\mathbf{e}_{\mathbf{k},-\lambda}$.

The transformation laws for the operators can now be read off from (54) and (55). Substituting (58) into (54) gives

$$\mathcal{R}a_{\mathbf{k}\lambda}^{\dagger}\mathcal{R}^{-1} = a_{-\mathbf{k}-\lambda}^{\dagger}. \tag{59}$$

* Here we designate the subscripts as vectors to avoid confusion.

† One should note carefully that this is the inverse of Eq. (36.9), for the case $k = 1$. In (36.9) we rotate the components of a tensor with respect to a fixed basis, whereas in (56) we rotate the basis itself.

Under the anti-unitary transformation (55), $\mathbf{e}_{\mathbf{k}\lambda} \rightarrow \mathbf{e}^*_{\mathbf{k}\lambda}$, and therefore

$$\Theta a^\dagger_{\mathbf{k}\lambda}\Theta^{-1} = -a^\dagger_{-\mathbf{k}\lambda}. \tag{60}$$

These transformation laws are consistent with the fact that the helicity is the eigenvalue of a pseudoscalar observable that is even under time reversal.•

53. Commutation Rules for the Free Fields—Complementarity

In the preceding section the corpuscular aspects of electromagnetic phenomena were emphasized. As we have already mentioned, this is only one of an infinite number of complementary descriptions. In fact, any complete set of solutions of the Helmholz equation can be used to decompose the vector potential into destruction and creation operators. We have already treated two representations of this type; one was based on the plane waves $\mathbf{e}_{\mathbf{k}\lambda}e^{i\mathbf{k}\cdot\mathbf{r}}$, and the other on the spherical wave functions $\mathbf{f}^\lambda_{jm}$. The spherical and plane wave representations are incompatible, i.e., $[a^\dagger_\lambda(\mathbf{k}), a_{jm\lambda'}(k')] \neq 0$. We are already familiar with these aspects of complementarity from the wave mechanics of a single nonrelativistic particle.

Because the electromagnetic field possesses an infinite number of degrees of freedom it is also possible to formulate the theory in terms of a very different set of dynamical variables, the field operators themselves. The field operators $\mathcal{E}$ and $\mathcal{H}$ are linear in the vector potential, and by the same token, linear in the destruction and creation operators. The field operators therefore fail to commute with the natural observables of the photon picture, the occupation numbers. To be more precise, the field operators only have off-diagonal matrix elements in the representation in which the photon occupation numbers are diagonal. In a state with a definite number of photons the expectation value of any field strength vanishes, and it is only meaningful to discuss fluctuations in the field strengths.* Conversely, in a state where one of the field strengths has a precise value the number of photons will not have a well-defined value. These remarks lead to a much deeper insight into the concept of complementarity than we were able to attain in our study of systems with a finite number of degrees of freedom.

If one wishes to set up the representation in which the field strengths

* The analogy to the one-dimensional oscillator should be noted. The field strengths are similar to q and p. These observables do not assume precise values in the stationary states of the oscillator, and it is the latter states that are analogous to states of the electromagnetic field with a definite number of photons.

have precise values one must know the commutation rules between the various field operators. We begin by computing the commutation rules for the components of the vector potential. With the help of Eqs. (52.12) and (52.3) we find

$$[A_i(\mathbf{r}_1, t_1), A_j(\mathbf{r}_2, t_2)] = -2i \sum_{\mathbf{k}} \frac{\hbar c^2}{\omega \Omega} t_{ij} e^{i\mathbf{k}\cdot(\mathbf{r}_1 - \mathbf{r}_2)} \sin \omega(t_1 - t_2), \tag{1}$$

where

$$t_{ij} = \sum_{\alpha=1}^{2} \epsilon^i_{\mathbf{k}\alpha} \epsilon^j_{\mathbf{k}\alpha}, \tag{2}$$

and $\epsilon^i_{\mathbf{k}\alpha}$ is the ith Cartesian component of the linear polarization vector. We observe that t_{ij} is a second rank symmetric tensor that can only depend on the components of $\mathbf{k}$. The only tensors of this type are δ_{ij} and $k_i k_j$, multiplied by arbitrary functions of $|\mathbf{k}|$. Hence t_{ij} must have the form $a\,\delta_{ij} + b k_i k_j / k^2$. But by definition $\Sigma_i t_{ii} = 2$, whence $3a + b = 2$. Furthermore $\Sigma_\alpha (\mathbf{k} \cdot \boldsymbol{\epsilon}_{\mathbf{k}\alpha}) \epsilon^j_{\mathbf{k}\alpha} = \Sigma_i k_i t_{ij} = 0 = (a + b) k_j$, or $a = -b$. Consequently

$$t_{ij} = \delta_{ij} - \frac{k_i k_j}{k^2}. \tag{3}$$

The sum over $\mathbf{k}$ in (1) can be converted into an integral by means of (7.44); as a result (1) becomes

$$[A_i(\mathbf{r}_1, t_1), A_j(\mathbf{r}_2, t_2)] = -4\pi i \hbar c \int \frac{d^3k}{(2\pi)^3} e^{i\mathbf{k}\cdot\mathbf{r}} \frac{\sin kct}{k} \left(\delta_{ij} - \frac{k_i k_j}{k^2} \right), \tag{4}$$

where $\mathbf{r} = \mathbf{r}_1 - \mathbf{r}_2$, $t = t_1 - t_2$. There is no problem in evaluating this integral. However $\mathbf{A}$ is not of direct physical concern and so we shall first put (4) into a more appropriate form for computing the $\mathcal{E} - \mathcal{E}$ commutator. In an obvious notation we have

$$[\mathcal{E}_i(1), \mathcal{E}_j(2)] = -4\pi i \hbar c \frac{1}{c^2} \frac{\partial^2}{\partial t_1 \, \partial t_2} \int \frac{d^3k}{(2\pi)^3} e^{i\mathbf{k}\cdot\mathbf{r}} \frac{\sin kct}{k} \left(\delta_{ij} - \frac{k_i k_j}{k^2} \right). \tag{5}$$

But

$$\frac{1}{c^2} \frac{\partial^2}{\partial t_1 \, \partial t_2} \int \frac{d^3k}{(2\pi)^3} e^{i\mathbf{k}\cdot\mathbf{r}} \frac{\sin kc(t_1 - t_2)}{k} \frac{k_i k_j}{k^2} = \int \frac{d^3k}{(2\pi)^3} e^{i\mathbf{k}\cdot\mathbf{r}} \frac{\sin kct}{k} k_i k_j$$

$$= \frac{\partial^2}{\partial x_{1i} \, \partial x_{2j}} \int \frac{d^3k}{(2\pi)^3} e^{i\mathbf{k}\cdot\mathbf{r}} \frac{\sin kct}{k}.$$

Hence (5) becomes

$$[\mathcal{E}_i(1), \mathcal{E}_j(2)] = -4\pi i \hbar c \left[\delta_{ij} \frac{1}{c^2} \frac{\partial^2}{\partial t_1 \, \partial t_2} - \frac{\partial^2}{\partial x_{1i} \, \partial x_{2j}} \right] \int \frac{d^3k}{(2\pi)^3} e^{i\mathbf{k}\cdot\mathbf{r}} \frac{\sin kct}{k}.$$

The remaining integral is easily evaluated:

$$D(r, t) \equiv -\int \frac{d^3k}{(2\pi)^3} e^{i\mathbf{k}\cdot\mathbf{r}} \frac{1}{k} \sin kct$$
$$= -\frac{1}{4\pi^2 r}\int_{-\infty}^{\infty} dk \sin kr \sin kct$$
$$= \frac{1}{4\pi r}[\delta(r + ct) - \delta(r - ct)]. \tag{6}$$

Our final result for the commutator (first obtained by Jordan and Pauli in 1928) is

$$[\mathcal{E}_i(1), \mathcal{E}_j(2)] = 4\pi i\hbar c \left\{\delta_{ij}\frac{1}{c^2}\frac{\partial^2}{\partial t_1\,\partial t_2} - \frac{\partial^2}{\partial x_{1i}\,\partial x_{2j}}\right\} D(|\mathbf{r}_1 - \mathbf{r}_2|, t_1 - t_2). \tag{7}$$

The essential property of the function $D(\mathbf{r}_1 - \mathbf{r}_2, t_1 - t_2)$ is that it vanishes everywhere except on the light cone joining points *1* and *2*. Therefore (7) states that *one can simultaneously measure the electric field strength at two space-time points, provided these points cannot be connected by a light signal.* This is a very sensible result from the intuitive point of view. Note that it implies that *one can measure* $\mathcal{E}(\mathbf{r}, t)$ *throughout all of space at a fixed instant t, at least in principle.*

The remaining commutation rules are derived in the same way, and one finds

$$[\mathcal{H}_i(1), \mathcal{H}_j(2) = [\mathcal{E}_i(1), \mathcal{E}_j(2)], \tag{8}$$
$$[\mathcal{E}_i(1), \mathcal{H}_i(2)] = 0, \tag{9}$$

and

$$[\mathcal{E}_i(1), \mathcal{H}_j(2)] = 4\pi i\hbar c\epsilon_{ijk}\frac{\partial^2}{\partial t_1\,\partial x_{2k}} D(1,2). \tag{10}$$

The first statement in italics following (7) therefore applies to any pair of field operators that cannot be connected by a light signal.

Let us examine (7) more closely for equal times, i.e., $t_1 = t_2$. We note that $D(r, t) = -D(r, -t)$, and therefore

$$\lim_{t\to 0}\frac{\partial^2}{\partial t^2} D(r, t) = 0, \qquad \lim_{t\to 0} D(r, t) = 0.$$

Hence

$$[\mathcal{E}_i(\mathbf{r}_1, t), \mathcal{E}_j(\mathbf{r}_2, t)] = 0; \tag{11}$$

on the other hand, only the first time derivative appears in (10), and therefore $\mathcal{E}$ and $\mathcal{H}$ fail to commute when $\mathbf{r}_1 = \mathbf{r}_2$ at equal times. Thus the electric field $\mathcal{E}(\mathbf{r}, t)$ over a constant time surface (i.e., all space at an instant of time) is a commuting set of observables. The same statement holds for $\mathcal{H}$. In fact it can be shown that *the set of operators* $\mathcal{E}$ *(or* $\mathcal{H}$*) at an instant of time is a complete commuting set.* One may therefore introduce a representation in which $\mathcal{E}(\mathbf{r})$ has a precise value, but in this representation the photon occupation numbers will have no diagonal elements.* This is then the other side of Prof. Bohr's "complementary coin."

We may now make contact with the uncertainty relations concerning the measurement of space-time averaged field strengths derived in Sec. 3. To do so we recall from Sec. 24 that the uncertainty relation for two observables A_1 and A_2 is $\Delta A_1\, \Delta A_2 \gtrsim \frac{1}{2}|\langle[A_1, A_2]\rangle|$. We apply this to the operator

$$\mathcal{E}_i(V_1) = \frac{1}{V_1}\int_{V_1} dV_1\, \mathcal{E}_i(\mathbf{r}_1, t_1),$$

and a similarly defined operator $\mathcal{E}_j(V_2)$, where $V_i = \Omega_i T_i$ is a volume in 4-space. Let V_1 be completely in the future with respect to V_2. Then (7) informs us that

$$\Delta\mathcal{E}_x(V_1)\, \Delta\mathcal{E}_y(V_2) \gtrsim \frac{\hbar}{2}\left|\int \frac{dV_1\, dV_2}{V_1 V_2} \frac{\partial^2}{\partial x_1\, \partial y_2} \frac{\delta(t - r/c)}{r}\right|, \tag{12}$$

where $t = t_1 - t_2$, and $r = |\mathbf{r}_1 - \mathbf{r}_2|$. This agrees with (3.6) except for a factor of $\frac{1}{2}$, which is within the error of the definition of uncertainty products used in Sec. 3.

The present derivation of (12) contributes an important check on the logical consistency of the entire theoretical structure that we have set up. It will be recalled that this uncertainty relation was originally derived in Sec. 3 by a rather intuitive argument that used as its basic ingredient the uncertainty relation $\Delta p\, \Delta q \gtrsim \hbar$ for the canonical variables pertaining to the test charges used in measuring the field strength. At that time we did not have a mathematical theory capable of describing either particles or an electromagnetic field, but we were able to conclude that if the particles manifest a wave-particle duality, the fields must also display "complementary" characteristics. We then went on to set up a mathematical theory that incorporated the uncertainty principle for "particles," i.e., for objects that, in the correspondence principle limit, behave like Newtonian point-particles. Finally, by using a rather formal analogy, we quantized the field. We have now closed the circle by showing that

* This representation plays an important role in phenomena where many photons are involved. See R. Glauber, *Phys. Rev.* 130, 2529; 131, 2766 (1963).

this quantization procedure leads to precisely the same uncertainty product for the fields as we originally deduced from very general and qualitative considerations.

PROBLEMS

1. Let $\mathbf{M}_o$ be the orbital angular momentum operator of the electromagnetic field, and $|\mathbf{k}\rangle$ a one-photon state of arbitrary polarization. Verify that

$$\langle \mathbf{k}|\hat{\mathbf{k}} \cdot \mathbf{M}_0|\mathbf{k}\rangle = 0.$$

(Adopt a definition of $\mathbf{M}_o$ that guarantees that its vacuum expectation value vanishes.)

2. Show that

$$[\mathcal{E}_i(\mathbf{r}, t), \mathcal{H}_i(\mathbf{r}', t')] = 0.$$

3. By employing the commutation rules for the field strengths given in Sec. 53, show that

$$\mathbf{M} = \frac{1}{8\pi c}\int d^3r\, [\mathbf{r} \times (\mathcal{E} \times \mathcal{H} - \mathcal{H} \times \mathcal{E})]$$

is the operator that generates rotations.

IX

Time-Dependent Perturbation Theory

Transitions between various stationary states of a system occur when it is forced by an externally applied disturbance. In an isolated system—i.e., having a time-independent Hamiltonian H—transitions will also take place if the initial state is not an eigenstate of H. Most collision and decay processes fall under the latter heading.

The formulation of approximation methods for time-dependent phenomena is rather difficult, and we therefore approach the problem in several steps. In this chapter we shall first present a modernized version of a method originally due to Dirac. This approach depends on a straightforward expansion in powers of the perturbing interaction. Not all interactions are weak, however, and in certain circumstances even weak interactions can lead to large transition probabilities. This fact is illustrated by the simple though important example of magnetic resonance (Sec. 55). In any resonance or decay phenomenon the total transition probability sooner or later becomes comparable to unity, and a naive application of perturbation theory fails. Throughout most of this chapter we shall confine the discussion to collision phenomena where perturbation theory, or rather obvious generalizations thereof, gives accurate results. The more difficult theory of spontaneous decay and resonance scattering, and applications to electromagnetic processes, will be developed in Chapt. X.

54. The Interaction Picture

As we have just said, time-dependent problems can, broadly speaking, be separated into two classes: (a) the Hamiltonian is explicitly time dependent as, for example, in the case where an atom is exposed to a time dependent external field; (b) the Hamiltonian is time independent, but the state initially prepared by the experimenter is not a stationary state. In actual practice collision problems belong to (b), although one can use stationary state methods in scattering theory by employing certain idealizations that have already been discussed in detail in Sec. 12.

The theoretical treatment of cases (a) and (b) is actually very similar. One can always transform a Hamiltonian describing (b) into a problem involving a time-dependent H, and it is often convenient to do this. To see how this comes about, consider the Hamiltonian

$$H = H_0 + H'. \tag{1}$$

In the Schrödinger picture the states satisfy

$$\left(i\hbar\frac{\partial}{\partial t} - H\right)|a';t\rangle_S = 0, \tag{2}$$

and all observables A are independent of t. Let us now assume that H_0 describes two noninteracting systems and H' their mutual interaction, which we assume to be of finite range. In a collision problem we begin with an eigenstate belonging to H_0, because the two systems are initially separated in space and H' is therefore inoperative, up to say $t = 0$. We can therefore obtain $|a';t\rangle_S$ when $t < 0$ from $(i\hbar\,\partial_t - H_0)|a';t\rangle_S = 0$, which we usually know how to solve. To take advantage of this information it is convenient to transform the equations, and for this purpose we introduce the ket

$$|a';t\rangle_I = e^{iH_0t/\hbar}|a';t\rangle_S. \tag{3}$$

Note that $|a';t\rangle_I$ would be t-independent if H' were really zero. If we substitute $|a';t\rangle_S = e^{-iH_0t/\hbar}|a';t\rangle_I$ into (2), and premultiply by $e^{iH_0t/\hbar}$, we obtain

$$\left(i\hbar\frac{\partial}{\partial t} - H_I'(t)\right)|a';t\rangle_I = 0, \tag{4}$$

where

$$H_I'(t) = e^{iH_0t/\hbar}H'e^{-iH_0t/\hbar}. \tag{5}$$

Thus (4) is now the Schrödinger equation of a system having an explicitly t-dependent Hamiltonian, which is case (a) above. Finally we note that when we use (3)–(5) all observables have the t-dependence

$$A_I(t) = e^{iH_0t/\hbar}Ae^{-iH_0t/\hbar}. \tag{6}$$

Equations (3)–(6) define the *interaction picture;* it is distinguished from the Heisenberg and Schrödinger pictures in that *both* the states and the observables move in time.

The states $|a't\rangle_I$ for different values of t are, as always, related by a unitary operator U_I:

$$|a';t_2\rangle_I = U_I(t_2, t_1)|a';t_1\rangle_I. \tag{7}$$

The relation between U_I and U_S can be found by substituting into (3), viz.

$$U_S(t_2, t_1) = e^{-iH_0t_2/\hbar}U_I(t_2, t_1)e^{iH_0t_1/\hbar}.$$

An expansion for U_I in powers of H' can be obtained from (4) by the method of (28.16)–(28.20):

$$U_I(t_2, t_1) = 1 - \frac{i}{\hbar}\int_{t_1}^{t_2} H_I'(t)U_I(t, t_1)\,dt \tag{8}$$

$$= 1 + \sum_{n=1}^{\infty}\left(\frac{1}{i\hbar}\right)^n \int_{t_1}^{t_2} dt_1' \int_{t_1}^{t_1'} dt_2' \cdots \int_{t_1}^{t'_{n-1}} dt_n'\, H_I(t_1') \cdots H_I'(t_n'). \tag{9}$$

This last expression is called the Dyson expansion. In principle it provides us with a complete solution of the basic problem in time-dependent perturbation theory. In practice it is usually impossible to compute more than a few terms in the expansion, and what is worse, in many circumstances the expansion converges very slowly or not at all. There are also important interactions (beta decay, interatomic forces in liquids and gases) where essentially every term in (9) is separately infinite, even though U_S surely exists. In spite of these shortcomings, the Dyson expansion has been used with great success in quantum electrodynamics, quantum statistics, and other important areas of physics. A great deal is known about the structure of general terms in the series, and many ingenious methods for summing subsequences of terms are known. By far the most important tool in this type of analysis is the Feynman diagram. As we shall not make much use of the interaction picture in this book, we refer the interested reader to the literature * for treatments of this aspect of time-dependent perturbation theory.

•Under very special circumstances the integral equation (8) can be solved in closed form. The simplest situation of this type occurs when $[H_I'(t_1), H_I'(t_2)] = 0$; in this case one can integrate the differential equation for U_I directly (recall (28.21)):

$$U_I(t_2, t_1) = \exp\left\{\frac{1}{i\hbar}\int_{t_1}^{t_2} H_I(t)\, dt\right\}. \tag{10}$$

A complete solution can also be found when

$$[H_I'(t_1), H_I'(t_2)] = C(t_1, t_2), \tag{11}$$

where $C(t_1, t_2)$ is an arbitrary c-number. The emission of radiation by a prescribed current treated in Sec. 59 is actually an example of this situation. The trick† is to construct yet another unitary transformation beyond the interaction picture, which is chosen so as to transform the Hamiltonian into a c-number. Thus we put

$$|a'; t\rangle_{II} = e^{iW(t)/\hbar}|a'; t\rangle_I, \tag{12}$$

and then (4) becomes

$$e^{iW(t)/\hbar}\left(i\hbar\frac{\partial}{\partial t} - H_I'(t)\right)e^{-iW(t)/\hbar}|a'; t\rangle_{II} = 0,$$

or

$$\left(i\hbar\frac{\partial}{\partial t} - H_{II}'(t)\right)|a'; t\rangle_{II} = 0, \tag{13}$$

with

$$H_{II}'(t) = e^{iW(t)/\hbar}\left(H_I'(t) - i\hbar\frac{\partial}{\partial t}\right)e^{-iW(t)/\hbar}. \tag{14}$$

Since $\dot{W}(t)$ need not commute with W, H_{II}' is *not* just $e^{iW/\hbar}(H_I' - \dot{W})e^{-iW/\hbar}$. To see this explicitly we expand the exponentials:

$$\begin{aligned} H_{II}'(t) = H_I'(t) &+ \frac{i}{\hbar}[W(t), H_I'(t)] + \frac{1}{2!}\left(\frac{i}{\hbar}\right)^2 [W(t), [W(t), H_I'(t)]] + \cdots \\ &- \left\{\dot{W}(t) + \frac{i}{2\hbar}[W(t), \dot{W}(t)] + \frac{1}{3!}\left(\frac{i}{\hbar}\right)^2 [W(t), [W(t), \dot{W}(t)]] + \cdots\right\}. \end{aligned} \tag{15}$$

* See Bjorken and Drell, Nozières, Schweber.

† J. Schwinger, *Phys. Rev.* **75**, 651 (1949).

Now we note that if W (and therefore $\dot{W}$) is some linear functional of H_I', all the triple, quadruple, . . . , commutators vanish since the double commutator is already a c-number, i.e., (11). In fact, let us choose

$$\dot{W}(t) = H_I'(t), \tag{16}$$

or

$$W(t) = \int^t dt'\, H_I'(t'), \tag{17}$$

where the lower limit of integration is left open for the moment. Putting (16) and (17) into (15), we therefore have

$$H_{II}'(t) = \frac{i}{2\hbar}\int^t dt'\,[H_I'(t'), H_I'(t)] = \frac{i}{2\hbar}\int^t dt' C(t', t), \tag{18}$$

and so we have successfully constructed a c-number Hamiltonian. Hence we can solve (13) immediately:

$$|a't_2\rangle_{II} = \exp\left\{-\frac{i}{\hbar}\int_{t_1}^{t_2} dt'\, H_{II}'(t')\right\}|a't_1\rangle_{II}. \tag{19}$$

Thus

$$\begin{aligned}|a't_2\rangle_I &= e^{-iW(t_2)/\hbar}|a't_2\rangle_{II}\\ &= e^{-iW(t_2)/\hbar}\exp\left\{-\frac{i}{\hbar}\int_{t_1}^{t_2} H_{II}'(t')\,dt'\right\} e^{iW(t_1)/\hbar}|a't_1\rangle_I.\end{aligned}$$

Since W commutes with H_{II}', our final result is

$$U_I(t_2, t_1) = \exp\left\{-\frac{i}{\hbar}\int_{t_1}^{t_2} H_I'(t')\,dt'\right\}\exp\left\{-\frac{i}{\hbar}\int_{t_1}^{t_2} H_{II}'(t')\,dt'\right\}. \tag{20}$$

The integration limit in (17) has therefore dropped out, as it should.•

55. Magnetic Resonance

Here we shall apply time-dependent unitary transformations to an example of considerable practical interest. As a by-product, we shall also come to appreciate some of the limitations of time-dependent perturbation theory.

55.1 INTERACTION OF A SPIN WITH AN OSCILLATING MAGNETIC FIELD

The example in question is the determination of the magnetic moment of an atom or nucleus by observing the resonant absorption of energy from an oscillating magnetic field.

Consider first the motion of a spin in a static magnetic field $\mathcal{H}_0\hat{\mathbf{u}}_z$, where $\hat{\mathbf{u}}_z$ is a unit vector in the z-direction. If the system has a magnetic moment $\boldsymbol{\mu} = \gamma \mathbf{J}$, where $\mathbf{J}$ is the angular momentum operator in units of $\hbar$, the Hamiltonian is

$$H_0 = -\hbar\Omega J_z, \tag{1}$$

where Ω is the Larmor frequency, i.e.,

$$\hbar\Omega = \gamma \mathcal{H}_0. \tag{2}$$

In the Heisenberg picture, the angular momentum moves according to

$$\mathbf{J}(t) = e^{-i\Omega t J_z} \mathbf{J} e^{i\Omega t J_z}. \tag{3}$$

This is merely a clockwise rotation of $\mathbf{J}$ about $\hat{\mathbf{u}}_z$, precisely as in classical mechanics, and as one can see explicitly from

$$\begin{aligned} J_x(t) &= J_x \cos \Omega t + J_y \sin \Omega t, \\ J_y(t) &= J_y \cos \Omega t - J_x \sin \Omega t. \end{aligned} \tag{4}$$

In magnetic resonance experiments, the system is also subjected to an alternating magnetic field perpendicular to $\hat{\mathbf{u}}_z$. If the amplitude of this oscillating field is $2\lambda\mathcal{H}_0$, the full Hamiltonian is *

$$H = -\hbar\Omega[J_z + 2\lambda J_x \cos \omega t]. \tag{5}$$

In practice, the amplitude of the oscillating field is small compared to the static field, i.e., $\lambda \ll 1$, and therefore the dominant factor in the motion is the uniform clockwise precession expressed by (4). As we shall see in a moment, it is therefore very convenient to express the transverse field in terms of two rotating fields (see Fig. 55.1). This is possible since one can write

$$\hat{\mathbf{u}}_x \cos \omega t = \tfrac{1}{2}[\hat{\mathbf{u}}_c(t) + \hat{\mathbf{u}}_a(t)], \tag{6}$$

where $\hat{\mathbf{u}}_c$ and $\hat{\mathbf{u}}_a$ are unit vectors which rotate about $\hat{\mathbf{u}}_z$ with frequency ω in the clockwise and anticlockwise sense, respectively:

$$\begin{aligned} \hat{\mathbf{u}}_c(t) &= \hat{\mathbf{u}}_x \cos \omega t - \hat{\mathbf{u}}_y \sin \omega t, \\ \hat{\mathbf{u}}_a(t) &= \hat{\mathbf{u}}_x \cos \omega t + \hat{\mathbf{u}}_y \sin \omega t. \end{aligned} \tag{7}$$

Thus

$$H = -\hbar\Omega\{J_z + \lambda[\mathbf{J} \cdot \hat{\mathbf{u}}_c(t) + \mathbf{J} \cdot \hat{\mathbf{u}}_a(t)]\}. \tag{8}$$

The advantage of this break-up is that the clockwise rotating field $\lambda\mathcal{H}_0\hat{\mathbf{u}}_c(t)$ will exert a (small) torque on the spin, which keeps in step with

* The parameters λ, ω, and Ω are, without loss of generality, assumed to be positive throughout.

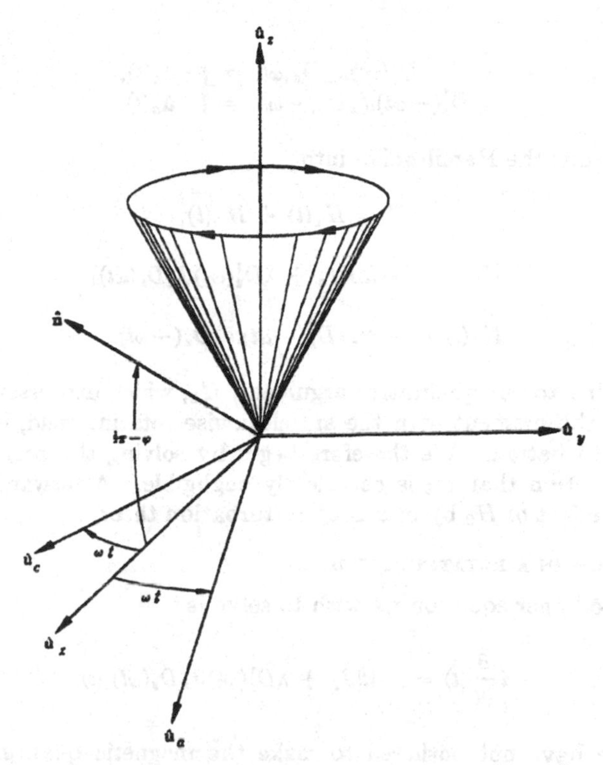

Fig. 55.1. The cone shows the precession of the classical angular momentum vector in the static field $\mathcal{H}_0\hat{\mathbf{u}}_z$. The clockwise and anticlockwise fields point along $\hat{\mathbf{u}}_c(t)$ and $\hat{\mathbf{u}}_a(t)$, respectively.

the torque due to the static field $\mathcal{H}_0\hat{\mathbf{u}}_z$ if the frequency ω is chosen to equal the Larmor frequency. In contrast to this, the counterclockwise rotating field $\lambda\mathcal{H}_0\hat{\mathbf{u}}_a(t)$ can never exert a torque in phase with that of $\mathcal{H}_0\hat{\mathbf{u}}_z$. One would therefore expect that the clockwise rotating field can lead to very dramatic perturbations when $\omega \approx \Omega$, whereas the effect of $\lambda\mathcal{H}_0\hat{\mathbf{u}}_a(t)$ should average out for all values of ω. This qualitative picture will be borne out by the calculations that follow.

We shall see that the computations are much more transparent when we express the rotating perturbations in terms of unitary operators. Let us introduce the operator that generates rotations about the z-axis

$$D_z(\omega t) = e^{-i\omega t J_z}; \tag{9}$$

then *

$$D_z^\dagger(\omega t)J_x D_z(\omega t) = \mathbf{J}\cdot\hat{\mathbf{u}}_c(t), \tag{10}$$
$$D_z^\dagger(-\omega t)J_x D_z(-\omega t) = \mathbf{J}\cdot\hat{\mathbf{u}}_a(t). \tag{11}$$

We then split the Hamiltonian into

$$H = H_1(t) + H_2(t), \tag{12}$$

where

$$H_1(t) = -\hbar\Omega[J_z + \lambda D_z^\dagger(\omega t)J_x D_z(\omega t)], \tag{13}$$

and

$$H_2(t) = -\hbar\Omega\lambda D_z^\dagger(-\omega t)J_x D_z(-\omega t). \tag{14}$$

According to our qualitative argument, H_2, which expresses the interaction of the moment with the anticlockwise rotating field, is always a small perturbation. We therefore begin by solving the problem under the assumption that H_2 is completely negligible. Afterwards we shall treat the effect of H_2 by means of perturbation theory.

55.2 MOTION IN A ROTATING FIELD

The Schrödinger equation we wish to solve is

$$i\frac{\partial}{\partial t}|t\rangle = -\Omega[J_z + \lambda D_z^\dagger(\omega t)J_x D_z(\omega t)]|t\rangle, \tag{15}$$

where we have not bothered to make the magnetic quantum number explicit in $|t\rangle$ (note that the Hilbert space in question is finite, and spanned by the $(2j+1)$ kets $|jm\rangle$). Since the perturbation in (15) rotates about the z-axis with circular frequency ω, we can make the field static by transforming to a rotating coordinate frame. In that frame the problem is trivially solved, and the solution may then be transformed back into the fixed laboratory frame. The transformation just described is effected by introducing the ket

$$|t\rangle_R \equiv D_z(\omega t)|t\rangle. \tag{16}$$

Because

$$D_z(\omega t)i\frac{\partial}{\partial t}|t\rangle = \left(i\frac{\partial}{\partial t} - \omega J_z\right)|t\rangle_R,$$

* This formula appears to disagree with (36.5), which claims that (10) gives the anticlockwise rotation, and not the clockwise. This contradiction is removed if one recalls that (36.5) gives the components of the rotated vector in a fixed coordinate system, whereas (10) gives the components of a fixed vector in the moving coordinate frame $\hat{\mathbf{u}}_c(t)$.

the transformed Schrödinger equation is

$$i \frac{\partial}{\partial t} |t\rangle_R = [(\omega - \Omega)J_z - \lambda\Omega J_x]|t\rangle_R. \tag{17}$$

Therefore

$$|t\rangle_R = e^{-i[(\omega-\Omega)J_z - \lambda\Omega J_x]t}|0\rangle, \tag{18}$$

and the solution of the original equation is

$$|t\rangle = e^{i\omega t J_z} e^{-i[(\omega-\Omega)J_z - \lambda\Omega J_x]t}|0\rangle, \tag{19}$$

where we used (16). The time-dependent operator appearing in (18) also corresponds to a rotation, since we can put

$$(\omega - \Omega)J_z - \lambda\Omega J_x = \Delta \mathbf{J} \cdot \hat{\mathbf{n}}, \tag{20}$$

where

$$\Delta^2 = (\omega - \Omega)^2 + (\lambda\Omega)^2, \tag{21}$$

$$\hat{\mathbf{n}} = \hat{\mathbf{u}}_z \cos\varphi + \hat{\mathbf{u}}_x \sin\varphi, \tag{22}$$

$$\tan\varphi = -\lambda\Omega/(\omega - \Omega); \tag{23}$$

$\hat{\mathbf{n}}$ and φ are shown in Fig. 55.1. The operator

$$D_{\hat{\mathbf{n}}}(\Delta t) = e^{-i\hat{\mathbf{n}}\cdot\mathbf{J}\Delta t} \tag{24}$$

allows us to write (19) as

$$|t\rangle = D_z^\dagger(\omega t) D_{\hat{\mathbf{n}}}(\Delta t)|0\rangle. \tag{25}$$

We can now compute the quantity of central interest, namely the probability $p_j(m'm; t)$ that the system be found in the state $|jm'\rangle$ if it was initially (i.e., at $t = 0$) in the state $|jm\rangle$. According to (25) the probability amplitude in question is

$$\langle jm'|jm; t\rangle = e^{im'\omega t}\langle jm'|D_{\hat{\mathbf{n}}}(\Delta t)|jm\rangle,$$

or

$$p_j(m'm; t) = |d^{(j)}_{m'm}(\beta)|^2, \tag{26}$$

where β is the second Euler angle corresponding to a rotation of Δt about the axis $\hat{\mathbf{n}}$, and $d^{(j)}_{m'm}$ is the function which appears in the $(2j + 1)$-dimensional irreducible representation of the rotation group (see (34.6)).

Let us evaluate the transition probability explicitly for $j = \frac{1}{2}$.* Since

$$e^{-\frac{1}{2}i\hat{\mathbf{n}}\cdot\boldsymbol{\sigma}\Delta t} = \cos\tfrac{1}{2}\Delta t - i\boldsymbol{\sigma}\cdot\hat{\mathbf{n}}\sin\tfrac{1}{2}\Delta t,$$

we see that the off-diagonal elements of $p_{\frac{1}{2}}(m'm; t)$ are obtained from

$$D_{\hat{\mathbf{n}}}(\Delta t)\Big|_{\text{off}} = -i\sigma_x \sin\varphi \sin\tfrac{1}{2}\Delta t,$$

* The result for the general case follows from (26), with β being determined from $\sin\frac{1}{2}\beta = (\lambda\Omega/\Delta)\sin\frac{1}{2}\Delta t$.

and the diagonal elements from

$$D_{\hat{n}}(\Delta t)\Big|_{\text{diag}} = \cos \tfrac{1}{2}\Delta t - i\sigma_z \cos\varphi \sin \tfrac{1}{2}\Delta t.$$

Therefore

$$p_{\frac{1}{2}}(m' \neq m; t) = \sin^2 \varphi \sin^2 \tfrac{1}{2}\Delta t,$$

and

$$p_{\frac{1}{2}}(\pm\tfrac{1}{2}, \pm\tfrac{1}{2}; t) = |\cos \tfrac{1}{2}\Delta t \mp i \cos\varphi \sin \tfrac{1}{2}\Delta t|^2.$$

By eliminating φ in favor of the other parameters we find

$$p_{\frac{1}{2}}(m' \neq m; t) = \frac{(\lambda\Omega)^2}{(\Omega - \omega)^2 + (\lambda\Omega)^2} \sin^2 \tfrac{1}{2}\Delta t, \tag{27}$$

and

$$p_{\frac{1}{2}}(m = m'; t) = 1 - \frac{(\lambda\Omega)^2}{(\Omega - \omega)^2 + (\lambda\Omega)^2} \sin^2 \tfrac{1}{2}\Delta t. \tag{28}$$

These results confirm our expectations: when the frequency ω of the rotating field equals the Larmor frequency Ω, the perturbation of the system is very large. If ω is varied, the transition probability (27) goes through a resonance of width $\lambda\Omega$ at $\omega = \Omega$; this resonance is seen to be very narrow if λ is small. By measuring the energy absorbed by a spin system (which is proportional to the time average of (27) over a period $1/\Delta$) one can therefore measure Ω, and since $\mathcal{H}_0$ is assumed known, one thereby determines the magnetic moment. In actual practice, there are many complications due to the interaction of the spin system with its environment,* but the essentials of the magnetic resonance technique are already contained in the simple example discussed here.

Were we to solve (15) by treating the rotating field as a perturbation, we would obtain the transition probabilities as a power series in λ. For example, the first term of the expansion (54.9) leads to

$$p_{\frac{1}{2}}(m' \neq m; t) = \left(\frac{\lambda\Omega}{\Omega - \omega}\right)^2 \sin^2 \tfrac{1}{2}(\Omega - \omega)t, \tag{29}$$

as is clear if one keeps only the leading term of (27) as $\lambda \to 0$. Equation (29) is a good approximation to the exact result (27) *provided* $|\Omega - \omega| \gg \lambda\Omega$, i.e., provided the applied field has a frequency which falls *outside* the resonance. *In the vicinity of the resonance the perturbation theory result is totally wrong, no matter how small* λ *may be.* In particular, when $\omega = \Omega$, Eq. (29) becomes

$$p_{\frac{1}{2}}(m' \neq m; t) = \tfrac{1}{4}(\lambda\Omega t)^2, \tag{29'}$$

which leads to probabilities in excess of unity when $t > 2/\lambda\Omega$. Such

* See Slichter for a comprehensive treatment of these questions.

breakdowns of perturbation theory usually occur when the frequency of a disturbance matches one of the natural frequencies of the system.

The effect of the anticlockwise rotating field can be obtained from the preceding formulas by replacing Ω by $-\Omega$. Thus, for example,

$$p_{\frac{1}{2}}(m \neq m'; t) = \frac{(\Omega\lambda)^2}{(\Omega + \omega)^2 + (\Omega\lambda)^2} \sin^2\{\tfrac{1}{2}\sqrt{(\Omega + \omega)^2 + (\lambda\Omega)^2}\, t\}. \quad (30)$$

Inasmuch as this expression never resonates, our calculations confirm the earlier intuitive discussion concerning the strikingly different effects of the two rotating fields.

55.3 MOTION IN A LINEARLY POLARIZED FIELD

•In the laboratory one works with a linearly polarized field. We had surmised that the field component which rotates in the anticlockwise sense can be treated perturbatively, and have just seen that this argument is supported by (30). We shall now discuss this perturbation in more detail.

The complete Schrödinger equation reads

$$i\hbar \frac{\partial}{\partial t} |t\rangle = [H_1(t) + H_2(t)]|t\rangle, \quad (31)$$

where the Hamiltonian has been split as in (12)–(14). Since we already know how to solve for the motion caused by $H_1(t)$, we make the *Ansatz*

$$|t\rangle = U_1(t)|t\rangle_0, \quad (32)$$

where

$$\left[i\hbar \frac{\partial}{\partial t} - H_1(t)\right] U_1(t) = 0. \quad (33)$$

According to (25)

$$U_1(t) = D_z^\dagger(\omega t) D_{\hat{n}}(\Delta t). \quad (34)$$

Substituting (32) into (31), we find

$$i\hbar \frac{\partial}{\partial t} |t\rangle_0 = U_1^\dagger(t) H_2(t) U_1(t) |t\rangle_0, \quad (35)$$

i.e., we have gone into the interaction picture appropriate to $H_1(t)$. To first order in H_2,

$$|t\rangle_0 = \left\{1 - \frac{i}{\hbar} \int_0^t U_1^\dagger(t') H_2(t') U_1(t')\, dt'\right\} |0\rangle. \quad (36)$$

The solution of (31) to first order in H_2 is therefore

$$|t\rangle = U_1(t)[1 + iF(t)]|0\rangle, \quad (37)$$

where

$$F(t) = -\frac{1}{\hbar} \int_0^t U_1^\dagger(t') H_2(t') U_1(t')\, dt'. \quad (38)$$

Equations (34) and (14) give

$$F(t) = \lambda\Omega \int_0^t D_{\hat{\mathbf{a}}}^\dagger(\Delta t') D_z(2\omega t') J_x D_z^\dagger(2\omega t') D_{\hat{\mathbf{a}}}(\Delta t')\, dt'. \tag{39}$$

The reason for the appearance of the operator $D_z(2\omega t')$ is that in the moving coordinate system the counterclockwise field rotates twice as rapidly as in the laboratory frame. Upon carrying out the z-rotation in the integrand of (39) we obtain $J_x \to \mathbf{J} \cdot \hat{\mathbf{u}}_a(2t)$, and therefore

$$F(t) = \lambda\Omega \int_0^t D_{\hat{\mathbf{a}}}^\dagger(\Delta t')(J_x \cos 2\omega t' + J_y \sin 2\omega t') D_{\hat{\mathbf{a}}}(\Delta t')\, dt'. \tag{40}$$

Although it is easy (but very tedious) to evaluate (40) exactly, there would be little sense in doing so. We are really only interested in the situation near resonance; furthermore, (40) already embodies an expansion in powers of λ, and there is therefore no point in keeping higher powers of λ. If we are sufficiently close to the resonance to satisfy $|\omega - \Omega| \ll \lambda\Omega$, (20) informs us that we can make the approximation

$$D_{\hat{\mathbf{a}}}(\Delta t) \simeq e^{i\lambda\Omega J_x t}. \tag{41}$$

With this simplification we can reduce (40) to

$$F(t) = \lambda\Omega \int_0^t \{J_x \cos 2\Omega t' + [J_y \cos \lambda\Omega t' + J_z \sin \lambda\Omega t'] \sin 2\Omega t'\}\, dt'.$$

Retaining only the leading term as $\lambda \to 0$ we obtain

$$1 + iF(t) = 1 + \tfrac{1}{2}i\lambda[J_x \sin 2\Omega t + J_y(1 - \cos 2\Omega t)] \tag{42}$$

as the correction factor to the time evolution operator in the immediate vicinity of the resonance. The matrix elements of $F(t)$ are seen to be uniformly small if $\lambda j \ll 1$, where j is the length of the spin in question. Our earlier intuitive argument regarding the unimportance of the counterclockwise field is therefore confirmed.*

56. *Transitions in a Continuum: Scattering*

In the preceding section we treated a problem where the unperturbed Hamiltonian had a discrete spectrum, and the perturbation was explicitly time dependent. The situation where the spectrum is continuous and the perturbation is independent of time is encountered much more frequently in physics. All scattering processes, whether they be elastic or

* In resonance experiments the really significant question is whether the counterclockwise field changes the shape and position of the resonance. This constitutes a rather difficult problem which has been solved by F. Bloch and A. Siegert, *Phys. Rev.* **57**, 522 (1940); they show that the resonance is shifted by an amount of order λ^2.

inelastic, belong to this category, as do phenomena associated with the emission of radiation or the decay of unstable particles. This section is devoted to derivations of some rather general formulas for collision cross sections. In the remainder of this chapter we shall apply these results to a variety of important problems where perturbative approximations are legitimate.

56.1 STATIONARY COLLISION STATES AND THE *T*-MATRIX

In the theory of potential scattering presented in Sec. 12, we saw that the stationary collision states played an important role in the realistic (i.e., time-dependent) formulation of the scattering process. We shall follow a similar route here, and first derive some important properties of stationary collision states.

Let $|t\rangle$ be a solution of the complete Schrödinger equation,

$$\left(i\hbar\frac{\partial}{\partial t} - H\right)|t\rangle = 0. \tag{1}$$

The Hamiltonian $H = H_0 + H'$ is assumed to be time independent. Here H_0 is the kinetic energy of the two colliding systems together with their separate internal energies, and H' the interaction between them. We envisage the situation where for early times $|t\rangle$ represents the two systems spatially separated and proceeding towards each other with a fairly well-defined momentum and in precisely defined internal states. In the collision of two atoms, the incident state would therefore be the product of two packets, one for each atom, describing their center-of-mass motion and their internal quantum state. We may therefore say that for early times $|t\rangle$ tends to $|t\rangle_0$, where, by hypothesis,

$$\left(i\hbar\frac{\partial}{\partial t} - H_0\right)|t\rangle_0 = 0. \tag{2}$$

The state $|t\rangle_0$ is constructed so as to assure that for $t \approx 0$ the systems are within range of each other. Although $|t\rangle_0$ coincides with $|t\rangle$ for $t \lll 0$, the former is not a solution of (1) once H' comes into play.

We may now construct an integral equation that incorporates (1) and the initial condition. As usual, we write

$$\left(i\hbar\frac{\partial}{\partial t} - H_0\right)|t\rangle = H'|t\rangle, \tag{3}$$

and introduce a Green's function (it is really an operator) that satisfies

$$\left(i\hbar\frac{\partial}{\partial t} - H_0\right)G(t-t') = \delta(t-t'). \tag{4}$$

In particular, we use the retarded boundary condition, i.e., $G_+(t - t') = 0$ if $t < t'$. Then

$$G_+(t - t') = -\frac{i}{\hbar}\theta(t - t')e^{-iH_0(t-t')/\hbar}, \tag{5}$$

where $\theta(x)$ is the unit step function. The desired integral equation is then

$$|t^{(+)}\rangle = |t\rangle_0 + \int_{-\infty}^{\infty} G_+(t - t')H'|t'^{(+)}\rangle\, dt'. \tag{6}$$

The notation $|t^{(+)}\rangle$ is intended to emphasize that the solution of (6) will tend to the free wave packet $|t\rangle_0$ as $t \to -\infty$, and therefore has outgoing (and not incoming) scattered waves.

Before proceeding with the formal development we should remark on a tacit assumption already incorporated into the discussion. We have split H into two pieces, and that is the crux of the matter, because our formulation is not legitimate if H' does not have a well-defined range. Thus in any collision between neutral atoms (Van der Waals forces) the split makes sense. The same holds for nuclear problems. The Coulomb interaction would seem to violate the criterion, but only in a trivial way, and (6) actually is correct for this case. There is a very important class of interactions where this breakup does not work, however, namely the interaction of charged particles with the radiation field. The stationary states of a *single* charged particle are not eigenstates of H(matter) + H(field), because, loosely speaking, the charge is affected by the fluctuating field strengths present in the electromagnetic vacuum state (recall the first footnote of Sec. 53). We shall make some further remarks concerning this question in the section on emission and absorption of radiation (Sec. 61).

In spite of the limitations just alluded to, (6) is able to describe a broad class of problems. Unlike the discussion of Sec. 12, it is not restricted to elastic scattering.

As we have said, we shall require those special states that are actually eigenstates of H. These states shall be denoted by $|a^{(+)}, t\rangle = |a^{(+)}\rangle e^{-iE_a t/\hbar}$. The symbol a incorporates all the quantum numbers necessary for the specification of the state, i.e., the initial momenta of the colliding particles, as well as their internal states. The corresponding stationary state belonging to H_0 is written as $|a, t\rangle = |a\rangle e^{-iE_a t/\hbar}$. Because we are concerned with states belonging to the continuous spectrum we must assign the same energy eigenvalue E_a to both $|a\rangle$ and $|a^{(+)}\rangle$. Returning to (6) we have

$$|a^{(+)}\rangle = |a\rangle - \frac{i}{\hbar}\int_{-\infty}^{0} e^{i(H_0-E_a)t/\hbar}H'|a^{(+)}\rangle\, dt. \tag{7}$$

The integrand oscillates indefinitely as $t \to -\infty$, reflecting the fact that we are now dealing with states of infinite norm instead of wave packets. But we shall always use the states $|a^{(+)}\rangle$ as mathematical aids in the construction of wave packets that were free packets in the *past*. We therefore define the singular integral by inserting the factor $e^{\epsilon t/\hbar}$ into the integrand of (7), where ϵ is a positive infinitesimal. This being done, we can integrate (7) to obtain

$$|a^{(+)}\rangle = |a\rangle + \frac{1}{E_a - H_0 + i\epsilon} H'|a^{(+)}\rangle. \tag{8}$$

This will be recognized as the Lippmann-Schwinger equation, (30.39), previously derived from the time-independent Schrödinger equation by requiring the Green's function in coordinate space to have outgoing radial waves. The argument just presented here shows that the outgoing wave condition is equivalent to the condition that the state vector tends to a free packet in the past.

Let us assume that the vectors $\{|a\rangle\}$ are a complete set of eigenstates of H_0. We may then write

$$|a^{(+)}\rangle = |a\rangle + \sum_{a'} \frac{1}{E_a - E_{a'} + i\epsilon} |a'\rangle\langle a'|H'|a^{(+)}\rangle, \tag{9}$$

instead of (8). In Sec. 30.4 we saw that the coefficient of each plane wave state in the expansion of the scattering term $(E_a - H_0 + i\epsilon)^{-1}H'|a^{(+)}\rangle$ was proportional to the scattering amplitude. We shall soon show that this is still true in the case where inelastic scattering is also possible. We therefore define an operator T with matrix elements *

$$T_{ba} = \langle b|H'|a^{(+)}\rangle \equiv \langle b|T|a\rangle. \tag{10}$$

Here $|b\rangle$ is an eigenstate of H_0; eigenstates of H will always be supplied with a superscript $(+)$. The operator T will play an important role in this chapter, and even more so in Chapt. XIII. It is usually called the transition matrix (or T-matrix for short), because it gives the amplitude for the transition $|a\rangle \to |b\rangle$. In terms of T, (9) reads

$$|a^{(+)}\rangle = |a\rangle + \sum_{a'} \frac{1}{E_a - E_{a'} + i\epsilon} |a'\rangle T_{a'a} \tag{11}$$

$$= \left(1 + \frac{1}{E_a - H_0 + i\epsilon} T\right)|a\rangle. \tag{12}$$

* Compare this equation with (12.19). Observe that in our present discussion the roles of the plane wave $\phi_{\mathbf{k}}$, the scattering wave function $\psi_{\mathbf{k}}$, and the potential V are played by $|b\rangle$, $|a^{(+)}\rangle$, and H', respectively.

In the applications discussed in this chapter we shall only require perturbative approximations to scattering amplitudes. For this purpose we require an equation for T in terms of H'. This equation is obtained immediately by multiplying (11) by H', and taking the scalar product with $\langle b|$:

$$T_{ba} = H'_{ba} + \sum_{a'} \frac{H'_{ba'}T_{a'a}}{E_a - E_{a'} + i\epsilon}, \tag{13}$$

or equivalently,

$$T = H' + H' \frac{1}{E_a - H_0 + i\epsilon} T. \tag{14}$$

The iterative solution to this equation is obviously

$$T = H' + H' \frac{1}{E_a - H_0 + i\epsilon} H' + \cdots. \tag{15}$$

It should be noted that these equations for T are only correct when T acts on a ket of energy E_a. For this reason the notation $T(E_a)$ is often used.

56.2 RELATIONSHIP BETWEEN THE CROSS SECTION AND THE T-MATRIX

We are now in a position to evaluate the result of a collision experiment. Nothing essential is lost if we consider a heavy target located at the origin of our coordinate system. Readers who have understood the development that follows can, at their leisure, provide a derivation free of this restriction.

An accelerator does not produce pure states, and we shall therefore describe the beam by a density matrix.* One might suppose that such a description would lead to a complicated and obscure formalism, but the opposite is really the case. As we shall see, a realistic treatment actually yields a more visualizable result than one attains by restricting oneself to pure states.

Before the collision we describe the state of the system by the density matrix $\rho_A(t)$, where

$$\rho_A(t) = \sum_{\mathbf{p}_a\mathbf{p}_{a'}} e^{-iE_at/\hbar}|a\rangle\langle\mathbf{p}_a|\rho_A|\mathbf{p}_{a'}\rangle\langle a'|e^{iE_{a'}t/\hbar}. \tag{16}$$

By construction, $\rho_A(0)$ describes the situation where the target is immersed in the beam, but does *not* interact with it. The matrix ele-

* For another formulation of collision theory in terms of mixtures see E. H. Wichmann, *Am. J. Phys.* **33**, 20 (1965).

ments of ρ_A depend only on the momenta $\mathbf{p}_a$ of the noninteracting states $|a\rangle$, because we assume that in the initial configuration the target and projectile are each in one definite quantum state of the internal motion. The fact that ρ_A projects onto these internal states has not been made explicit in (16), and is to be understood henceforth. The incident beam is supposed to have a rather well-defined momentum $\mathbf{p}_A$, and $\langle \mathbf{p}|\rho_A|\mathbf{p}'\rangle$ therefore vanishes unless $\mathbf{p}$ and $\mathbf{p}'$ are within Δp_A of $\mathbf{p}_A$. We shall have more to say about ρ_A in a moment.

Equation (16) only holds before the collision, i.e., when t is sufficiently negative. The correct density matrix for all values of t is obtained by replacing the noninteracting states $|a\rangle$ by the collision states $|a^{(+)}\rangle$:

$$\rho_A^{(+)}(t) = \sum_{\mathbf{p}_a\mathbf{p}_{a'}} e^{-iE_at/\hbar}|a^{(+)}\rangle\langle \mathbf{p}_a|\rho_A|\mathbf{p}_{a'}\rangle\langle a'^{(+)}|e^{iE_{a'}t/\hbar}. \tag{17}$$

Let $\Lambda_B(t)$ be the projection operator onto the state $|B, t\rangle$ defined by the detection apparatus. (Here we have purposely ignored the inessential complications implied by the fact that a realistic detector does not define a pure state.) When detection occurs the products of the collision process are spatially separated, and $|B, t\rangle$ therefore satisfies * $(i\hbar\,\partial_t - H_0)|B, t\rangle = 0$. The probability that a transition from $\rho_A(t)$ to $|B, t\rangle$ has taken place is given by

$$P_{BA}(t) = \operatorname{tr} \Lambda_B(t)\rho_A^{(+)}(t). \tag{18}$$

It is actually somewhat easier (and just as useful) to evaluate the transition rate dP/dt. A formula for $\dot{P}$ is readily derived if we recall the "Liouville" equation (28.33): $i\hbar\dot{\rho}^{(+)} = [H, \rho^{(+)}]$. The density matrix Λ_B obeys the equation $i\hbar\dot{\Lambda}_B = [H_0, \Lambda_B]$, because it describes the noninteracting system. If we then use the fact that $\operatorname{tr}[X, Y] = 0$ for any pair of (bounded) operators X and Y, we find that

$$\hbar\dot{P}_{BA}(t) = 2 \operatorname{Im} \operatorname{tr}[\Lambda_B(t)H'\rho_A^{(+)}(t)]. \tag{19}$$

It will suffice to evaluate $\dot{P}_{BA}$ in the case where $|B, t\rangle$ is one of the stationary states $|b, t\rangle$ of the free Hamiltonian H_0. The trace in (19) can then be evaluated with the help of (10) and (17):

$$\hbar\dot{P}_{bA}(t) = 2 \operatorname{Im} \sum_{\mathbf{p}_a\mathbf{p}_{a'}} e^{-i(E_a-E_{a'})t/\hbar}T_{ba}\langle \mathbf{p}_a|\rho_A|\mathbf{p}_{a'}\rangle\langle a'^{(+)}|b\rangle. \tag{20}$$

* Collisions in which the *non*interacting reaction products possess a *different* Hamiltonian from H_0 can also occur. Such processes are called rearrangement collisions; an example is $d + d \rightarrow He^3 + n$. The theory of such processes is deferred to Chapt. XIII.

According to (9)

$$\langle b|a'^{(+)}\rangle = \delta_{ba'} + \frac{T_{ba'}}{E_{a'} - E_b + i\epsilon}. \tag{21}$$

The δ-function in (21) requires $\mathbf{p}_{a'} = \mathbf{p}_b$. In collision experiments one always detects scattered particles moving at a finite angle with respect to the direction $\hat{\mathbf{p}}_A$ of the incident beam. Hence the δ-function in (21) does not contribute to (20), which therefore reads

$$\hbar \dot{P}_{bA}(t) = 2\,\mathrm{Im} \sum_{\mathbf{p}_a\mathbf{p}_{a'}} e^{-i(E_a - E_{a'})t/\hbar} \frac{T_{ba}T^*_{ba'}}{E_{a'} - E_b - i\epsilon} \langle \mathbf{p}_a|\rho_A|\mathbf{p}_{a'}\rangle. \tag{22}$$

We shall assume that the momentum spread Δp_A of the incident beam is sufficiently small so that the variation of the amplitude T_{ba} with $\mathbf{p}_a$ can be neglected in (22). Hence we may write

$$\hbar \dot{P}_{bA}(t) = -i|T_{bA}|^2 \sum_{\mathbf{p}_a\mathbf{p}_{a'}} e^{-i(E_a - E_{a'})t/\hbar} \langle \mathbf{p}_a|\rho_A|\mathbf{p}_{a'}\rangle \times \left\{ \frac{1}{E_{a'} - E_b - i\epsilon} - \frac{1}{E_a - E_b + i\epsilon} \right\}. \tag{23}$$

By T_{bA} we here mean any typical matrix element T_{ba} between momentum eigenstates such that $\mathbf{p}_a$ falls inside the (very narrow) momentum spread of the incident beam.

Now we must analyze the density matrix in more detail. Consider the real function $f(\mathbf{R}, \mathbf{P})$ defined by

$$\langle \mathbf{P} + \tfrac{1}{2}\mathbf{Q}|\rho|\mathbf{P} - \tfrac{1}{2}\mathbf{Q}\rangle = \int_\Omega d^3R\, e^{-i\mathbf{Q}\cdot\mathbf{R}/\hbar} f(\mathbf{R}, \mathbf{P}), \tag{24}$$

where Ω is the usual periodicity volume. The probability of finding a particle of momentum $\mathbf{P}$ in the mixture ρ is obtained by setting $\mathbf{Q} = 0$ in (24):

$$\langle \mathbf{P}|\rho|\mathbf{P}\rangle = \int d^3R\, f(\mathbf{R}, \mathbf{P}). \tag{25}$$

The probability density of finding a particle at the spatial point $\mathbf{R}$ is

$$\begin{aligned} n(\mathbf{R}) = \langle \mathbf{R}|\rho|\mathbf{R}\rangle &= \sum_{\mathbf{p}\mathbf{p}'} \langle \mathbf{R}|\mathbf{p}\rangle\langle \mathbf{p}|\rho|\mathbf{p}'\rangle\langle \mathbf{p}'|\mathbf{R}\rangle \\ &= \Omega^{-1} \sum_{\mathbf{P}\mathbf{Q}} e^{i\mathbf{R}\cdot\mathbf{Q}/\hbar} \langle \mathbf{P} + \tfrac{1}{2}\mathbf{Q}|\rho|\mathbf{P} - \tfrac{1}{2}\mathbf{Q}\rangle \\ &= \sum_{\mathbf{P}} f(\mathbf{R}, \mathbf{P}). \end{aligned} \tag{26}$$

Thus $f(\mathbf{R}, \mathbf{P})$ has the remarkable property that when it is integrated over the spatial variable **R** it gives the probability distribution in momentum space, whereas its integral over all momenta gives the probability distribution in coordinate space. In classical statistics one defines a distribution function that prescribes the joint probability that a molecule has, simultaneously, a certain position and momentum, i.e., the probability distribution in phase space. Because of this analogy $f(\mathbf{R}, \mathbf{P})$ is also called a (Wigner) phase space distribution function. The uncertainty relations between coordinates and momenta imply that in general our $f(\mathbf{R}, \mathbf{P})$ cannot be such a joint probability distribution, and it is hardly surprising that $f(\mathbf{R}, \mathbf{P})$ is not positive definite for an arbitrary mixture. However, when the mixture in question is composed of wave packets traveling along well-defined trajectories,* the negative terms in f are negligible, and the naive interpretation of the Wigner distribution function as a joint probability in phase space is permissible. This is certainly the case when we deal with free particles collimated in a beam whose transverse dimensions are large compared to the mean de Broglie wavelength. In the discussion that follows we shall therefore adopt the classical interpretation, and say that $f(\mathbf{R}, \mathbf{P})$ is the phase-space distribution of the incident beam at the time when it overlaps with the target.

Let us now return to the evaluation of (23). In the exponential we replace $E_a - E_{a'} = (p_a^2 - p_{a'}^2)/2m$ by $\mathbf{P} \cdot \mathbf{Q}/m$. We shall also assume that the spread Δp_A is sufficiently narrow to permit the replacement of E_a and $E_{a'}$ by $E_a^0 + P^2/2m$ in the denominators,† where E_a^0 is the energy of $|a\rangle$ when $\mathbf{p}_a = 0$. Inserting the representation (24) into (23), recalling (7.42), and using $(x - i\epsilon)^{-1} - (x + i\epsilon)^{-1} = 2\pi i\, \delta(x)$, we obtain

$$\dot{P}_{bA}(t) = \frac{2\pi\Omega}{\hbar} |T_{bA}|^2 \sum_{\mathbf{P}} \delta\left(E_b - \frac{P^2}{2m} - E_a^0\right) f(-\mathbf{P}t/m, \mathbf{P}).$$

Because of the narrow momentum spread in the beam, the coordinate argument, $-\mathbf{P}t/m$, can be replaced by $-\mathbf{v}_A t$, where $\mathbf{v}_A \equiv \mathbf{p}_A/m$ is the mean incident velocity. Our final result for the transition probability

* The precise meaning of this phrase is the following. Let $\{\chi_n(\mathbf{R})\}$ define the representation in which ρ is diagonal:

$$\langle \mathbf{R}|\rho|\mathbf{R}'\rangle = \sum_n p_n \chi_n(\mathbf{R}) \chi_n^*(\mathbf{R}'),$$

with $\Sigma p_n = 1$. Then the assertion concerning $f(\mathbf{R}, \mathbf{P})$ is correct when the $\chi_n(\mathbf{R})$ can be approximated by semiclassical wave functions.

† This means that we ignore the spreading of the wave packets. The correction is similar to the one in the equation preceding (12.25).

per unit time is therefore

$$\dot{P}_{bA}(t) = \frac{2\pi\Omega}{\hbar}|T_{bA}|^2 \sum_{\mathbf{P}} \delta(E_b - E_a^0 - P^2/2m) f(-\mathbf{v}_A t, \mathbf{P}). \tag{27}$$

This formula has a simple intuitive interpretation. Let us assume that for fixed $\mathbf{P}$, $f(\mathbf{R}, \mathbf{P})$ has a fairly constant value when $\mathbf{R}$ lies inside a cylindrical region, and falls off rapidly when $\mathbf{R}$ moves out of this region. The target lies at the center of the cylinder, and the symmetry axis is along the direction of the incident momentum. The appearance of the spatial argument $-\mathbf{v}_A t$ in (27) merely states that the transition rate starts to grow when the *front* of the beam pulse hits the target, attains a fairly constant value while the pulse traverses the target, and then falls off to zero once more. The vanishing of the x- and y-components of $\mathbf{R}$ in (27) is the statement that the target is located at the origin of the coordinate system. The δ-function in (27) selects out the portion of the incident beam that has the energy of the state selected by the detection apparatus; this δ-function therefore insures that the total energy is conserved in the collision.

The probability $P_{bA}(\infty)$ that a transition to the state $|b\rangle$ has occurred during the collision is obtained by integrating * over $\infty > t > -\infty$. This integral can be converted into an integral along the z-axis:

$$P_{bA}(\infty) = \frac{2\pi}{\hbar}\frac{\Omega}{v_A}|T_{bA}|^2 \sum_{\mathbf{P}} \delta(E_b - E_a^0 - P^2/2m) \int_{-\infty}^{\infty} dz\, f(00z, \mathbf{P}). \tag{28}$$

A realistic detector has a nonzero momentum resolution Δp_B centered at $\mathbf{p}_B$, and the transition probability actually measured is therefore

$$P_{BA}(\infty) \equiv \sum_{p_b \in \Delta p_B} P_{bA}(\infty) \underset{\Omega\to\infty}{\longrightarrow} \frac{\Omega}{(2\pi\hbar)^3}\int_{\Delta p_B} d^3p_b\, P_{bA}(\infty). \tag{29}$$

The differential $\Omega/(2\pi\hbar)^3\, d^3p$ may be written as $\rho_B\, dE_b$, with

$$\rho_B = \frac{\Omega}{(2\pi\hbar)^3} p_B^2 \left(\frac{\partial p_b}{\partial E_b}\right)_{p_B} d\Omega_B, \tag{30}$$

where $d\Omega_B$ is the element of solid angle specifying the orientation of $\mathbf{p}_B$. As we see, ρ_B is the number of final states per unit energy. The quantity $(\rho_B/\Omega d\Omega_B)$ is proportional to the number of final states per element of

* One can, of course, compute $P_{bA}(\infty)$ directly from (18), without going through the seemingly absurd motions of first computing dP/dt, and then integrating over t. The advantage of our approach is that it avoids the evaluation of the Wigner distribution function subsequent to the collision.

volume in phase space, and therefore ρ_B is often referred to as the phase space factor.

Let us assume that the detector's energy resolution $\Delta p_B(\partial E_b/\partial p_b)$ exceeds the energy spread $\Delta p_A(\partial E_a/\partial p_a)$ of the incident beam. The integration stipulated by (29) then removes the energy δ-function in (28), and the subsequent integration over $\mathbf{P}$ can be carried out with the help of (26). The differential transition probability (i.e., the probability for a transition into $d\Omega_B$) is therefore

$$dP_{BA} = \frac{2\pi}{\hbar}\frac{\Omega}{v_A}\rho_B|T_{BA}|^2 \int_{-\infty}^{\infty} dz\, n(00z),$$

where T_{BA} stands for any typical T-matrix element between states within the beam and those accepted by the detector. The quantity $\int dz\, n(00z)$ is the probability per unit area that a particle be incident along the z-axis. By definition, the differential cross section is the transition probability divided by the incident probability per unit area, or

$$d\sigma_{BA} = \frac{2\pi}{\hbar}\frac{\Omega}{v_A}\rho_B|T_{BA}|^2. \tag{31}$$

This is a very important formula. It provides the detailed connection between the matrix elements of T and the quantity measured in the laboratory.

The derivation of (31) makes it quite clear that when the final state $|b\rangle$ is specified by several momentum variables $\mathbf{p}_{bi}$ (e.g., $\mathrm{e} + \mathrm{d} \rightarrow \mathrm{e} + \mathrm{n} + \mathrm{p}$), one must integrate (28) over all these momenta, and instead of (30) one obtains *

$$\rho_B = \frac{1}{dE_B}\prod_{i=1}^{N}\frac{\Omega}{(2\pi\hbar)^3}\,d^3p_{bi}, \qquad E_B \equiv \sum_{i=1}^{N} E_{bi}. \tag{32}$$

Applications of these formulas will be given in the remainder of this chapter.

56.3 THE GOLDEN RULE

If one computes the transition probability between energy eigenstates, i.e., $P_{ba}(t) = |\langle b,t|a^{(+)},t\rangle|^2$, one can construct very concise though rather obscure derivations of (31). Derivations of a similar nature can also be carried out with the help of the interaction picture. The upshot of

* Observe that $d\sigma$ always has the dimensions of an area. The number of differentials that appear on the right-hand side of (31) depends on the number N of final momenta. When $N = 1$, there is only an element of solid angle; when $N = 2$ there is a second element of solid angle, and also one energy (or momentum) differential.

such arguments is that the transition rate between energy eigenstates is

$$``\dot{P}_{ba}" = \frac{2\pi}{\hbar} |T_{ba}|^2 \delta(E_b - E_a). \tag{33}$$

In the lowest approximation to T, i.e., $T \simeq H'$, (33) reads

$$``\dot{P}_{ba}" = \frac{2\pi}{\hbar} |H'_{ba}|^2 \delta(E_b - E_a). \tag{34}$$

In order to get a sensible (i.e., finite) number from (34), one sums over final states having an energy in the immediate vicinity of E_b, and thereby defines a total transition rate to this entire group of states by

$$\dot{P}_{Ba} = \frac{2\pi}{\hbar} |H'_{ba}|^2 \rho_B. \tag{35}$$

This formula of Dirac has given such faithful and meritorious service that Fermi called it the Golden Rule. We shall also use this terminology.

The cross section is determined from (35) by dividing by the incident flux. As the initial state is now a plane wave state normalized to one in a box of volume Ω, the flux is given by v_a/Ω. The cross section is therefore

$$d\sigma_{Ba} = \frac{2\pi}{\hbar} \frac{\Omega}{v_a} |H'_{ba}|^2 \rho_B. \tag{36}$$

This agrees with (31) when the Born approximation $T \simeq H'$ is made in the latter expression.

57. *Collision Phenomena in the Born Approximation*

We shall now apply the Golden Rule to a variety of collision phenomena. Our aim here is to illustrate the power and limitations of the technique, and to establish its relationship to the treatment of elastic scattering given in Chapt. III.

At the outset we shall study a system that does not really occur in nature, but which does constitute a prototype for more complex processes that abound in atomic and nuclear physics. The model consists of two spinless and distinguishable particles designated by α and β which interact via a central potential $V(|\mathbf{r}_\alpha - \mathbf{r}_\beta|)$. Beyond this force there is also a fixed, attractive, central field that only acts on β. The full Hamiltonian is therefore

$$H = \frac{p_\alpha^2}{2m_\alpha} + \frac{p_\beta^2}{2m_\beta} + V(r_{\alpha\beta}) + U(r_\beta), \tag{1}$$

where we have chosen the origin at the symmetry point of the external potential U. The latter potential is assumed strong enough so as to possess a number of bound states,

$$\left(\frac{p_\beta^2}{2m_\beta} + U(\mathbf{r}_\beta)\right)|n\rangle = E_n|n\rangle, \tag{2}$$

whose wave functions will be designated by

$$\varphi_n(\mathbf{r}_\beta) \equiv \langle \mathbf{r}_\beta | n \rangle. \tag{3}$$

Let X_n stand for nth bound state of β. The following processes are then possible,* provided the incident energy of α is sufficiently high:

$$\alpha + X_0 \rightarrow \alpha + X_0, \tag{4}$$
$$\alpha + X_0 \rightarrow \alpha + X_n, \tag{5}$$
$$\alpha + X_0 \rightarrow \alpha + \beta. \tag{6}$$

We shall call these elastic scattering, inelastic scattering, and production, respectively. The system we are discussing could therefore serve as a model for electron scattering by a nucleus,† the counterparts to (4)–(6) being

$$e^- + X_Z^A \rightarrow e^- + X_Z^A,$$
$$e^- + X_Z^A \rightarrow e^- + X_Z^{A*},$$
$$e^- + X_Z^A \rightarrow e^- + p + X_{Z-1}^{A-1*},$$

where X_Z^A stands for a nucleus of the indicated mass number and charge, and the asterisk designates an excited state.

In applying the Golden Rule to processes (4) and (5), it is clear that the perturbing Hamiltonian H' is just $V(r_{\alpha\beta})$. The production process is more involved, and we shall therefore treat it separately.

57.1 ELASTIC AND INELASTIC SCATTERING

To compute the transition rate for the scattering processes (4) and (5), it is only necessary to know the various factors that enter into the formula

* If V is sufficiently attractive to bind the α-β system, the pick-up reaction $\alpha + X_0 \rightarrow (\alpha + \beta)_{\text{bound}}$ is also possible.

† See R. Hofstadter, *Ann. Rev. Nucl. Sci.* 7, 23 (1957). In these experiments the electrons are ultrarelativistic, and some modifications of the formulas that we shall derive are required.

(recall (56.35))

$$\dot{P}_{Ba} = \frac{2\pi}{\hbar}\,\rho_B|\langle b|V(r_{\alpha\beta})|a\rangle|^2. \tag{7}$$

The states $|a\rangle$ and $|b\rangle$ are eigenstates of the unperturbed Hamiltonian

$$H_0 = \frac{p_\alpha^2}{2m_\alpha} + \frac{p_\beta^2}{2m_\beta} + U(r_\beta), \tag{8}$$

and are therefore products of states for α and β separately. Thus $|a\rangle$ contains α in a plane wave state of momentum $\hbar\mathbf{k}_\alpha$ and β in the ground state $|0\rangle$, while $|b\rangle$ is a product of another plane wave for α with momentum $\hbar\mathbf{k}'_\alpha$, and β in the bound state $|n\rangle$, where $n = 0$ for elastic and $n \neq 0$ for inelastic scattering. The energies of these states are

$$E_a = E_\alpha + E_0, \tag{9}$$
$$E_b = E'_\alpha + E_n, \tag{10}$$

where $E_\alpha = (\hbar k_\alpha)^2/2m_\alpha$, etc. Our continuum wave functions satisfy periodic boundary conditions on a cube of volume Ω:

$$\langle \mathbf{r}_\alpha|\mathbf{k}_\alpha\rangle = \Omega^{-\frac{1}{2}}e^{i\mathbf{k}_\alpha\cdot\mathbf{r}_\alpha}. \tag{11}$$

These states are orthonormal,

$$\langle \mathbf{k}_\alpha|\mathbf{k}'_\alpha\rangle = \frac{1}{\Omega}\int d^3r_\alpha\, e^{-i(\mathbf{k}_\alpha-\mathbf{k}_\alpha')\cdot\mathbf{r}_\alpha} = \delta_{\mathbf{k}_\alpha,\mathbf{k}'_\alpha}, \tag{12}$$

and $\mathbf{k}_\alpha$ only takes on the discrete values listed in (7.40). The matrix element in (7) is therefore

$$\langle b|V|a\rangle = \frac{1}{\Omega}\int d^3r_\alpha\, d^3r_\beta\, e^{-i\mathbf{k}_\alpha'\cdot\mathbf{r}_\alpha}\varphi_n^*(\mathbf{r}_\beta)V(r_{\alpha\beta})\varphi_0(\mathbf{r}_\beta)e^{i\mathbf{k}_\alpha\cdot\mathbf{r}_\alpha}. \tag{13}$$

Here it is already clear that our method only makes sense if V is sufficiently regular to possess a Fourier transform, a condition which is familiar to us from Sec. 13. Let us then introduce

$$\tilde{V}(k) = \int_\Omega e^{i\mathbf{k}\cdot\mathbf{r}}V(r)\,d^3r, \tag{14}$$

and its inverse

$$V(r) = \frac{1}{\Omega}\sum_{\mathbf{k}} e^{-i\mathbf{k}\cdot\mathbf{r}}\tilde{V}(k). \tag{15}$$

In terms of $\tilde{V}(k)$, (13) reduces to

$$\langle b|V|a\rangle = \frac{1}{\Omega}\,\tilde{V}(q)F_n(\mathbf{q}), \tag{16}$$

where

$$\hbar \mathbf{q} = \hbar(\mathbf{k}_\alpha - \mathbf{k}'_\alpha) \tag{17}$$

is the momentum transferred to α during the collision, and

$$F_n(\mathbf{q}) = \int d^3r\, e^{i\mathbf{q}\cdot\mathbf{r}} \varphi_n^*(\mathbf{r})\varphi_0(\mathbf{r}). \tag{18}$$

Finally we must compute $\rho_B\, dE_b$, the number of final states in the interval $(E_b, E_b + dE_b)$. The number of states in the interval $(\mathbf{k}'_\alpha, \mathbf{k}'_\alpha + d\mathbf{k}'_\alpha)$ is $\Omega\, d^3k'_\alpha/(2\pi)^3$, and therefore

$$\rho_B = \frac{\Omega\, d^3k'_\alpha}{(2\pi)^3\, dE'_\alpha}. \tag{19}$$

But $dE'_\alpha = (\hbar^2/m_\alpha)k'_\alpha\, dk'_\alpha$ and $d^3k'_\alpha = k'^2_\alpha\, dk'_\alpha\, d\Omega'_\alpha$, where $d\Omega'_\alpha$ is the element of solid angle for the scattered particle. Hence (19) becomes

$$\rho_B = \frac{\Omega}{(2\pi)^3} \frac{m_\alpha k'_\alpha}{\hbar^2}\, d\Omega'_\alpha. \tag{20}$$

The transition rate is therefore

$$\dot{P}(\mathbf{k}_\alpha \to \mathbf{k}'_\alpha, n) = \frac{1}{\Omega} \frac{m_\alpha k'_\alpha}{(2\pi)^2 \hbar^3}\, d\Omega'_\alpha |\tilde{V}(q)|^2 |F_n(\mathbf{q})|^2. \tag{21}$$

Here it must be noted that k'_α is fixed by energy conservation, because

$$k'_\alpha = \sqrt{k_\alpha^2 - \gamma^2} \leqslant k_\alpha, \tag{22}$$

where $\hbar^2\gamma^2/2m_\alpha \equiv E_n - E_0$ is the excitation energy, and the equality only holds for elastic scattering. The momentum transfer is related to the scattering angle θ by

$$q^2 = 4k_\alpha^2 \sin^2 \tfrac{1}{2}\theta - 2k_\alpha(\sqrt{k_\alpha^2 - \gamma^2} - k_\alpha)\cos\theta - \gamma^2. \tag{23}$$

When the incident energy is large compared to the excitation energy, $k_\alpha \gg \gamma$, and (23) may be approximated by

$$q^2 = 4k_\alpha^2 \sin^2 \tfrac{1}{2}\theta + \gamma^2(\cos\theta - 1) + \tfrac{1}{4}(\gamma^4/k_\alpha^2)\cos\theta. \tag{23'}$$

The quantity of direct experimental interest is not the transition rate, but the cross section, which is obtained from (21) by dividing by the incident flux, $\hbar k_\alpha/m_\alpha\Omega$. The differential cross section is therefore

$$\frac{d\sigma_n^B}{d\Omega'_\alpha} = \frac{k'_\alpha}{k_\alpha}\left(\frac{m_\alpha}{2\pi\hbar^2}\right)^2 |\tilde{V}(q)|^2 |F_n(\mathbf{q})|^2. \tag{24}$$

Observe that this result is independent of the volume Ω, as it must be.

Let us first make contact with our earlier treatment of elastic scattering. The elastic cross section is

$$\frac{d\sigma_{el}^B}{d\Omega_\alpha'} = \left(\frac{m_\alpha}{2\pi\hbar^2}\right)^2 |\tilde{V}(q)|^2 |F_0(\mathbf{q})|^2. \tag{25}$$

This formula almost agrees with the Born approximation result obtained in (13.2). The most important difference arises from

$$F_0(\mathbf{q}) = \int e^{i\mathbf{q}\cdot\mathbf{r}} |\varphi_0(\mathbf{r})|^2 \, d^3r, \tag{26}$$

the so-called *elastic form factor.* Now we note that

$$F_0(0) = 1. \tag{27}$$

In fact, if R is the r.m.s. radius of the ground state,

$$R^2 = \int r^2 |\varphi_0(\mathbf{r})|^2 \, d^3r, \tag{28}$$

then $F_0(q) \approx 1$ for $qR \ll 1$, because (26), when expanded in powers of q, gives

$$F_0(q) \underset{q\to 0}{\longrightarrow} 1 - \tfrac{1}{6}q^2R^2 \tag{29}$$

if φ_0 is an s-state. Thus (25) agrees with (13.2) when $qR \ll 1$, provided the reduced mass μ is replaced by m_α in (13.2). That is to say, near the forward direction (25) equals the Born approximation cross section for scattering off an infinitely heavy (or rigidly fixed) β-particle. Actually β moves about in a region of radius $\sim R$, but this fact is only noticeable in the cross section when the monentum transfer is comparable to or larger than $\hbar/R$ (see Fig. 57.1).

Now that we have established contact with the elastic Born approximation, it is crystal clear that the Golden Rule is only applicable if the potential V is weak in the sense of the criteria deduced in our earlier discussion of the Born approximation. If V is strong, the plane waves for particle α are appreciably distorted, and (25) no longer holds. It is, however, very easy to modify (25) to a form that is much closer to the truth under many circumstances. We have already noted that (25) can be written as

$$\frac{d\sigma_{el}^B}{d\Omega_\alpha'} = \sigma^B(q) |F_0(\mathbf{q})|^2,$$

where σ^B is the Born approximation to the differential cross section for scattering through momentum transfer $\hbar q$ from a fixed target. The modification alluded to then consists of replacing σ^B by σ_{el}, the exact

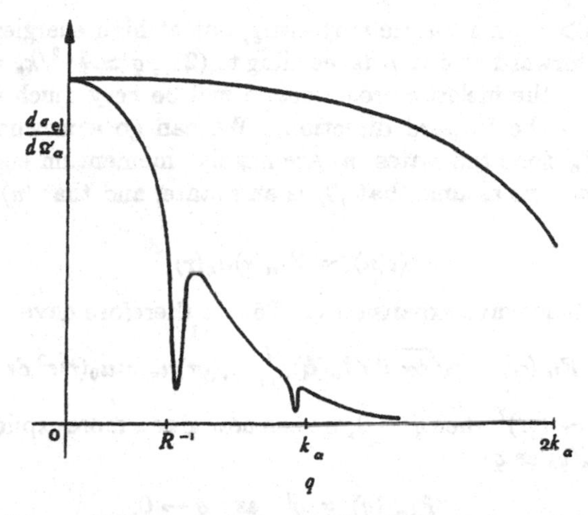

Fig. 57.1. Influence of the form factor on the elastic cross section. The upper curve shows the elastic cross section for scattering from a rigidly fixed particle which, as we recall from Sec. 12, is trivially related to the center-of-mass cross section when the target has a finite mass. The lower curve shows the modification due to $F_0(q)$. If the probability distribution $|\varphi_0(\mathbf{r}_\beta)|^2$ has a fairly well-defined edge, diffraction minima of the type illustrated will occur. Such diffraction effects are observed in electron scattering by complex nuclei (see Hofstadter, *loc. cit.*).

elastic cross section off a rigidly fixed β-particle:

$$\frac{d\sigma_{\text{el}}^{\text{imp}}}{d\Omega_\alpha'} = \sigma_{\text{el}}(q)|F_0(\mathbf{q})|^2; \tag{30}$$

this is the so-called *impulse approximation.**

Let us now study the inelastic cross section, i.e., (24) when $n \neq 0$. The great difference between the inelastic and elastic cross section is due to the very different behaviors of the form factors when $n \neq 0$. Since the different bound-state wave functions are orthogonal, (18) immediately yields

$$F_n(0) = 0 \qquad (n \neq 0). \tag{31}$$

Therefore the inelastic cross section vanishes for zero momentum transfer.

* Our discussion is somewhat oversimplified here. In the Born approximation the cross section is the product of a known kinematical factor, and a function of q only. The exact two-body cross section is not so simple, and the step from (29) to (30) is therefore not completely straightforward. See G. F. Chew, *Phys. Rev.* **80**, 196 (1950); Goldberger and Watson, pp. 683–690.

Of course $q > 0$ for inelastic scattering, but at high energies q is almost zero in the forward direction (according to (23), $q \simeq \frac{1}{2}\gamma^2/k_\alpha$ when $\theta = 0$), and therefore the inelastic cross section will be very much smaller than the elastic in the forward direction. We can go somewhat further in analyzing F_n since the states $|n\rangle$ are angular momentum eigenfunctions. For simplicity we assume that $|0\rangle$ is an s-state, and that $|n\rangle$ has angular momentum l:

$$\langle \mathbf{r}|n\rangle = Y_{lm}(\hat{\mathbf{r}})u_n(r).$$

Using the plane wave expansion (11.25) we therefore have

$$F_{lm}(\mathbf{q}) = \sqrt{4\pi}\, i^l Y^*_{lm}(\hat{\mathbf{q}}) \int_0^\infty j_l(qr)u_n(r)u_0(r)r^2\, dr. \tag{32}$$

Since $j_l(qr) \sim (qr)^l$ when $q \to 0$, we can now give a more explicit form than (31) for $F_{lm}(q)$ as $q \to 0$:

$$F_{lm}\,(q) \propto q^l \quad \text{as} \quad q \to 0. \tag{33}$$

In practice one does not distinguish final states which only differ in the magnetic quantum number m, and therefore the observed cross section involves a sum on m. The addition theorem tells us that

$$\sum_m |Y_{lm}(\hat{\mathbf{q}})|^2 = \frac{2l+1}{4\pi}.$$

Thus we find that the relevant cross section is given by

$$\frac{d\sigma_l}{d\Omega'_\alpha} = (2l+1)\frac{k'_\alpha}{k_\alpha}\sigma^B(q)\left|\int_0^\infty j_l(qr)u_n(r)u_0(r)r^2\, dr\right|^2. \tag{34}$$

Because of (33) and (23′), the angular distribution near the forward direction will be governed by θ^{2l} at high energy (i.e., $k_\alpha \gg \gamma$).

57.2 PRODUCTION

The production process $\alpha + X_0 \to \alpha + \beta$ is more complicated because the perturbation, from the point of view of the final state, contains both $V(r_{\alpha\beta})$ and $U(r_\beta)$. This can be seen most clearly in terms of the inverse (or time-reversed) capture reaction $\alpha + \beta \to \alpha + X_0$, which is admittedly hard to produce in the laboratory. Nevertheless, in treating $\alpha + \beta \to \alpha + X_0$ by the Golden Rule we would certainly consider $V + U$ as H'. We shall not be able to resolve this ambiguity here.* Instead, we shall consider only those high-energy collisions in which β emerges with such high momentum as to make its interaction in the final state with U

* See, however, Chapt. XIII, and Goldberger and Watson, Sec. 5.4.

negligible. In this situation we are justified in using plane wave states for all the particles in both initial and final states. The matrix element in (7) then becomes

$$\langle b|V|a\rangle = \Omega^{-\frac{3}{2}} \int d^3r_\alpha\, d^3r_\beta\, e^{-i(\mathbf{k}'_\alpha\cdot\mathbf{r}_\alpha+\mathbf{k}_\beta\cdot\mathbf{r}_\beta)} V(r_{\alpha\beta})\varphi_0(r_\beta)e^{i\mathbf{k}_\alpha\cdot\mathbf{r}_\alpha},$$

where $\hbar\mathbf{k}_\beta$ is the final momentum of β. The same calculation which reduces (13) to (16) gives us

$$\langle b|V|a\rangle = \left(\frac{2\pi}{\Omega}\right)^{\frac{3}{2}} \tilde{V}(q) f_0(\mathbf{q}-\mathbf{k}_\beta), \tag{35}$$

where f_0 is the ground state wave function in momentum space:

$$f_0(k) = \frac{1}{(2\pi)^{\frac{3}{2}}} \int e^{-i\mathbf{k}\cdot\mathbf{r}} \varphi_0(r)\, d^3r. \tag{36}$$

The density of states factor ρ_B is more complicated now (see (56.32)) since there are two continua to contend with:

$$\rho_B = \frac{\Omega^2}{(2\pi)^6} \frac{d^3k'_\alpha\, d^3k_\beta}{dE_b}, \tag{37}$$

where

$$E_b = \frac{\hbar^2 k_\alpha'^2}{2m_\alpha} + \frac{\hbar^2 k_\beta^2}{2m_\beta} = \frac{\hbar^2 k_\alpha^2}{2m_\alpha} + E_0. \tag{38}$$

The choice of variables that one uses to describe the final state is dictated by the experimental arrangement. Perhaps the simplest case occurs when one measures the orientation of $\mathbf{k}'_\alpha$ and $\mathbf{k}_\beta$, and the energy of one of the particles, say β. For fixed E_β, $dE_b = dE'_\alpha$, and (37) reduces to

$$\rho_B = \frac{\Omega^2}{(2\pi)^6} \frac{m_\alpha k'_\alpha}{\hbar^2} d\Omega'_\alpha \frac{m_\beta k_\beta}{\hbar^2} dE_\beta\, d\Omega_\beta. \tag{39}$$

The differential cross section is then obtained by dividing the transition rate by the flux—recall that this means that σ is an area, no matter how complex the final state. Putting all the pieces together, we therefore have

$$\frac{d\sigma^B}{dE_\beta\, d\Omega'_\alpha\, d\Omega_\beta} = \left(\frac{m_\alpha}{2\pi\hbar^2}\right)^2 \frac{k'_\alpha}{k_\alpha} \frac{m_\beta k_\beta}{\hbar^2} |\tilde{V}(q)|^2 |f_0(\mathbf{k}_\alpha - \mathbf{k}_\beta - \mathbf{k}'_\alpha)|^2,$$

which we can write more compactly in terms of the cross section for scattering off a static potential $V(r)$ as

$$\frac{d\sigma^B}{dE_\beta\, d\Omega'_\alpha\, d\Omega_\beta} = \sigma^B(q) \frac{k'_\alpha}{k_\alpha} \frac{m_\beta k_\beta}{\hbar^2} |f_0(\mathbf{k}_\alpha - \mathbf{k}_\beta - \mathbf{k}'_\alpha)|^2. \tag{40}$$

It should be remembered that the various kinematic quantities in this formula are restricted by energy conservation. We also note that (40) has the dimension area $\times$ (energy)$^{-1}$, because f_0 has the dimension (length)$^{\frac{3}{2}}$.

Aside from σ^B, the dominant factor in determining the cross section is

$$|f_0(\mathbf{k}_\alpha - \mathbf{k}_\beta - \mathbf{k}'_\alpha)|^2 = \frac{1}{(2\pi)^3}\left|\int e^{-i(\mathbf{k}_\alpha - \mathbf{k}_\beta - \mathbf{k}'_\alpha)\cdot\mathbf{r}}\varphi_0(r)\, d^3r\right|^2. \quad (41)$$

This is just the probability for finding the momentum $\hbar(\mathbf{k}_\alpha - \mathbf{k}_\beta - \mathbf{k}'_\alpha)$ in the ground state. This probability is normalized to

$$\int d^3k\, |f_0(k)|^2 = 1. \quad (42)$$

In view of this it would perhaps be clearer to write

$$d\sigma^B = \sigma^B(q)\frac{k'_\alpha}{k_\alpha}\, d\Omega'_\alpha |f_0(\mathbf{k}_\alpha - \mathbf{k}_\beta - \mathbf{k}'_\alpha)|^2\, d^3k_\beta \quad (43)$$

instead of (40).

The appearance in the cross section of the ground state momentum distribution has a very simple intuitive interpretation. This interpretation is based on the approximation that underlies the derivation of (43), namely our neglect of the force due to the potential U on the knock-on particle β. This approximation is tantamount to assuming that α scatters off a free gas of β particles with the momentum distribution appropriate to the ground state, i.e., $|f_0(\mathbf{k})|^2$. The cross section would then be the product of essentially two factors: (a) the cross section for free space α-β scattering, and (b) the probability of finding a β particle in the target with a momentum $\hbar\mathbf{k}$ that satisfies the conservation law $\mathbf{k} + \mathbf{k}_\alpha = \mathbf{k}'_\alpha + \mathbf{k}_\beta$. This is precisely the form of (43); the kinematic factor k'_α/k_α is due to our definition of $\sigma^B(q)$ as the cross section for scattering off a β particle with $m_\beta = \infty$. If we had not neglected the influence of U on the emerging β particle, its final momentum would not have been uniquely correlated to $\mathbf{k}_\alpha$, $\mathbf{k}'_\alpha$, and its "initial" momentum in the ground state, because it could then scatter off U during and subsequent to the primary α-β collision. We can actually establish a rough criterion for the validity of (43) on the basis of this intuitive picture. What is clearly necessary (see Fig. 57.2) is that the momentum transferred to β by the α-β collision should greatly exceed the momentum imparted during the same time by the potential U. Let $\bar{F}$ be the average force due to U, i.e.,

$$\bar{F} = -\int |\varphi_0(r)|^2 \frac{\partial U}{\partial r}\, d^3r,$$

and τ the α-β collision time. Then we require $\tau|\bar{F}|$ to be much smaller than the momentum transferred to β, which is of order $\hbar k_\beta$ at high energies. The time τ is of order $m_\alpha r_V/\hbar k_\alpha$, where r_V is the range of V. Therefore we must have

$$\frac{|\bar{F}|m_\alpha r_V}{\hbar^2 k_\alpha k_\beta} \ll 1 \tag{44}$$

if the effect of U is to be negligible *during* the α-β collision. In addition, we must also make sure that U does not distort the outgoing β wave function appreciably, which means that the criterion concerning the Born approximation for the elastic scattering off U by a β particle having momentum $\hbar k_\beta$ must be satisfied as well.

The impulse approximation for the production process again consists of replacing σ^B in (43) by the exact α-β scattering cross section. The

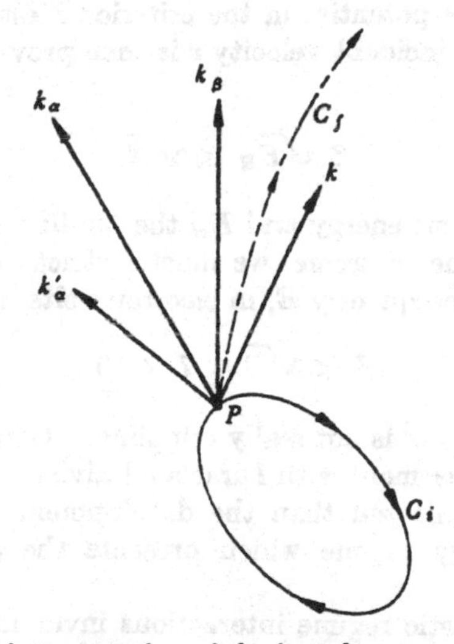

Fig. 57.2. A classical interpretation of the impulse approximation. Before the appearance of the impinging particle α, the target particle β follows the orbit C_i. If the projectile has a sufficiently high velocity $\hbar k_\alpha/m_\alpha$, the collision occurs at a well-defined instant at which time $\mathbf{r}_\beta(t)$ is at P, and $m_\beta\dot{\mathbf{r}}_\beta = \hbar\mathbf{k}$, with $\mathbf{k}$ being shown in the sketch. After the encounter, but before β has escaped from the induence of U, the two particles have the momenta $\hbar\mathbf{k}_\beta$ and $\hbar\mathbf{k}'_\alpha$, respectively. These momenta are related by $\mathbf{k} + \mathbf{k}_\alpha = \mathbf{k}'_\alpha + \mathbf{k}_\beta$, provided the collision time τ is sufficiently short to permit the neglect of the impulse $\tau\langle\partial U/\partial r_\beta\rangle$. After the collision the ejected particle follows the orbit C_f. This orbit can be approximated by the linear trajectory determined by the point P and the momentum $\hbar k_\beta$ if the conditions discussed in the text are satisfied, i.e., when C_f is nearly straight.

conditions involving the potential U discussed in the preceding paragraph must also be met if the impulse approximation is to be valid. The impulse approximation differs from the Golden Rule result (43) in that it does not require the α-β potential V to be weak. Since (44) and the Born condition for β scattering off U both improve with increasing energy, it is clear that the impulse approximation must become exact in the limit $k_\alpha \to \infty$, $k_\beta \to \infty$.

57.3 COLLISIONS OF FAST ELECTRONS WITH ATOMS

As a practical application, we consider the scattering of energetic electrons by complex atoms. By "fast" and "energetic" we mean that the Born approximation should be valid. The field to which the incident electron is subjected is strongest when it is well inside the K-shell, where the potential energy is Ze^2/r. Here Z is the atomic number and $\mathbf{r}$ the location of the incident electron relative to the nucleus. To be on the safe side, we use this potential in the criterion * established in Sec. 13. A lower limit on the incident velocity v is then provided by $Z \ll 137v/c$, or equivalently†

$$Z\sqrt{E_H} \ll \sqrt{E}, \tag{45}$$

where E is the incident energy and E_H the binding energy of hydrogen, 13.5 ev. At the other extreme, we must restrict ourselves to nonrelativistic electrons; consequently E, in electron volts, must satisfy

$$4Z \ll \sqrt{E} \ll 7 \times 10^2. \tag{46}$$

The restriction $E \ll mc^2$ is not really crippling. Once the Born approximation is made, a treatment with Dirac's relativistic theory is not fundamentally more complicated than the development that follows below. It is the low energy regime which presents the greatest theoretical difficulties.

In the nonrelativistic regime interactions involving the electron spin are small, and will therefore be neglected. The nuclear recoil is an even smaller correction, and we shall simply treat the nucleus as infinitely heavy. We shall also ignore the fact that the incident electron and those in the atom are indistinguishable. This is permissible as long as one

* We assume that the atom is neutral originally. The potential seen by the electron therefore falls off rapidly at large distances, and the difficulties associated with the infinite range of the Coulomb interaction do not arise.

† The Born approximation actually leads to considerably more accurate inelastic cross sections at low energies than one would surmise from this inequality. See Bethe.

scattered electron has most of the incident energy. This remark can be understood as follows. The correct scattering amplitude is the difference between two terms, each of which can be computed as if the electrons were distinguishable (recall Sec. 43.2). In one of these terms it is the incident electron that emerges with a great deal of energy, while either the electrons initially in the atom remain bound or else one of them is ejected into a free state where the kinetic energy is low compared to E. The other term in the antisymmetrized amplitude describes the process where one of the electrons initially in the atom emerges with high momentum, whereas the incident electron ends up in a state whose energy is small compared to E. The latter term—the exchange amplitude—is very small provided one of the electrons in the final state carries off the lion's share of the total energy.*

In using the Born approximation we must evaluate the matrix element

$$M_\nu(\mathbf{q}) = \frac{1}{\Omega}\int d^3r\, e^{i\mathbf{q}\cdot\mathbf{r}} \left\langle \nu \left| \sum_{j=1}^{Z} \frac{e^2}{|\mathbf{r}-\mathbf{x}_j|} - \int d^3r' \frac{Ze^2N(r')}{|\mathbf{r}-\mathbf{r}'|} \right| 0 \right\rangle, \tag{47}$$

where $|0\rangle$ is the ground state of the atom, $|\nu\rangle$ the state excited in the collision, $\mathbf{x}_j$ the position operator of the jth atomic electron, $\hbar\mathbf{q}$ the momentum transfer suffered by the incident electron, and $N(r)$ the nuclear charge distribution, normalized so that $\int N(r)\, d^3r = 1$. Here we have assumed that the nucleus is an inert object that is not excited by the electron. This is an excellent approximation, because we are concerned with incident energies that are small on the scale of nuclear energies, even though they are large by atomic standards. The reduction of (47) follows the same route as we took in going from (13) to (16). Using

$$\frac{1}{4\pi|\mathbf{x}-\mathbf{y}|} = \frac{1}{\Omega}\sum_{\mathbf{k}} \frac{e^{-i\mathbf{k}\cdot(\mathbf{x}-\mathbf{y})}}{k^2},$$

we obtain

$$M_\nu(\mathbf{q}) = \frac{4\pi Ze^2}{\Omega q^2}\left[F_\nu(\mathbf{q}) - \delta_{\nu 0}F_N(q)\right]. \tag{48}$$

The second term, which only contributes to elastic scattering, is due to the nucleus. The quantity

$$F_N(q) = \int e^{i\mathbf{q}\cdot\mathbf{r}} N(r)\, d^3r \tag{49}$$

is the nuclear form factor. The first term in (48) is the electronic form

* For more detailed discussion, see Prob. 2, and especially Drukarev, p. 73.

factor:

$$ZF_\nu(\mathbf{q}) = \left\langle \nu \middle| \sum_{j=1}^{Z} e^{i\mathbf{q}\cdot\mathbf{x}_j} \middle| 0 \right\rangle. \tag{50}$$

Henceforth we shall call $F_N(q) - F_0(q)$ the elastic form factor of the atom.

Let us first consider elastic scattering. For sufficiently small momentum transfers we can use the expansion (29) for the form factors. Thus

$$F_N(q) \simeq 1 - \tfrac{1}{6}q^2R_N^2 \qquad (q \ll R_N^{-1}),$$

where R_N is the r.m.s. radius of the nucleus, which is of order $Z^{\frac{1}{3}} \times 10^{-13}$ cm. The elastic form factor of the electronic charge distribution behaves like

$$F_0(q) \simeq 1 - \tfrac{1}{6}q^2R_e^2 \qquad (q \ll R_e^{-1}),$$

where

$$R_e^2 = \frac{1}{Z}\sum_{j=1}^{Z} \langle 0|\mathbf{x}_j^2|0\rangle.$$

R_e, the r.m.s. radius of the electronic charge distribution, is of order $Z^{-\frac{1}{3}}a_0$, where $a_0 = 0.528 \times 10^{-8}$ cm is the Bohr radius. Near the forward direction the elastic amplitude will therefore be a constant,

$$M_0(\mathbf{q}) \underset{q\to 0}{\longrightarrow} -\frac{4\pi Ze^2}{6\Omega} R_e^2.$$

When the momentum transfer grows beyond $1/R_e$, the electronic form factor decreases rapidly.* When q lies in the interval

$$R_e^{-1} \ll q \ll R_N^{-1},$$

we can make the approximations $F_0(q) \simeq 0$, $F_N(q) \simeq 1$. Under these circumstances the elastic cross section is given accurately by the Rutherford formula for scattering off a point nucleus with charge Ze. Once q approaches R_N^{-1}, the cross section will fall below the Rutherford formula because the deviation of the nuclear form factor from unity is then noticeable. The finite extent of the nuclear charge distribution cannot be revealed by a nonrelativistic electron, however, and strictly speaking we should put $F_N(q) = 1$ throughout. But it turns out that the relativistic theory leads to an atomic elastic form factor which is essentially the same as the one we have found here.† Consequently Fig. 57.3, which

* For example, in hydrogen $F_0(q) = (1 + \frac{1}{4}q^2a_0^2)^{-2}$.

† Cf., e.g., Hofstadter, *loc. cit.* When the momentum transfer $\hbar q$ exceeds Mc, where M is the mass of the nucleon, there are also important contributions due to scattering by the nucleon's magnetic moment.

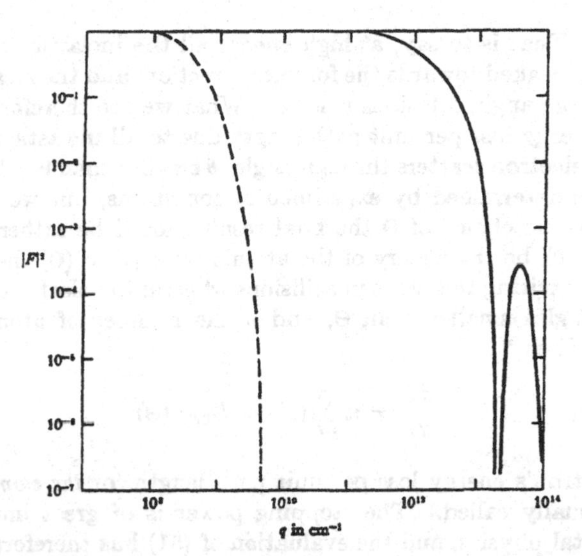

Fig. 57.3. Elastic form factors for carbon. The dashed line is the square of the electronic form factor F_0, and the full line the square of the nuclear form factor F_N. The latter has a diffraction minimum because the nuclear charge distribution has a fairly well-defined edge. The electronic charge distribution is rather diffuse, and F_0 does not display diffraction minima. F_N has been taken from Hofstadter (*loc. cit.*), Figs. 2 and 4.

is based on our present considerations, displays the most important features of elastic electron scattering as it is actually observed.

57.4 STOPPING POWER

We now turn to inelastic atomic scattering. This is a vast subject,* and we shall merely touch on one aspect of it. An electron traversing a substance loses energy by exciting the atoms it passes. It may also slow down by emitting light (*bremsstrahlung*), but this is only a significant energy loss mechanism for extremely relativistic electrons. Consequently the range and energy loss per unit path length of a nonrelativistic electron traveling through matter are determined by the inelastic atomic cross sections.

When a fast electron traverses a bubble or cloud chamber, or a photographic emulsion, it usually follows a rather well-defined trajectory because the overwhelming majority of collisions involve small momentum

* Cf., e.g., H. B. W. Massey, *Encyclopedia of Physics* **36** (Springer, Berlin, 1956); Massey and Burhop; Mott and Massey; Drukarev.

transfers. That is to say, at high energy all the inelastic cross sections are sharply peaked towards the forward direction, and the mean free path between large angle collisions is long. What we are therefore interested in is the energy loss per unit path length due to all inelastic processes in which the electron scatters through angles θ smaller than Θ. To a certain extent Θ is determined by experimental conditions, but we expect that with a sensible choice of Θ the final result should be rather insensitive to it. Let E_ν be the energy of the atomic state $|\nu\rangle$, $\sigma_\nu(\Theta)$ the total cross section for exciting this state in collisions wherein the electron is scattered through angles smaller than Θ, and n the number of atoms per unit volume. Then *

$$\frac{dE}{dx} = n \sum_\nu (E_\nu - E_0)\sigma_\nu(\Theta) \tag{51}$$

is the electron's energy loss per unit path length, or *the stopping power*, as it is usually called. The stopping power is of great importance in experimental physics, and the evaluation of (51) has therefore received a good deal of attention throughout the history of atomic and nuclear physics.†

Let us first obtain an expression for $\sigma_\nu(\Theta)$. The derivation of a differential cross section from $M_\nu(\mathbf{q})$ has already been given in Sec. 57.1, and we therefore quote the result without proof:

$$\frac{d\sigma_\nu}{d\Omega'} = 4Z^2a_0^2\left(\frac{k'_\nu}{k}\right)(q_\nu a_0)^{-4}|F_\nu(\mathbf{q}_\nu)|^2. \tag{52}$$

Here we have placed the suffix ν on k' and $\mathbf{q}$ to remind us that the momentum of the scattered electron, and therefore the momentum transfer, depends on the excitation energy. The cross section $\sigma_\nu(\Theta)$ is then given by

$$\sigma_\nu(\Theta) = \int_0^{2\pi} d\phi \int_0^\Theta \sin\theta\, d\theta\, (d\sigma_\nu/d\Omega'). \tag{53}$$

For the purpose of evaluating dE/dx, it is best to use $k_\nu k'_\nu \sin\theta\, d\theta = q_\nu\, dq_\nu$ to convert (53) into an integral over momentum transfer.

* This formula assumes that the amplitudes for scattering off different atoms add incoherently. The validity of this assumption is discussed in Jackson, Sec. 13.4.

† The quantum theory of energy loss is due to Bethe (1930). For more recent developments see J. Ashkin and H. A. Bethe in *Experimental Nuclear Physics*, Vol. I, edited by E. Segrè (Wiley, New York, 1953); A. Dalgarno in *Atomic and Molecular Processes*, edited by D. R. Bates (Academic Press, New York, 1962); G. Knop and W. Paul in *Alpha-, Beta-, and Gamma-Ray Spectroscopy*, edited by K. Siegbahn (North Holland, Amsterdam, 1965). A very informative discussion can be found in N. Bohr, *Det. Kgl. dansk. Vid. Selskab* **18**, No. 8 (1948).

The formula for the stopping power then reads *

$$\frac{dE}{dx} = \frac{8\pi n}{k^2 a_0^2} \sum_\nu (E_\nu - E_0) \int_{Q_\nu}^{\bar{q}_\nu} |\langle 0|\rho_{\mathbf{q}}|\nu\rangle|^2 \frac{dq}{q^3}, \tag{54}$$

where the limits Q_ν and $\bar{q}_\nu$ are determined by setting $\theta = 0$ and $\theta = \Theta$ in (23), respectively, and the definition of $\rho_{\mathbf{q}}$ follows from comparing (52) and (50):

$$\rho_{\mathbf{q}} = \sum_{j=1}^{Z} e^{-i\mathbf{q}\cdot\mathbf{x}_j}. \tag{55}$$

As we see, $\rho_{\mathbf{q}}$ is the operator that describes the Fourier decomposition of the electronic charge density.

The excitation spectrum of an atom is enormously complex, and when $E_\nu - E_0$ exceeds the ionization potential, it is infinitely degenerate. Hence a brute-force evaluation of the sum over excited states must be avoided at all costs. Following Bethe, we shall exploit the identity

$$\langle[\mathcal{O}, H]\mathcal{O}^\dagger\rangle_0 = \sum_\nu (E_\nu - E_0)|\langle 0|\mathcal{O}|\nu\rangle|^2, \tag{56}$$

where H is the Hamiltonian of the atom, $\langle\cdot\ \cdot\ \cdot\rangle_0$ the expectation value in $|0\rangle$, and $\mathcal{O}$ any operator that does not depend on the index ν. (To prove (56), one merely inserts the complete set $\{|\nu\rangle\}$ between $[\mathcal{O}, H]$ and $\mathcal{O}^\dagger$.) If it were permissible to interchange the summation and integration in (54), we could use (56) and thereby evaluate dE/dx. Unfortunately the limits of the q-integration depend on ν. The evaluation of (54) is therefore not quite so straightforward, and some approximations must be made.

At the end of this section we shall show that for those transitions that contribute significantly to the stopping power, the limits of integration satisfy

$$\bar{q}_\nu \simeq 2k \sin \tfrac{1}{2}\Theta \equiv q_{max}, \tag{57}$$

and

$$Q_\nu \ll R_\nu^{-1}, \tag{58}$$

provided the Born approximation criterion (45) is satisfied. Here R_ν is the characteristic radius of the charge distribution due to those atomic electrons that given the dominant contribution to transitions with

* Here we assume that the ground state $|0\rangle$ is an S-state, in which case the sum over all excited states belonging to one energy eigenvalue E_ν produces a contribution to (54) that depends only on $|\mathbf{q}|$. When the atom has nonzero angular momentum, the formula for dE/dx also contains an average over the initial magnetic substates of the target, and the dependence on the orientation of $\mathbf{q}$ again disappears.

excitation energy $E_\nu - E_0$. Equations (57) and (58) allow us to simplify the expression for dE/dx. First we observe that q_{max} is independent of ν. We therefore break the integral in (54) into two pieces: $Q_\nu < q < q_0$ and $q_0 < q < q_{max}$, where q_0 is also independent of ν, and $q_0 R_\nu$ small compared to unity. Our identity (56) can be used in the large-q integral. In the integral over small q's we can, because of (58), approximate $\rho_{\mathbf{q}}$ by $\Sigma_j(1 - i\mathbf{q}\cdot\mathbf{x}_j)$. We must then evaluate

$$\frac{dE}{dx} = \frac{8\pi n}{k^2 a_0^2}(J_1 + J_2), \tag{59}$$

where

$$J_1 = \int_{q_0}^{q_{max}} \langle[\rho_{\mathbf{q}}, H]\rho_{-\mathbf{q}}\rangle_0\, q^{-3}\, dq, \tag{60}$$

and

$$J_2 = \sum_\nu (E_\nu - E_0) \int_{Q_\nu}^{q_0} |\langle 0|\mathbf{q}\cdot\mathbf{D}|\nu\rangle|^2 q^{-3}\, dq. \tag{61}$$

Here $e\mathbf{D} \equiv e\Sigma_j\mathbf{x}_j$ is the dipole moment operator of the electrons in the target atom. The spherical symmetry * of the atom's ground state allows us to simplify these expressions for J_1 and J_2. The expectation value in (60) is unaffected by the substitution $\mathbf{q} \to -\mathbf{q}$, and is also real. Hence

$$J_1 = \tfrac{1}{2}\int_{q_0}^{q_{max}} \langle[[\rho_{\mathbf{q}}, H], \rho_{-\mathbf{q}}]\rangle_0\, q^{-3}\, dq. \tag{62}$$

By the same token, (61) cannot depend on the orientation of $\mathbf{q}$. We may therefore choose the z-direction along $\mathbf{q}$, and write

$$J_2 = \sum_\nu (E_\nu - E_0)|\langle 0|D_z|\nu\rangle|^2 \ln(q_0/Q_\nu) \tag{63}$$

instead of (61).

The evaluation of J_1 can be carried out in closed form because $\rho_{\mathbf{q}}$ commutes with the Coulomb interaction. With the help of (6.12) we find

$$\begin{aligned}[\rho_{\mathbf{q}}, H] &= \frac{i\hbar}{2m}\sum_{j=1}^{Z}\left(\mathbf{p}_j\cdot\frac{\partial\rho_{\mathbf{q}}}{\partial\mathbf{x}_j} + \frac{\partial\rho_{\mathbf{q}}}{\partial\mathbf{x}_j}\cdot\mathbf{p}_j\right)\\ &= \frac{\hbar\mathbf{q}}{2m}\cdot\sum_j(\mathbf{p}_j e^{-i\mathbf{q}\cdot\mathbf{x}_j} + e^{-i\mathbf{q}\cdot\mathbf{x}_j}\mathbf{p}_j),\end{aligned}$$

and therefore

$$[[\rho_{\mathbf{q}}, H], \rho_{-\mathbf{q}}] = \hbar^2 Z q^2/m. \tag{64}$$

The integration in (62) is therefore elementary:

$$J_1 = (\hbar^2 Z/2m)\ln(q_{max}/q_0). \tag{65}$$

* Our results do not depend on this assumption provided an average over the initial states is carried out. Equation (62) also follows from time reversal.

Next we turn to J_2. The quantity Q_ν is given by (23′) with $\theta = 0$: $Q_\nu = \frac{1}{2}\gamma^2/k = (E_\nu - E_0)/\hbar v$, where v is the incident velocity. Thus

$$J_2 = \sum_\nu (E_\nu - E_0)|\langle 0|D_z|\nu\rangle|^2 \ln\left(\frac{\hbar v q_0}{E_\nu - E_0}\right).$$

Upon inserting I/I into the last logarithm, we obtain

$$J_2 = \tfrac{1}{2}\langle[[D_z, H], D_z]\rangle_0 \ln(\hbar v q_0/I) + \sum_\nu (E_\nu - E_0)|\langle 0|D_z|\nu\rangle|^2 \ln[I/(E_\nu - E_0)].$$

The double commutator is a special case of (64); its value is $\hbar^2 Z/m$. Combining our results for J_1 and J_2, and substituting into (59), we have

$$\frac{dE}{dx} = \frac{8\pi n}{k^2 a_0^2}\left\{\frac{\hbar^2 Z}{2m}\ln\left(\frac{\hbar v q_{\max}}{I}\right) + \sum_\nu (E_\nu - E_0)|\langle 0|D_z|\nu\rangle|^2 \ln[I/(E_\nu - E_0)]\right\}. \tag{66}$$

It is very comforting that q_0, the magnitude of $\mathbf{q}$ below which we approximated $\rho_\mathbf{q}$ by $Z - i\mathbf{q}\cdot\mathbf{D}$, has vanished from this formula. This disappearance of q_0 is not accidental; it occurs because the value of the double commutator (64) does not change when one makes the dipole approximation for $\rho_\mathbf{q}$. The fact that I is arbitrary and therefore at our disposal allows us to carry out a formal simplification by requiring the last term in (66) to vanish, i.e., by defining I as

$$\ln I = (2m/\hbar^2 Z)\sum_\nu (E_\nu - E_0)|\langle 0|D_z|\nu\rangle|^2 \ln(E_\nu - E_0). \tag{67}$$

Our final result is therefore *

$$\frac{dE}{dx} = \frac{4\pi n Z\hbar^2}{k^2 a_0^2 m}\ln\left(\frac{\hbar v q_{\max}}{I}\right) = \frac{4\pi n Z e^4}{m v^2}\ln\left(\frac{2mv^2 \sin\frac{1}{2}\Theta}{I}\right). \tag{68}$$

Let us be quite clear that this expression for dE/dx represents a great deal of progress beyond (54). In the latter equation the result of the sum over states is an unknown *function* of both the incident velocity v and the maximum scattering angle Θ. Our new expression contains only one unknown, the velocity-independent *parameter* I. The whole complexity of the atom's excitation spectrum has been buried in I, and (67)

* Observe that $\hbar$ only appears implicitly in this expression (through the definition of I). A classical derivation of (68), in terms of a slightly different parameter I, was first given by Bohr in 1915. For a comparison of the classical and quantum theories, see Bohr, *loc. cit.*, and Jackson, Chapt. 13.

even gives us a formula with which we can, in principle, compute it theoretically. In practice I is usually determined experimentally.

We must still investigate the circumstances under which the integration limits $\bar{q}_\nu$ and Q_ν satisfy (57) and (58). For this purpose we must estimate the maximum excitation energy as a function of momentum transfer. A simple model of the atom will suffice for this estimate. We suppose that the electrons in the target are dynamically independent, i.e., that the atomic wave function is a product of Z one-electron wave functions. In the Born approximation the collision amplitude is proportional to the matrix element of $\rho_{\mathbf{q}}$. As we see from (55), each term in $\rho_{\mathbf{q}}$ involves the coordinates of only a *single* target electron. If the atomic wave functions are of the product type, only *one* target electron changes its state during the collision, and the excitation energy of the atom is, within these approximations, equal to the change of energy of this one electron.* This one-electron excitation energy will be largest when a K-shell electron is ejected into the continuum during the course of the collision.

We shall analyze the validity of Eqs. (57) and (58) for transitions originating in the K-shell. The reader can easily extend the argument to the more loosely bound shells. In this case R_ν is the radius of the K-shell, that is, $R_\nu \sim a_0 Z^{-1}$. The initial momentum of the ejected electron is therefore at most of order $Z\hbar/a_0$. In a collision of momentum transfer $\hbar q$ the maximum final momentum of the ejected electron is thus $\sim\hbar(q + Za_0^{-1})$, or

$$(E_\nu - E_0)_{\max} \sim Z^2 E_H + \frac{\hbar^2}{2m}(q + Za_0^{-1})^2. \tag{69}$$

We note that $\gamma \ll k$ when (45) is satisfied, where γ is defined in (22) *et seq.* The maximum value of q as a function of scattering angle is then obtained by substituting (69) into (23′):

$$q_{\max}(\theta) \simeq k\left[\theta^2 + \left(\frac{(E_\nu - E_0)_{\max}}{2E}\right)^2\right]^{\frac{1}{2}}, \tag{70}$$

where $(E_\nu - E_0)_{\max}$ is to be eliminated in favor of $q_{\max}(\theta)$ by means of (69), and we have assumed that $\theta^2 \ll 1$.

Consider first $\theta = 0$. The derivation of Bethe's formula (68) presupposed that (58) is satisfied, where $Q_\nu \lesssim q_{\max}(0)$. We may now use (69) and (70) to find the values of Z for which (58) holds. When $q_{\max}(0)R_\nu \ll 1$, $q_{\max}(0) \ll Za_0^{-1}$, because $R_\nu \sim Z^{-1}a_0$. Hence (69) re-

* As we shall see in Chapt. XI, a rather accurate theory of atomic spectra can be based on product-type wave functions. The argument that follows should therefore be sufficiently reliable for our purpose.

duces to $(E_\nu - E_0)_{max} \sim 2Z^2E_H$, or $q_{max}(0)R_\nu \sim Z(E_H/E)^{\frac{1}{2}}$. Thus (58) is satisfied when the Born criterion (45) is met. The expansion of ρ_q in J_2 is therefore legitimate when the Born approximation is valid.

If $q_{max}(\Theta)$ is to equal q_{max} as defined by (57), we must have $\Theta \gg (E_\nu - E_0)_{max}/2E$. But $(E_\nu - E_0)_{max}/2E$ is of order Z^2E_H/E, which, by hypothesis, is always small compared to unity [cf. (45)]. It therefore follows that (57) is valid provided Θ is not too small (this does not preclude $\Theta \ll 1$).

The energy loss suffered by heavy charged particles, such as mesons, protons, or alphas, is also of great interest. The evaluation of dE/dx for heavy particles only requires a few minor modifications. The approximations we have made (especially the small-q expansion of ρ_q) are actually much more accurate in this case because a nuclear particle can have an energy that is immense on the atomic scale and still be thoroughly nonrelativistic. Thus for protons (46) is to be replaced by $4Z \ll \sqrt{E} \ll 3 \times 10^4$. The derivation of dE/dx is sketched in Prob. 3, and the result is found to be

$$\frac{dE}{dx} = \frac{4\pi n Z Z'^2 e^4}{mv^2} \ln\left(\frac{mv^2\Theta}{I}\right), \tag{71}$$

where Z' is the charge of the heavy particle. For all practical purposes one should replace Θ by the maximum scattering angle $2m/M$ in this formula; once this is done, dE/dx has the remarkable property of being independent of the mass of the projectile. By measuring the stopping power one can therefore determine the velocity of the particle in question.

Equation (71) gives a rather accurate description of the data on energy loss of heavy particles; the empirical values of I are roughly $13Z$ ev for the lighter elements.* This is to be compared with the theoretical value of 14.9 ev for hydrogen found by Bethe by evaluating (67).

58. The Photo-Effect in Hydrogen

The photo-effect in hydrogen constitutes a very interesting example of a perturbation problem. In this process a photon of momentum $\hbar\mathbf{k}$ impinges on the hydrogen ground state and ejects the electron into a continuum state of momentum $\hbar\mathbf{k}'$. Because we only have a nonrelativistic

* For detailed comparisons between theory and experiment see the references quoted on p. 458.

theory of the electron, we must confine ourselves to photon energies $\hbar\omega$, small compared to mc^2. Under these circumstances we can ignore the electron spin, and treat the proton as if its mass M were infinite.

Instead of plunging directly into the calculation, we first discuss some of the qualitative features of the phenomenon. It is not difficult to see that these features change markedly with the photon energy. The lowest energy at which the reaction is possible is called the threshold energy. To within terms of order E_0/Mc^2, the threshold occurs at $\hbar\omega = E_0$, where $-E_0$ is the binding energy of the hydrogen ground state. We must now distinguish between two regimes:

$$(\hbar\omega - E_0) \lesssim E_0, \tag{1}$$

and

$$\hbar\omega \gg E_0. \tag{2}$$

In regime (1) the kinetic and potential energies of the ejected electron are of comparable magnitude, and it is therefore not permissible to ignore the influence of the Coulomb interaction on the final state. On the other hand, when the inequality (2) is satisfied, the electron in the final state can be treated as approximately free. Hence a straightforward Born approximation calculation will do in the high-energy regime, but we must do better near threshold. Fortunately a familiar simplification occurs at low energy: The interaction is confined to the lowest partial wave permitted by angular momentum conservation. The binding energy is $\frac{1}{2}\alpha^2mc^2$; near threshold the photon wavelength $\lambda\!\!\!^{-}$ is therefore of order a_0/α, where $1/\alpha = 137$, i.e., immense compared to the size of the atom. Hence for incident energies small compared to E_0/α, only states having a total angular momentum $j = 1$ will interact appreciably. (We recall from Sec. 52.3 that one-photon states with zero angular momentum do not exist. Consequently the lowest total angular momentum is $j = 1$, because the target is in an s-state.)

It is now clear how one must set about doing these calculations. In the high-energy regime we can approximate the final electron state by a plane wave. As in (57.43), we expect the angular distribution at high energy to be proportional to the momentum distribution in the hydrogen ground state. Near threshold we shall have to describe the final state with the p-wave Coulomb continuum eigenfunctions derived in Sec. 17. There is no problem in joining these two calculations together, because the threshold calculation should work up to $\hbar\omega < E_0/\alpha$, and the plane wave approximation for the final state is also valid in this intermediate energy range.

The interaction H' that describes the coupling of the electromagnetic field to the electron is given by the **A**-dependent portion of (53.27).

There are two terms, one linear and the other quadratic in $\mathbf{A}$. The linear term has a nonzero matrix element between the incident one-photon state and the vacuum, but the quadratic term does not. Hence the $\mathbf{A}^2$-term only contributes to the photo-effect if H' is treated to second (or higher) order. But in the following paragraph we shall show that the linear part of H' is a small perturbation, and we may therefore delete the quadratic term completely.* Consequently the perturbation of concern to us is

$$H' = -\frac{e}{mc}\mathbf{p}\cdot\mathbf{A}(\mathbf{x}), \tag{3}$$

where $\mathbf{p}$ and $\mathbf{x}$ are the momentum and position operators of the electron. [In writing (3) we have used the fact that $\nabla\cdot\mathbf{A} = 0$.]

Let us estimate the magnitude of H'. For a photon of wavelength $\lambda\!\!\!^{-}$, the electric field $\mathcal{E}$ can be estimated from the energy density, $\hbar\omega \sim \lambda\!\!\!^{-3}\mathcal{E}^2$. Thus $|\mathbf{A}|$ is of order $c\mathcal{E}/\omega \sim \omega\sqrt{\hbar/c}$. At high energy the final state is a momentum eigenstate, and $|\mathbf{p}|$ in (3) can then be replaced by $\sqrt{2mW}$, where $W = \hbar\omega - E_0$ is the excitation energy. As we are only interested in an order-of-magnitude estimate at the moment, we shall also use this value of $|\mathbf{p}|$ in the low energy regime. Thus we estimate that the magnitude of H' is given by

$$\frac{H'}{W} \sim \alpha^{\frac{1}{2}}\sqrt{2\frac{\hbar\omega}{W}\frac{\hbar\omega}{mc^2}}. \tag{4}$$

The characteristic excitation energies in the process are of order W, and H' may therefore be treated in first order throughout. Hence in using the Golden Rule we need only evaluate matrix elements of H' between eigenstates of the unperturbed Hamiltonian $H_\gamma + H_H$, where H_γ and H_H are the Hamiltonians of the free electromagnetic field and the hydrogen atom, respectively.

58.1 HIGH ENERGIES

It is obvious that the high-energy regime (2) is the easier one to deal with. Here our only task is the evaluation of

$$\langle\mathbf{k}'; 0|H'|G; 1_{\mathbf{k}\lambda}\rangle, \tag{5}$$

where the ket and bra are both products of one-electron states and eigenstates of H_γ. Thus $|G; 1_{\mathbf{k}\lambda}\rangle = a^\dagger_{\mathbf{k}\lambda}|G; 0\rangle$ represents the electron in the hydrogen ground state and one photon of momentum $\hbar\mathbf{k}$ and helicity λ, while $|\mathbf{k}'; 0\rangle$ is simultaneously the electromagnetic vacuum state and a

* An order-of-magnitude estimate of the $\mathbf{A}^2$-term is given in Sec. 61.1.

one-electron state of momentum $\hbar\mathbf{k}'$. To evaluate (5) we insert the Fourier expansion (52.12) of the vector potential into the matrix element. Only the destruction operators in $\mathbf{A}$ can contribute to (5). If we then use (52.34) in (52.12), and $\langle 0|a_{\mathbf{k}'\lambda'}a^\dagger_{\mathbf{k}\lambda}|0\rangle = \delta_{\mathbf{k}\mathbf{k}'}\delta_{\lambda\lambda'}$, we obtain

$$\begin{aligned}\langle \mathbf{k}'; 0|H'|G; 1_{\mathbf{k}\lambda}\rangle &= -\sqrt{\frac{hc^2}{\omega\Omega}}\,\frac{e}{mc}\langle \mathbf{k}'|\mathbf{e}_{\mathbf{k}\lambda}\cdot\mathbf{p}e^{i\mathbf{k}\cdot\mathbf{r}}|G\rangle \\ &= -\sqrt{\frac{hc^2}{\omega\Omega}}\,\frac{e\hbar}{mc}\sqrt{\frac{(2\pi)^3}{\Omega}}\,(\mathbf{e}_{\mathbf{k}\lambda}\cdot\mathbf{k}')f_G(\mathbf{k}-\mathbf{k}'), \end{aligned} \tag{6}$$

where

$$f_G(q) = (2\pi)^{-\frac{3}{2}}\int e^{i\mathbf{q}\cdot\mathbf{r}}\langle \mathbf{r}|G\rangle\, d^3r$$

is the momentum-space wave function of the hydrogen ground state, and $\langle \mathbf{r}|G\rangle = (\pi a_0^3)^{-\frac{1}{2}}e^{-r/a_0}$. An elementary integration yields

$$f_G(q) = \frac{2\sqrt{2}}{\pi}\,\frac{a_0^{\frac{3}{2}}}{(1+q^2a_0^2)^2}. \tag{7}$$

The differential cross section is obtained from (6) by substituting into the Golden Rule. The density of final states is given by (57.20), while the incident flux of photons is c/Ω. Hence

$$\frac{d\sigma_\lambda}{d\Omega_e} = \frac{4\pi^2e^2k'}{\omega cm}\,|(\mathbf{e}_{\mathbf{k}\lambda}\cdot\mathbf{k}')f_G(\mathbf{k}-\mathbf{k}')|^2, \tag{8}$$

where $d\Omega_e$ is the element of solid angle for the photo-electron. That this result actually has the dimension of an area can be seen by using (7), which leads to

$$\frac{d\sigma_\lambda}{d\Omega_e} = 64\alpha a_0^2\left(\frac{E_0}{\hbar\omega}\right)\frac{k'|\mathbf{e}_{\mathbf{k}\lambda}\cdot\mathbf{k}'|^2a_0^3}{[1+(\mathbf{k}-\mathbf{k}')^2a_0^2]^4}. \tag{9}$$

The vectors $\mathbf{k}$ and $\mathbf{k}'$ are not independent because energy conservation requires

$$\hbar ck = E_0 + (\hbar k')^2/2m. \tag{10}$$

Well above threshold we can ignore E_0 in (10) and use the approximate relationship

$$k' \simeq k\sqrt{2mc/\hbar k}. \tag{11}$$

This equation shows that the momentum of the photo-electron is much larger than the momentum of the incident photon in the non-relativistic high-energy regime. Equation (11) may also be written as $k'a_0 \simeq \sqrt{\hbar\omega/E_0}$.

The angular distribution is determined by the expression

$$\frac{|\mathbf{e}_{\mathbf{k}\lambda} \cdot \mathbf{k}'|^2}{[1 + (\mathbf{k} - \mathbf{k}')^2 a_0^2]^4}. \tag{12}$$

As we have just seen, $[1 + (\mathbf{k} - \mathbf{k}')^2 a_0^2]^{-4}$ is proportional to the momentum distribution in the ground state of hydrogen evaluated at the momentum determined by the free space momentum conservation law. The appearance of the momentum distribution has precisely the

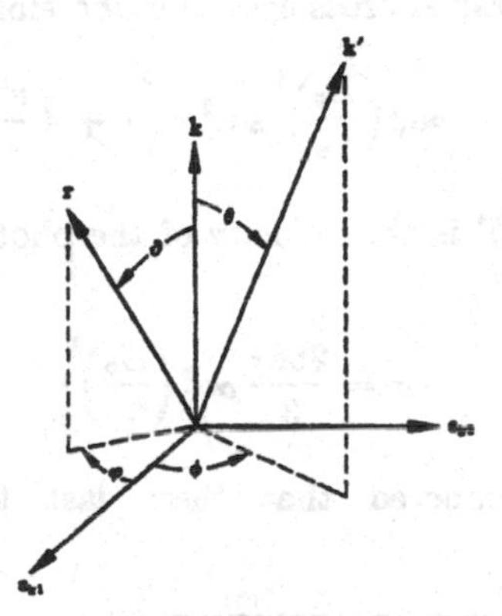

Fig. 58.1. Polarization and propagation vectors in the photo-effect. The vector r and the associated angles only enter into the threshold calculation of Sec. 58.2.

same explanation here as in (57.43) *et seq.* The numerator in (12) arises from the vector coupling $\mathbf{A} \cdot \mathbf{p}$ characteristic of electromagnetic interactions, and has no counterpart in the scalar interactions discussed in Sec. 57. We recall that $\mathbf{A}$ is a transverse field, i.e., $\mathbf{e}_{\mathbf{k}\lambda}$ is perpendicular to $\mathbf{k}$. As a consequence the differential cross section *vanishes* in the forward direction, where $\mathbf{k}'$ and $\mathbf{k}$ are parallel. This fact can also be understood from angular momentum conservation. In the initial state the *total* angular momentum along $\mathbf{k}$ is ± 1. If we expand the final electron state in terms of spherical harmonics $Y_{lm}(\theta, \phi)$ with $\mathbf{k}$ as the z-axis, then m is necessarily ± 1. (The angles θ and ϕ are defined in Fig. 58.1.) But $Y_{l,\pm 1}(0, \phi) = 0$, and the vanishing of the angular distribution in the forward direction is therefore a general property of the photo-effect cross section, and does not depend on the plane wave approximation for the final state.*

For most purposes one only needs the cross section averaged over both helicity states. We then require $\frac{1}{2}\Sigma_\lambda|\mathbf{e}_{\mathbf{k}\lambda}\cdot\mathbf{k}'|^2 = \frac{1}{2}\Sigma_\alpha(\boldsymbol{\varepsilon}_{\mathbf{k}\alpha}\cdot\mathbf{k}')^2$, where we have switched to the linear polarization vectors [cf. (52.31)]. A glance at Fig. 58.1 shows that this last sum is just $\frac{1}{2}k'^2\sin^2\theta$. As we already noted after Eq. (11), at the energy at which the present calculation is valid $k'a_0 \gg 1$ and $k' \gg k$, and the denominator in (12) can therefore be approximated by $k'^2a_0^2[1 - 2(k/k')\cos\theta] \simeq (\hbar\omega/E_0)[1 - \alpha(\hbar\omega/E_0)^{\frac{1}{2}}\cos\theta]$. The polarization-averaged cross section is therefore †

$$\frac{d\sigma}{d\Omega_e} = 32\alpha a_0^2\left(\frac{E_0}{\hbar\omega}\right)^{\frac{7}{2}}\sin^2\theta\left(1 + 4\frac{v'}{c}\cos\theta\right), \tag{13}$$

where $v' = \alpha c(\hbar\omega/E_0)^{\frac{1}{2}}$ is the velocity of the photo-electron. The total cross section is

$$\sigma = \frac{256\pi}{3}\alpha a_0^2\left(\frac{E_0}{\hbar\omega}\right)^{\frac{7}{2}}. \tag{14}$$

It should be remembered that these last two formulas require $E_0 \ll \hbar\omega \ll mc^2$.

58.2 THE CROSS SECTION NEAR THRESHOLD

When the photon energy $\hbar\omega$ is comparable to the binding energy E_0, we require a formula for the transition amplitude that is exact insofar as the Coulomb interaction is concerned. At first sight it would seem that there is no difficulty in using the Golden Rule for this purpose: All that we need is the matrix element of the perturbing interaction H' between $|G; 1_{\mathbf{k}\lambda}\rangle$ and a state with no photons and one electron in a Coulomb scattering state. While this statement is perfectly correct, it is somewhat ambiguous because it does not tell us precisely which Coulomb scattering state to use. Clearly it is a state characterized by an incident wave of momentum $\hbar\mathbf{k}'$, but the boundary conditions which are to be imposed on this state are not self-evident.

The correct boundary condition can be inferred from an intuitive

* If the electron spin and its coupling to the radiation field are taken into account, $d\sigma/d\Omega_e$ does not vanish in the forward direction because angular momentum can be conserved by means of a spin-flip. The spin-flip cross section is smaller than (9) by an over-all factor of $\hbar\omega/mc^2$.

† In comparing this result with Bethe and Salpeter, Eq. 70.5, one might conclude that there is a discrepancy of a factor of 2. This is not so: our cross section is an average over both helicity states, whereas that of Bethe and Salpeter is not.

argument.* In the photo-effect the counter selects an electron which has the momentum $\hbar\mathbf{k}'$ *after* it has emerged from the atom. We recall from Eq. (56.8) *et seq.* that the collision states $|a^{(+)}\rangle$, with outgoing wave boundary conditions, are an idealized description of the situation where the particle is in a plane wave state *before* the collision. It is therefore clear that states of this type are not appropriate final states in the photo-effect. If we return to the argument starting with (56.3), and seek a solution of the Schrödinger equation that tends to a free packet in the *future* instead of the past, we must incorporate the step function $\theta(t' - t)$, instead of $\theta(t - t')$, into the Green's function. The stationary states corresponding to this "final" (as compared to initial) condition satisfy a Lippmann-Schwinger equation with $+i\epsilon$ replaced by $-i\epsilon$, a fact which we already established with the help of time reversal in Sec. 39.2. We recall that the solutions of this time-reversed Lippmann-Schwinger equation possess *incoming*, not outgoing, scattered waves. On the basis of this argument we conclude that the transition amplitude for the photo-effect is

$$\langle \mathbf{k}'^{(-)}; 0|H'|G; 1_{\mathbf{k}\lambda}\rangle \equiv A(\mathbf{k}'; \mathbf{k}\lambda), \tag{15}$$

where the electron scattering state $|\mathbf{k}'^{(-)}\rangle$ satisfies

$$|\mathbf{k}'^{(-)}\rangle = |\mathbf{k}'\rangle + \frac{1}{E' - K - i\epsilon} V|\mathbf{k}'^{(-)}\rangle. \tag{16}$$

Here K is the electron's kinetic energy, V the Coulomb interaction, and $E' = (\hbar\mathbf{k}')^2/2m$. When the Coulomb potential is ignored in the final state, (15) reduces to (5).

The electromagnetic portion of (15) is evaluated as in the first step of (6):

$$A(\mathbf{k}'; \mathbf{k}\lambda) = -\sqrt{\frac{hc^2}{\omega\Omega}}\frac{e}{mc}\langle \mathbf{k}'^{(-)}|\mathbf{e}_{\mathbf{k}\lambda}\cdot\mathbf{p}e^{i\mathbf{k}\cdot\mathbf{x}}|G\rangle. \tag{17}$$

Neither the ket nor the bra in this last matrix element is an eigenstate of $\mathbf{p}$, and the second step of (6) cannot be repeated. We recall that when $\hbar\omega \ll E_0/\alpha$, $ka_0 \ll 1$. In regime (1) we may therefore approximate the factor $e^{i\mathbf{k}\cdot\mathbf{x}}$ by one. We are then faced with the evaluation of $\langle \mathbf{k}'^{(-)}|\mathbf{e}_{\mathbf{k}\lambda}\cdot\mathbf{p}|G\rangle$. The operator $\mathbf{p}$ is a polar vector, and $|G\rangle$ is an s-state of even parity. Consequently only the portion of the final state which has angular momentum one and odd parity contributes when $e^{i\mathbf{k}\cdot\mathbf{x}} \simeq 1$.†

* A formal proof that (15) is the exact amplitude to first order in H' can be found in Chapt. XIII.

† This is an example of the famous dipole selection rule which dominates atomic spectroscopy. For a detailed discussion see Sec. 61.3.

This conclusion agrees with the qualitative argument presented in the introduction to this section, and shows that only the p-wave Coulomb continuum eigenfunctions will enter into our calculation.

It is advantageous to eliminate $\mathbf{p}$ from $\langle \mathbf{k}'^{(-)}|\mathbf{e}_{\mathbf{k}\lambda}\cdot\mathbf{p}|G\rangle$. This can be done by an ancient device based on the observation that $\mathbf{x}$ commutes with the Coulomb interaction. Hence $[\mathbf{x}, H_H] = i\hbar\mathbf{p}/m$. Furthermore,

$$\begin{aligned}\langle \mathbf{k}'^{(-)}|[\mathbf{x}, H_H]|G\rangle &= -(E_0 + E')\langle \mathbf{k}'^{(-)}|\mathbf{x}|G\rangle \\ &= -\hbar\omega\langle \mathbf{k}'^{(-)}|\mathbf{x}|G\rangle. \end{aligned} \tag{18}$$

We have therefore reduced (17) to

$$A(\mathbf{k}'; \mathbf{k}\lambda) = -ie\sqrt{\frac{\hbar\omega}{\Omega}}\int d^3r\,\langle \mathbf{k}'^{(-)}|\mathbf{r}\rangle \mathbf{e}_{\mathbf{k}\lambda}\cdot\mathbf{r}\langle r|G\rangle. \tag{19}$$

Now we must relate $\langle \mathbf{r}|\mathbf{k}'^{(-)}\rangle$ to the Coulomb wave functions. In Sec. 17 we only discussed eigenfunctions with outgoing wave boundary conditions, states that we now designate by $|\mathbf{k}'^{(+)}\rangle$. As we know, these states are related to $|\mathbf{k}'^{(-)}\rangle$ by time reversal: $\Theta|-\mathbf{k}'^{(+)}\rangle = |\mathbf{k}'^{(-)}\rangle$, or

$$\langle \mathbf{k}'^{(-)}|\mathbf{r}\rangle = \langle \mathbf{r}|-\mathbf{k}'^{(+)}\rangle. \tag{20}$$

Except for a slight difference of normalization, $\langle \mathbf{r}|\mathbf{k}'^{(+)}\rangle$ is given by (17.19). This difference comes about because our kets $|\mathbf{k}'^{(\pm)}\rangle$ are normalized to unity in a box of volume Ω, whereas (17.19) has δ-function normalization. Hence

$$\begin{aligned}\langle \mathbf{k}'^{(-)}|\mathbf{r}\rangle &= \sqrt{8\pi^3/\Omega}\,\psi_{-\mathbf{k}'}(\mathbf{r}) \\ &= \sqrt{4\pi/\Omega}\sum_{l=0}^{\infty}\sqrt{2l+1}\,i^l C_l(k'; r)Y_{l0}(\pi-\beta),\end{aligned}$$

where β is the angle between $\mathbf{r}$ and $\mathbf{k}'$. With the help of the addition theorem (35.7) we can write this as

$$\langle \mathbf{k}'^{(-)}|\mathbf{r}\rangle = 4\pi\Omega^{-\frac{1}{2}}\sum_{lm} i^{-l}C_l(k'; r)Y_{lm}(\theta\phi)Y^*_{lm}(\vartheta\varphi),$$

where the angles ϑ and φ are shown in Fig. 58.1, and we have also used $Y_{l0}(\pi-\beta) = (-1)^l Y_{l0}(\beta)$.

The angular integration in (19) is very simple if $\mathbf{e}_{\mathbf{k}\lambda}\cdot\mathbf{r}$ is expressed in terms of spherical harmonics. According to (52.31) and (10.20), $\mathbf{r}\cdot\mathbf{e}_{\mathbf{k}\pm1} = \mp 2^{-\frac{1}{2}}(x \pm iy) = r(4\pi/3)^{\frac{1}{2}}Y_{1,\pm1}(\vartheta\varphi)$, whence

$$\langle \mathbf{k}'^{(-)}|\mathbf{x}\cdot\mathbf{e}_{\mathbf{k}\lambda}|G\rangle = -i\left(\frac{64\pi^3}{3\Omega}\right)^{\frac{1}{2}} Y_{1\lambda}(\theta\phi)\int_0^\infty C_1(k'; r)r\langle r|G\rangle r^2\,dr.$$

Here we see quite explicitly that only p-states with $m = \pm 1$ can be excited.

After substituting the explicit forms for Y_{1m} and $\langle r|G\rangle$, we obtain

$$\frac{d\sigma_\lambda}{d\Omega_e} = 2\alpha a_0^2\,(ka_0)(\hbar\omega/E_0)|I(k')|^2 \sin^2\theta, \tag{21}$$

where

$$I(k') = a_0^{-4}\int_0^\infty r^3 C_1(k';r)e^{-r/a_0}\,dr. \tag{22}$$

Note that (21) does not depend on the helicity of the incident photon.

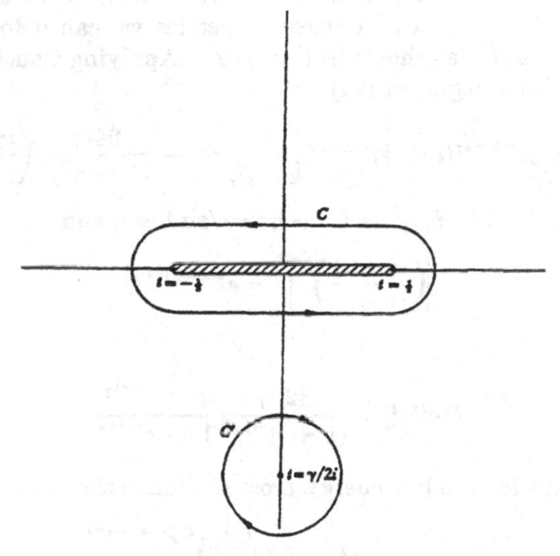

Fig. 58.2. Contours pertaining to the evaluation of $I(k')$.

An analytic expression for $I(k')$ can be found by using the integral representation for the Coulomb wave function. For our present purpose the most convenient representation is obtained by making the substitution $s = 2ik'r(t - \frac{1}{2})$ in (17.32), and combining the resulting expression with (17.24):

$$C_1(k';r) = \frac{e^{\frac{1}{2}\pi\gamma}\Gamma(2 - i\gamma)}{2\pi(2k'r)^2}\int_C dt\, e^{2ik'rt}(t - \tfrac{1}{2})^{-2-i\gamma}(t + \tfrac{1}{2})^{-2+i\gamma}.$$

The integration path C is shown in Fig. 58.2, and

$$\frac{1}{\gamma} = k'a_0 \equiv \sqrt{\frac{\hbar\omega}{E_0} - 1} \tag{23}$$

First we perform the r-integration of (22):

$$\int_0^\infty e^{2ik'rt}e^{-r/a_0}r\,dr = -\frac{a_0^2\gamma^2}{4(t+\frac{1}{2}i\gamma)^2}.$$

We are then left with the integral

$$I(k') = \frac{e^{\frac{1}{2}\pi\gamma}\Gamma(2-i\gamma)\gamma^4}{16i}\int_C \frac{dt}{2\pi i}\frac{(t-\frac{1}{2})^{-2-i\gamma}(t+\frac{1}{2})^{-2+i\gamma}}{(t+\frac{1}{2}i\gamma)^2}. \tag{24}$$

The only singularities of the integrand are the branch cut shown in Fig. 58.2, and a second-order pole at $t = -\frac{1}{2}i\gamma$. As $|t| \to \infty$, the integrand falls off as $|t|^{-6}$. Because of these properties we can deform the integration path into C', as shown in Fig. 58.2. Applying Cauchy's theorem we find that the integral in (24) equals

$$-\frac{d}{dt}(t-\tfrac{1}{2})^{-2-i\gamma}(t+\tfrac{1}{2})^{-2+i\gamma}\Big|_{t=-\frac{1}{2}i\gamma} = -\frac{64i\gamma}{(1+\gamma^2)^3}\left(\frac{i\gamma-1}{i\gamma+1}\right)^{i\gamma}.$$

The identities $|\Gamma(2-i\gamma)|^2 = (1+\gamma^2)\pi\gamma/\sinh\pi\gamma$, and

$$\left|\left(\frac{i\gamma-1}{i\gamma+1}\right)^{i\gamma}\right|^2 = e^{-4\gamma\cot^{-1}\gamma},$$

then lead to

$$|I(k')|^2 = \frac{32\pi\gamma}{(1+\gamma^{-2})^5}\frac{e^{-4\gamma\cot^{-1}\gamma}}{1-e^{-2\pi\gamma}}.$$

Our final result for the low-energy cross section is therefore

$$\frac{d\sigma_\lambda}{d\Omega_e} = 64\pi\alpha a_0^2\sin^2\theta\left(\frac{E_0}{\hbar\omega}\right)^4\frac{e^{-4\gamma\cot^{-1}\gamma}}{1-e^{-2\pi\gamma}}. \tag{25}$$

This formula gives the exact cross section (to first order in H', of course) as long as $\hbar\omega \ll E_0/\alpha$.

Let us verify that (25) joins smoothly onto the high-energy result (13) at intermediate energies. This intermediate regime is characterized by $\hbar\omega \sim O(E_0/\alpha)$, where $\gamma \ll 1$. By expanding (25) about $\gamma = 0$ we obtain

$$\frac{d\sigma_\lambda}{d\Omega_e} \simeq 32\alpha a_0^2\sin^2\theta\left(\frac{E_0}{\hbar\omega}\right)^{\frac{7}{2}}, \tag{26}$$

which agrees with (13) except for the small term proportional to v'/c. [When $\hbar\omega \sim O(E_0/\alpha)$, $v'/c \sim O(\sqrt{\alpha})$.] This small term represents the leading contribution of final states having an angular momentum $j > 1$, and cannot be reproduced by the p-wave calculation we have just done.

The behavior of (25) near threshold is exceptionally interesting. Here $\gamma \to \infty$, whence

$$\frac{d\sigma_\lambda}{d\Omega_e} = 64\pi\alpha a_0^2 e^{-4} \sin^2\theta \left[1 - \frac{8}{3}\left(\frac{\hbar\omega - E_0}{E_0}\right) + \cdots\right], \tag{27}$$

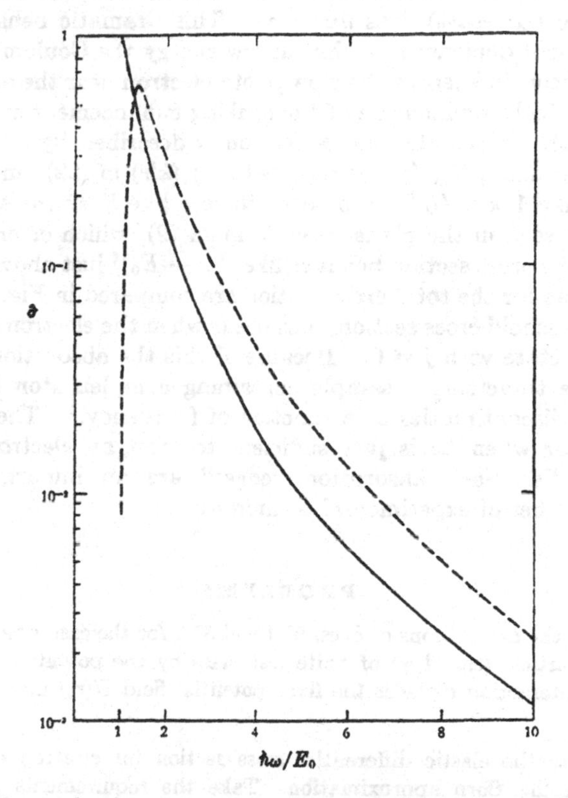

Fig. 58.3. The total cross section for the photo-effect in hydrogen near threshold. The quantity $\hat{\sigma}$ is the total cross section divided by the exact result at threshold, i.e., $(512\pi^2/3)\alpha a_0^2 e^{-4} = 6.31 \times 10^{-18}$ cm^2. The solid curve is the exact p-wave calculation, while the broken curve represents the result of integrating the plane wave formula (9) over $d\Omega_e$. The latter cross section is given by $(e^4/2\pi)(E_0/\hbar\omega)^5\gamma^{-3}$ in the units defined above. One observes that the plane wave calculation gives the correct trend when $\hbar\omega$ is above $\sim 2E_0$. If one continues the calculation to higher energies than shown here, one finds that the quantitative discrepancy visible in the graph at $\hbar\omega \sim 10E_0$ persists up to energies of order E_0/α. In particular, the factor $\exp(-4\gamma \cot^{-1}\gamma)$ approaches its asymptotic value of 1 rather slowly.

when $(\hbar\omega - E_0) \ll E_0$. (Note that e stands for the base of the natural logarithms, not the electronic charge.) The remarkable feature of (27) is that it is *finite at threshold.* The cross section is the product of two factors: the phase space factor $k'm/(2\pi)^3\hbar^2$, and the square of the matrix element $A(\mathbf{k}'; \mathbf{k}\lambda)$ given by (15). At threshold the phase space factor vanishes like k', and our finite result (27) therefore implies that $A(\mathbf{k}'; \mathbf{k}\lambda)$ *diverges* like $(\hbar\omega - E_0)^{-\frac{1}{4}}$ as $\hbar\omega \to E_0$. This dramatic behavior of the matrix element demonstrates that at low energy the Coulomb attraction is very effective in keeping the slow photo-electron near the origin. This is the "physical" explanation of the striking enhancement of the overlap integral (22). When the photo-electron is described by a plane wave, on the other hand, $C_1(k'; r)$ is replaced by $j_1(k'r)$ in (22), and instead of diverging like $1/k'^{\frac{1}{2}}$, $I(k')$ then tends to zero like k' at threshold. This can also be seen in the plane wave formula (9), which erroneously predicts that the cross section behaves like $(\hbar\omega - E_0)^{\frac{3}{2}}$ just above threshold. Our formulas for the total cross section are compared in Fig. 58.3.

Finite threshold cross sections also occur when the electron is originally bound in a state with $j \neq 0$. Because of this the absorption coefficient for photons traversing a sample containing complex atoms displays a number of discontinuities as a function of frequency.* These discontinuities occur when $\hbar\omega$ is just sufficient to eject an electron from the various shells. Such absorption "edges" are an important ingredient in a number of experimental techniques.

PROBLEMS

1. Repeat the calculations of Secs. 57.1 and 57.2 for the case where β is bound to a third particle (called γ) of finite rest mass by the potential $U(|\mathbf{r}_\beta - \mathbf{r}_\gamma|)$. (This β-γ interaction replaces the fixed potential field $U(r_\beta)$ introduced in the text.)

2. Evaluate the elastic differential cross section for electron scattering by hydrogen in the Born approximation. Take the requirements of the Pauli principle into account, and determine the error committed when the identity of the electrons is ignored. Compute numerical values of the correction as a function of incident energy and scattering angle.

3. Consider the scattering of a particle of charge $Z'e$ and mass $M \gg m$ by a neutral atom. The velocity v of this projectile is assumed to be large enough for the Born approximation to be applicable.

(a) Show that in an inelastic collision, the momentum transfer is always small compared to Mv, or to be precise, $\hbar q_{max} = 2Mv(m/M)$. Also show that except for exceedingly small scattering angles θ, $q\hbar \sim Mv\theta$.

* Cf., e.g., Bethe and Salpeter, Fig. 34.

(b) Show that the differential cross section for exciting the νth atomic state is given by (57.52) multiplied by Z'^2.

(c) Derive Eq. (57.71).

4. Consider the scattering of a fast electron by an atom in the angular region where the momentum transfer is independent of excitation energy, i.e., $\theta \gg Z^2 E_H/E$. Define the inelastic cross section as

$$d\sigma_{\text{in}} = \sum_{\nu>0} d\sigma_\nu.$$

(a) Show that

$$\frac{d\sigma_{\text{in}}}{dq} = 8\pi \left(\frac{e^2}{\hbar v}\right)^2 \frac{1}{q^3} \left\{ Z - Z^2 |F_0(q)|^2 + \left\langle \sum_{j \neq k} e^{i\mathbf{q}\cdot(\mathbf{x}_j - \mathbf{x}_k)} \right\rangle_0 \right\},$$

where we have approximated the nuclear form factor by its value at $q = 0$.

(b) Apply this result to the helium atom. Assume that the ground state is adequately described by the product wave function of Sec. 48, and show that

$$\frac{d\sigma_{\text{in}}}{dq} = 16\pi \left(\frac{e^2}{\hbar v}\right)^2 \frac{1}{q^3} [1 - |F_0(q)|^2].$$

5. Consider the collision of a fast electron with hydrogen resulting in an ionization process where the secondary electron emerges with low energy. The identity of the electrons can then be ignored.

(a) Show that the differential cross section is given by

$$d\sigma = 4 \left(\frac{k'}{k}\right) a_0^{-2} q^{-4}\, d^3\kappa\, d\Omega' \left| \int \psi_{-\kappa}(\mathbf{r}) e^{i\mathbf{q}\cdot\mathbf{r}} \langle \mathbf{r} | G \rangle \, d^3r \right|^2,$$

where $\hbar\boldsymbol{\kappa}$ is the momentum of the slow electron, $\hbar\mathbf{q}$ the momentum transfer suffered by the fast electron, $\psi_\kappa(\mathbf{r})$ the Coulomb continuum function given in (17.19), and $\hbar\mathbf{k}$ and $\hbar\mathbf{k}'$ are the momenta of the fast electron before and after the collision, respectively.

(b) Consider the special case of scattering through small angles such that $qa_0 \ll 1$. Show that under these circumstances

$$d\sigma = \frac{8}{\pi} \left(\frac{k'}{k}\right) \frac{a_0^2}{q^2} \cos^2\chi\, |I(\kappa)|^2\, d^3\kappa\, d\Omega',$$

where I is the function defined by (58.22), and χ is the angle between $\boldsymbol{\kappa}$ and $\mathbf{q}$.

(c) Show that for all values of q the projection of the angular momentum of the secondary electron along $\mathbf{q}$ vanishes.

Appendix

A detailed compilation of the fundamental constants can be found in E. R. Cohen and J. W. DuMond, *Encyclopedia of Physics*, Vol. 35, Springer-Verlag, Berlin (1957). Here we confine ourselves to those combinations of the natural constants that occur most frequently in atomic and nuclear physics. The numbers given below are unaffected by the present experimental errors. If higher accuracy is desired, one should consult Cohen and DuMond, *loc. cit.*

(a) *Energy Conversion Factors*

$$\begin{aligned} 1 \text{ ev} &= 1.6021 \times 10^{-12} \text{ erg} \\ &= 2.418 \times 10^{14} \text{ cycles/sec} \\ 1^\circ \text{ Kelvin} &= 8.617 \times 10^{-5} \text{ ev} \end{aligned}$$

(b) *Rest Masses*

Particle	Mev	Units of m_e
Electron	0.51098	1
Muon	105.65	206.8
π^0	135.0	264.2
$\pi^\pm$	139.6	273.2
Proton	938.2	1836
Neutron	939.5	1839
Deuteron	1875.5	3670

(c) *Constants Related to the Electron*

Fine structure constant:

$$\alpha = \frac{e^2}{\hbar c} = 7.297 \times 10^{-3} = \frac{1}{137.04}$$

Bohr radius:

$$a_0 = \frac{\hbar^2}{me^2} = 0.5292 \times 10^{-8}\ \text{cm}$$

Compton wavelength:

$$\lambda\!\!\!^{-}_C = \frac{\hbar}{mc} = \alpha a_0 = 3.862 \times 10^{-11}\ \text{cm}$$

Classical radius:

$$r_0 = \frac{e^2}{mc^2} = \alpha^2 a_0 = 2.818 \times 10^{-13}\ \text{cm}$$

Binding energy of electron by infinitely heavy proton in nonrelativistic approximation:

$$E_0 = \tfrac{1}{2}\alpha^2 mc^2 = e^2/2a_0 = \hbar^2/2ma_0^2 = 13.605\ \text{ev}$$

Ionization energy of hydrogen: 13.598 ev
Bohr magneton:

$$\mu_0 = e\hbar/2mc = 0.5788 \times 10^{-8}\ \text{ev/gauss}$$

(d) *Magnetic Moments*

electron	0.5795×10^{-8} ev/gauss
proton	$2.7928\mu_N$
neutron	$-1.9128\mu_N$

where μ_N is the nuclear magneton:

$$\mu_N = e\hbar/2m_p c = 0.3152 \times 10^{-11}\ \text{ev/gauss}$$

(e) *Energy-Area Units, $\hbar^2/2m$*

electron	3.810×10^{-16} cm² ev
proton	20.75×10^{-26} cm² Mev

(f) *de Broglie Wavelengths, $\hbar/p$*

electron	$\dfrac{1.952 \times 10^{-8}\ \text{cm}}{\sqrt{E(\text{ev})}}$
proton	$\dfrac{4.555 \times 10^{-13}\ \text{cm}}{\sqrt{E(\text{Mev})}}$

Bibliography

Bethe, H. A., and R. Jackiw, *Intermediate Quantum Mechanics* (2nd Edition), W. A. Benjamin, New York, 1966.

Bethe, H. A., and P. Morrison, *Elementary Nuclear Theory*, John Wiley, New York, 1956.

Bethe, H. A., and E. E. Salpeter, *Quantum Mechanics of One- and Two-Electron Atoms*, Springer-Verlag, Berlin, 1957.

Bjorken, J. D., and S. D. Drell, *Relativistic Quantum Fields*, McGraw-Hill, New York, 1965.

Blatt, J. M., and V. F. Weisskopf, *Theoretical Nuclear Physics*, John Wiley, New York, 1952.

Boerner, H., *Darstellungen von Gruppen*, Springer-Verlag, Berlin, 1955.

Bohm, D., *Quantum Theory*, Prentice-Hall, New York, 1951.

Bohr, N., *Atomic Physics and Human Knowledge*, John Wiley, New York, 1958.

Born, M., and E. Wolf, *Principles of Optics*, Pergamon Press, New York, 1959.

Condon, E. U., and G. H. Shortley, *Theory of Atomic Spectra*, Cambridge University Press, New York, 1935.

Courant, R., and D. Hilbert, *Methods of Mathematical Physics*, Interscience, New York, 1953.

Dalitz, R. H., *Strange Particles and Strong Interactions*, Oxford University Press, New York, 1962.

de Benedetti, S., *Nuclear Interactions*, John Wiley, New York, 1964.

de-Shalit, A., and I. Talmi, *Nuclear Shell Theory*, Academic Press, New York, 1963.

Dirac, P. A. M., *Quantum Mechanics* (3rd Edition), Oxford University Press, New York, 1947.

Drukarev, G. F., *The Theory of Electron-Atom Collisions*, Academic Press, New York, 1965.

Edmonds, A., *Angular Momentum in Quantum Mechanics*, Princeton University Press, Princeton, New Jersey, 1957.

Fano, U., and G. Racah, *Irreducible Tensorial Sets*, Academic Press, New York, 1959.

Gel'fand, I. M., and G. E. Shilov, *Generalized Functions*, Vol. 1, Academic Press, New York, 1964.

Gel'fand, I. M., R. A. Minlos, and Z. Ya. Shapiro, *Representations of the Rotation and Lorentz Groups*, Macmillan, New York, 1963.

Goldberger, M. L., and K. M. Watson, *Collision Theory*, John Wiley, New York, 1964.

Goldstein, H., *Classical Mechanics*, Addison-Wesley, Reading, Massachusetts, 1950.

Hamermesh, M., *Group Theory*, Addison-Wesley, Reading, Massachusetts, 1962.

Hamilton, J., *Theory of Elementary Particles*, Oxford University Press, New York, 1959.

Heisenberg, W., *The Physical Principles of the Quantum Theory*, Dover, New York, 1950.

Heitler, W., *The Quantum Theory of Radiation* (3rd Edition), Oxford University Press, New York, 1954.

Jackson, J. D., *Classical Electrodynamics*, John Wiley, New York, 1962.

Källén, G., *Elementary Particle Physics*, Addison-Wesley, Reading, Massachusetts, 1964.

Kemble, E. C., *The Fundamental Principles of Quantum Mechanics*, McGraw-Hill, New York, 1937.

Kronig, R. deL., *Band Spectra and Molecular Structure*, Cambridge University Press, New York, 1930.

Landau, L. D., *Classical Theory of Fields*, Addison-Wesley, Reading, Massachusetts, 1951. [Referred to as *CTF*]

Landau, L. D., and E. M. Lifshitz, *Quantum Mechanics*, Addison-Wesley, Reading, Massachusetts, 1958. [Referred to as *QM*]

Lighthill, M., *Introduction to Fourier Analysis and Generalized Functions*, Cambridge University Press, New York, 1958.

Mackey, G. W., *The Mathematical Foundations of Quantum Mechanics*, W. A. Benjamin, New York, 1963.

Mandelstam, S., and W. Yourgrau, *Variational Principles in Dynamics and Quantum Theory*, Pitman, New York, 1960.

Massey, H. S. W., and E. H. S. Burhop, *Electronic and Ionic Impact Phenomena*, Oxford University Press, New York, 1952.

Merzbacher, E., *Quantum Mechanics*, John Wiley, New York, 1961.

Messiah, A., *Quantum Mechanics*, Interscience, New York, 1961.
Møller, C., *The Theory of Relativity*, Oxford University Press, New York, 1952.
Morse, P. M., and H. Feshbach, *Methods of Theoretical Physics*, McGraw-Hill, New York, 1953.
Mott, N. F., and H. S. W. Massey, *The Theory of Atomic Collisions* (2nd Edition), Oxford University Press, New York, 1949.
von Neumann, J., *Mathematical Foundations of Quantum Mechanics*, Princeton University Press, Princeton, New Jersey, 1955.
Nozières, P., *The Theory of Interacting Fermi Systems*, W. A. Benjamin, New York, 1964.
Pauli, W., *Die allgemeine Prinzipien der Wellenmechanik*, Encyclopedia of Physics, Vol. 5, Springer, Berlin, 1958.
Pauling, L., and E. B. Wilson, *Introduction to Quantum Mechanics*, McGraw-Hill, New York, 1935.
Ramsay, N. F., *Molecular Beams*, Oxford University Press, New York, 1956.
Riesz, F., and B. Sz. Nagy, *Functional Analysis*, Frederick Ungar, New York, 1955.
Sakurai, J. J., *Invariance Principles and Elementary Particles*, Princeton University Press, Princeton, New Jersey, 1964.
Schwartz, L., *Théorie des Distributions*, Hermann, Paris, 1957/59.
Schweber, S. S., *Relativistic Quantum Field Theory*, Row, Peterson, Evanston, Illinois, 1961.
Schwinger, J., *Quantum Electrodynamics*, Dover, New York, 1958.
Schwinger, J., *On Angular Momentum*, NYO-3071, Technical Information Services, Oak Ridge, Tennessee, 1952. [Referred to as *AM*]
Slichter, C. P., *Principles of Magnetic Resonance*, Harper and Row, New York, 1963.
Streater, R. F., and A. S. Wightman, *PCT, Spin and Statistics, and All That*, W. A. Benjamin, New York, 1964.
Thirring, W., *Principles of Quantum Electrodynamics*, Academic Press, New York, 1958.
Tinkham, M., *Group Theory and Quantum Mechanics*, McGraw-Hill, New York, 1964.
Vulikh, B. Z., *Functional Analysis for Scientists and Technologists*, Addison-Wesley, Reading, Massachusetts, 1963.
van der Waerden, B. L., *Die gruppentheoretische Methode in der Quantenmechanik*, Springer, Berlin, 1932.
Whittaker, E. T., *History of the Theories of Aether and Electricity*, Thomas Nelson, London, 1951.
Wigner, E. P., *Group Theory and Its Applications to Quantum Mechanics*, Academic Press, New York, 1959.
Wilson, R., *The Nucleon-Nucleon Interaction*, Interscience, New York, 1963.

Index